BASIC ELECTROMAGNETICS
WITH APPLICATIONS

NANNAPANENI NARAYANA RAO

Department of Electrical Engineering
University of Illinois at Urbana-Champaign

PRENTICE-HALL, INC., *Englewood Cliffs, New Jersey*

MICROWAVES AND FIELDS SERIES
Nathan Marcuvitz, Editor

© 1972 by Prentice-Hall, Inc.
Englewood Cliffs, New Jersey

ISBN-13-060236-1

10 9 8 7 6 5 4 3 2 1

Library of Congress Catalog Card Number: 72-000041

Printed in the United States of America

Drawings by GEORGE E. MORRIS

Prentice-Hall International, Inc., *London*
Prentice-Hall of Australia, Pty. Ltd., *Sydney*
Prentice-Hall of Canada, Ltd., *Toronto*
Prentice-Hall of India Private Ltd., *New Delhi*
Prentice-Hall of Japan, Inc., *Tokyo*

To

My Wife and Children
Sarojini and Vanaja, Durgaprasad, and Hariprasad

CONTENTS

v

3   THE STATIC MAGNETIC FIELD   134

4   THE ELECTROMAGNETIC FIELD   193

Preface

This book is designed for an introductory undergraduate course in electromagnetics. In view of the rapid growth of the several specialized branches of electrical engineering, a student may not have the opportunity to take advanced courses in field theory. On the other hand, electromagnetics is one of the fundamental subjects having a wide variety of applications, as evidenced by its place in the undergraduate core curriculum. Hence, a thorough understanding of the basic concepts of electromagnetics must be imparted to the electrical engineering student in the introductory course itself.

To facilitate the aforementioned task, an attempt is made in this book to present the basic field theory at an introductory level and at the same time in sufficient depth to establish the concepts firmly in the student's mind and to enable the interested student to use the advanced books without having to relearn the subject or reorient his understanding of the concepts. This is done by combining the classical approach of introducing field theory with statics and the modern approach of emphasizing dynamics to develop Maxwell's equations and the associated constitutive relations in a gradual manner and finally use them to discuss several applications. A number of worked-out examples are distributed throughout the book to illustrate and, in some cases, extend the various concepts and to aid the student's grasp of the subject matter.

The book does not presuppose knowledge of vector analysis. Chapter 1 contains the discussion of coordinate systems and vector analysis necessary and sufficient for the remaining chapters. Other mathematical tools such as the Dirac delta function and the phasor technique are introduced wherever necessary.

Chapters 2 and 3 are devoted to static electric and magnetic fields, respectively, in free space. Starting with Coulomb's and Ampere's laws in chapters 2 and 3, respectively, Maxwell's equations for static fields are introduced in a logical manner. The coverage of static magnetic field in chapter 3 is as much detailed as the coverage of static electric field in chapter 2 unlike the traditional mode of presentation in which the electric field topics are emphasized.

Chapter 4 is devoted to the electromagnetic field in free space. Maxwell's equations for time-varying fields are introduced. Energy storage in electric and magnetic fields and power flow in electromagnetic field are discussed. The use of phasor technique in dealing with sinusoidally time-varying vector fields is illustrated. Maxwell's equations and the power and energy relations are then specialized for sinusoidally time-varying fields.

The discussion in chapters 2, 3, and 4 is in terms of the field vectors **E** and **B**. Chapter 5 is devoted to the study of fields in the presence of materials. The interaction between fields and charges in materials is discussed in terms of equivalent charge and current distributions which are related to the fields and act as though they were situated in free space, thereby entering into Maxwell's equations. By defining field vectors **D** and **H** and relating them to **E** and **B**, respectively, Maxwell's equations for free space developed in chapters 2, 3, and 4 are generalized so that they can be used for material media as well as for free space. The power and energy relations developed in chapter 4 are also generalized for material media. Boundary conditions are derived for the fields.

Chapter 6 serves as an introduction to the applications of Maxwell's equations. A variety of topics providing a continuous coverage from statics to electromagnetic waves via quasistatics and distributed circuits are discussed. The presentation is oriented towards introducing the fundamental concepts leading to and associated with the applications. For example, the circuit parameters conductance, capacitance, and inductance are introduced simultaneously so that the student can better appreciate the development of the frequency behavior of a physical structure made up of two parallel conductors leading to the concept of a distributed circuit. Yet another example is the introduction of waveguides by starting with uniform plane waves incident obliquely on a perfect conductor, which provides a physical understanding of the waveguiding phenomenon.

There is enough material in this book for a two-semester course. However, by deemphasizing certain topics and omitting certain other topics, it is possible to use this text for a one-semester course. In the latter case, the student can read the remaining material by himself with the aid of the answers to the odd-numbered problems included at the end of the book. The many example problems throughout and the numerous homework problems at the end of each chapter make this book especially suitable for a course oriented towards problem solving.

This text is based on lecture notes prepared for courses taught since 1965 at the University of Illinois at Urbana-Champaign and earlier at the University of Washington. I am indebted to Professor E. C. Jordan at the University of Illinois and Professor A. V. Eastman at the University of Washington for their help in several instances without which this book would not have materialized.

Urbana, Illinois   *N. Narayana Rao*

1

VECTOR ANALYSIS

Vector analysis is a shorthand notation by means of which we perform mathematical manipulations with quantities which have associated with them not only magnitude but also direction in space. Such quantities are known as vectors, in contrast to scalars which have only magnitude associated with them. Force and velocity are examples of vectors. Mass and length are examples of scalars. The electric and magnetic fields are examples of vectors. Voltage and current are examples of scalars. Since this book is concerned with electric and magnetic fields, it is necessary that we first learn the notation and certain rules of vector analysis. To distinguish vector quantities from scalar quantities, we use boldface type: **A**. Graphically, the vector **A** is represented by a line whose length is equal to the magnitude of **A**, denoted $|\mathbf{A}|$ or simply A, and with an arrowhead at the end of the line pointing toward the direction of **A**. If the top of the page is taken to be pointing toward the north, then Figs. 1.1(a), (b), and (c) represent vectors **A**, **B**, and **C** directed north, northeast, and west-northwest, respectively.

1.1   Some Simple Rules

a. Equality of Vectors.

Two vectors **A** and **B** are equal if and only if their magnitudes as well as directions are the same.

1

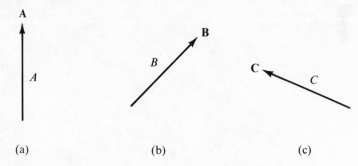

Fig. 1.1. Graphical representation of vectors.

b. Addition and Subtraction of Vectors.

Two vectors **A** and **B** are added by placing the beginning of one vector at the tip of the other as shown in Figs. 1.2(a) and (b). The sum vector is then obtained by joining the beginning of the first vector to the tip of the second vector. This rule is also known as the parallelogram law since, if we consider the two vectors as the adjacent sides of a parallelogram with their beginnings at a common point O as shown in Fig. 1.2(c), the sum vector is then given by the diagonal of the parallelogram drawn from the corner O to the opposite corner. From Figs. 1.2(a) and (b), it is clear that vector addition is commutative, that is,

$$\mathbf{A} + \mathbf{B} = \mathbf{B} + \mathbf{A} \tag{1-1}$$

Subtraction is a special case of addition. If we want to subtract a vector **B** from a vector **A**, we first construct the vector $(-\mathbf{B})$, which has the same magnitude as that of **B** but opposite direction, and then add it to **A**, that is,

$$\mathbf{A} - \mathbf{B} = \mathbf{A} + (-\mathbf{B}) \tag{1-2}$$

The graphical construction pertinent to (1-2) is shown in Fig. 1.3(a). If we decide to obtain $\mathbf{A} - \mathbf{B}$ from the construction of a parallelogram with **A** and **B** as the adjacent sides emanating from the common point O similar to that in Fig. 1.2(c), then the construction of Fig. 1.3(b) indicates that

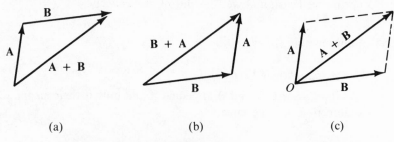

Fig. 1.2. Addition of two vectors.

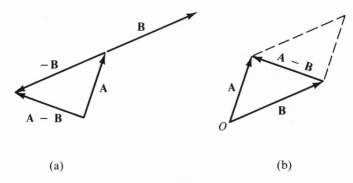

(a)   (b)

Fig. 1.3. Vector subtraction.

$\mathbf{A} - \mathbf{B}$ is given by the diagonal of the parallelogram drawn from the tip of $\mathbf{B}$ to the tip of $\mathbf{A}$. Finally, the constructions of Fig. 1.4 illustrate that vector addition is associative, that is,

$$\mathbf{A} + (\mathbf{B} + \mathbf{C}) = (\mathbf{A} + \mathbf{B}) + \mathbf{C} \tag{1-3}$$

c. Multiplication and Division by a Scalar.

When a vector $\mathbf{A}$ is multiplied by a scalar m, it is equivalent to adding $\mathbf{A}$ or $(-\mathbf{A})$ a total of m times, depending upon whether m is positive or negative. Hence the direction of $m(\mathbf{A})$ is the same as or opposite to that of $\mathbf{A}$, depending upon whether m is positive or negative, whereas the magnitude of $m(\mathbf{A})$ is $|m|$ times the magnitude of $\mathbf{A}$. Thus

$$|m(\mathbf{A})| = |m||\mathbf{A}| = |m|A \tag{1-4}$$

$$\text{Direction of } m(\mathbf{A}) = \begin{cases} \text{direction of } \mathbf{A} & \text{if } m > 0 \\ \text{direction of } (-\mathbf{A}) & \text{if } m < 0 \end{cases} \tag{1-5}$$

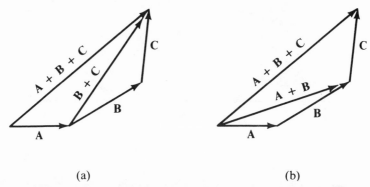

(a)   (b)

Fig. 1.4. Illustrating the associative property of vector addition.

Division by a scalar is, of course, a special case of multiplication, that is, to divide a vector by m we multiply it by $1/m$.

d. Unit Vector.

If we divide a vector **A** by its magnitude A, we obtain a vector whose magnitude is unity and whose direction is the same as the direction of **A**. The resulting vector is called the "unit vector" in the direction of **A** and is denoted i_A. Thus

$$i_A = \frac{\mathbf{A}}{|\mathbf{A}|} = \frac{\mathbf{A}}{A} \tag{1-6}$$

Unit vectors play a very important role in vector analysis, as we will find throughout this book.

e. Scalar or Dot Product of Two Vectors.

The scalar or dot product of two vectors **A** and **B** is a scalar quantity of value equal to the product of the magnitudes of **A** and **B** and the cosine of the angle between **A** and **B**. It is represented by a dot between **A** and **B**. Thus

$$\mathbf{A} \cdot \mathbf{B} = |\mathbf{A}||\mathbf{B}| \cos \alpha = AB \cos \alpha \tag{1-7}$$

where α is the angle between **A** and **B**. Noting that

$$\mathbf{A} \cdot \mathbf{B} = AB \cos \alpha = A(B \cos \alpha) = B(A \cos \alpha) \tag{1-8}$$

we see from the constructions of Fig. 1.5 that the dot-product operation consists of multiplying the magnitude of one vector by the scalar obtained by projecting the second vector onto the first vector. This suggests that the dot product is useful for problems such as finding the work done in displacing a mass. The dot-product operation is commutative since

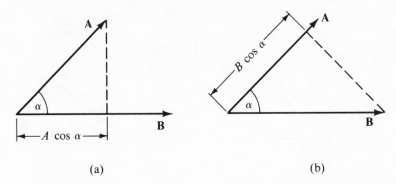

(a)   (b)

Fig. 1.5. Showing that the dot product of **A** and **B** is the product of the magnitude of one vector and the projection of the second vector onto the first vector.

$$\mathbf{B} \cdot \mathbf{A} = |\mathbf{B}||\mathbf{A}| \cos \alpha = |\mathbf{A}||\mathbf{B}| \cos \alpha = \mathbf{A} \cdot \mathbf{B} \qquad (1\text{-}9)$$

Furthermore, the distributive property also holds, that is,

$$\mathbf{A} \cdot (\mathbf{B} + \mathbf{C}) = \mathbf{A} \cdot \mathbf{B} + \mathbf{A} \cdot \mathbf{C} \qquad (1\text{-}10)$$

To prove the distributive property, we note from the construction shown in Fig. 1.6 that the projection of $(\mathbf{B} + \mathbf{C})$ onto $\mathbf{A}$ is equal to the sum of the projections of $\mathbf{B}$ and $\mathbf{C}$ onto $\mathbf{A}$. It follows from this that (1-10) is correct.

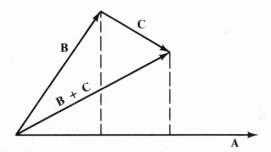

Fig. 1.6. For proving the distributive property of the dot-product operation.

f. Vector or Cross Product of Two Vectors.

In contrast to the dot product, the vector or cross product of two vectors $\mathbf{A}$ and $\mathbf{B}$ is another vector whose magnitude is the product of the magnitudes of $\mathbf{A}$ and $\mathbf{B}$ and the sine of the angle α between $\mathbf{A}$ and $\mathbf{B}$ and whose direction is the direction of advance of a right-hand screw as it is turned from $\mathbf{A}$ towards $\mathbf{B}$ through the angle α, as shown in Fig. 1.7(a). Thus

$$\mathbf{A} \times \mathbf{B} = |\mathbf{A}||\mathbf{B}| \sin \alpha \, \mathbf{i}_N = AB \sin \alpha \, \mathbf{i}_N \qquad (1\text{-}11)$$

where $\mathbf{i}_N$ is the unit vector in the direction of advance of a right-hand screw as it is turned from $\mathbf{A}$ towards $\mathbf{B}$ through α. For example, if vector $\mathbf{A}$ is a unit vector directed eastward and vector $\mathbf{B}$ is a unit vector directed northward, then a right-hand screw advances upward as it is turned from east towards north through the 90° angle so that $\mathbf{A} \times \mathbf{B}$ has a magnitude $(1)(1)(\sin 90°)$ or unity and is directed upward. Alternatively, if we decide to turn the right-hand screw from east toward north through the 270° angle, we note that the screw advances downward. There is no inconsistency, however, since the product $|\mathbf{A}||\mathbf{B}| \sin \alpha$ is then equal to $(1)(1)(\sin 270°)$ or -1. When the minus sign is associated with the direction of advance of the screw, the direction of $\mathbf{A} \times \mathbf{B}$ becomes upward.

From the constructions of Figs. 1.7(a) and (b) it follows that

$$\mathbf{B} \times \mathbf{A} = -\mathbf{A} \times \mathbf{B} \qquad (1\text{-}12)$$

so that the commutative law does not hold for the cross product. Similarly,

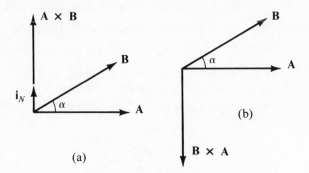

Fig. 1.7. Cross-product operations for two vectors **A** and **B**.

the associative law does not hold for the cross product, that is,

$$(A \times B) \times C \neq A \times (B \times C) \qquad (1\text{-}13)$$

This can be demonstrated very easily by considering a particular case in which the three vectors **A**, **B**, and **C** are unit vectors directed eastward, northward, and southward, respectively, as shown in Fig. 1.8. Then $(A \times B)$ is the unit vector directed upward. $(A \times B) \times C$ is the unit vector directed eastward, that is, **A**. On the other hand, $(B \times C)$ is equal to zero and hence $A \times (B \times C)$ is equal to zero. Thus the associative law does not hold. That the distributive law,

$$A \times (B + C) = A \times B + A \times C \qquad (1\text{-}14)$$

holds will be proved in an example after we discuss the scalar triple product.

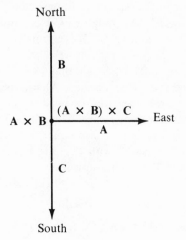

Fig. 1.8. For demonstrating that the associative law does not hold for the cross-product operation.

The cross-product operation is very convenient to define unit vectors. Thus a unit vector perpendicular to both **A** and **B** is, according to (1-11), given by

$$\mathbf{i}_N = \frac{\mathbf{A} \times \mathbf{B}}{|\mathbf{A}||\mathbf{B}|\sin\alpha} \tag{1-15}$$

g. Scalar Triple Product.

Another useful operation but of less importance is the scalar triple product $\mathbf{A} \cdot (\mathbf{B} \times \mathbf{C})$. Using the definitions of dot and cross products we have

$$
\begin{aligned}
\mathbf{A} \cdot (\mathbf{B} \times \mathbf{C}) &= |\mathbf{A}||\mathbf{B} \times \mathbf{C}|\cos(\text{angle between } \mathbf{A} \text{ and } \mathbf{B} \times \mathbf{C}) \\
&= |\mathbf{A}||\mathbf{B}||\mathbf{C}|\sin(\text{angle between } \mathbf{B} \text{ and } \mathbf{C}) \\
&\quad \times \cos(\text{angle between } \mathbf{A} \text{ and } \mathbf{B} \times \mathbf{C})
\end{aligned} \tag{1-16}
$$

From the construction of Fig. 1.9, we note that

$$
\begin{aligned}
\mathbf{A} \cdot (\mathbf{B} \times \mathbf{C}) &= ABC \sin\beta \cos\alpha = (A\cos\alpha)(BC\sin\beta) \\
&= \text{volume of the parallelepiped formed by } \mathbf{A}, \mathbf{B}, \text{ and } \mathbf{C}
\end{aligned} \tag{1-17}
$$

Thus the scalar triple product has the geometric meaning that it represents the volume of the parallelepiped formed by the three vectors. From con-

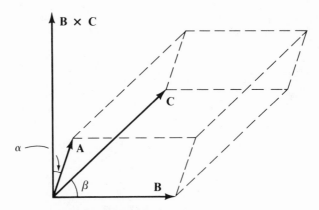

Fig. 1.9. Parallelepiped formed by **A**, **B**, and **C**.

structions similar to Fig. 1.9, it can be shown that $(\mathbf{A} \times \mathbf{B}) \cdot \mathbf{C}$ or $(\mathbf{C} \times \mathbf{A}) \cdot \mathbf{B}$ represent the same volume so that

$$\mathbf{A} \cdot (\mathbf{B} \times \mathbf{C}) = \mathbf{B} \cdot (\mathbf{C} \times \mathbf{A}) = \mathbf{C} \cdot (\mathbf{A} \times \mathbf{B}) \tag{1-18}$$

Also, the parentheses in the scalar triple product are unnecessary since, for example, $\mathbf{A} \cdot \mathbf{B} \times \mathbf{C}$ can mean only $\mathbf{A} \cdot (\mathbf{B} \times \mathbf{C})$ and not $(\mathbf{A} \cdot \mathbf{B}) \times \mathbf{C}$. This is so because $\mathbf{A} \cdot \mathbf{B}$ is a scalar and for a vector product, we need two vectors. Hence $(\mathbf{A} \cdot \mathbf{B}) \times \mathbf{C}$ is meaningless. It is therefore customary to omit the parentheses when writing a scalar triple product.

EXAMPLE 1-1. Vector **A** has a magnitude of 4 units and is directed towards the east. Vector **B** has a magnitude of 4 units and is oriented in a direction making an angle of 120° toward north from east. Vector **C** has a magnitude of 3 units and is directed 30° south of east. Find

(a) **A** + **B**

(b) 3**A** − 4**C**

(c) **A** + **B** − **C**

(d) **A** · **B**

(e) **B** × **C**

(f) **A** · **B** × **C**

(g) **A** × (**B** × **C**)

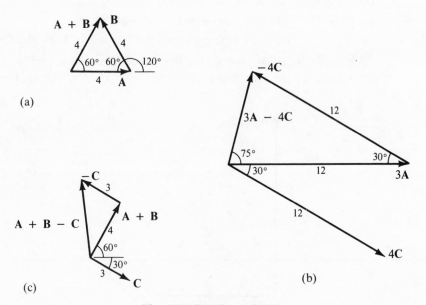

Fig. 1.10. For Example 1-1.

(a) From the construction of Fig. 1.10(a), (**A** + **B**) has a magnitude of 4 units and is directed 60° north of east.

(b) 3**A** = 12 units towards the east; 4**C** = 12 units directed 30° south of east. From the construction of Fig. 1.10(b), 3**A** − 4**C** has a magnitude of 24 cos 75° or 6·21 units and is directed 75° north of east.

(c) From the construction of Fig. 1.10(c), **A** + **B** − **C** has a magnitude of 5 units and is directed (60° + tan⁻¹ ¾) or 96°52′ north of east.

(d) **A** · **B** = |**A**||**B**| cos (angle between **A** and **B**) = (4)(4)(cos 120°) = −8.

(e) $|\mathbf{B} \times \mathbf{C}| = |\mathbf{B}||\mathbf{C}||\sin$ (angle between $\mathbf{B}$ and $\mathbf{C})| = (4)(3)(\sin 150°) =$ 6. The direction of $\mathbf{B} \times \mathbf{C}$ is the direction in which a right-hand screw advances when it is turned from $\mathbf{B}$ toward $\mathbf{C}$ through the angle 150°. This direction is downward. Thus, $\mathbf{B} \times \mathbf{C}$ has a magnitude of 6 units and is directed downward.

(f) $\mathbf{A} \cdot \mathbf{B} \times \mathbf{C} = |\mathbf{A}||\mathbf{B} \times \mathbf{C}|$ cos (angle between $\mathbf{A}$ and $\mathbf{B} \times \mathbf{C}) =$ $(4)(6)(\cos 90°) = 0$. This is consistent with the reasoning that, since all three vectors are in a plane, the area of the parallelogram formed by them is zero.

(g) $|\mathbf{A} \times (\mathbf{B} \times \mathbf{C})| = |\mathbf{A}||\mathbf{B} \times \mathbf{C}||\sin$ (angle between $\mathbf{A}$ and $\mathbf{B} \times \mathbf{C})| =$ $(4)(6)(\sin 90°) = 24$. The direction of $\mathbf{A} \times (\mathbf{B} \times \mathbf{C})$ is the direction in which a right-hand screw advances if it is turned from $\mathbf{A}$ toward $\mathbf{B} \times \mathbf{C}$ through the angle 90°, that is, from east to downward through the angle 90°. The screw advances towards the north. Thus $\mathbf{A} \times (\mathbf{B} \times \mathbf{C})$ has a magnitude of 24 units and is directed northward. ∎

EXAMPLE 1-2.   Show that $\mathbf{A} \times (\mathbf{B} + \mathbf{C}) = \mathbf{A} \times \mathbf{B} + \mathbf{A} \times \mathbf{C}$.
We will prove this equality by showing that

$$\mathbf{D} = \mathbf{A} \times (\mathbf{B} + \mathbf{C}) - \mathbf{A} \times \mathbf{B} - \mathbf{A} \times \mathbf{C} = 0$$

Taking the dot product of an arbitrary vector $\mathbf{E}$ and the vector $\mathbf{D}$ and using (1-10) and (1-18), we have

$$
\begin{aligned}
\mathbf{E} \cdot \mathbf{D} &= \mathbf{E} \cdot [\mathbf{A} \times (\mathbf{B} + \mathbf{C}) - \mathbf{A} \times \mathbf{B} - \mathbf{A} \times \mathbf{C}] \\
&= \mathbf{E} \cdot \mathbf{A} \times (\mathbf{B} + \mathbf{C}) - \mathbf{E} \cdot \mathbf{A} \times \mathbf{B} - \mathbf{E} \cdot \mathbf{A} \times \mathbf{C} \\
&= (\mathbf{B} + \mathbf{C}) \cdot \mathbf{E} \times \mathbf{A} - \mathbf{B} \cdot \mathbf{E} \times \mathbf{A} - \mathbf{C} \cdot \mathbf{E} \times \mathbf{A} \\
&= \mathbf{B} \cdot \mathbf{E} \times \mathbf{A} + \mathbf{C} \cdot \mathbf{E} \times \mathbf{A} - \mathbf{B} \cdot \mathbf{E} \times \mathbf{A} - \mathbf{C} \cdot \mathbf{E} \times \mathbf{A} = 0
\end{aligned}
$$

This result implies that $\mathbf{D}$ is either zero or perpendicular to $\mathbf{E}$. However, since $\mathbf{E}$ is an arbitrary vector, it can be chosen such that it is not perpendicular to $\mathbf{D}$, in which case $\mathbf{D}$ has to be zero for $\mathbf{E} \cdot \mathbf{D}$ to be zero. Thus $\mathbf{D}$ is equal to zero and hence the equality $\mathbf{A} \times (\mathbf{B} + \mathbf{C}) = \mathbf{A} \times \mathbf{B} + \mathbf{A} \times \mathbf{C}$ is correct. ∎

EXAMPLE 1-3.   Two unit vectors $\mathbf{i}_A$ and $\mathbf{i}_B$ drawn at a point are perpendicular to each other. A vector $\mathbf{C}$ is also drawn from the same point. Express $\mathbf{C}$ in terms of its component vectors along $\mathbf{i}_A$ and $\mathbf{i}_B$.
From Fig. 1.11, the projection of $\mathbf{C}$ onto the line along $\mathbf{i}_A$ is equal to $\mathbf{C} \cos \alpha = \mathbf{C} \cdot \mathbf{i}_A$. Hence the component vector of $\mathbf{C}$ along $\mathbf{i}_A$ is $(\mathbf{C} \cdot \mathbf{i}_A)\mathbf{i}_A$. Similarly, the component vector of $\mathbf{C}$ along $\mathbf{i}_B$ is $(\mathbf{C} \cdot \mathbf{i}_B)\mathbf{i}_B$. Since the component vectors form two adjacent sides of a rectangle whose diagonal is $\mathbf{C}$, as in Fig. 1.11, we have

$$\mathbf{C} = (\mathbf{C} \cdot \mathbf{i}_A)\mathbf{i}_A + (\mathbf{C} \cdot \mathbf{i}_B)\mathbf{i}_B$$

It also follows from Fig. 1.11 that

$$(\mathbf{C} \cdot \mathbf{i}_A)^2 + (\mathbf{C} \cdot \mathbf{i}_B)^2 = C^2$$

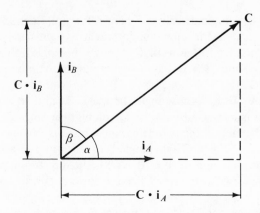

Fig. 1.11. Components of a vector along mutually perpendicular unit vectors.

Likewise, if we have three mutually perpendicular unit vectors $\mathbf{i}_A$, $\mathbf{i}_B$, and $\mathbf{i}_C$ drawn from a point, then the component vectors of a vector $\mathbf{D}$ along the unit vectors are $(\mathbf{D} \cdot \mathbf{i}_A)\mathbf{i}_A$, $(\mathbf{D} \cdot \mathbf{i}_B)\mathbf{i}_B$, and $(\mathbf{D} \cdot \mathbf{i}_C)\mathbf{i}_C$, respectively, so that

$$\mathbf{D} = (\mathbf{D} \cdot \mathbf{i}_A)\mathbf{i}_A + (\mathbf{D} \cdot \mathbf{i}_B)\mathbf{i}_B + (\mathbf{D} \cdot \mathbf{i}_C)\mathbf{i}_C$$

Furthermore,

$$(\mathbf{D} \cdot \mathbf{i}_A)^2 + (\mathbf{D} \cdot \mathbf{i}_B)^2 + (\mathbf{D} \cdot \mathbf{i}_C)^2 = D^2 \quad \blacksquare$$

1.2 Coordinate Systems

In the previous section we discussed some simple rules of vector analysis without involving any coordinate system. In physical problems, we cannot simply go on describing vectors by symbols **A**, **B**, **C**, and so on, if we wish to simplify the geometry associated with the mathematical operations using these vectors. We need to describe a vector in terms of component vectors along a set of reference directions such as east, north, and upward. Although several different coordinate systems are in existence, we will be interested only in three: (a) the cartesian, (b) the circular cylindrical or simply cylindrical, and (c) the spherical coordinate systems. Each coordinate system involves three surfaces which are mutually orthogonal. At any particular point, unit vectors can be drawn tangential to the curves of intersection of pairs of the three orthogonal surfaces. The three unit vectors drawn in this manner will be mutually perpendicular and will define the reference directions at that point. Once such reference directions are defined everywhere in space, we can represent vectors in terms of their component vectors along the reference directions and use them for performing vector operations. We will discuss each coordinate system separately and then summarize the details in the form of a table.

a. Cartesian Coordinate System.

For the cartesian coordinate system, the three mutually orthogonal surfaces are three planes. Let us consider three orthogonal planes which intersect at a particular point O which we will call the origin, as shown in Fig. 1.12(a). The three planes also define three straight lines which are the intersections of pairs of planes. These three straight lines are mutually perpendicular and form a set of coordinate axes which are denoted x, y, and z axes. Values of x, y, and z are measured from the origin so that the origin is taken as the reference point. We say that the coordinates of the origin are $(0, 0, 0)$, that is, $x = 0$, $y = 0$, and $z = 0$. Thus, if we consider the x axis, values of x on one side of the origin are positive and on the other side, they are negative. The direction of increasing values of x is indicated by an arrowhead. We will direct a unit vector $\mathbf{i}_x$ drawn from the origin in the direction of increasing values of x. By doing the same with the y and z axes, we define unit vectors $\mathbf{i}_y$ and $\mathbf{i}_z$ at O. Now, we note that we can choose the directions of increasing values of x, y, and z in two ways: (a) such that $\mathbf{i}_x \times \mathbf{i}_y = \mathbf{i}_z$ as in Fig. 1.12(a); or (b) such that $\mathbf{i}_y \times \mathbf{i}_x = \mathbf{i}_z$. The first is known as a right-hand coordinate system since, if a right-hand screw is turned from the direction of increasing values of x towards the direction of increasing values of y through the smaller angle 90°, it advances in the direction of increasing values of z. The second choice is known as a left-hand coordinate system since it requires a left-hand screw to advance in the direction of increasing values of z when turned from the direction of increasing values of x towards the direction of increasing values of y through the smaller angle 90°. By convention, the right-hand coordinate system is used.

Movement on the yz plane requires no displacement along the x direction; hence the value of x is constant on this plane. In particular, since the value of x at the origin is zero, this constant is zero. Also, the unit vector $\mathbf{i}_x$ is in the increasing x direction and hence is normal to this plane. Similarly, for the xz plane, $y = $ constant $= 0$ and $\mathbf{i}_y$ is normal to this plane; for the xy plane, $z = $ constant $= 0$ and $\mathbf{i}_z$ is normal to this plane. Any other point in space is now defined by the intersection of three planes parallel to the three planes defining the origin. Alternatively, we can displace the three planes $x = 0$, $y = 0$, and $z = 0$ along the coordinate axes (or unit vectors) perpendicular to them and obtain a new point of intersection. For example, by moving the $x = 0$ plane by one unit along the x axis, the $y = 0$ plane by three units along the y axis, and the $z = 0$ plane by four units along the z axis, we obtain a point of intersection whose coordinates are $(1, 3, 4)$, as shown in Fig. 1.12(b). On any plane parallel to the $x = 0$ plane, the value of x is constant and equal to its displacement from the $x = 0$ plane; on any plane parallel to the $y = 0$ plane, the value of y is constant and equal to its displacement from the $y = 0$ plane; and on any plane parallel to the $z = 0$ plane, the value of z is constant and equal to its displacement from the $z = 0$ plane. Thus the point $(1, 3, 4)$ is the intersection of the three planes $x = 1$, $y = 3$, and $z = 4$. These three planes also define three straight lines which are intersections of pairs of planes. Unit vectors $\mathbf{i}_x$, $\mathbf{i}_y$, and $\mathbf{i}_z$ can be drawn along these lines of intersections. These unit vectors are parallel to the

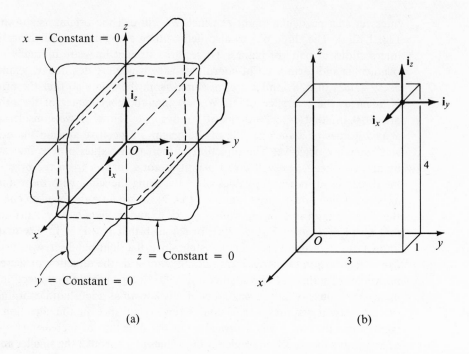

(a)

(b)

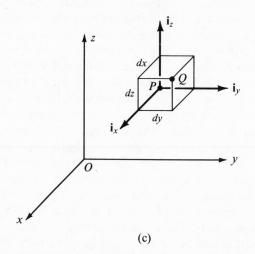

(c)

Fig. 1. 12. Cartesian coordinate system. (a) The three orthogonal planes defining the coordinate system. (b) Unit vectors at an arbitrary point. (c) Differential volume formed by incrementing the coordinates.

corresponding unit vectors at the origin since the lines of intersection are parallel to the x, y, and z axes. In general, an arbitrary point (a, b, c) is defined by the intersection of the three planes $x = a$, $y = b$, and $z = c$. Unit vectors $\mathbf{i}_x$, $\mathbf{i}_y$, and $\mathbf{i}_z$ are directed normal to these planes along increasing values of x, y, and z, respectively, and are parallel to the corresponding unit vectors at the origin. Thus, in the cartesian coordinate system, the directions of the unit vectors $\mathbf{i}_x$, $\mathbf{i}_y$, and $\mathbf{i}_z$ are everywhere the same as their directions at the origin.

Let us now consider two points $P(x, y, z)$ and $Q(x + dx, y + dy, z + dz)$, where Q is obtained by incrementing infinitesimally each coordinate from its value at P. The three orthogonal planes intersecting at P and the three orthogonal planes intersecting at Q define a rectangular box of edges dx, dy, and dz in the $\mathbf{i}_x$, $\mathbf{i}_y$, and $\mathbf{i}_z$ directions, respectively, as shown in Fig. 1.12(c). The differential displacements (or length elements) along the unit vectors $\mathbf{i}_x$, $\mathbf{i}_y$, and $\mathbf{i}_z$ in going from P to Q are therefore the same as the differential increments dx, dy, and dz of the coordinates x, y, and z, respectively. The vector displacement $d\mathbf{l}$ from P to Q is given by

$$d\mathbf{l} = dx\,\mathbf{i}_x + dy\,\mathbf{i}_y + dz\,\mathbf{i}_z \tag{1-19a}$$

The magnitude of this displacement is

$$dl = \sqrt{(dx)^2 + (dy)^2 + (dz)^2} \tag{1-19b}$$

The three displacements $dx\,\mathbf{i}_x$, $dy\,\mathbf{i}_y$, and $dz\,\mathbf{i}_z$ also define three surfaces of infinitesimal areas in the three planes intersecting at P. To take into account the orientation of the surface area, it is convenient to represent the area by a vector quantity whose magnitude is equal to the area and whose direction is that of the normal to the area. The three infinitesimal surfaces are then $\pm dy\,dz\,\mathbf{i}_x$, $\pm dz\,dx\,\mathbf{i}_y$, and $\pm dx\,dy\,\mathbf{i}_z$, where the $\pm$ sign takes into account two possible directions of normal to the surface. The infinitesimal volume of the box is $dx\,dy\,dz$. We will use the differential length elements, surface areas, and volume introduced here in later sections.

Any arbitrary surface is defined by an equation of the type

$$f(x, y, z) = 0 \tag{1-20}$$

where f denotes a function. Since an arbitrary curve is an intersection of two appropriate surfaces, it is defined by a pair of equations

$$f(x, y, z) = 0 \quad \text{and} \quad g(x, y, z) = 0 \tag{1-21}$$

where f and g are two different functions. Alternatively, a curve may be defined by three parametric equations

$$x = x(t) \quad y = y(t) \quad z = z(t) \tag{1-22}$$

where t is an independent parameter.

A vector drawn from the origin to an arbitrary point $P(x, y, z)$ is called the position vector defining the point P. It is denoted by the symbol $\mathbf{r}$.

Thus, in the cartesian coordinate system,

$$\mathbf{r} = x\mathbf{i}_x + y\mathbf{i}_y + z\mathbf{i}_z \tag{1-23}$$

b. Cylindrical Coordinate System.

For the cylindrical coordinate system, the three mutually orthogonal surfaces are a cylinder and two planes, as shown in Fig. 1.13(a). One of the planes is the same as the $z = $ constant plane in the cartesian coordinate

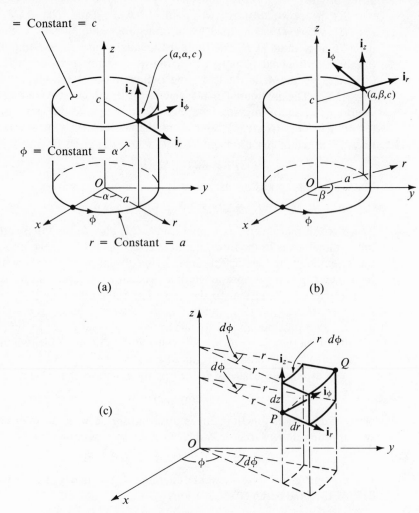

(a)   (b)

(c)

Fig. 1.13. Cylindrical coordinate system. (a) The three orthogonal surfaces defining the coordinate system (b) Unit vectors at an arbitrary point. (c) Differential volume formed by incrementing the coordinates.

system. The second plane is orthogonal to the $z = $ constant plane and hence contains the z axis. It makes an angle ϕ with a reference plane, conveniently chosen as the xz plane of the cartesian coordinate system. This plane is therefore defined by $\phi = $ constant. The third orthogonal surface, which is cylindrical, has the z axis as its axis. On such a cylindrical surface, the radial distance r from the z axis is a constant. Thus the three orthogonal surfaces defining the cylindrical coordinates of a point are given by $r = $ constant, $\phi = $ constant, and $z = $ constant. In particular, the origin is defined by $r = 0$, $\phi = 0$, and $z = 0$. Note that only two of the coordinates (r and z) are distances, whereas the third coordinate (ϕ) is an angle. Since the radius of a cylinder cannot be negative, the coordinate r varies only from 0 to ∞. Since one revolution of the $\phi = $ constant plane about the z axis sweeps the entire space, the coordinate ϕ varies from 0 to 2π. The coordinate z varies from $-\infty$ to $+\infty$ as in the cartesian coordinate system.

Through any arbitrary point (a, α, c) we can pass a cylinder $r = a$, a plane $\phi = \alpha$, and another plane $z = c$. These three orthogonal surfaces define three curves, mutually perpendicular at (a, α, c), two of which are straight lines and the third is a circle. We draw unit vectors $\mathbf{i}_r$, $\mathbf{i}_\phi$, and $\mathbf{i}_z$ tangential to these curves at (a, α, c) and directed toward increasing values of r, ϕ, and z, respectively, as shown in Fig. 1.13(a). It follows that $\mathbf{i}_r$, $\mathbf{i}_\phi$, and $\mathbf{i}_z$ are mutually perpendicular and normal to the surfaces $r = a, \phi = \alpha$, and $z = c$, respectively, at the point (a, α, c). If we now consider a point (a, β, c), this point is defined by the intersection of the surfaces $r = a, \phi = \beta$, and $z = c$. Three mutually perpendicular unit vectors $\mathbf{i}_r$, $\mathbf{i}_\phi$, and $\mathbf{i}_z$ can be drawn at the point (a, β, c) tangential to the curves of intersection of pairs of these surfaces and in the directions of increasing r, ϕ, and z, respectively, as shown in Fig. 1.13(b). However, we note that the unit vectors $\mathbf{i}_r$ and $\mathbf{i}_\phi$ at this point are not parallel to the corresponding unit vectors at the point (a, α, c). Thus, unlike the unit vectors in the cartesian coordinate system, the unit vectors $\mathbf{i}_r$ and $\mathbf{i}_\phi$ do not have the same directions at all points; that is, the directions of $\mathbf{i}_r$ and $\mathbf{i}_\phi$ are functions of the coordinates r and ϕ, whereas $\mathbf{i}_z$ remains uniform. We also note that a right-hand coordinate system defined by $\mathbf{i}_r \times \mathbf{i}_\phi = \mathbf{i}_z$ and a left-hand coordinate system defined by $\mathbf{i}_\phi \times \mathbf{i}_r = \mathbf{i}_z$ are possible. However, we will work with the right-hand coordinate system.

Let us now consider two points $P(r, \phi, z)$ and $Q(r + dr, \phi + d\phi, z + dz)$, where Q is obtained by incrementing infinitesimally each coordinate from its value at P. The three orthogonal surfaces intersecting at P and the three orthogonal surfaces intersecting at Q define a box which can be considered as a rectangular box since dr, $d\phi$, and dz are infinitesimally small. The sides of this box are made up of the differential length elements dr, $r\, d\phi$, and dz along the $\mathbf{i}_r$, $\mathbf{i}_\phi$, and $\mathbf{i}_z$ directions, respectively, as shown in Fig. 1.13(c). Thus the differential displacements along the unit vectors $\mathbf{i}_r$, $\mathbf{i}_\phi$, and $\mathbf{i}_z$ in going from P to Q are dr, $r\, d\phi$, and dz, respectively. We note that the differential displacement in the ϕ direction is not $d\phi$ but $r\,d\phi$. The vector displacement $d\mathbf{l}$ from P to Q is given by

$$d\mathbf{l} = dr\,\mathbf{i}_r + r\,d\phi\,\mathbf{i}_\phi + dz\,\mathbf{i}_z \qquad (1\text{-}24a)$$

The magnitude of this displacement is

$$dl = \sqrt{(dr)^2 + (r\,d\phi)^2 + (dz)^2} \qquad (1\text{-}24b)$$

The infinitesimal areas in the three surfaces intersecting at P are $\pm(r\,d\phi)(dz)\mathbf{i}_r$, $\pm(dr)(dz)\,\mathbf{i}_\phi$, and $\pm(r\,d\phi)(dr)\mathbf{i}_z$. Finally, the infinitesimal volume of the box is $(dr)(r\,d\phi)(dz) = r\,dr\,d\phi\,dz$. Equations similar to (1-20), (1-21), and (1-22) define arbitrary surfaces and curves. The position vector defining an arbitrary point $P(r, \phi, z)$ is given by

$$\mathbf{r} = r\mathbf{i}_r + z\mathbf{i}_z \qquad (1\text{-}25)$$

c. Spherical Coordinate System.

For the spherical coordinate system, the three mutually orthogonal surfaces are a sphere, a cone, and a plane, as shown in Fig. 1.14(a). The plane is the same as the $\phi =$ constant plane in the cylindrical coordinate system. The sphere is centered at the origin. On the surface of such a sphere, the radial distance r from the origin is constant and hence the sphere is defined by $r =$ constant. The spherical coordinate r should not be confused with the cylindrical coordinate r. When these two coordinates appear in the same expression, we will use subscripts c and s to distinguish between

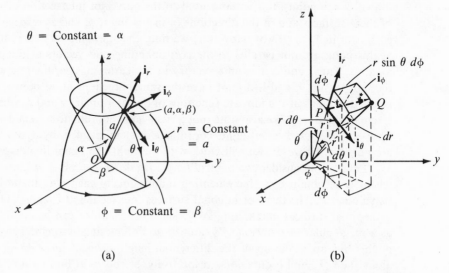

(a)   (b)

Fig. 1.14. Spherical coordinate system. (a) The three orthogonal surfaces defining the coordinate system. (b) Differential volume formed by incrementing the coordinates.

cylindrical and spherical. The cone has its vertex at the origin and its surface is symmetrical about the z axis, so that the angle θ which the conical surface makes with the z axis is constant. Thus the three orthogonal surfaces defining the spherical coordinates are given by $r = $ constant, $\theta = $ constant, and $\phi = $ constant. In particular, the origin is defined by $r = 0$, $\theta = 0$, and $\phi = 0$. Note that only one coordinate (r) is distance whereas the other two (θ and ϕ) are angles. Since the radius of a sphere cannot be negative, the coordinate r varies only from 0 to ∞. Likewise, it is sufficient if the coordinate θ is allowed to vary from 0 to π to cover the entire space. The coordinate ϕ varies from 0 to 2π as in the cylindrical coordinate system.

Through any arbitrary point (a, α, β) we can pass a sphere $r = a$, a cone $\theta = \alpha$, and a plane $\phi = \beta$. These three orthogonal surfaces define three curves, mutually perpendicular at (a, α, β). We draw unit vectors $\mathbf{i}_r$, $\mathbf{i}_\theta$, and $\mathbf{i}_\phi$ tangential to these curves at (a, α, β) and directed towards increasing values of r, θ, and ϕ, respectively, as shown in Fig. 1.14(a). It follows that $\mathbf{i}_r$, $\mathbf{i}_\theta$, and $\mathbf{i}_\phi$ are mutually perpendicular and normal to the surfaces $r = a$, $\theta = \alpha$, and $\phi = \beta$, respectively, at the point (a, α, β). By doing the same at another point, it may be seen that the directions of all three unit vectors $\mathbf{i}_r$, $\mathbf{i}_\theta$, and $\mathbf{i}_\phi$ are functions r, θ, and ϕ. We also note that a right-hand coordinate system defined by $\mathbf{i}_r \times \mathbf{i}_\theta = \mathbf{i}_\phi$ and a left-hand coordinate system defined by $\mathbf{i}_\theta \times \mathbf{i}_r = \mathbf{i}_\phi$ are possible. We will, however, work with the right-hand coordinate system.

Let us now consider two points $P(r, \theta, \phi)$ and $Q(r + dr, \theta + d\theta, \phi + d\phi)$, where Q is obtained by incrementing infinitesimally each coordinate from its value at P. The three orthogonal surfaces intersecting at P and the three orthogonal surfaces intersecting at Q define a box which can be considered as a rectangular box since dr, $d\theta$, and $d\phi$ are infinitesimally small. The sides of this box are made up of the differential length elements dr, $r\,d\theta$, and $r \sin \theta\,d\phi$ along the $\mathbf{i}_r$, $\mathbf{i}_\theta$, and $\mathbf{i}_\phi$ directions, respectively. Thus the differential displacements along the unit vectors $\mathbf{i}_r$, $\mathbf{i}_\theta$, and $\mathbf{i}_\phi$ in going from P to Q are dr, $r\,d\theta$, and $r \sin \theta\,d\phi$, respectively, as shown in Fig. 1.14(b). We note that the differential displacements in the θ and ϕ directions are $r\,d\theta$ and $r \sin \theta\,d\phi$ and not $d\theta$ and $d\phi$. The vector displacement $d\mathbf{l}$ from P to Q is given by

$$d\mathbf{l} = dr\,\mathbf{i}_r + r\,d\theta\,\mathbf{i}_\theta + r \sin \theta\,d\phi\,\mathbf{i}_\phi \tag{1-26a}$$

The magnitude of this displacement is

$$dl = \sqrt{(dr)^2 + (r\,d\theta)^2 + (r \sin \theta\,d\phi)^2} \tag{1-26b}$$

The infinitesimal areas in the three surfaces intersecting at P are $\pm(r\,d\theta)(r \sin \theta\,d\phi)\mathbf{i}_r$, $\pm(dr)(r \sin \theta\,d\phi)\mathbf{i}_\theta$, and $\pm(dr)(r\,d\theta)\mathbf{i}_\phi$. Finally, the infinitesimal volume of the box is $(dr)(r\,d\theta)(r \sin \theta\,d\phi) = r^2 \sin \theta\,dr\,d\theta\,d\phi$. Equations similar to (1-20), (1-21), and (1-22) define arbitrary surfaces and curves. The position vector defining an arbitrary point $P(r, \theta, \phi)$ is given by

$$\mathbf{r} = r\mathbf{i}_r \tag{1-27}$$

The various details discussed thus far in this section are summarized in Table 1.1.

TABLE 1.1. Summary of Details Pertinent to the Cartesian, Cylindrical, and Spherical Coordinate Systems

	Cartesian	*Cylindrical*	*Spherical*
Orthogonal Surfaces	three planes	a cylinder and two planes	a sphere, a cone, and a plane
Geometry	Fig. 1.12	Fig. 1.13	Fig. 1.14
Coordinates	x, y, z	r, ϕ, z	r, θ, ϕ
Unit Vectors	$\mathbf{i}_x, \mathbf{i}_y, \mathbf{i}_z$	$\mathbf{i}_r, \mathbf{i}_\phi, \mathbf{i}_z$	$\mathbf{i}_r, \mathbf{i}_\theta, \mathbf{i}_\phi$
Limits of Coordinates	$-\infty < x < \infty$ $-\infty < y < \infty$ $-\infty < z < \infty$	$0 < r < \infty$ $0 < \phi < 2\pi$ $-\infty < z < \infty$	$0 < r < \infty$ $0 < \theta < \pi$ $0 < \phi < 2\pi$
Differential Length Elements	$dx\,\mathbf{i}_x, dy\,\mathbf{i}_y, dz\,\mathbf{i}_z$	$dr\,\mathbf{i}_r, r\,d\phi\,\mathbf{i}_\phi, dz\,\mathbf{i}_z$	$dr\,\mathbf{i}_r, r\,d\theta\,\mathbf{i}_\theta,$ $r\sin\theta\,d\phi\,\mathbf{i}_\phi$
Differential Areas	$dx\,dy\,\mathbf{i}_z$ $dy\,dz\,\mathbf{i}_x$ $dz\,dx\,\mathbf{i}_y$	$r\,dr\,d\phi\,\mathbf{i}_z$ $r\,d\phi\,dz\,\mathbf{i}_r$ $dr\,dz\,\mathbf{i}_\phi$	$r\,dr\,d\theta\,\mathbf{i}_\phi$ $r^2\sin\theta\,d\theta\,d\phi\,\mathbf{i}_r$ $r\sin\theta\,dr\,d\phi\,\mathbf{i}_\theta$
Differential Volume	$dx\,dy\,dz$	$r\,dr\,d\phi\,dz$	$r^2\sin\theta\,dr\,d\theta\,d\phi$

Since any particular point in space can be defined by its coordinates in any one of the three coordinate systems, it is possible to derive relationships between the different sets of coordinates from simple considerations of geometry.

EXAMPLE 1-4. Express the cylindrical coordinates of a point in terms of its spherical coordinates.

From the construction of Fig. 1.15, the distance of point P (r_s, θ, ϕ) from the z axis is $r_s \sin \theta$. This is the radius of the cylinder passing through P and having the z axis as its axis. The height of point P above the xy plane is $r_s \cos \theta$. This is the value of z on the constant z plane passing through P. Thus the cylindrical coordinates (r_c, ϕ, z) of point P are

$$r_c = r_s \sin \theta$$
$$\phi = \phi$$
$$z = r_s \cos \theta$$

The various relationships between the different sets of coordinates obtained in this manner are summarized in Table 1.2. ∎

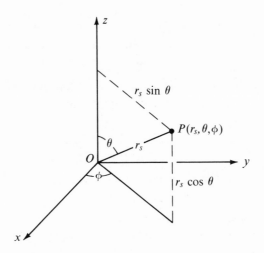

Fig. 1. 15. Conversion from spherical coordinates to cylindrical coordinates.

TABLE 1.2. Relationships Between Different Sets of Coordinates

	Cartesian x, y, z	Cylindrical r, ϕ, z	Spherical r, θ, ϕ
Cartesian x, y, z		$x = r \cos \phi$ $y = r \sin \phi$ $z = z$	$x = r \sin \theta \cos \phi$ $y = r \sin \theta \sin \phi$ $z = r \cos \theta$
Cylindrical r, ϕ, z	$r = \sqrt{x^2 + y^2}$ $\phi = \tan^{-1} \dfrac{y}{x}$ $z = z$		$r_c = r_s \sin \theta$ $\phi = \phi$ $z = r_s \cos \theta$
Spherical r, θ, ϕ	$r = \sqrt{x^2 + y^2 + z^2}$ $\theta = \tan^{-1} \dfrac{\sqrt{x^2 + y^2}}{z}$ $\phi = \tan^{-1} \dfrac{y}{x}$	$r_s = \sqrt{r_c^2 + z^2}$ $\theta = \tan^{-1} \dfrac{r_c}{z}$ $\phi = \phi$	

1.3 Components of Vectors

Once we set up a coordinate system and define unit vectors pertinent to that coordinate system, we can express vectors at any point in terms of their components along the unit vectors at that point and perform vector operations using the components. Let $\mathbf{i}_1, \mathbf{i}_2, \mathbf{i}_3$ be a set of mutually perpendicular vectors at a point P such that $\mathbf{i}_1 \times \mathbf{i}_2 = \mathbf{i}_3$, so that they can represent any one of the sets of unit vectors $(\mathbf{i}_x, \mathbf{i}_y, \mathbf{i}_z)$, $(\mathbf{i}_r, \mathbf{i}_\phi, \mathbf{i}_z)$, and $(\mathbf{i}_r, \mathbf{i}_\theta, \mathbf{i}_\phi)$ in the three different coordinate systems. Let $\mathbf{A}$, $\mathbf{B}$, and $\mathbf{C}$ be three vectors at the point

P. Then we have, from Example 1-3,

$$A = (A \cdot i_1)i_1 + (A \cdot i_2)i_2 + (A \cdot i_3)i_3$$
$$= A_1i_1 + A_2i_2 + A_3i_3 \tag{1-28}$$

where A_1, A_2, and A_3 are equal to $(A \cdot i_1)$, $(A \cdot i_2)$, and $(A \cdot i_3)$, respectively; that is, A_1, A_2, and A_3 are the components of A along i_1, i_2, and i_3, respectively. Similarly,

$$B = B_1i_1 + B_2i_2 + B_3i_3 \tag{1-29a}$$

$$C = C_1i_1 + C_2i_2 + C_3i_3 \tag{1-29b}$$

Now, we can perform the vector operations discussed in Section 1.1 as follows:

(a) *Equality of vectors:* Two vectors A and B are equal if and only if their respective components are equal; that is,

$$B_i = A_i, \qquad i = 1, 2, 3 \tag{1-30}$$

(b) *Magnitude of a vector:*

$$|A| = A = \sqrt{A_1^2 + A_2^2 + A_3^2} \tag{1-31}$$

(c) *Addition and subtraction of vectors:*

$$A + B = (A_1 + B_1)i_1 + (A_2 + B_2)i_2 + (A_3 + B_3)i_3 \tag{1-32a}$$

$$B - C = (B_1 - C_1)i_1 + (B_2 - C_2)i_2 + (B_3 - C_3)i_3 \tag{1-32b}$$

(d) *Multiplication and division by a scalar:*

$$m(A) = mA_1i_1 + mA_2i_2 + mA_3i_3 \tag{1-33a}$$

$$\frac{1}{m}(B) = \frac{B_1}{m}i_1 + \frac{B_2}{m}i_2 + \frac{B_3}{m}i_3 \tag{1-33b}$$

(e) *Unit vector:* The unit vector along the direction of a vector A is given by

$$i_A = \frac{A_1i_1 + A_2i_2 + A_3i_3}{\sqrt{A_1^2 + A_2^2 + A_3^2}} \tag{1-34}$$

(f) *Scalar or dot product of two vectors:*

$$A \cdot B = (A_1i_1 + A_2i_2 + A_3i_3) \cdot (B_1i_1 + B_2i_2 + B_3i_3)$$
$$= A_1B_1 + A_2B_2 + A_3B_3 \tag{1-35}$$

(g) *Vector or cross product of two vectors:*

$$A \times B = (A_1i_1 + A_2i_2 + A_3i_3) \times (B_1i_1 + B_2i_2 + B_3i_3)$$
$$= A_1B_2i_3 - A_1B_3i_2 - A_2B_1i_3 + A_2B_3i_1 + A_3B_1i_2 - A_3B_2i_1$$
$$= (A_2B_3 - A_3B_2)i_1 + (A_3B_1 - A_1B_3)i_2 + (A_1B_2 - A_2B_1)i_3$$
$$= \begin{vmatrix} i_1 & i_2 & i_3 \\ A_1 & A_2 & A_3 \\ B_1 & B_2 & B_3 \end{vmatrix} \tag{1-36}$$

(h) *Scalar triple product:*

$$\mathbf{A} \cdot \mathbf{B} \times \mathbf{C} = (A_1\mathbf{i}_1 + A_2\mathbf{i}_2 + A_3\mathbf{i}_3) \cdot \begin{vmatrix} \mathbf{i}_1 & \mathbf{i}_2 & \mathbf{i}_3 \\ B_1 & B_2 & B_3 \\ C_1 & C_2 & C_3 \end{vmatrix}$$

$$= \begin{vmatrix} A_1 & A_2 & A_3 \\ B_1 & B_2 & B_3 \\ C_1 & C_2 & C_3 \end{vmatrix} \tag{1-37}$$

EXAMPLE 1-5.   Find the dot and cross products of the unit vector $\mathbf{i}_{rc}$ at the point $P(r_c, \phi_c, z)$ and the unit vector $\mathbf{i}_\theta$ at the point $Q(r_s, \theta, \phi_s)$.

Since the unit vectors $\mathbf{i}_x$, $\mathbf{i}_y$, and $\mathbf{i}_z$ are uniform everywhere, we express $\mathbf{i}_{rc}$ at P and $\mathbf{i}_\theta$ at Q in terms of their components along $\mathbf{i}_x$, $\mathbf{i}_y$, and $\mathbf{i}_z$ and then perform the dot- and cross-product operations.

From the construction of Fig. 1.16 we have

$$\mathbf{i}_{rc} = \cos \phi_c \, \mathbf{i}_x + \sin \phi_c \, \mathbf{i}_y$$

$$\mathbf{i}_\theta = \cos \theta \cos \phi_s \, \mathbf{i}_x + \cos \theta \sin \phi_s \, \mathbf{i}_y - \sin \theta \, \mathbf{i}_z$$

Using (1-35) and (1-36) and simplifying, we then obtain

$$\mathbf{i}_{rc} \cdot \mathbf{i}_\theta = \cos \theta \cos (\phi_s - \phi_c)$$

$$\mathbf{i}_{rc} \times \mathbf{i}_\theta = \sin \theta \, \mathbf{i}_{\phi c} + \cos \theta \sin (\phi_s - \phi_c) \mathbf{i}_z$$

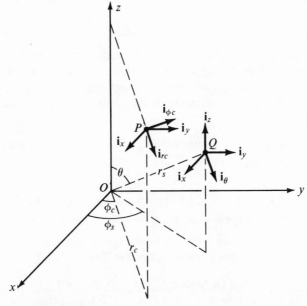

Fig. 1. 16. To find the dot product of $\mathbf{i}_{rc}$ at $P(r_c, \phi_c, z)$ and $\mathbf{i}_\theta$ at $Q(r_s, \theta, \phi_s)$.

If P and Q are the same points (r, θ, ϕ), then these results reduce to

$$\mathbf{i}_{rc} \cdot \mathbf{i}_\theta = \cos \theta$$

$$\mathbf{i}_{rc} \times \mathbf{i}_\theta = \sin \theta \, \mathbf{i}_\phi$$

Dot products and cross products between different unit vectors at the same point (r, θ, ϕ) are listed in Tables 1.3 and 1.4, respectively. ∎

TABLE 1.3. Dot Products of Unit Vectors at a Point (r, θ, ϕ)

	$\mathbf{i}_x$	$\mathbf{i}_y$	$\mathbf{i}_z$	$\mathbf{i}_{rc}$	$\mathbf{i}_\phi$	$\mathbf{i}_{rs}$	$\mathbf{i}_\theta$
$\mathbf{i}_x \cdot$	1	0	0	$\cos \phi$	$-\sin \phi$	$\sin \theta \cos \phi$	$\cos \theta \cos \phi$
$\mathbf{i}_y \cdot$		1	0	$\sin \phi$	$\cos \phi$	$\sin \theta \sin \phi$	$\cos \theta \sin \phi$
$\mathbf{i}_z \cdot$			1	0	0	$\cos \theta$	$-\sin \theta$
$\mathbf{i}_{rc} \cdot$				1	0	$\sin \theta$	$\cos \theta$
$\mathbf{i}_\phi \cdot$					1	0	0
$\mathbf{i}_{rs} \cdot$						1	0
$\mathbf{i}_\theta \cdot$							1

Table 1.4. Cross Products of Unit Vectors at a Point (r, θ, ϕ)

	$\mathbf{i}_x$	$\mathbf{i}_y$	$\mathbf{i}_z$	$\mathbf{i}_{rc}$	$\mathbf{i}_\phi$	$\mathbf{i}_{rs}$	$\mathbf{i}_\theta$
$\mathbf{i}_x \times$	0	$\mathbf{i}_z$	$-\mathbf{i}_y$	$\sin \phi \, \mathbf{i}_z$	$\cos \phi \, \mathbf{i}_z$	$\sin \theta \sin \phi \, \mathbf{i}_z - \cos \theta \, \mathbf{i}_y$	$\cos \theta \sin \phi \, \mathbf{i}_z + \sin \theta \, \mathbf{i}_y$
$\mathbf{i}_y \times$		0	$\mathbf{i}_x$	$-\cos \phi \, \mathbf{i}_z$	$\sin \phi \, \mathbf{i}_z$	$-\sin \theta \cos \phi \, \mathbf{i}_z + \cos \theta \, \mathbf{i}_x$	$-\cos \theta \cos \phi \, \mathbf{i}_z - \sin \theta \, \mathbf{i}_x$
$\mathbf{i}_z \times$			0	$\mathbf{i}_\phi$	$-\mathbf{i}_{rc}$	$\sin \theta \, \mathbf{i}_\phi$	$\cos \theta \, \mathbf{i}_\phi$
$\mathbf{i}_{rc} \times$				0	$\mathbf{i}_z$	$-\cos \theta \, \mathbf{i}_\phi$	$\sin \theta \, \mathbf{i}_\phi$
$\mathbf{i}_\phi \times$					0	$-\sin \theta \, \mathbf{i}_z + \cos \theta \, \mathbf{i}_{rc}$	$-\cos \theta \, \mathbf{i}_z - \sin \theta \, \mathbf{i}_{rc}$
$\mathbf{i}_{rs} \times$						0	$\mathbf{i}_\phi$
$\mathbf{i}_\theta \times$							0

Since any vector drawn at a point can be expressed in terms of its components along any one of the three sets of unit vectors, it is possible to derive relationships between the components of a vector in one coordinate system and the components of the same vector in another coordinate system.

EXAMPLE 1-6.   Express the component A_θ of a vector $\mathbf{A}$ in terms of its components A_x, A_y, and A_z.

$$
\begin{aligned}
A_\theta &= \mathbf{A} \cdot \mathbf{i}_\theta \\
&= (A_x \mathbf{i}_x + A_y \mathbf{i}_y + A_z \mathbf{i}_z) \cdot (\cos \theta \cos \phi \, \mathbf{i}_x + \cos \theta \sin \phi \, \mathbf{i}_y - \sin \theta \, \mathbf{i}_z) \\
&= A_x \cos \theta \cos \phi + A_y \cos \theta \sin \phi - A_z \sin \theta
\end{aligned}
$$

The various relationships derived in this manner between different components of a vector are summarized in Table 1.5. ∎

EXAMPLE 1-7.   Show that $(\mathbf{A} \times \mathbf{B}) \times \mathbf{C} = (\mathbf{A} \cdot \mathbf{C})\mathbf{B} - (\mathbf{B} \cdot \mathbf{C})\mathbf{A}$.

First, we note that the vector $\mathbf{A} \times \mathbf{B}$ is perpendicular to both vectors $\mathbf{A}$ and $\mathbf{B}$ and hence is normal to the plane containing $\mathbf{A}$ and $\mathbf{B}$. But the vector

TABLE 1.5. Relationships Between Components of a Vector in Different Coordinate Systems

	Cartesian A_x, A_y, A_z	Cylindrical A_r, A_ϕ, A_z	Spherical A_r, A_θ, A_ϕ
Cartesian A_x, A_y, A_z		$A_x = \dfrac{A_r x - A_\phi y}{\sqrt{x^2 + y^2}}$ $A_y = \dfrac{A_r y + A_\phi x}{\sqrt{x^2 + y^2}}$ $A_z = A_z$	$A_x = \dfrac{A_r x\sqrt{x^2 + y^2} + A_\theta xz - A_\phi y\sqrt{x^2 + y^2 + z^2}}{\sqrt{(x^2 + y^2)(x^2 + y^2 + z^2)}}$ $A_y = \dfrac{A_r y\sqrt{x^2 + y^2} + A_\theta yz + A_\phi x\sqrt{x^2 + y^2 + z^2}}{\sqrt{(x^2 + y^2)(x^2 + y^2 + z^2)}}$ $A_z = \dfrac{A_r z - A_\theta \sqrt{x^2 + y^2}}{\sqrt{x^2 + y^2 + z^2}}$
Cylindrical A_r, A_ϕ, A_z	$A_r = A_x \cos\phi + A_y \sin\phi$ $A_\phi = -A_x \sin\phi + A_y \cos\phi$ $A_z = A_z$		$A_{rc} = \dfrac{A_{rs} r_c + A_\theta z}{\sqrt{r_c^2 + z^2}}$ $A_\phi = A_\phi$ $A_z = \dfrac{A_{rs} z - A_\theta r_c}{\sqrt{r_c^2 + z^2}}$
Spherical A_r, A_θ, A_ϕ	$A_r = A_x \sin\theta\cos\phi + A_y\sin\theta\sin\phi + A_z\cos\theta$ $A_\theta = A_x\cos\theta\cos\phi + A_y\cos\theta\sin\phi - A_z\sin\theta$ $A_\phi = -A_x\sin\phi + A_y\cos\phi$	$A_{rs} = A_{rc}\sin\theta + A_z\cos\theta$ $A_\theta = A_{rc}\cos\theta - A_z\sin\theta$ $A_\phi = A_\phi$	

$(\mathbf{A} \times \mathbf{B}) \times \mathbf{C}$ is perpendicular to the vector $(\mathbf{A} \times \mathbf{B})$ as well as to the vector $\mathbf{C}$. Hence $(\mathbf{A} \times \mathbf{B}) \times \mathbf{C}$ lies in the plane containing $\mathbf{A}$ and $\mathbf{B}$. In view of this, $(\mathbf{A} \times \mathbf{B}) \times \mathbf{C}$ can be written as a superposition of two vectors proportional to $\mathbf{A}$ and $\mathbf{B}$; that is,

$$(\mathbf{A} \times \mathbf{B}) \times \mathbf{C} = m\mathbf{A} + n\mathbf{B}$$

To find m and n, we expand $(\mathbf{A} \times \mathbf{B}) \times \mathbf{C}$. Thus

$$
(\mathbf{A} \times \mathbf{B}) \times \mathbf{C} = \begin{vmatrix} \mathbf{i}_1 & \mathbf{i}_2 & \mathbf{i}_3 \\ (A_2B_3 - A_3B_2) & (A_3B_1 - A_1B_3) & (A_1B_2 - A_2B_1) \\ C_1 & C_2 & C_3 \end{vmatrix}
$$

$$
= (A_2C_2B_1 + A_3C_3B_1 - B_2C_2A_1 - B_3C_3A_1)\mathbf{i}_1
$$
$$
+ (A_1C_1B_2 + A_3C_3B_2 - B_1C_1A_2 - B_3C_3A_2)\mathbf{i}_2
$$
$$
+ (A_1C_1B_3 + A_2C_2B_3 - B_1C_1A_3 - B_2C_2A_3)\mathbf{i}_3
$$
$$
= (A_1C_1B_1 + A_2C_2B_1 + A_3C_3B_1 - B_1C_1A_1
$$
$$
- B_2C_2A_1 - B_3C_3A_1)\mathbf{i}_1
$$
$$
+ (A_1C_1B_2 + A_2C_2B_2 + A_3C_3B_2 - B_1C_1A_2
$$
$$
- B_2C_2A_2 - B_3C_3A_2)\mathbf{i}_2
$$
$$
+ (A_1C_1B_3 + A_2C_2B_3 + A_3C_3B_3 - B_1C_1A_3
$$
$$
- B_2C_2A_3 - B_3C_3A_3)\mathbf{i}_3
$$
$$
= (\mathbf{A} \cdot \mathbf{C})\mathbf{B} - (\mathbf{B} \cdot \mathbf{C})\mathbf{A}
$$

Similarly, it can be shown that

$$\mathbf{A} \times (\mathbf{B} \times \mathbf{C}) = (\mathbf{A} \cdot \mathbf{C})\mathbf{B} - (\mathbf{A} \cdot \mathbf{B})\mathbf{C} \quad \blacksquare$$

EXAMPLE 1-8.   Given

$$\mathbf{A} = 2\mathbf{i}_x - \mathbf{i}_z$$
$$\mathbf{B} = 2\mathbf{i}_x - \mathbf{i}_y + 2\mathbf{i}_z$$
$$\mathbf{C} = 2\mathbf{i}_x - 3\mathbf{i}_y + \mathbf{i}_z$$

We wish to perform several operations with these vectors as follows:

(a) $\mathbf{A} + \mathbf{B} = (2\mathbf{i}_x - \mathbf{i}_z) + (2\mathbf{i}_x - \mathbf{i}_y + 2\mathbf{i}_z) = 4\mathbf{i}_x - \mathbf{i}_y + \mathbf{i}_z$

(b) $\mathbf{B} - \mathbf{C} = (2\mathbf{i}_x - \mathbf{i}_y + 2\mathbf{i}_z) - (2\mathbf{i}_x - 3\mathbf{i}_y + \mathbf{i}_z) = 2\mathbf{i}_y + \mathbf{i}_z$

(c) $\mathbf{A} + \mathbf{B} - \mathbf{C} = \mathbf{A} + (\mathbf{B} - \mathbf{C}) = (2\mathbf{i}_x - \mathbf{i}_z) + (2\mathbf{i}_y + \mathbf{i}_z) = 2\mathbf{i}_x + 2\mathbf{i}_y$

(d) $|\mathbf{B}| = \sqrt{2^2 + (-1)^2 + 2^2} = 3$

(e) $\mathbf{i}_B = \dfrac{\mathbf{B}}{|\mathbf{B}|} = \dfrac{2\mathbf{i}_x - \mathbf{i}_y + 2\mathbf{i}_z}{3} = \dfrac{2}{3}\mathbf{i}_x - \dfrac{1}{3}\mathbf{i}_y + \dfrac{2}{3}\mathbf{i}_z$

(f) $\mathbf{A} \cdot \mathbf{B} = (2\mathbf{i}_x - \mathbf{i}_z) \cdot (2\mathbf{i}_x - \mathbf{i}_y + 2\mathbf{i}_z) = 4 + 0 - 2 = 2$

(g) Cosine of the angle between $\mathbf{A}$ and $\mathbf{B} = \dfrac{\mathbf{A} \cdot \mathbf{B}}{|\mathbf{A}||\mathbf{B}|} = \dfrac{2}{3\sqrt{5}}$

(h) $\mathbf{B} \times \mathbf{C} = \begin{vmatrix} \mathbf{i}_x & \mathbf{i}_y & \mathbf{i}_z \\ 2 & -1 & 2 \\ 2 & -3 & 1 \end{vmatrix} = 5\mathbf{i}_x + 2\mathbf{i}_y - 4\mathbf{i}_z$

(i) Sine of the angle between $\mathbf{B}$ and $\mathbf{C} = \dfrac{|\mathbf{B} \times \mathbf{C}|}{|\mathbf{B}||\mathbf{C}|} = \dfrac{\sqrt{45}}{3\sqrt{14}} = \sqrt{\dfrac{5}{14}}$

(j) $\mathbf{B} \cdot \mathbf{C} \times \mathbf{A} = \begin{vmatrix} 2 & -1 & 2 \\ 2 & -3 & 1 \\ 2 & 0 & -1 \end{vmatrix} = 14$

(k) $\mathbf{A} \times (\mathbf{B} \times \mathbf{C}) = \begin{vmatrix} \mathbf{i}_x & \mathbf{i}_y & \mathbf{i}_z \\ 2 & 0 & -1 \\ 5 & 2 & -4 \end{vmatrix} = 2\mathbf{i}_x + 3\mathbf{i}_y + 4\mathbf{i}_z$

(l) Components of $\mathbf{B}$ in spherical coordinates at $(1, \pi/2, \pi)$

$$B_r = B_x \sin \theta \cos \phi + B_y \sin \theta \sin \phi + B_z \cos \theta$$

$$= 2 \sin \frac{\pi}{2} \cos \pi - 1 \sin \frac{\pi}{2} \sin \pi + 2 \cos \frac{\pi}{2} = -2$$

$$B_\theta = B_x \cos \theta \cos \phi + B_y \cos \theta \sin \phi - B_z \sin \theta$$

$$= 2 \cos \frac{\pi}{2} \cos \pi - 1 \cos \frac{\pi}{2} \cos \pi - 2 \sin \frac{\pi}{2} = -2$$

$$B_\phi = -B_x \sin \phi + B_y \cos \phi = -2 \sin \pi - 1 \cos \pi = 1$$

(m) By using a vector product, find any vector perpendicular to $\mathbf{B}$. We can consider the unit vector $\mathbf{i}_x$ for simplicity. Then

$$\mathbf{D} = \mathbf{B} \times \mathbf{i}_x = \begin{vmatrix} \mathbf{i}_x & \mathbf{i}_y & \mathbf{i}_z \\ 2 & -1 & 2 \\ 1 & 0 & 0 \end{vmatrix} = 2\mathbf{i}_y + \mathbf{i}_z$$

We can verify that $\mathbf{D}$ is indeed perpendicular to $\mathbf{B}$ by showing that

$$\mathbf{B} \cdot \mathbf{D} = (2\mathbf{i}_x - \mathbf{i}_y + 2\mathbf{i}_z) \cdot (2\mathbf{i}_y + \mathbf{i}_z) = 0 - 2 + 2 = 0 \quad \blacksquare$$

1.4 Scalar and Vector Fields

A mathematical function or a graphical sketch constructed so as to describe the variation of a quantity in a given region is said to represent the "field" of that quantity associated with that region. We distinguish between scalar and vector fields, depending upon whether the quantity of interest is a scalar or a vector. We will first discuss scalar fields or functions. A simple example of scalar function is one by means of which we attempt to describe how the depth d of water in a lake varies from point to point on the lake surface. Assuming the lake surface to be plane, we first set up a two-dimensional coordinate system to define each and every point on the surface by a set of coordinates (x, y) with respect to a chosen origin. To each set of coordinates (x, y) we assign a number for d, which represents the depth of water beneath the point defined by that set of coordinates. The coordinates (x, y) are the independent variables and the depth d is the dependent variable. The function $d(x, y)$ represents the depth field associated with points on the surface of the lake.

If we join points in the xy plane for which the depth is equal to a particular constant, we obtain a curve known as a constant-depth contour. Similarly, by joining the points which have associated with them the same depth value but different from the previous constant, we obtain a different constant-depth contour. In this manner we can draw several constant-depth contours with convenient increments ranging from zero depth to the greatest depth, as shown in Fig. 1.17. The constant-depth contours provide a graphical representation of the depth field $d(x, y)$.

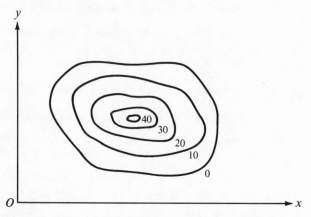

Fig. 1.17. Sketch of a two-dimensional scalar field $d(x, y)$ showing contours of constant values of d.

To add one more dimension to the scalar field, let us consider the temperature field associated with points inside a room. We can set up a coordinate system to define the location of each and every point inside the room with respect to a chosen origin. However, we will need all three coordinates (x, y, z) in this case instead of just two coordinates as in the previous example. To each set of coordinates (x, y, z) we assign a number which represents the temperature T at the point defined by that set of (x, y, z). The coordinates (x, y, z) are the independent variables and the temperature T is the dependent variable. The function $T(x, y, z)$ represents the temperature field. If we join points in the coordinate system for which the temperature is equal to a particular constant, we obtain a surface which is known as a constant-temperature or isothermal surface. Similarly, by joining the points which have associated with them the same temperature value but different from the previous constant, we obtain a different isothermal surface. In this manner we describe the temperature field in the room by a set of isothermal surfaces.

The addition of time t as an independent variable introduces one more dimension to the problem. The temperature at each and every point in the room varies with time in general so that the discussion in the preceding paragraph is valid only for fixed times or for the special case in which the temperature does not vary with time. In the latter case the temperature field in the room is said to be "static." In the general case, however, the temperature distributions measured throughout the room at two times t_1 and t_2 can be different so that the shapes of the isothermal surfaces representing the same constant temperatures at the two times can be different. Mathematically, we need two different functions of (x, y, z) to describe the temperature fields at these two times. To generalize this statement, since t is a continuous independent variable, T is a function of four independent variables x, y, z, and t. Thus we describe the time-varying temperature field in the room by a function $T(x, y, z, t)$.

The same concepts can be used to describe vector fields. However, in the case of vector quantities, we need to describe not only how the magnitude of the vector varies as a function of the independent variables but also how the direction of the vector varies. Hence, if we wish to describe the variation of a vector as a function of position in three-dimensional space and also of time, we associate a set of two numbers to each possible combination (x, y, z, t) or (r, ϕ, z, t) or (r, θ, ϕ, t), depending upon the coordinate system used, where one of the two numbers represents the magnitude and the other, the direction of the vector. More conveniently, since the variation of the unit vectors in each coordinate system is completely known (the unit vectors are independent of time), it is sufficient if we describe how each component of the vector of interest varies with (x, y, z, t) or (r, ϕ, z, t) or (r, θ, ϕ, t). Thus we have reduced the problem of describing a vector field to one of describing the component scalar fields. Mathematically, we write

$$\mathbf{F}(x, y, z, t) = F_x(x, y, z, t)\mathbf{i}_x + F_y(x, y, z, t)\mathbf{i}_y + F_z(x, y, z, t)\mathbf{i}_z \qquad (1\text{-}38)$$

$$\mathbf{F}(r, \phi, z, t) = F_r(r, \phi, z, t)\mathbf{i}_r + F_\phi(r, \phi, z, t)\mathbf{i}_\phi + F_z(r, \phi, z, t)\mathbf{i}_z \qquad (1\text{-}39)$$

$$\mathbf{F}(r, \theta, \phi, t) = F_r(r, \theta, \phi, t)\mathbf{i}_r + F_\theta(r, \theta, \phi, t)\mathbf{i}_\theta + F_\phi(r, \theta, \phi, t)\mathbf{i}_\phi \qquad (1\text{-}40)$$

where $\mathbf{F}$ is the vector of interest and remembering that the unit vectors $\mathbf{i}_r$ and $\mathbf{i}_\phi$ in cylindrical coordinates and $\mathbf{i}_r$, $\mathbf{i}_\theta$, and $\mathbf{i}_\phi$ in spherical coordinates are themselves known functions of the coordinates.

EXAMPLE 1-9.   Consider a circular disk of radius a rotating with a constant angular velocity ω about an axis passing normally through its center. It is desired to describe the linear velocity vector field associated with the points on the disk.

We can choose the center of the disk as the origin and set up a two-dimensional coordinate system. We have a choice of the coordinates (x, y) or the coordinates (r, ϕ). Since the linear velocity of a point is equal to the product of the angular velocity and the distance from the axis about which the disk rotates, we note that points equidistant from the center have the same magnitude of velocity. Also, the velocity is directed everywhere in the angular direction. This suggests the use of (r, ϕ) coordinate system. Then, at a point (r, ϕ) the velocity magnitude is $r\omega$ and its direction is $\mathbf{i}_\phi$, as shown in Fig. 1.18(a). Thus the expression for the linear velocity vector field is given as

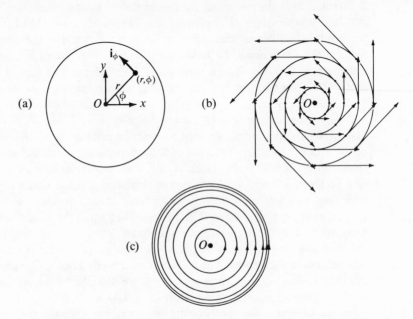

Fig. 1.18. (a) Rotating disk. (b) Field of the linear velocity vector associated with the points on the rotating disk. (c) Same as (b) with the arrows omitted and the density of direction lines used to indicate the magnitude variation.

$$\mathbf{v}(r, \phi) = v_r(r, \phi)\mathbf{i}_r + v_\phi(r, \phi)\mathbf{i}_\phi = r\omega\mathbf{i}_\phi \qquad \text{for } r < a \qquad (1\text{-}41)$$

The constant-magnitude contours are circles centered at the origin and having radii proportional to the magnitudes. The velocity direction is everywhere tangential to these circles. One way of pictorially representing the vector field is by drawing at several points vectors whose lengths are equal to the radii of the circles passing through those points and hence proportional to the velocity magnitudes at those points and whose directions are everywhere along $\mathbf{i}_\phi$, as shown in Fig. 1.18(b). For this field, these vectors are everywhere tangential to the constant-magnitude contours (circles) passing through those points; that is, the constant-magnitude contours are also the curves along which points on the disk move as the disk rotates. Such curves are known as direction lines since they indicate the direction of the vector field. The constant-magnitude contours and direction lines are not the same curves for a general vector field. The pictorial representation of Fig. 1.18(b) can be simplified by omitting the vectors and simply placing arrowheads along the circles, that is, the direction lines, as shown in Fig. 1.18(c). Also, by decreasing the spacing between the direction lines as r increases, the density of direction lines is used to indicate the magnitude variation. This is the common procedure adapted for graphically depicting a vector field. In Chapter 2 we will discuss a procedure for obtaining the equations for the direction lines from the field expressions. ∎

1.5 Differentiation of Vectors

In calculus, we have learned the rules for differentiation of scalar functions. If f is a function of x, then the derivative of f with respect to x is

$$\frac{df}{dx} = \lim_{\Delta x \to 0} \frac{f(x + \Delta x) - f(x)}{\Delta x} \tag{1-42}$$

If f is a function of (x, y, z), then the partial derivative of f with respect to x is

$$\frac{\partial f}{\partial x} = \lim_{\Delta x \to 0} \frac{f(x + \Delta x, y, z) - f(x, y, z)}{\Delta x} \tag{1-43}$$

and the differential increase in f from a point (x, y, z) to a neighboring point $(x + dx, y + dy, z + dz)$ is

$$df = \frac{\partial f}{\partial x} dx + \frac{\partial f}{\partial y} dy + \frac{\partial f}{\partial z} dz \tag{1-44}$$

where $\partial f/\partial y$ and $\partial f/\partial z$ are given by expressions similar to (1-43).

Differentiation of vector functions is defined in the same manner as differentiation of scalar functions. Let us consider a vector function $\mathbf{A}(x, y, z)$. The differential increment in $\mathbf{A}$ from a point (x, y, z) to a neighboring point $(x + dx, y + dy, z + dz)$ is

$$d\mathbf{A} = \frac{\partial \mathbf{A}}{\partial x} dx + \frac{\partial \mathbf{A}}{\partial y} dy + \frac{\partial \mathbf{A}}{\partial z} dz \tag{1-45}$$

where

$$\frac{\partial \mathbf{A}}{\partial x} = \lim_{\Delta x \to 0} \frac{\mathbf{A}(x + \Delta x, y, z) - \mathbf{A}(x, y, z)}{\Delta x} \tag{1-46}$$

$$\frac{\partial \mathbf{A}}{\partial y} = \lim_{\Delta y \to 0} \frac{\mathbf{A}(x, y + \Delta y, z) - \mathbf{A}(x, y, z)}{\Delta y} \tag{1-47}$$

$$\frac{\partial \mathbf{A}}{\partial z} = \lim_{\Delta z \to 0} \frac{\mathbf{A}(x, y, z + \Delta z) - \mathbf{A}(x, y, z)}{\Delta z} \tag{1-48}$$

Since $[\mathbf{A}(x + \Delta x, y, z) - \mathbf{A}(x, y, z)]$ is a vector, the derivative $\partial\mathbf{A}/\partial x$ is a vector which, in general, is oriented in a direction different from that of **A**. Similarly, $\partial\mathbf{A}/\partial y$ and $\partial\mathbf{A}/\partial z$ are vectors which, in general, are oriented in directions different from that of **A**. If we express a vector function in terms of its component vector functions in cartesian coordinates, that is, if

$$\mathbf{A}(x, y, z) = A_x(x, y, z)\mathbf{i}_x + A_y(x, y, z)\mathbf{i}_y + A_z(x, y, z)\mathbf{i}_z \qquad (1\text{-}49)$$

then

$$d\mathbf{A} = dA_x\,\mathbf{i}_x + dA_y\,\mathbf{i}_y + dA_z\,\mathbf{i}_z \qquad (1\text{-}50)$$

since $\mathbf{i}_x, \mathbf{i}_y$ and $\mathbf{i}_z$ are independent of x, y and z.

EXAMPLE 1-10.   The unit vector $\mathbf{i}_r$ in cylindrical coordinates is independent of r and z but not ϕ. Hence $\partial\mathbf{i}_r/\partial r = d\mathbf{i}_r/\partial z = 0$ but $\partial\mathbf{i}_r/\partial\phi \neq 0$. We wish to find $\partial\mathbf{i}_r/\partial\phi$ in two ways: (a) from first principles, and (b) by using (1-50).

(a) By definition,

$$\frac{\partial\mathbf{i}_r}{\partial\phi} = \lim_{\Delta\phi\to 0} \frac{\mathbf{i}_r(r, \phi + \Delta\phi, z) - \mathbf{i}_r(r, \phi, z)}{\Delta\phi} \qquad (1\text{-}51)$$

To deduce the right side of (1-51), consider the unit vectors $\mathbf{i}_{r1}$ and $\mathbf{i}_{r2}$ at the points $P(r, \phi, z)$ and $Q(r, \phi + \Delta\phi, z)$, as shown in Fig. 1.19. Then we can write

$$\mathbf{i}_{r2} = (\mathbf{i}_{r2} \cdot \mathbf{i}_{r1})\mathbf{i}_{r1} + (\mathbf{i}_{r2} \cdot \mathbf{i}_\phi)\mathbf{i}_\phi = \cos\Delta\phi\,\mathbf{i}_{r1} + \sin\Delta\phi\,\mathbf{i}_\phi \qquad (1\text{-}52)$$

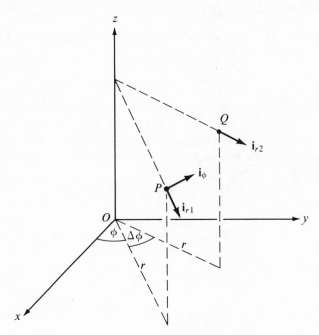

Fig. 1.19. For the evaluation of $\partial\mathbf{i}_r/\partial\phi$.

where $\mathbf{i}_\phi$ is the unit vector in the ϕ direction at point P. Using (1-52), we have

$$\mathbf{i}_r(r, \phi + \Delta\phi, z) - \mathbf{i}_r(r, \phi, z) = \mathbf{i}_{r2} - \mathbf{i}_{r1}$$
$$= (\cos \Delta\phi - 1)\mathbf{i}_{r1} + \sin \Delta\phi\, \mathbf{i}_\phi \qquad (1\text{-}53)$$

Substituting (1-53) into (1-51), we get

$$\frac{\partial \mathbf{i}_r}{\partial \phi} = \lim_{\Delta\phi \to 0} \frac{(\cos \Delta\phi - 1)\mathbf{i}_{r1} + \sin \Delta\phi\, \mathbf{i}_\phi}{\Delta\phi} = \mathbf{i}_\phi \qquad (1\text{-}54)$$

(b) To use (1-50), we first note that since $\mathbf{i}_r$ is only a function of ϕ, $\partial \mathbf{i}_r/\partial\phi$ is the same as $d\mathbf{i}_r/d\phi$. Expressing $\mathbf{i}_r$ in terms of $\mathbf{i}_x$, $\mathbf{i}_y$, and $\mathbf{i}_z$, we have

$$\mathbf{i}_r = \cos \phi\, \mathbf{i}_x + \sin \phi\, \mathbf{i}_y \qquad (1\text{-}55)$$

Then, from (1-50), we obtain

$$d\mathbf{i}_r = d(\cos \phi)\, \mathbf{i}_x + d(\sin \phi)\, \mathbf{i}_y$$
$$= (-\sin \phi\, \mathbf{i}_x + \cos \phi\, \mathbf{i}_y)d\phi = d\phi\, \mathbf{i}_\phi \qquad (1\text{-}56)$$

or

$$\frac{\partial \mathbf{i}_r}{\partial \phi} = \frac{d\mathbf{i}_r}{d\phi} = \mathbf{i}_\phi$$

which agrees with (1-54). Partial derivatives of unit vectors obtained in this manner are listed in Table 1.6. ∎

TABLE 1.6. Partial Derivatives of Unit Vectors; All Derivatives Not Listed in the Table are Zero

	∂x	∂y	∂z	∂r_c	$\partial \phi$	$\partial \theta$
$\partial \mathbf{i}_{rc}/$	$\dfrac{-\sin \phi}{r_c} \mathbf{i}_\phi$	$\dfrac{\cos \phi}{r_c} \mathbf{i}_\phi$	0	0	$\mathbf{i}_\phi$	0
$\partial \mathbf{i}_\phi/$	$\dfrac{\sin \phi}{r_c} \mathbf{i}_{rc}$	$\dfrac{-\cos \phi}{r_c} \mathbf{i}_{rc}$	0	0	$-\mathbf{i}_{rc}$	0
$\partial \mathbf{i}_{rs}/$	$\dfrac{1}{r_s}(-\sin \phi\, \mathbf{i}_\phi$ $+ \cos \theta \cos \phi\, \mathbf{i}_\theta)$	$\dfrac{1}{r_s}(\cos \phi\, \mathbf{i}_\phi$ $+ \cos \theta \sin \phi\, \mathbf{i}_\theta)$	$\dfrac{-\sin \theta}{r_s} \mathbf{i}_\theta$	$\dfrac{\cos \theta}{r_s} \mathbf{i}_\theta$	$\sin \theta\, \mathbf{i}_\phi$	$\mathbf{i}_\theta$
$\partial \mathbf{i}_\theta/$	$\dfrac{\cot \theta}{r_s}(-\sin \phi\, \mathbf{i}_\phi$ $- \sin \theta \cos \phi\, \mathbf{i}_{rs})$	$\dfrac{\cot \theta}{r_s}(\cos \phi\, \mathbf{i}_\phi$ $- \sin \theta \sin \phi\, \mathbf{i}_{rs})$	$\dfrac{\sin \theta}{r_s} \mathbf{i}_{rs}$	$\dfrac{-\cos \theta}{r_s} \mathbf{i}_{rs}$	$\cos \theta\, \mathbf{i}_\phi$	$-\mathbf{i}_{rs}$

Expressions similar to (1-50) are not true in cylindrical and spherical coordinates; that is,

$$d\mathbf{A} \neq dA_r\, \mathbf{i}_r + dA_\phi\, \mathbf{i}_\phi + dA_z\, \mathbf{i}_z \qquad (1\text{-}57)$$

$$d\mathbf{A} \neq dA_r\, \mathbf{i}_r + dA_\theta\, \mathbf{i}_\theta + dA_\phi\, \mathbf{i}_\phi \qquad (1\text{-}58)$$

To derive the correct expressions for these two coordinate systems, we make use of the differentiation rule,

$$d(f\mathbf{A}) = f\, d\mathbf{A} + df\, \mathbf{A} \qquad (1\text{-}59)$$

where f is a scalar function. Thus if

$$\mathbf{A} = A_r\mathbf{i}_r + A_\phi\mathbf{i}_\phi + A_z\mathbf{i}_z$$

we have

$$dA = d(A_r\mathbf{i}_r) + d(A_\phi\mathbf{i}_\phi) + d(A_z\mathbf{i}_z)$$
$$= A_r\, d\mathbf{i}_r + dA_r\, \mathbf{i}_r + A_\phi\, d\mathbf{i}_\phi + dA_\phi\, \mathbf{i}_\phi + dA_z\, \mathbf{i}_z \qquad (1\text{-}60)$$

Similarly, if

$$\mathbf{A} = A_r\mathbf{i}_r + A_\theta\mathbf{i}_\theta + A_\phi\mathbf{i}_\phi$$

we have

$$dA = A_r\, d\mathbf{i}_r + dA_r\, \mathbf{i}_r + A_\theta\, d\mathbf{i}_\theta + dA_\theta\, \mathbf{i}_\theta + A_\phi\, d\mathbf{i}_\phi + dA_\phi\, \mathbf{i}_\phi \qquad (1\text{-}61)$$

Finally, if $\mathbf{A}$ is also a function of t in addition to x, y, z, we have

$$dA = dA_x\, \mathbf{i}_x + dA_y\, \mathbf{i}_y + dA_z\, \mathbf{i}_z \qquad (1\text{-}50)$$

where

$$dA_i = \frac{\partial A_i}{\partial x}\, dx + \frac{\partial A_i}{\partial y}\, dy + \frac{\partial A_i}{\partial z}\, dz + \frac{\partial A_i}{\partial t}\, dt \qquad i = x, y, z \qquad (1\text{-}62)$$

Rules for the differentiation of dot and cross products of vectors are as follows:

$$d(\mathbf{A} \cdot \mathbf{B}) = d\mathbf{A} \cdot \mathbf{B} + \mathbf{A} \cdot d\mathbf{B} \qquad (1\text{-}63)$$

$$d(\mathbf{A} \times \mathbf{B}) = d\mathbf{A} \times \mathbf{B} + \mathbf{A} \times d\mathbf{B} \qquad (1\text{-}64)$$

1.6 The Gradient

Gradient is an operation performed on a scalar function which results in a vector function. The magnitude of this vector function at any point in the region of the scalar field is the maximum rate of increase of the scalar function at that point. The direction of the vector function at that point is the direction in which this maximum rate of increase occurs. To illustrate this concept mathematically, let us consider a scalar function $V(x, y, z)$ which is single-valued everywhere so that it is differentiable. From calculus, we can express the change in V from a point (x, y, z) to another point $(x + dx, y + dy, z + dz)$ an infinitesimal distance away from it as

$$dV = \frac{\partial V}{\partial x}\, dx + \frac{\partial V}{\partial y}\, dy + \frac{\partial V}{\partial z}\, dz$$

$$= \left(\frac{\partial V}{\partial x}\, \mathbf{i}_x + \frac{\partial V}{\partial y}\, \mathbf{i}_y + \frac{\partial V}{\partial z}\, \mathbf{i}_z\right) \cdot (dx\, \mathbf{i}_x + dy\, \mathbf{i}_y + dz\, \mathbf{i}_z) \qquad (1\text{-}65)$$

$$= \nabla V \cdot d\mathbf{l}$$

where the symbol ∇ stands for "del" and is a vector operator defined as

$$\nabla = \frac{\partial}{\partial x}\, \mathbf{i}_x + \frac{\partial}{\partial y}\, \mathbf{i}_y + \frac{\partial}{\partial z}\, \mathbf{i}_z \qquad (1\text{-}66)$$

When "del" operates on a scalar function, the operation is known as evaluating the gradient of the scalar function; that is, ∇V is the gradient of V. Thus

$$\nabla V = \frac{\partial V}{\partial x}\mathbf{i}_x + \frac{\partial V}{\partial y}\mathbf{i}_y + \frac{\partial V}{\partial z}\mathbf{i}_z \qquad (1\text{-}67)$$

The vector $d\mathbf{l}$ is the infinitesimal displacement vector drawn from the point (x, y, z) to the point $(x + dx, y + dy, z + dz)$.

To discuss the physical significance of ∇V, let us consider a surface containing a point $P(x_0, y_0, z_0)$ on which the scalar function is constant and equal to $V(x_0, y_0, z_0) = V_0$. If we now consider another point $Q(x_0 + dx, y_0 + dy, z_0 + dz)$ on the same surface and an infinitesimal distance away from P, dV between these two points is zero since V remains constant throughout this surface. It follows from (1-65) that for the vector $d\mathbf{l}$ drawn from P to Q on this surface,

$$\nabla V \cdot d\mathbf{l} = 0 \qquad (1\text{-}68)$$

and hence ∇V is perpendicular to $d\mathbf{l}$. But since (1-68) is true for all points Q on the constant V surface surrounding P, ∇V must be normal to all possible infinitesimal displacement vectors drawn away from P on the constant V surface, and hence it is normal to that surface. Actually, it is sufficient for ∇V to be normal to any two different infinitesimal displacement vectors drawn away from P on the constant V surface to conclude that ∇V is normal to that surface. Thus we can reach the general conclusion that the gradient of a scalar function at any point is directed normal to the surface passing through that point and on which the value of the scalar function is a constant. Designating $\mathbf{i}_n$ as the unit vector normal to the constant V surface, we then have

$$\nabla V = |\nabla V|\,\mathbf{i}_n \qquad (1\text{-}69)$$

Let us now consider two adjacent surfaces of constant V equal to V_0 and $V_0 + dV$, respectively, as shown in Fig. 1.20. Let P and Q be points on the V_0 and $V_0 + dV$ surfaces, respectively, and let $d\mathbf{l}$ be the displacement vector drawn from P to Q. Then, since dV is infinitesimally small and hence the two surfaces are infinitesimally close to each other, we have, according to (1-65),

$$dV = (\nabla V)_{\text{at } P} \cdot d\mathbf{l} = |\nabla V|\,\mathbf{i}_n \cdot dl\,\mathbf{i}_l \qquad (1\text{-}70)$$

where we have substituted the right side of (1-69) for $(\nabla V)_{\text{at } P}$ and expressed $d\mathbf{l}$ as $dl\,\mathbf{i}_l$. From (1-70), we have

$$\left(\frac{dV}{dl}\right)_{\text{at } P} = |\nabla V|\,\mathbf{i}_n \cdot \mathbf{i}_l = |\nabla V|\cos\alpha \qquad (1\text{-}71)$$

where α is the angle between $\mathbf{i}_n$ and $\mathbf{i}_l$. But since the maximum value of $\cos\alpha$ is unity, the maximum value of $(dV/dl)_{\text{at } P}$ is equal to $|\nabla V|$ and it occurs for α equal to zero, that is, for the case in which $\mathbf{i}_l = \mathbf{i}_n$. Thus $|\nabla V|$ is indeed the maximum rate of increase of V and it occurs in the direction normal to the constant V surface, consistent with the conclusion of the previous paragraph.

EXAMPLE 1-11. Consider the scalar function $V(x, y, z) = xy$. Obtain a unit vector normal to the constant V surface of value 2 at the point $(2, 1, 0)$ in two ways:

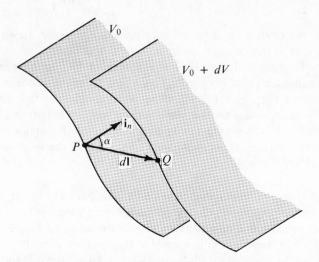

Fig. 1.20. Surfaces of constant V equal to V_0 and $V_0 + dV$, respectively, for evaluating ∇V.

(a) by using the cross product of two vectors which are tangential to the surface at that point; (b) by using the concept of the gradient of a scalar function. What is the maximum rate of increase of the scalar function at the point $(2, 1, 0)$?

(a) The constant V surface of value 2 is given by

$$xy = 2$$

A cut section of this surface is shown in Fig. 1.21(a) and its cross section in the xy plane is repeated in Fig. 1.21(b). The unit vector $\mathbf{i}_z$ is tangential to the surface everywhere. Hence it is sufficient if we find another vector tangential to the surface at $(2, 1, 0)$ so that we can take the cross product of these two vectors to find a unit vector normal to the surface. For simplicity, we can find the tangential unit vector $\mathbf{i}_t$ lying in the xy plane. To find the components of $\mathbf{i}_t$ along $\mathbf{i}_x$ and $\mathbf{i}_y$, we need the angle α which the tangent to the curve $xy = 2$ in Fig. 1.21(b) makes with the x axis. Noting that the curve is defined by $y = 2/x$, we obtain

$$\left(\frac{dy}{dx}\right)_{2,1} = \left(-\frac{2}{x^2}\right)_{2,1} = -\frac{1}{2}$$

Hence $\tan \alpha = \frac{1}{2}$ or $\alpha = 26.6°$. Now,

$$\mathbf{i}_t = \cos \alpha \, \mathbf{i}_x - \sin \alpha \, \mathbf{i}_y = \frac{2}{\sqrt{5}}\mathbf{i}_x - \frac{1}{\sqrt{5}}\mathbf{i}_y$$

The unit vector normal to the surface $xy = 2$ at $(2, 1, 0)$ is then given by

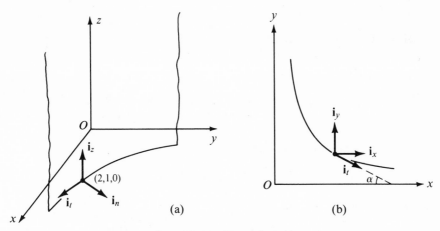

Fig. 1.21. For Example 1-11.

$$\mathbf{i}_n = \mathbf{i}_z \times \mathbf{i}_t = \begin{vmatrix} \mathbf{i}_x & \mathbf{i}_y & \mathbf{i}_z \\ 0 & 0 & 1 \\ \dfrac{2}{\sqrt{5}} & -\dfrac{1}{\sqrt{5}} & 0 \end{vmatrix} = \frac{1}{\sqrt{5}}\mathbf{i}_x + \frac{2}{\sqrt{5}}\mathbf{i}_y$$

(b) The direction of the gradient of a scalar is normal to the surface of constant value of the scalar. Hence by evaluating the gradient of the given scalar function at the point $(2, 1, 0)$ we can find the required unit vector.

$$\mathbf{\nabla}V = \frac{\partial(xy)}{\partial x}\mathbf{i}_x + \frac{\partial(xy)}{\partial y}\mathbf{i}_y = y\mathbf{i}_x + x\mathbf{i}_y$$

$$(\mathbf{\nabla}V)_{2,1,0} = \mathbf{i}_x + 2\mathbf{i}_y$$

This vector is normal to the surface $xy = 2$. To find the unit vector we divide it by its magnitude which is $\sqrt{5}$. Thus

$$\mathbf{i}_n = \frac{1}{\sqrt{5}}\mathbf{i}_x + \frac{2}{\sqrt{5}}\mathbf{i}_y$$

which agrees with the result of part (a). The maximum rate of increase of the scalar function at $(2, 1, 0)$ is the magnitude of the gradient. Hence it is equal to $\sqrt{5}$. ∎

Equation (1-67) gives the gradient of a scalar in cartesian coordinates. We can similarly consider cylindrical coordinates and write the following steps:

$$\begin{aligned} dV &= \frac{\partial V}{\partial r}\,dr + \frac{\partial V}{\partial \phi}\,d\phi + \frac{\partial V}{\partial z}\,dz \\ &= \left(\frac{\partial V}{\partial r}\mathbf{i}_r + \frac{\partial V}{\partial \phi}\mathbf{i}_\phi + \frac{\partial V}{\partial z}\mathbf{i}_z\right) \cdot (dr\,\mathbf{i}_r + d\phi\,\mathbf{i}_\phi + dz\,\mathbf{i}_z) \end{aligned} \qquad (1\text{-}72)$$

But, in cylindrical coordinates,

$$dl = dr\, \mathbf{i}_r + r\, d\phi\, \mathbf{i}_\phi + dz\, \mathbf{i}_z \qquad (1\text{-}24a)$$

Hence we have to modify (1-72) as

$$
\begin{aligned}
dV &= \frac{\partial V}{\partial r}\, dr + \frac{1}{r}\frac{\partial V}{\partial \phi}\, r\, d\phi + \frac{\partial V}{\partial z}\, dz \\
&= \left(\frac{\partial V}{\partial r}\mathbf{i}_r + \frac{1}{r}\frac{\partial V}{\partial \phi}\mathbf{i}_\phi + \frac{\partial V}{\partial z}\mathbf{i}_z\right) \cdot (dr\, \mathbf{i}_r + r\, d\phi\, \mathbf{i}_\phi + dz\, \mathbf{i}_z) \qquad (1\text{-}73) \\
&= \nabla V \cdot d\mathbf{l}
\end{aligned}
$$

Thus, in cylindrical coordinates,

$$\nabla V = \frac{\partial V}{\partial r}\mathbf{i}_r + \frac{1}{r}\frac{\partial V}{\partial \phi}\mathbf{i}_\phi + \frac{\partial V}{\partial z}\mathbf{i}_z \qquad (1\text{-}74)$$

Similarly, in spherical coordinates,

$$\nabla V = \frac{\partial V}{\partial r}\mathbf{i}_r + \frac{1}{r}\frac{\partial V}{\partial \theta}\mathbf{i}_\theta + \frac{1}{r \sin \theta}\frac{\partial V}{\partial \phi}\mathbf{i}_\phi \qquad (1\text{-}75)$$

EXAMPLE 1-12.   Find the rate at which the scalar function $V = r^2 \sin 2\phi$, in cylindrical coordinates, increases in the direction of the vector $\mathbf{A} = \mathbf{i}_r + \mathbf{i}_\phi$ at the point $(2, \pi/4, 0)$.

Evidently, the required quantity is $\nabla V \cdot \mathbf{A}/|\mathbf{A}|$ evaluated at $(2, \pi/4, 0)$.

$$
\begin{aligned}
\nabla V &= \frac{\partial}{\partial r}(r^2 \sin 2\phi)\, \mathbf{i}_r + \frac{1}{r}\frac{\partial}{\partial \phi}(r^2 \sin 2\phi)\, \mathbf{i}_\phi + \frac{\partial}{\partial z}(r^2 \sin 2\phi)\, \mathbf{i}_z \\
&= 2r \sin 2\varphi\, \mathbf{i}_r + 2r \cos 2\phi\, \mathbf{i}_\phi \\
\nabla V \cdot \mathbf{A} &= (2r \sin 2\phi\, \mathbf{i}_r + 2r \cos 2\phi\, \mathbf{i}_\phi) \cdot (\mathbf{i}_r + \mathbf{i}_\phi) \\
&= 2r \sin 2\phi + 2r \cos 2\phi \\
\frac{\nabla V \cdot \mathbf{A}}{|\mathbf{A}|} &= \sqrt{2}\, r \sin 2\phi + \sqrt{2}\, r \cos 2\phi
\end{aligned}
$$

Finally, the rate of increase of V along $\mathbf{A}$ at the point $(2, \pi/4, 0)$ is equal to $2\sqrt{2}$. ∎

1.7   Volume, Surface, and Line Integrals

In the study of electromagnetic fields, we repeatedly encounter three types of integrals: (a) the volume integral, (b) the surface integral, and (c) the line integral. We will discuss each of these separately and provide some examples for evaluating them.

a. The Volume Integral.

If the density of a quantity is specified throughout a certain volume, we make use of volume integral to evaluate the amount of that quantity in that volume. For example, let us assume that the density of mass ρ of a body is known as a function of the coordinates (x, y, z), (r, ϕ, z), or (r, θ, ϕ). To obtain the total mass contained in the volume occupied by the body, we divide the volume into a number of infinitesimal volumes $dv_1, dv_2, dv_3, \ldots$. Within each infinitesimal volume, the density may be considered to be constant so that the mass contained within each volume is given by the product of the infinitesimal volume and the density in that volume. The total mass m is then the sum of these several masses, that is,

$$m = \rho_1 \, dv_1 + \rho_2 \, dv_2 + \rho_3 \, dv_3 + \cdots = \sum_i \rho_i \, dv_i \qquad (1\text{-}76)$$

where ρ_i is the density associated with the volume dv_i. Equation (1-76) gives only the approximate value of m since the density within each infinitesimal volume is not quite constant. However, it becomes exact in the limit that dv_i tends to zero (i.e., shrinks to a point) in which case the summation becomes an integral

$$m = \int_V \rho \, dv \qquad (1\text{-}77)$$

where the integration is performed throughout the volume V of the body, as indicated by the letter V associated with the integral sign. The integral on the right side of (1-77) is known as a volume integral. The volume integral is a triple integral since dv is the product of three differential lengths.

EXAMPLE 1-13.   The density of mass of a spherical body of radius a centered at the origin is given by

$$\rho(r, \theta, \phi) = \frac{\rho_0}{r}$$

where ρ_0 is a constant. It is desired to find the mass m of the spherical body.

The differential volume dv in spherical coordinates is $r^2 \sin \theta \, dr \, d\theta \, d\phi$. Substituting for ρ and dv in (1-77) and introducing the limits for the three variables r, θ, and ϕ, we have

$$m = \int_{r=0}^{a} \int_{\theta=0}^{\pi} \int_{\phi=0}^{2\pi} \frac{\rho_0}{r} r^2 \sin \theta \, dr \, d\theta \, d\phi$$
$$= 2\pi \rho_0 a^2 \quad \blacksquare$$

b. The Surface Integral.

If the density of flow of a fluid or, in general, the flux density of any physical quantity is specified over a certain surface, we make use of surface integrals to evaluate the amount of the flux of that quantity crossing that surface. For example, let us assume that the density of current at all points on a particular surface S is known. Since current is due to the flow of charges, the current density at a point has magnitude and direction and hence is a vector. Let us denote the current density vector as $\mathbf{J}$. To obtain the current crossing the surface S, we first divide the surface into several infinitesimal areas of magnitudes $dS_1, dS_2, dS_3, \ldots$. Since each of these areas is very,

very small in magnitude, we can treat them as plane surfaces and define their orientations by their corresponding normal vectors $\mathbf{i}_{n1}, \mathbf{i}_{n2}, \mathbf{i}_{n3}, \ldots$. Furthermore, we can consider the current density vector associated with each area to be constant.

Let us then consider one infinitesimal area $dS_1\, \mathbf{i}_{n1}$ and its associated current density vector $\mathbf{J}_1$ as shown in Fig. 1.22. Let the angle between $\mathbf{i}_{n1}$

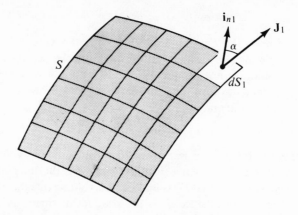

Fig. 1.22. Division of a surface S into several infinitesimal areas to evaluate the flux of a vector $\mathbf{J}$ crossing the surface.

and $\mathbf{J}_1$ be α. Then the projection of the area dS_1 onto a plane normal to the current density vector $\mathbf{J}_1$ has an area $dS_1 \cos \alpha$. The current crossing this projected area and hence the current crossing the surface $dS_1\, \mathbf{i}_{n1}$ is equal to $|\mathbf{J}_1|\, dS_1 \cos \alpha$, or $\mathbf{J}_1 \cdot dS_1\, \mathbf{i}_{n1}$. Similarly, we can obtain the currents flowing through all the other infinitesimal surfaces and add them up to give the total current I as

$$I = \mathbf{J}_1 \cdot dS_1\, \mathbf{i}_{n1} + \mathbf{J}_2 \cdot dS_2\, \mathbf{i}_{n2} + \mathbf{J}_3 \cdot dS_3\, \mathbf{i}_{n3} + \cdots$$
$$= \sum_k \mathbf{J}_k \cdot dS_k\, \mathbf{i}_{nk} = \sum_k \mathbf{J}_k \cdot d\mathbf{S}_k \tag{1-78}$$

where $d\mathbf{S}_k = dS\, \mathbf{i}_{nk}$. Equation (1-78) is approximate since the assumption of constant current density vector for any infinitesimal surface is true only in the limit that the area of that surface tends to zero (i.e., shrinks to a point). In this limit, the summation in (1-78) becomes an integral, giving us

$$I = \int_S \mathbf{J} \cdot d\mathbf{S} = \int_S \mathbf{J} \cdot \mathbf{i}_n \, dS \tag{1-79}$$

where the integration is performed over the entire surface S. The integral on the right side of (1-79) is known as a surface integral. The surface integral is a double integral since dS is the product of two differential lengths. If the surface is closed, we call it a closed surface integral and write it with a small circle associated with the integral sign, as follows:

$$I = \oint_S \mathbf{J} \cdot d\mathbf{S} \qquad (1\text{-}80)$$

Also, the normal vectors to the differential areas comprising the closed surface are then usually chosen to be pointing away from the volume bounding that surface so that the closed surface integral represents the flux emanating from the volume.

EXAMPLE 1-14.   In a certain region, the current density vector is given by

$$\mathbf{J} = 3x\mathbf{i}_x + (y - 3)\mathbf{i}_y + (2 + z)\mathbf{i}_z \quad \text{amp/m}^2$$

Find the total current flowing out of the surface of the box bounded by the five planes $x = 0$, $y = 0$, $y = 2$, $z = 0$, and $3x + z = 3$.

With reference to Fig. 1.23, we will consider the normal vector to be always pointing out of the box so that $\int \mathbf{J} \cdot d\mathbf{S}$ gives the current flowing out of the surface.

For the surface $x = 0$, $d\mathbf{S} = -dy\,dz\,\mathbf{i}_x$, $\mathbf{J} = (y - 3)\mathbf{i}_y + (2 + z)\mathbf{i}_z$.

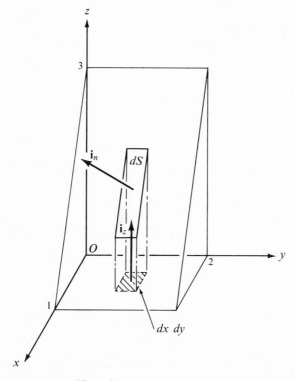

Fig. 1.23. For Example 1-14.

$$\mathbf{J} \cdot d\mathbf{S} = 0$$

$$\int \mathbf{J} \cdot d\mathbf{S} = 0$$

For the surface $y = 0$, $d\mathbf{S} = -dz\, dx\, \mathbf{i}_y$, $\mathbf{J} = 3x\mathbf{i}_x - 3\mathbf{i}_z + (2 + z)\mathbf{i}_z$.

$$\mathbf{J} \cdot d\mathbf{S} = 3\, dz\, dx$$

$$\int \mathbf{J} \cdot d\mathbf{S} = \int_{x=0}^{1} \int_{z=0}^{3-3x} 3\, dz\, dx = \frac{9}{2}$$

For the surface $y = 2$, $d\mathbf{S} = dz\, dx\, \mathbf{i}_y$, $\mathbf{J} = 3x\mathbf{i}_x - \mathbf{i}_y + (2 + z)\mathbf{i}_z$.

$$\mathbf{J} \cdot d\mathbf{S} = -1\, dz\, dx$$

$$\int \mathbf{J} \cdot d\mathbf{S} = \int_{x=0}^{1} \int_{z=0}^{3-3x} (-1)\, dz\, dx = -\frac{3}{2}$$

For the surface $z = 0$, $d\mathbf{S} = -dx\, dy\, \mathbf{i}_z$, $\mathbf{J} = 3x\mathbf{i}_x + (y - 3)\mathbf{i}_y + 2\mathbf{i}_z$.

$$\mathbf{J} \cdot d\mathbf{S} = -2\, dx\, dy$$

$$\int \mathbf{J} \cdot d\mathbf{S} = \int_{x=0}^{1} \int_{y=0}^{2} (-2)\, dx\, dy = -4$$

For the surface $3x + z = 3$,

$$\mathbf{i}_n = \frac{\nabla(3x + z)}{|\nabla(3x + z)|} = \frac{3\mathbf{i}_x + \mathbf{i}_z}{\sqrt{3^2 + 1^2}} = \frac{3}{\sqrt{10}}\mathbf{i}_x + \frac{1}{\sqrt{10}}\mathbf{i}_z$$

From $d\mathbf{S}\, \mathbf{i}_n \cdot \mathbf{i}_z = dx\, dy$, we have

$$dS = \frac{dx\, dy}{\mathbf{i}_n \cdot \mathbf{i}_z} = \sqrt{10}\, dx\, dy$$

$$d\mathbf{S} = (3\mathbf{i}_x + \mathbf{i}_z)\, dx\, dy$$

$$\mathbf{J} = 3x\mathbf{i}_x + (y - 3)\mathbf{i}_y + (5 - 3x)\mathbf{i}_z$$

$$\mathbf{J} \cdot d\mathbf{S} = (9x + 5 - 3x)\, dx\, dy = (6x + 5)\, dx\, dy$$

$$\int \mathbf{J} \cdot d\mathbf{S} = \int_{x=0}^{1} \int_{y=0}^{2} (6x + 5)\, dx\, dy = 16$$

Finally, adding the values of $\int \mathbf{J} \cdot d\mathbf{S}$ for the five surfaces, we obtain the total current flowing out of the box to be $0 + \frac{9}{2} - \frac{3}{2} - 4 + 16 = 15$ amp. ▮

c. Line Integral.

The line integral consists of evaluating along a specified path the integral of the dot product of a vector and the differential displacement vector tangential to the path. For example, let us consider a path from point a to point b, as shown in Fig. 1.24 in a region of a known force vector field $\mathbf{F}$. To find the total work done by the force from a to b, we divide the path from a to

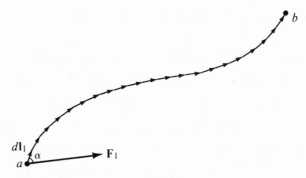

Fig. 1.24. Division of a curve into several infinitesimal segments to evaluate the work done by the force vector **F** along the curve.

b into a number of segments of infinitesimal lengths $dl_1, dl_2, dl_3, \ldots$. Since the length of each segment is very, very small, we can treat these segments as straight lines and define their orientations by the corresponding differential vectors $d\mathbf{l}_1, d\mathbf{l}_2, d\mathbf{l}_3, \ldots$. Within each segment, we can consider the force vector to be constant.

Let us then consider one infinitesimal segment $d\mathbf{l}_1$ and its associated force vector $\mathbf{F}_1$. Let the angle between $d\mathbf{l}_1$ and $\mathbf{F}_1$ be α. The component of the force $\mathbf{F}_1$ along the direction of $d\mathbf{l}_1$ is equal to $F_1 \cos \alpha$. Hence the work done by $\mathbf{F}_1$ along $d\mathbf{l}_1$ is equal to $(F_1 \cos \alpha)(dl_1)$, or $\mathbf{F}_1 \cdot d\mathbf{l}_1$. Similarly, we can obtain the work done by the forces for all the other infinitesimal segments and add them up to give the total work W done from a to b as

$$W = \mathbf{F}_1 \cdot d\mathbf{l}_1 + \mathbf{F}_2 \cdot d\mathbf{l}_2 + \mathbf{F}_3 \cdot d\mathbf{l}_3 + \cdots = \sum_i \mathbf{F}_i \cdot d\mathbf{l}_i \qquad (1\text{-}81)$$

Equation (1-81) is approximate since the assumption of constant force vector for any infinitesimal segment is true only in the limit that the length of that segment tends to zero (i.e., shrinks to a point). In this limit, the summation in (1-81) becomes an integral, giving us

$$W = \int_a^b \mathbf{F} \cdot d\mathbf{l} \qquad (1\text{-}82)$$

where the integration is performed along the path from a to b. The integral on the right side of (1-82) is known as a line integral. For the case of the force vector, it represents the work done by the force field. For other vectors it will have different meanings. When the line integral is evaluated around a closed path C, it is known as the "circulation" around that path and we write it with a small circle associated with the integral sign, as follows:

$$W = \oint_C \mathbf{F} \cdot d\mathbf{l} \qquad (1\text{-}83)$$

EXAMPLE 1-15.   Find the work done by the force vector

$$\mathbf{F} = y\mathbf{i}_x - x\mathbf{i}_y$$

around the closed path *abcdefga* shown in Fig. 1.25.

The work done by the force vector is $\oint_{abcdefga} \mathbf{F} \cdot d\mathbf{l}$. This integral consists of seven parts which will be evaluated independently. First, we note that

$$\mathbf{F} \cdot d\mathbf{l} = (y\mathbf{i}_x - x\mathbf{i}_y) \cdot (dx\,\mathbf{i}_x + dy\,\mathbf{i}_y) = y\,dx - x\,dy$$

Along path *ab*, $y = x^2$, $dy = 2x\,dx$, $\mathbf{F} \cdot d\mathbf{l} = -x^2\,dx$.

$$\int_a^b \mathbf{F} \cdot d\mathbf{l} = -\int_{x=0}^{-1} x^2\,dx = \frac{1}{3}$$

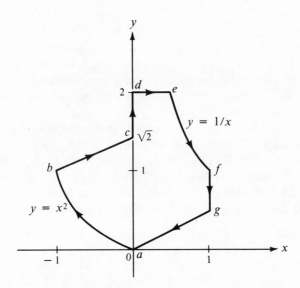

Fig. 1.25. For Example 1-15.

Along path *bc*, $y = (\sqrt{2} - 1)x + \sqrt{2}$, $dy = (\sqrt{2} - 1)\,dx$

$$\mathbf{F} \cdot d\mathbf{l} = \sqrt{2}\,dx.$$

$$\int_b^c \mathbf{F} \cdot d\mathbf{l} = \int_{x=-1}^{0} \sqrt{2}\,dx = \sqrt{2}$$

Along path *cd*, $x = 0$, $dx = 0$, $\mathbf{F} \cdot d\mathbf{l} = 0$.

$$\int_c^d \mathbf{F} \cdot d\mathbf{l} = 0$$

Along path *de*, $y = 2$, $dy = 0$, $\mathbf{F} \cdot d\mathbf{l} = 2\,dx$.

$$\int_d^e \mathbf{F} \cdot d\mathbf{l} = \int_{x=0}^{1/2} 2\,dx = 1$$

Along path ef, $y = 1/x$, $dy = -(1/x^2)\, dx$, $\mathbf{F} \cdot d\mathbf{l} = (2/x)dx$.

$$\int_e^f \mathbf{F} \cdot d\mathbf{l} = \int_{x=1/2}^1 \frac{2}{x}\, dx = 2 \ln 2$$

Along path fg, $x = 1$, $dx = 0$, $\mathbf{F} \cdot d\mathbf{l} = -dy$.

$$\int_f^g \mathbf{F} \cdot d\mathbf{l} = -\int_{y=1}^{1/2} dy = \frac{1}{2}$$

Along path ga, $x = 2y$, $dx = 2\, dy$, $\mathbf{F} \cdot d\mathbf{l} = 0$.

$$\int_g^a \mathbf{F} \cdot d\mathbf{l} = 0$$

Finally, adding the values of $\int \mathbf{F} \cdot d\mathbf{l}$ for the seven paths, we obtain the total work done to be $\frac{1}{3} + \sqrt{2} + 0 + 1 + 2 \ln 2 + \frac{1}{2} + 0 \approx 4.634$. The fact that the integrals along the paths cd and ga are zero is obvious if we note that $\mathbf{F} = y\mathbf{i}_x - x\mathbf{i}_y = -r\mathbf{i}_\phi$. Thus the force vector is everywhere tangential to the circle with the center at the origin and, since cd and ga are radial to the origin, $\mathbf{F} \cdot d\mathbf{l} \equiv 0$ for these paths. Hence $\int \mathbf{F} \cdot d\mathbf{l}$ is zero for the paths cd and ga. ∎

Integration of vectors is performed by expressing the integrand in terms of its components in cartesian coordinates, thereby reducing the problem to one of evaluating three scalar integrals. Thus, for example,

$$\int \mathbf{A}\, dm = \int (A_x\mathbf{i}_x + A_y\mathbf{i}_y + A_z\mathbf{i}_z)\, dm$$

$$= \left(\int A_x\, dm\right)\mathbf{i}_x + \left(\int A_y\, dm\right)\mathbf{i}_y + \left(\int A_z\, dm\right)\mathbf{i}_z \qquad (1\text{-}84)$$

where dm stands for dv, dS, or dl, depending upon whether the integration is over a volume, surface, or along a line, respectively. Similar expressions using the components in cylindrical and spherical coordinate systems are not correct since some or all of the unit vectors in these coordinate systems are functions of the coordinates. For example, the magnitude of the sum of two component vectors along the unit vector $\mathbf{i}_r$ at two different points is not, in general, the sum of the magnitudes of the two vectors since the two components are directed in different directions. For that matter, the direction of the sum of the two component vectors is not the direction of either of the component vectors. Thus

$$\int \mathbf{A}\, dm \neq \left(\int A_r\, dm\right)\mathbf{i}_r + \left(\int A_\phi\, dm\right)\mathbf{i}_\phi + \left(\int A_z\, dm\right)\mathbf{i}_z \qquad (1\text{-}85a)$$

$$\int \mathbf{A}\, dm \neq \left(\int A_r\, dm\right)\mathbf{i}_r + \left(\int A_\theta\, dm\right)\mathbf{i}_\theta + \left(\int A_\phi\, dm\right)\mathbf{i}_\phi \qquad (1\text{-}85b)$$

The integrand must, in general, be expressed as the sum of its component vectors along $\mathbf{i}_x$, $\mathbf{i}_y$, and $\mathbf{i}_z$ for correct results.

1.8 Divergence and the Divergence Theorem

In Section 1.7 we introduced the concept of the surface integral. Let us consider a closed surface S enclosing a volume V in a region in which the current density vector $\mathbf{J}$ is specified. Then the amount of current emanating from this volume is given by

$$I = \oint_S \mathbf{J} \cdot d\mathbf{S} \tag{1-86}$$

where the integration is performed over the closed surface S. If we let this volume shrink to an infinitesimal value Δv, we obtain an infinitesimal amount of current flowing out of the surface ΔS bounding Δv. In the limit that we let the volume shrink to a point, the current emanating from the point may tend to zero. On the other hand, since the volume occupied by the point is zero, the ratio of the current emanating from the point to the volume occupied by the point can be nonzero; that is, although the quantity $\oint_{\Delta S} \mathbf{J} \cdot d\mathbf{S}$ may tend to zero in the limit $\Delta v \to 0$, the quantity

$$\frac{\oint_{\Delta S} \mathbf{J} \cdot d\mathbf{S}}{\Delta v}$$

can approach a nonzero value in the limit $\Delta v \to 0$. The quantity

$$\frac{\oint_{\Delta S} \mathbf{J} \cdot d\mathbf{S}}{\Delta v}$$

is the amount of current, or the flux of the quantity whose density vector is represented by $\mathbf{J}$, per unit volume emanating from the infinitesimal volume Δv. The value that this quantity approaches as Δv tends to zero (i.e., shrinks to a point) is known as the divergence of the vector $\mathbf{J}$. The divergence of $\mathbf{J}$ is represented as the dot product of the vector operator ∇ and the vector $\mathbf{J}$, that is, as $\nabla \cdot \mathbf{J}$. Thus

$$\nabla \cdot \mathbf{J} = \lim_{\Delta v \to 0} \frac{\oint_{\Delta S} \mathbf{J} \cdot d\mathbf{S}}{\Delta v} \tag{1-87}$$

Since the surface integral results in a scalar, the divergence of a vector is a scalar. It is the flux emanating per unit volume as the volume shrinks to a point. Hence the concept of divergence is valid at a point.

To make use of the concept of divergence of a vector, we need to derive expressions for it in terms of the components of the vector in different coordinate systems. Let us choose the cylindrical coordinate system for this purpose. The method of deriving the required expressions consists of following exactly the steps involved in the definition of divergence. First we choose an infinitesimal volume at an arbitrary point $P(r_0, \phi_0, z_0)$, as shown in Fig. 1.26. The infinitesimal volume is formed by the surfaces $r = r_0, r = r_0 + dr$,

$\phi = \phi_0$, $\phi = \phi_0 + d\phi$, $z = z_0$, and $z = z_0 + dz$. The resulting differential surfaces 1, 2, 3, 4, 5, and 6 are given by $-r_0\, d\phi\, dz\, \mathbf{i}_r$, $(r_0 + dr)\, d\phi\, dz\, \mathbf{i}_r$, $-dr\, dz\, \mathbf{i}_\phi$, $dr\, dz\, \mathbf{i}_\phi$, $-r_0\, d\phi\, dr\, \mathbf{i}_z$, and $r_0\, d\phi\, dr\, \mathbf{i}_z$, respectively. Expressing $\mathbf{J}$ in terms of its components in cylindrical coordinates, we have

$$\mathbf{J} = J_r\mathbf{i}_r + J_\phi\mathbf{i}_\phi + J_z\mathbf{i}_z \tag{1-88}$$

The next step is to evaluate the integral of $\mathbf{J} \cdot d\mathbf{S}$ over the surface bounding the differential volume. We do this by evaluating the surface integrals over the six surfaces separately and then adding them up. Over

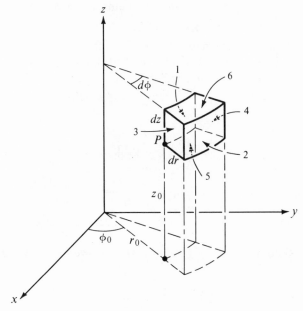

Fig. 1.26. For obtaining the expression for the divergence of a vector in cylindrical coordinates.

each surface, we can assume that $\mathbf{J}$ is constant since the surface area is infinitesimal. Only one of the three components of $\mathbf{J}$ will contribute to the flux crossing a particular surface since the other two components are tangential. Thus the flux leaving the volume from any surface is simply the product of the surface area and the normal component of the $\mathbf{J}$ vector evaluated on that surface or its negative, depending upon whether that component is directed out of or into the volume. In this manner we obtain

flux leaving the volume from surface 1 $= -[J_r]_{r=r_0}\, r_0\, d\phi\, dz$ \qquad (1-89)

and

flux leaving the volume from surface 2 $= [J_r]_{r=r_0+dr}(r_0 + dr)\, d\phi\, dz$

$$\tag{1-90}$$

From (1-89) and (1-90) we have

net flux out of the volume due to surfaces 1 and 2

$$= [J_r]_{r=r_0+dr}(r_0 + dr)\, d\phi\, dz - [J_r]_{r=r_0} r_0\, d\phi\, dz$$

$$= \{[rJ_r]_{r=r_0+dr} - [rJ_r]_{r=r_0}\}\, d\phi\, dz \tag{1-91a}$$

Similarly,

net flux out of the volume due to surfaces 3 and 4

$$= \{[J_\phi]_{\phi=\phi_0+d\phi} - [J_\phi]_{\phi=\phi_0}\}\, dr\, dz \tag{1-91b}$$

and

net flux out of the volume due to surfaces 5 and 6

$$= \{[J_z]_{z=z_0+dz} - [J_z]_{z=z_0}\}\, r_0\, dr\, d\phi \tag{1-91c}$$

The total flux emanating from the differential volume is the sum of the expressions on the right sides of (1-91a), (1-91b), and (1-91c). Adding these three expressions and dividing by the differential volume,

$$\Delta v = r_0\, dr\, d\phi\, dz \tag{1-92}$$

we obtain

$$\frac{\oint_{\Delta S} \mathbf{J} \cdot d\mathbf{S}}{\Delta v} = \frac{[rJ_r]_{r=r_0+dr} - [rJ_r]_{r=r_0}}{r_0\, dr}$$

$$+ \frac{[J_\phi]_{\phi=\phi_0+d\phi} - [J_\phi]_{\phi=\phi_0}}{r_0\, d\phi} \tag{1-93}$$

$$+ \frac{[J_z]_{z=z_0+dz} - [J_z]_{z=z_0}}{dz}$$

By taking the limit of (1-93) as $\Delta v \to 0$, we obtain $\mathbf{\nabla} \cdot \mathbf{J}$ at $P(r_0, \phi_0, z_0)$ as

$$[\mathbf{\nabla} \cdot \mathbf{J}]_{(r_0, \phi_0, z_0)} = \lim_{\Delta v \to 0} \frac{\oint_{\Delta S} \mathbf{J} \cdot d\mathbf{S}}{\Delta v}$$

$$= \lim_{dr \to 0} \frac{[rJ_r]_{r=r_0+dr} - [rJ_r]_{r=r_0}}{r_0\, dr} + \lim_{d\phi \to 0} \frac{[J_\phi]_{\phi=\phi_0+d\phi} - [J_\phi]_{\phi=\phi_0}}{r_0\, d\phi}$$

$$+ \lim_{dz \to 0} \frac{[J_z]_{z=z_0+dz} - [J_z]_{z=z_0}}{dz} \tag{1-94}$$

$$= \frac{1}{r_0}\left[\frac{\partial}{\partial r}(rJ_r)\right]_{r=r_0} + \frac{1}{r_0}\left[\frac{\partial J_\phi}{\partial \phi}\right]_{\phi=\phi_0} + \left[\frac{\partial J_z}{\partial z}\right]_{z=z_0}$$

$$= \left[\frac{1}{r}\frac{\partial}{\partial r}(rJ_r) + \frac{1}{r}\frac{\partial J_\phi}{\partial \phi} + \frac{\partial J_z}{\partial z}\right]_{(r_0, \phi_0, z_0)}$$

Now, since (1-94) is valid for any (r_0, ϕ_0, z_0), we can generalize (1-94) by stating that at any point (r, ϕ, z),

$$\mathbf{\nabla} \cdot \mathbf{J} = \frac{1}{r}\frac{\partial}{\partial r}(rJ_r) + \frac{1}{r}\frac{\partial J_\phi}{\partial \phi} + \frac{\partial J_z}{\partial z} \tag{1-95}$$

Similar expressions for the divergence can be derived in the cartesian and spherical coordinate systems by repeating the procedure followed for the cylindrical coordinate system. The resulting expressions are as follows:

Cartesian coordinates:

$$\mathbf{V} \cdot \mathbf{J} = \frac{\partial J_x}{\partial x} + \frac{\partial J_y}{\partial y} + \frac{\partial J_z}{\partial z} \tag{1-96}$$

Spherical coordinates:

$$\mathbf{V} \cdot \mathbf{J} = \frac{1}{r^2}\frac{\partial}{\partial r}(r^2 J_r) + \frac{1}{r\sin\theta}\frac{\partial}{\partial\theta}(\sin\theta\, J_\theta) + \frac{1}{r\sin\theta}\frac{\partial J_\phi}{\partial\phi} \tag{1-97}$$

By rewriting (1-96) as

$$\mathbf{V} \cdot \mathbf{J} = \left(\frac{\partial}{\partial x}\mathbf{i}_x + \frac{\partial}{\partial y}\mathbf{i}_y + \frac{\partial}{\partial z}\mathbf{i}_z\right) \cdot (J_x\mathbf{i}_x + J_y\mathbf{i}_y + J_z\mathbf{i}_z) \tag{1-98}$$

we note why the divergence of **J** is written as $\mathbf{V} \cdot \mathbf{J}$.

We will now derive a theorem which relates the closed surface integral $\oint_S \mathbf{J} \cdot d\mathbf{S}$ to a volume integral evaluated in the volume V bounded by S. To do this, let us divide the volume V into a large number of infinitesimal volumes $dv_1, dv_2, dv_3, \ldots$ having surfaces $\Delta S_1, \Delta S_2, \Delta S_3, \ldots$, respectively. For each infinitesimal volume, we can assume $\mathbf{V} \cdot \mathbf{J}$ to be uniform and equal to the value it approaches in the limit the volume shrinks to a point. According to definition, $\mathbf{V} \cdot \mathbf{J}$ is the flux of the quantity, represented by **J**, per unit volume in the limit that the volume shrinks to a vanishingly small value. Now, let us consider one of the infinitesimal volumes dv_1 with its associated surface ΔS_1 and vector $\mathbf{J}_1$. The total flux emanating from this volume is equal to $(\mathbf{V} \cdot \mathbf{J})_1\, dv_1$ where $(\mathbf{V} \cdot \mathbf{J})_1$ is the value of $\mathbf{V} \cdot \mathbf{J}$ evaluated in that volume. But, from the concept of surface integral, this flux is also equal to $\oint_{\Delta S_1} \mathbf{J}_1 \cdot d\mathbf{S}$. Thus

$$(\mathbf{V} \cdot \mathbf{J})_1\, dv_1 = \oint_{\Delta S_1} \mathbf{J}_1 \cdot d\mathbf{S} \tag{1-99}$$

By writing similar expressions for all the other infinitesimal volumes and adding them up, we obtain

$$\begin{aligned}(\mathbf{V} \cdot \mathbf{J})_1\, dv_1 &+ (\mathbf{V} \cdot \mathbf{J})_2\, dv_2 + (\mathbf{V} \cdot \mathbf{J})_3\, dv_3 + \cdots \\ &= \oint_{\Delta S_1} \mathbf{J}_1 \cdot d\mathbf{S} + \oint_{\Delta S_2} \mathbf{J}_2 \cdot d\mathbf{S} + \oint_{\Delta S_3} \mathbf{J}_3 \cdot d\mathbf{S} + \cdots\end{aligned} \tag{1-100}$$

But the right side of (1-100) is equal to $\oint_S \mathbf{J} \cdot d\mathbf{S}$, since contributions from all the surfaces and portions of the surfaces inside the boundary of the volume V cancel, leaving a net integral over the surface bounding the volume V. Equation (1-100) then becomes

$$\sum_i (\mathbf{V} \cdot \mathbf{J})_i\, dv_i = \oint_S \mathbf{J} \cdot d\mathbf{S} \tag{1-101}$$

Equation (1-101) is approximate since the assumption of uniform $\mathbf{V} \cdot \mathbf{J}$

inside any infinitesimal volume is true only in the limit that the volume shrinks to a point. In this limit, the summation in (1-101) becomes an integral, giving us

$$\int_V (\mathbf{\nabla} \cdot \mathbf{J}) \, dv = \oint_S \mathbf{J} \cdot d\mathbf{S} \tag{1-102}$$

where the integration is performed throughout the volume V bounded by S. The result represented by (1-102) is known as the divergence theorem. It permits the replacement of a surface integration by a volume integration and vice versa.

EXAMPLE 1-16.   In Example 1-14 we evaluated a surface integral to find the current flowing out of a box. It is now desired to compute the same quantity by using the divergence theorem and performing a volume integration.

According to the divergence theorem (1-102), the current flowing out of the box of Fig. 1.23 is $\int_V (\mathbf{\nabla} \cdot \mathbf{J}) \, dv$, where V is the volume of the box and $\mathbf{J}$ is the current density vector specified in Example 1-14. For this current density vector, the divergence is equal to 5. Hence

$$\oint_S \mathbf{J} \cdot d\mathbf{S} = \int_V (\mathbf{\nabla} \cdot \mathbf{J}) \, dv = \int_V 5 \, dv = 5 \int_V dv$$
$$= 5(\text{volume of the box})$$
$$= 5 \times \left(\frac{1 \times 2 \times 3}{2} \right) = 15 \text{ amp}$$

This result agrees with the result of Example 1-14. ∎

1.9   Curl and Stokes' Theorem

In Section 1.7 we introduced the concept of circulation or line integral around a closed path. Let us consider an infinitesimal area ΔS in the field of a vector $\mathbf{F}$ and orient it such that the circulation $\oint \mathbf{F} \cdot d\mathbf{l}$ around the periphery ΔC of the area is a maximum. Let $\mathbf{i}_n$ be the unit vector normal to the area for that particular orientation. Then we define a vector quantity known as the "curl" of $\mathbf{F}$, having the symbol "del cross" as

$$\mathbf{\nabla} \times \mathbf{F} = \lim_{\Delta S \to 0} \frac{\oint_{\Delta C} \mathbf{F} \cdot d\mathbf{l}}{\Delta S} \, \mathbf{i}_n \tag{1-103}$$

We note that in the limit $\Delta S \to 0$, although $\oint_{\Delta C} \mathbf{F} \cdot d\mathbf{l}$ may tend to zero, $\mathbf{\nabla} \times \mathbf{F}$ can be nonzero. The line integral in (1-103) is evaluated by traversing the perimeter of the area ΔS on the side of the unit vector $\mathbf{i}_n$ in such a direction that the area is on the left, as shown in Fig. 1.27. This is the same as the direction in which a right-hand screw turns as it advances in the direction of the normal vector. Just as the divergence of a vector is associated with a point in space, the curl of a vector is also associated with a point in space, in view of the limit on the right side of (1-103). Whereas the divergence of a vector is a scalar, the curl of a vector is another vector. The magnitude of this vector is the circulation per unit area as the area shrinks to a point, maintaining in this process an orientation of the area such that the circulation around its periphery is a maximum. The direction of the vector is the direction which the normal vector to the area assumes as the area shrinks to the point.

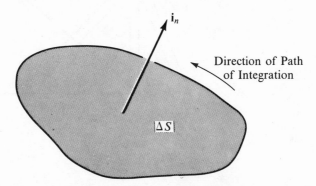

Fig. 1.27. Illustrating the sense of traversal around the periphery of area ΔS to evaluate the line integral in (1-103).

Later in this section we will explore the physical significance of curl, but first let us obtain the expressions for the curl of a vector in terms of the components of the vector. To do this, we first wish to show that the components of the curl of a vector at a point are simply the circulations per unit area at that point with the areas oriented normal to the corresponding coordinate axes.

Let us consider an infinitesimal plane area ABC, as shown in Fig. 1.28, such that its normal vector $\mathbf{i}_n$ is oriented in an arbitrary direction in the field of the vector $\mathbf{F}$. We can write

$$\oint_{ABCA} \mathbf{F} \cdot d\mathbf{l} = \oint_{ABOA} \mathbf{F} \cdot d\mathbf{l} + \oint_{BCOB} \mathbf{F} \cdot d\mathbf{l} + \oint_{CAOC} \mathbf{F} \cdot d\mathbf{l} \qquad (1\text{-}104)$$

since the contribution to the integrals on the right side from the paths between O and A, O and B, and O and C cancel. Dividing both sides of (1-104) by the area ABC, we get

$$\frac{\oint_{ABCA} \mathbf{F} \cdot d\mathbf{l}}{\text{area } ABC} = \frac{\oint_{ABOA} \mathbf{F} \cdot d\mathbf{l}}{\text{area } ABC} + \frac{\oint_{BCOB} \mathbf{F} \cdot d\mathbf{l}}{\text{area } ABC} + \frac{\oint_{CAOC} \mathbf{F} \cdot d\mathbf{l}}{\text{area } ABC} \qquad (1\text{-}105)$$

With the relationships

$$\text{area } AOB = (\text{area } ABC)\mathbf{i}_n \cdot \mathbf{i}_z \qquad (1\text{-}106a)$$

$$\text{area } BOC = (\text{area } ABC)\mathbf{i}_n \cdot \mathbf{i}_x \qquad (1\text{-}106b)$$

$$\text{area } COA = (\text{area } ABC)\mathbf{i}_n \cdot \mathbf{i}_y \qquad (1\text{-}106c)$$

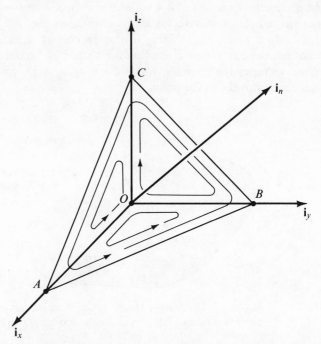

Fig. 1.28. For showing that the components of the curl of a vector at a point are the circulations per unit area at that point with the areas oriented normal to the corresponding coordinate axes.

(1-105) can be written as

$$\frac{\oint_{ABCA} \mathbf{F} \cdot d\mathbf{l}}{\text{area } ABC} = \frac{\oint_{ABOA} \mathbf{F} \cdot d\mathbf{l}}{\text{area } AOB} \mathbf{i}_n \cdot \mathbf{i}_z + \frac{\oint_{BCOB} \mathbf{F} \cdot d\mathbf{l}}{\text{area } BOC} \mathbf{i}_n \cdot \mathbf{i}_x + \frac{\oint_{CAOC} \mathbf{F} \cdot d\mathbf{l}}{\text{area } COA} \mathbf{i}_n \cdot \mathbf{i}_y$$

$$= \mathbf{i}_n \cdot \left(\frac{\oint_{ABOA} \mathbf{F} \cdot d\mathbf{l}}{\text{area } AOB} \mathbf{i}_z + \frac{\oint_{BCOB} \mathbf{F} \cdot d\mathbf{l}}{\text{area } BOC} \mathbf{i}_x + \frac{\oint_{CAOC} \mathbf{F} \cdot d\mathbf{l}}{\text{area } COA} \mathbf{i}_y \right)$$

$$(1\text{-}107)$$

Taking the limit of both sides of (1-107) as the area $ABC \longrightarrow 0$, we have

$$\lim_{ABC \to 0} \frac{\oint_{ABCA} \mathbf{F} \cdot d\mathbf{l}}{\text{area } ABC} = \mathbf{i}_n \cdot \left(\lim_{AOB \to 0} \frac{\oint_{ABOA} \mathbf{F} \cdot d\mathbf{l}}{\text{area } AOB} \mathbf{i}_z \right.$$

$$\left. + \lim_{BOC \to 0} \frac{\oint_{BCOB} \mathbf{F} \cdot d\mathbf{l}}{\text{area } BOC} \mathbf{i}_x + \lim_{COA \to 0} \frac{\oint_{CAOC} \mathbf{F} \cdot d\mathbf{l}}{\text{area } COA} \mathbf{i}_y \right)$$

$$(1\text{-}108)$$

The magnitude of $\nabla \times \mathbf{F}$ is the maximum possible value of

$$\lim_{ABC \to 0} \frac{\oint_{ABCA} \mathbf{F} \cdot d\mathbf{l}}{\text{area } ABC}$$

that is, the maximum possible value of the quantity on the left side of (1-108). The maximum value of this quantity occurs when the orientation of $\mathbf{i}_n$ coincides with the direction of the vector inside the parentheses on the right side of (1-108). It then follows that this maximum value is the magnitude of the vector inside the parentheses. Hence the vector inside the parentheses on the right side of (1-108) is indeed $\mathbf{\nabla} \times \mathbf{F}$. Thus the components of $\mathbf{\nabla} \times \mathbf{F}$ are simply the circulations per unit area at the point of interest with the areas oriented normal to the corresponding unit vectors and as these areas are shrunk to the point. Although the foregoing proof is carried out for the cartesian coordinate system, it is obvious that it is valid for any orthogonal coordinate system since the unit vectors in Fig. 1.28 can be replaced by any orthogonal set of unit vectors.

We will now derive the expressions for the components of $\mathbf{\nabla} \times \mathbf{F}$. Let us choose the spherical coordinate system for this purpose. To obtain the r component, we consider an infinitesimal area *abcd* normal to the unit vector $\mathbf{i}_r$ at point $P(r, \theta, \phi)$, as shown in Fig. 1.29(a). From our experience in deriving the expressions for the divergence in Section 1.8, there is no need to consider a point (r_0, θ_0, ϕ_0) and then generalize the result. Expressing $\mathbf{F}$ in terms of its components in spherical coordinates, we have

$$\mathbf{F} = F_r \mathbf{i}_r + F_\theta \mathbf{i}_\theta + F_\phi \mathbf{i}_\phi \tag{1-109}$$

Then we evaluate the circulation of $\mathbf{F}$ around the path *abcd* in Fig. 1.29(a), divide the circulation by the area of *abcd*, and find the limit of the resulting

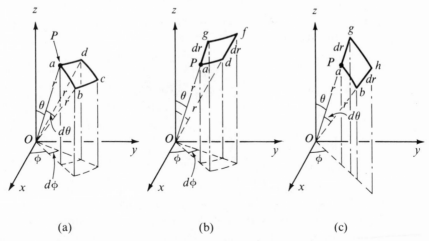

| (a) | (b) | (c) |

Fig. 1.29. For obtaining the expression for the curl of a vector in spherical coordinates.

quantity as the area *abcd* tends to zero. The circulation of **F** around *abcd* is the sum of four line integrals, evaluated along the four sides of the area *abcd*. For each side, we can assume that **F** is constant since the lengths are infinitesimal. Only one of the three components of **F** contribute to the line integral involving any particular side since the other two components are normal to the path. Thus the line integral along any side is simply the product of the length of the side and the tangential component of the **F** vector evaluated along that side or its negative, depending upon whether the component is directed along or opposite to the path of integration. In this manner, we obtain

$$\int_a^b \mathbf{F} \cdot d\mathbf{l} = [F_\theta]_\phi \, r \, d\theta \tag{1-110}$$

$$\int_b^c \mathbf{F} \cdot d\mathbf{l} = [F_\phi]_{\theta+d\theta} \, r \sin{(\theta + d\theta)} \, d\phi \tag{1-111}$$

$$\int_c^d \mathbf{F} \cdot d\mathbf{l} = -[F_\theta]_{\phi+d\phi} \, r \, d\theta \tag{1-112}$$

$$\int_d^a \mathbf{F} \cdot d\mathbf{l} = -[F_\phi]_\theta \, r \sin{\theta} \, d\phi \tag{1-113}$$

From (1-110)–(1-113) we have

$$\oint_{abcda} \mathbf{F} \cdot d\mathbf{l} = \int_a^b \mathbf{F} \cdot d\mathbf{l} + \int_b^c \mathbf{F} \cdot d\mathbf{l} + \int_c^d \mathbf{F} \cdot d\mathbf{l} + \int_d^a \mathbf{F} \cdot d\mathbf{l}$$

$$= \{[F_\theta]_\phi - [F_\theta]_{\phi+d\phi}\} r \, d\theta$$

$$+ \{[F_\phi]_{\theta+d\theta} \sin{(\theta + d\theta)} - [F_\phi]_\theta \sin{\theta}\} r \, d\phi \tag{1-114}$$

$$= \{[F_\theta]_\phi - [F_\theta]_{\phi+d\phi}\} r \, d\theta$$

$$+ \{[F_\phi \sin{\theta}]_{\theta+d\theta} - [F_\phi \sin{\theta}]_\theta\} r \, d\phi$$

Dividing both sides of (1-114) by area $abcd = r^2 \sin{\theta} \, d\theta \, d\phi$ and taking the limit as the area tends to zero, we have

$$\lim_{abcd \to 0} \frac{\oint_{abcda} \mathbf{F} \cdot d\mathbf{l}}{\text{area } abcd} = \lim_{\substack{d\theta \to 0 \\ d\phi \to 0}} \frac{\{[F_\theta]_\phi - [F_\theta]_{\phi+d\phi}\} r \, d\theta}{r^2 \sin{\theta} \, d\theta \, d\phi}$$

$$+ \lim_{\substack{d\theta \to 0 \\ d\phi \to 0}} \frac{\{[F_\phi \sin{\theta}]_{\theta+d\theta} - [F_\phi \sin{\theta}]_\theta\} r \, d\phi}{r^2 \sin{\theta} \, d\theta \, d\phi} \tag{1-115}$$

$$= -\frac{1}{r \sin{\theta}} \frac{\partial F_\theta}{\partial \phi} + \frac{1}{r \sin{\theta}} \frac{\partial}{\partial \theta} (\sin{\theta} \, F_\phi)$$

Similarly, to derive the θ component of $\nabla \times \mathbf{F}$, we consider an infinitesimal area *adfg* normal to the unit vector $\mathbf{i}_\theta$ at point $P(r, \theta, \phi)$, as shown in Fig. 1.29(b), and evaluate the circulation around the periphery of this area. Following in the same manner as for the r component of $(\nabla \times \mathbf{F})$, we have

$$\oint_{adfga} \mathbf{F} \cdot d\mathbf{l} = [F_\phi]_r \, r \sin \theta \, d\phi + [F_r]_{\phi+d\phi} \, dr$$
$$- [F_\phi]_{r+dr}(r + dr) \sin \theta \, d\phi - [F_r]_\phi \, dr \qquad (1\text{-}116)$$
$$= \{[rF_\phi]_r - [rF_\phi]_{r+dr}\} \sin \theta \, d\phi$$
$$+ \{[F_r]_{\phi+d\phi} - [F_r]_\phi\} \, dr$$

Noting that area $adfg = r \sin \theta \, dr \, d\phi$, we obtain

$$\lim_{adfg \to 0} \frac{\oint_{adfga} \mathbf{F} \cdot d\mathbf{l}}{\text{area } adfg} = \lim_{\substack{dr \to 0 \\ d\phi \to 0}} \frac{\{[rF_\phi]_r - [rF_\phi]_{r+dr}\} \sin \theta \, d\phi}{r \sin \theta \, dr \, d\phi}$$
$$+ \lim_{\substack{dr \to 0 \\ d\phi \to 0}} \frac{\{[F_r]_{\phi+d\phi} - [F_r]_\phi\} \, dr}{r \sin \theta \, dr \, d\phi} \qquad (1\text{-}117)$$
$$= -\frac{1}{r} \frac{\partial}{\partial r}(rF_\phi) + \frac{1}{r \sin \theta} \frac{\partial F_r}{\partial \phi}$$

Finally, to obtain the ϕ component of $\nabla \times \mathbf{F}$, we choose an infinitesimal area $aghb$ normal to the unit vector $\mathbf{i}_\phi$ at point $P(r, \theta, \phi)$, as shown in Fig. 1.29(c), and evaluate the circulation around the periphery of this area. Following in the same manner as for the r and θ components of $\nabla \times \mathbf{F}$, we have

$$\oint_{aghba} \mathbf{F} \cdot d\mathbf{l} = [F_r]_\theta \, dr + [F_\theta]_{r+dr}(r + dr) \, d\theta - [F_r]_{\theta+d\theta} \, dr - [F_\theta]_r r \, d\theta$$
$$= \{[F_r]_\theta - [F_r]_{\theta+d\theta}\} \, dr \qquad (1\text{-}118)$$
$$+ \{[rF_\theta]_{r+dr} - [rF_\theta]_r\} \, d\theta$$

Noting that area $aghb = r \, dr \, d\theta$, we obtain

$$\lim_{aghb \to 0} \frac{\oint_{aghba} \mathbf{F} \cdot d\mathbf{l}}{\text{area } aghb} = \lim_{\substack{dr \to 0 \\ d\theta \to 0}} \frac{\{[F_r]_\theta - [F_r]_{\theta+d\theta}\} \, dr}{r \, dr \, d\theta}$$
$$+ \lim_{\substack{dr \to 0 \\ d\theta \to 0}} \frac{\{[rF_\theta]_{r+dr} - [rF_\theta]_r\} \, d\theta}{r \, dr \, d\theta} \qquad (1\text{-}119)$$
$$= -\frac{1}{r} \frac{\partial F_r}{\partial \theta} + \frac{1}{r} \frac{\partial}{\partial r}(rF_\theta)$$

Thus, in the spherical coordinate system, we note that

$$\nabla \times \mathbf{F} = \frac{1}{r \sin \theta} \left[\frac{\partial}{\partial \theta}(\sin \theta \, F_\phi) - \frac{\partial F_\theta}{\partial \phi} \right] \mathbf{i}_r$$
$$+ \frac{1}{r} \left[\frac{1}{\sin \theta} \frac{\partial F_r}{\partial \phi} - \frac{\partial}{\partial r}(rF_\phi) \right] \mathbf{i}_\theta + \frac{1}{r} \left[\frac{\partial}{\partial r}(rF_\theta) - \frac{\partial F_r}{\partial \theta} \right] \mathbf{i}_\phi$$
$$= \begin{vmatrix} \dfrac{\mathbf{i}_r}{r^2 \sin \theta} & \dfrac{\mathbf{i}_\theta}{r \sin \theta} & \dfrac{\mathbf{i}_\phi}{r} \\[2mm] \dfrac{\partial}{\partial r} & \dfrac{\partial}{\partial \theta} & \dfrac{\partial}{\partial \phi} \\[2mm] F_r & rF_\theta & r \sin \theta \, F_\phi \end{vmatrix} \qquad (1\text{-}120)$$

Similar expressions for the curl can be derived in the cartesian and cylindrical coordinate systems by repeating the procedure followed for the spherical coordinate system. The resulting expressions are as follows:

Cartesian coordinates:

$$\mathbf{V} \times \mathbf{F} = \begin{vmatrix} \mathbf{i}_x & \mathbf{i}_y & \mathbf{i}_z \\ \dfrac{\partial}{\partial x} & \dfrac{\partial}{\partial y} & \dfrac{\partial}{\partial z} \\ F_x & F_y & F_z \end{vmatrix} \tag{1-121}$$

Cylindrical coordinates:

$$\mathbf{V} \times \mathbf{F} = \begin{vmatrix} \dfrac{\mathbf{i}_r}{r} & \mathbf{i}_\phi & \dfrac{\mathbf{i}_z}{r} \\ \dfrac{\partial}{\partial r} & \dfrac{\partial}{\partial \phi} & \dfrac{\partial}{\partial z} \\ F_r & rF_\phi & F_z \end{vmatrix} \tag{1-122}$$

The form of the right side of (1-121) explains why the curl of $\mathbf{F}$ is written as $\mathbf{V} \times \mathbf{F}$.

Let us now discuss briefly the physical significance of curl. To do this, we will use the concept of the curl meter or the paddle wheel as suggested by Skilling (see bibliography). Consider a stream of rectangular cross section carrying water in the z direction, as shown in Fig. 1.30(a). Assume the velocity $\mathbf{v}$ of the water to be independent of height but increasing uniformly from a value of zero at the banks to a maximum value of v_0 at the center. Thus

$$\mathbf{v} = \begin{cases} v_0 \dfrac{y}{a} \mathbf{i}_z & \text{for } 0 < y < a \\[2mm] v_0 \dfrac{2a - y}{a} \mathbf{i}_z & \text{for } a < y < 2a \end{cases} \tag{1-123}$$

The curl of the velocity vector is given by

$$\begin{aligned} \mathbf{V} \times \mathbf{v} &= \begin{vmatrix} \mathbf{i}_x & \mathbf{i}_y & \mathbf{i}_z \\ \dfrac{\partial}{\partial x} & \dfrac{\partial}{\partial y} & \dfrac{\partial}{\partial z} \\ 0 & 0 & v_z \end{vmatrix} \\[2mm] &= \dfrac{\partial v_z}{\partial y} \mathbf{i}_x - \dfrac{\partial v_z}{\partial x} \mathbf{i}_y \\[2mm] &= \begin{cases} \dfrac{v_0}{a} \mathbf{i}_x & 0 < y < a \\[2mm] -\dfrac{v_0}{a} \mathbf{i}_x & a < y < 2a \end{cases} \end{aligned} \tag{1-124}$$

Sketches of v_z and $(\mathbf{V} \times \mathbf{v})_x$ are shown in Figs. 1.30(b) and (c), respectively. Now, let us consider a frictionless paddle wheel having negligible influence on the velocity of the water and introduce it into the water with its shaft

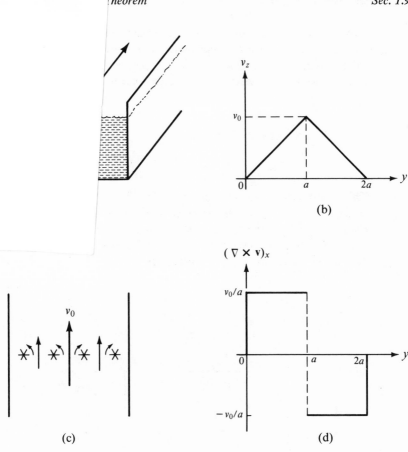

(b)

(c)   (d)

Fig. 1.30. For explaining the physical significance of curl using the paddle-wheel device.

vertical, that is, parallel to the x axis. It will turn in the counterclockwise direction on the left side of the center of the stream and in the clockwise direction on the right side of the center, as shown in Fig. 1.30(d). Moreover, since the velocity differential is independent of y, it will turn at the same rate independent of y. In exactly the midstream, it will not turn since the velocities on either side are equal and are in the same direction. Now, if we examine the graph of $(\nabla \times \mathbf{v})_x$ and compare it with the action of the paddle wheel, the physical meaning of curl is apparent. It signifies the ability of the vector field to rotate the paddle wheel. If we insert the paddle wheel horizontally, that is, along the z axis or along the y axis or in any other direction parallel to the yz plane, it will not rotate since the top and bottom plates are hit with the same force, thus indicating that the curl for this field has no horizontal component, as indeed the expression (1-124) shows. The curl has nothing to do with curvature or curling flow as the name might imply. We have already seen in the example just discussed that a vector field whose direction lines are straight lines has a nonzero curl. Likewise, it is possible to have vector fields whose direction lines are curved but with zero curl. As an example, consider the field given in cylindrical coordinates by

$$\mathbf{F} = \frac{1}{r}\mathbf{i}_\phi \tag{1-125}$$

For this vector field, (1-122) gives

$$\mathbf{V} \times \mathbf{F} = \begin{cases} 0 & \text{everywhere except at } r = 0 \\ \infty\,\mathbf{i}_z & \text{at } r = 0 \end{cases} \tag{1-126}$$

This can be explained by referring to Fig. 1.31. Although the magnitude of the force on the right side of the center of the paddle wheel is less than on the left side, there are more blades hit by the force on the right side, thereby keeping the paddle wheel still. At $r = 0$, however, there is circular motion of the fluid which turns the paddle wheel.

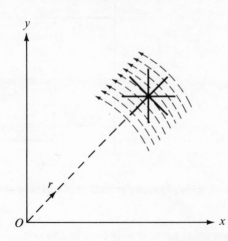

Fig. 1.31. An exaggerated picture of a paddle wheel in the field $(1/r)\mathbf{i}_\phi$.

Two important identities involving curl are

$$\mathbf{V} \cdot \mathbf{V} \times \mathbf{F} \equiv 0 \tag{1-127}$$

$$\mathbf{V} \times \mathbf{V}V \equiv 0 \tag{1-128}$$

The first identity states that the divergence of any vector which can be expressed as the curl of another vector is zero, whereas the second identity states that the curl of any vector which can be expressed as the gradient of a scalar is zero. These relations can be derived simply by carrying out the vector operations indicated by the left sides of (1-127) and (1-128). Conversely, if the divergence of a vector is zero, it can be expressed as the curl of a vector and if the curl of a vector is zero, it can be expressed as the gradient of a scalar.

We will now derive a theorem which relates the closed line integral

$\oint \mathbf{F} \cdot d\mathbf{l}$ to a surface integral evaluated over any surface bounded by the closed path. To do this, let us consider in the field of the vector $\mathbf{F}$ a contour C which is the boundary of a surface S, not necessarily plane, as shown in Fig. 1.32. Let us divide the surface S into a large number of infinitesimal areas dS_1, dS_2, dS_3, ... bounded by contours ΔC_1, ΔC_2, ΔC_3, ..., respectively. For each infinitesimal area, we can assume $\mathbf{\nabla} \times \mathbf{F}$ to be uniform and equal to the value it approaches in the limit the area shrinks to a point. According to definition, $\mathbf{\nabla} \times \mathbf{F}$ is the maximum circulation of $\mathbf{F}$ per unit area at a point. If an infinitesimal area dS is oriented such that its normal vector is in the direction of $\mathbf{\nabla} \times \mathbf{F}$, the circulation around the periphery of that infinitesimal area is $(\mathbf{\nabla} \times \mathbf{F}) \, dS$. If the infinitesimal area has some other orientation, we have to take the component of $\mathbf{\nabla} \times \mathbf{F}$ along the normal vector to that area and multiply by the area to obtain the circulation.

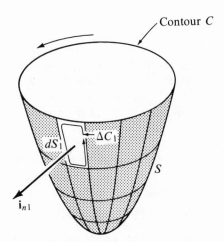

Contour C

dS_1

ΔC_1

S

$\mathbf{i}_{n1}$

Fig. 1.32. Division of a surface S bounded by a contour C into a number of infinitesimal areas to derive Stokes' theorem.

Let us then consider one of the infinitesimal areas dS_1 with its associated vector $\mathbf{F}_1$. The circulation around the contour ΔC_1 bounding this infinitesimal area is equal to $(\mathbf{\nabla} \times \mathbf{F})_1 \cdot \mathbf{i}_{n1} \, dS_1$, where $\mathbf{i}_{n1}$ is the unit normal vector to dS_1 oriented in accordance with the convention shown in Fig. 1.27 and $(\mathbf{\nabla} \times \mathbf{F})_1$ is the value of $\mathbf{\nabla} \times \mathbf{F}$ evaluated over that area. But, from the concept of line integral, this circulation is also equal to $\oint_{\Delta C_1} \mathbf{F}_1 \cdot d\mathbf{l}$. Thus

$$(\mathbf{\nabla} \times \mathbf{F})_1 \cdot \mathbf{i}_{n1} \, dS_1 = \oint_{\Delta C_1} \mathbf{F}_1 \cdot d\mathbf{l} \qquad (1\text{-}129)$$

By writing similar expressions for all the other infinitesimal areas and adding them up, we obtain

$$(\mathbf{\nabla} \times \mathbf{F})_1 \cdot \mathbf{i}_{n1} \, dS_1 + (\mathbf{\nabla} \times \mathbf{F})_2 \cdot \mathbf{i}_{n2} \, dS_2 + (\mathbf{\nabla} \times \mathbf{F})_3 \cdot \mathbf{i}_{n3} \, dS_3 + \cdots$$

$$= \oint_{\Delta C_1} \mathbf{F}_1 \cdot d\mathbf{l} + \oint_{\Delta C_2} \mathbf{F}_2 \cdot d\mathbf{l} + \oint_{\Delta C_3} \mathbf{F}_3 \cdot d\mathbf{l} + \cdots \qquad (1\text{-}130)$$

But the right side of (1-130) is equal to $\oint_C \mathbf{F} \cdot d\mathbf{l}$, since contributions from all the contours and portions of the contours inside the periphery of the surface S cancel, leaving a net integral around the periphery. Equation (1-130) then becomes

$$\sum_j (\nabla \times \mathbf{F})_j \cdot \mathbf{i}_{nj}\, dS_j = \oint_C \mathbf{F} \cdot d\mathbf{l} \qquad (1\text{-}131)$$

Equation (1-131) is approximate since the assumption of uniform $\nabla \times \mathbf{F}$ over any infinitesimal area is true only in the limit that the area shrinks to zero. In this limit, the summation in (1-131) becomes an integral, giving us

$$\int_S (\nabla \times \mathbf{F}) \cdot \mathbf{i}_n\, dS = \oint_C \mathbf{F} \cdot d\mathbf{l} \qquad (1\text{-}132)$$

or

$$\int_S (\nabla \times \mathbf{F}) \cdot d\mathbf{S} = \oint_C \mathbf{F} \cdot d\mathbf{l} \qquad (1\text{-}133)$$

where we have absorbed the unit vector $\mathbf{i}_n$ into the vector $d\mathbf{S}$. The result represented by (1-132) is known as Stokes' theorem. It permits the replacement of a line integration by a surface integration and vice versa. In (1-132) and (1-133), the sense of traversal around C must be such that the area on the side of the normal vector $\mathbf{i}_n$ is on the left.

EXAMPLE 1-17.   In Example 1-15 we used line integration to evaluate the work done by a force vector around a closed path. It is now desired to compute the same quantity by performing a surface integration.

According to Stokes' theorem, the work done by the force vector is $\int_S (\nabla \times \mathbf{F}) \cdot \mathbf{i}_n\, dS$, where S is the surface bounded by the closed path and $\mathbf{F}$ is the force vector specified in Example 1-15. For this force vector, the curl is equal to $-2\mathbf{i}_z$. The normal vector $\mathbf{i}_n$ must be chosen such that it is on the left side while traversing the path specified in Example 1-15. Hence $\mathbf{i}_n = -\mathbf{i}_z$, and

$$\int_{abcdefg} (\nabla \times \mathbf{F}) \cdot \mathbf{i}_n\, dS = \int_{abcdefg} (-2\mathbf{i}_z) \cdot (-\mathbf{i}_z)\, dS$$

$$= 2(\text{area } abcdefg)$$

$$\text{area } abcdefg = \frac{2}{3} + \frac{\sqrt{2}-1}{2} + \frac{1}{2} + \frac{1}{4} + \ln 2 = 2.317$$

Thus

$$\oint_{abcdefga} \mathbf{F} \cdot d\mathbf{l} = \int_{abcdefg} (\nabla \times \mathbf{F}) \cdot \mathbf{i}_n\, dS = 2(2.317) = 4.634$$

This result agrees with the result of Example 1-15.   ∎

1.10   The Laplacian

In Sections 1.6, 1.8, and 1.9 we introduced gradient, divergence, and curl, respectively. Gradient is an operation performed only on scalar functions, whereas divergence and curl are operations performed only on vector functions. In this section we will introduce another operation, known as the Laplacian, which is performed both on scalar and vector functions.

a. The Laplacian of a Scalar.

The Laplacian of a scalar function V is defined as the divergence of the gradient of V. The gradient of V is a vector and the divergence of a vector is a scalar. Hence the Laplacian of a scalar results in a scalar. The Laplacian operation has the symbol ∇^2. Thus

$$\nabla^2 V = \nabla \cdot \nabla V \qquad (1\text{-}134)$$

In cartesian coordinates,

$$\begin{aligned}
\nabla^2 V &= \left(\frac{\partial}{\partial x}\mathbf{i}_x + \frac{\partial}{\partial y}\mathbf{i}_y + \frac{\partial}{\partial z}\mathbf{i}_z\right) \cdot \left(\frac{\partial V}{\partial x}\mathbf{i}_x + \frac{\partial V}{\partial y}\mathbf{i}_y + \frac{\partial V}{\partial z}\mathbf{i}_z\right) \\
&= \frac{\partial^2 V}{\partial x^2} + \frac{\partial^2 V}{\partial y^2} + \frac{\partial^2 V}{\partial z^2}
\end{aligned} \qquad (1\text{-}135)$$

Similarly, expressions for $\nabla^2 V$ can be derived in other coordinate systems. These expressions are as follows:

Cylindrical coordinates:

$$\nabla^2 V = \frac{1}{r}\frac{\partial}{\partial r}\left(r\frac{\partial V}{\partial r}\right) + \frac{1}{r^2}\frac{\partial^2 V}{\partial \phi^2} + \frac{\partial^2 V}{\partial z^2} \qquad (1\text{-}136)$$

Spherical coordinates:

$$\nabla^2 V = \frac{1}{r^2}\frac{\partial}{\partial r}\left(r^2\frac{\partial V}{\partial r}\right) + \frac{1}{r^2 \sin\theta}\frac{\partial}{\partial \theta}\left(\sin\theta\frac{\partial V}{\partial \theta}\right) + \frac{1}{r^2 \sin^2\theta}\frac{\partial^2 V}{\partial \phi^2} \quad (1\text{-}137)$$

b. The Laplacian of a Vector.

The Laplacian of a vector $\mathbf{A}$ is defined as the gradient of divergence of $\mathbf{A}$ minus the curl of curl of $\mathbf{A}$; that is,

$$\nabla^2 \mathbf{A} = \nabla(\nabla \cdot \mathbf{A}) - \nabla \times \nabla \times \mathbf{A} \qquad (1\text{-}138)$$

Expansions for $\nabla^2 \mathbf{A}$ in different coordinate systems can be obtained by carrying out the operations on the right side of (1-138) and simplifying the resulting expressions. The results are as follows:

Cartesian coordinates:

$$\nabla^2 \mathbf{A} = (\nabla^2 A_x)\mathbf{i}_x + (\nabla^2 A_y)\mathbf{i}_y + (\nabla^2 A_z)\mathbf{i}_z \qquad (1\text{-}139)$$

Cylindrical coordinates:

$$\nabla^2 \mathbf{A} = \left(\nabla^2 A_r - \frac{A_r}{r^2} - \frac{2}{r^2} \frac{\partial A_\phi}{\partial \phi} \right) \mathbf{i}_r$$

$$+ \left(\nabla^2 A_\phi - \frac{A_\phi}{r^2} + \frac{2}{r^2} \frac{\partial A_r}{\partial \phi} \right) \mathbf{i}_\phi \qquad (1\text{-}140)$$

$$+ (\nabla^2 A_z) \mathbf{i}_z$$

Spherical coordinates:

$$\nabla^2 \mathbf{A} = \left[\nabla^2 A_r - \frac{2}{r^2} A_r - \frac{2}{r^2 \sin \theta} \frac{\partial}{\partial \theta} (A_\theta \sin \theta) - \frac{2}{r^2 \sin \theta} \frac{\partial A_\phi}{\partial \phi} \right] \mathbf{i}_r$$

$$+ \left(\nabla^2 A_\theta - \frac{A_\theta}{r^2 \sin^2 \theta} + \frac{2}{r^2} \frac{\partial A_r}{\partial \theta} - \frac{2 \cos \theta}{r^2 \sin^2 \theta} \frac{\partial A_\phi}{\partial \phi} \right) \mathbf{i}_\theta \qquad (1\text{-}141)$$

$$+ \left(\nabla^2 A_\phi - \frac{A_\phi}{r^2 \sin^2 \theta} + \frac{2}{r^2 \sin \theta} \frac{\partial A_r}{\partial \phi} + \frac{2 \cos \theta}{r^2 \sin^2 \theta} \frac{\partial A_\theta}{\partial \phi} \right) \mathbf{i}_\phi$$

1.11 Some Useful Vector Relations

In this section we will summarize the important vector relations discussed in this chapter and present additional useful vector identities. The following notation is used:

$\mathbf{i}_1, \mathbf{i}_2, \mathbf{i}_3 =$ set of mutually perpendicular unit vectors forming a right-hand coordinate system.

$u_1, u_2, u_3 =$ set of three orthogonal coordinates.

$dl_1, dl_2, dl_3 =$ differential displacements along the direction of the unit vectors $\mathbf{i}_1, \mathbf{i}_2, \mathbf{i}_3$, respectively.

$h_1, h_2, h_3 = dl_1/du_1, dl_2/du_2, dl_3/du_3$ known as the metric coefficients.

Table 1.7 summarizes u_1, u_2, u_3 and h_1, h_2, h_3 for the three coordinate systems.

TABLE 1.7. Coordinates and Metric Coefficients for the Three Coordinate Systems

	u_1	u_2	u_3	h_1	h_2	h_3
Cartesian	x	y	z	1	1	1
Cylindrical	r	ϕ	z	1	r	1
Spherical	r	θ	ϕ	1	r	$r \sin \theta$

We will denote the components of a vector $\mathbf{A}$ as A_1, A_2, and A_3 so that

$$\mathbf{A} = A_1 \mathbf{i}_1 + A_2 \mathbf{i}_2 + A_3 \mathbf{i}_3$$

The following general relations can be written:

$$\mathbf{A} \cdot \mathbf{B} = A_1 B_1 + A_2 B_2 + A_3 B_3 \qquad (1\text{-}142)$$

$$\mathbf{A} \times \mathbf{B} = \begin{vmatrix} \mathbf{i}_1 & \mathbf{i}_2 & \mathbf{i}_3 \\ A_1 & A_2 & A_3 \\ B_1 & B_2 & B_3 \end{vmatrix} \qquad (1\text{-}143)$$

$$\nabla V = \frac{1}{h_1}\frac{\partial V}{\partial u_1}\mathbf{i}_1 + \frac{1}{h_2}\frac{\partial V}{\partial u_2}\mathbf{i}_2 + \frac{1}{h_3}\frac{\partial V}{\partial u_3}\mathbf{i}_3 \tag{1-144}$$

$$\nabla \cdot \mathbf{J} = \frac{1}{h_1 h_2 h_3}\left[\frac{\partial}{\partial u_1}(h_2 h_3 J_1) + \frac{\partial}{\partial u_2}(h_3 h_1 J_2) + \frac{\partial}{\partial u_3}(h_1 h_2 J_3)\right] \tag{1-145}$$

$$\nabla \times \mathbf{F} = \begin{vmatrix} \dfrac{\mathbf{i}_1}{h_2 h_3} & \dfrac{\mathbf{i}_2}{h_3 h_1} & \dfrac{\mathbf{i}_3}{h_1 h_2} \\[2mm] \dfrac{\partial}{\partial u_1} & \dfrac{\partial}{\partial u_2} & \dfrac{\partial}{\partial u_3} \\[2mm] h_1 F_1 & h_2 F_2 & h_3 F_3 \end{vmatrix} \tag{1-146}$$

$$\nabla^2 V = \frac{1}{h_1 h_2 h_3}\left[\frac{\partial}{\partial u_1}\left(\frac{h_2 h_3}{h_1}\frac{\partial V}{\partial u_1}\right) + \frac{\partial}{\partial u_2}\left(\frac{h_3 h_1}{h_2}\frac{\partial V}{\partial u_2}\right) + \frac{\partial}{\partial u_3}\left(\frac{h_1 h_2}{h_3}\frac{\partial V}{\partial u_3}\right)\right] \tag{1-147}$$

We will now list some useful vector identities. U and V are scalar functions whereas $\mathbf{A}$, $\mathbf{B}$, $\mathbf{C}$, and $\mathbf{D}$ are vectors.

$$\mathbf{A} \cdot \mathbf{B} \times \mathbf{C} = \mathbf{B} \cdot \mathbf{C} \times \mathbf{A} = \mathbf{C} \cdot \mathbf{A} \times \mathbf{B}$$

$$\mathbf{A} \times (\mathbf{B} \times \mathbf{C}) = \mathbf{B}(\mathbf{A} \cdot \mathbf{C}) - \mathbf{C}(\mathbf{A} \cdot \mathbf{B})$$

$$(\mathbf{A} \times \mathbf{B}) \times \mathbf{C} = \mathbf{B}(\mathbf{A} \cdot \mathbf{C}) - \mathbf{A}(\mathbf{B} \cdot \mathbf{C})$$

$$\mathbf{A} \times (\mathbf{B} \times \mathbf{C}) + \mathbf{B} \times (\mathbf{C} \times \mathbf{A}) + \mathbf{C} \times (\mathbf{A} \times \mathbf{B}) = 0$$

$$(\mathbf{A} \times \mathbf{B}) \cdot (\mathbf{C} \times \mathbf{D}) = (\mathbf{A} \cdot \mathbf{C})(\mathbf{B} \cdot \mathbf{D}) - (\mathbf{B} \cdot \mathbf{C})(\mathbf{A} \cdot \mathbf{D})$$

$$(\mathbf{A} \times \mathbf{B}) \times (\mathbf{C} \times \mathbf{D}) = (\mathbf{A} \times \mathbf{B} \cdot \mathbf{D})\mathbf{C} - (\mathbf{A} \times \mathbf{B} \cdot \mathbf{C})\mathbf{D}$$

$$\nabla(U + V) = \nabla U + \nabla V$$

$$\nabla \cdot (\mathbf{A} + \mathbf{B}) = \nabla \cdot \mathbf{A} + \nabla \cdot \mathbf{B}$$

$$\nabla \times (\mathbf{A} + \mathbf{B}) = \nabla \times \mathbf{A} + \nabla \times \mathbf{B}$$

$$\nabla(UV) = U \nabla V + V \nabla U$$

$$\nabla \cdot (U\mathbf{A}) = \mathbf{A} \cdot \nabla U + U \nabla \cdot \mathbf{A}$$

$$\nabla(\mathbf{A} \cdot \mathbf{B}) = \mathbf{A} \times (\nabla \times \mathbf{B}) + \mathbf{B} \times (\nabla \times \mathbf{A}) + (\mathbf{A} \cdot \nabla)\mathbf{B} + (\mathbf{B} \cdot \nabla)\mathbf{A}$$

$$\nabla \cdot (\mathbf{A} \times \mathbf{B}) = \mathbf{B} \cdot \nabla \times \mathbf{A} - \mathbf{A} \cdot \nabla \times \mathbf{B}$$

$$\nabla \times (U\mathbf{A}) = \nabla U \times \mathbf{A} + U \nabla \times \mathbf{A}$$

$$\nabla \times (\mathbf{A} \times \mathbf{B}) = \mathbf{A} \nabla \cdot \mathbf{B} - \mathbf{B} \nabla \cdot \mathbf{A} + (\mathbf{B} \cdot \nabla)\mathbf{A} - (\mathbf{A} \cdot \nabla)\mathbf{B}$$

$$\nabla \cdot \nabla \times \mathbf{A} = 0$$

$$\nabla \times \nabla U = 0$$

$$\nabla \times \nabla \times \mathbf{A} = \nabla(\nabla \cdot \mathbf{A}) - \nabla^2 \mathbf{A}$$

$$\frac{d}{dt}(U\mathbf{A}) = U\frac{d\mathbf{A}}{dt} + \frac{dU}{dt}\mathbf{A}$$

$$\frac{d}{dt}(\mathbf{A} \cdot \mathbf{B}) = \mathbf{A} \cdot \frac{d\mathbf{B}}{dt} + \mathbf{B} \cdot \frac{d\mathbf{A}}{dt}$$

$$\frac{d}{dt}(\mathbf{A} \times \mathbf{B}) = \mathbf{A} \times \frac{d\mathbf{B}}{dt} + \mathbf{B} \times \frac{d\mathbf{A}}{dt}$$

PROBLEMS

1.1.   For the vectors of Example 1-1, perform the following operations:

(a) $\mathbf{A} - \mathbf{B}$.

(b) $\mathbf{B}/4 + \mathbf{C}/3$.

(c) $\mathbf{A} - \mathbf{B} + \mathbf{C}$.

(d) $\mathbf{B} \cdot \mathbf{C}$.

(e) $\mathbf{C} \times \mathbf{A}$.

(f) $(\mathbf{A} + \mathbf{B}) \cdot \mathbf{C}$.

(g) $(\mathbf{A} - \mathbf{B}) \times \mathbf{C}$.

(h) $(\mathbf{B}/4 + \mathbf{C}/3) \cdot (3\mathbf{A} - 4\mathbf{C})$.

(i) $(\mathbf{B}/4 + \mathbf{C}/3) \times (3\mathbf{A} - 4\mathbf{C})$.

(j) $\mathbf{B} \cdot \mathbf{C} \times \mathbf{A}$.

(k) $\mathbf{C} \cdot \mathbf{A} \times \mathbf{B}$.

(l) $\mathbf{A} \times (\mathbf{B} \times \mathbf{C})$.

(m) $\mathbf{B} \times (\mathbf{C} \times \mathbf{A})$.

1.2.   (a) Show that the area of a triangle having vectors $\mathbf{A}$ and $\mathbf{B}$ as two of its sides is equal to $\frac{1}{2}|\mathbf{A} \times \mathbf{B}|$.

(b) Show that the volume of a tetrahedron formed by three vectors $\mathbf{A}$, $\mathbf{B}$, and $\mathbf{C}$ originating from a point is equal to $\frac{1}{6}|\mathbf{A} \cdot \mathbf{B} \times \mathbf{C}|$.

1.3.   A triangle is formed by three vectors $\mathbf{A}$, $\mathbf{B}$, and $\mathbf{C}$ such that $\mathbf{C} = \mathbf{B} - \mathbf{A}$ and hence $\mathbf{C} \cdot \mathbf{C} = (\mathbf{B} - \mathbf{A}) \cdot (\mathbf{B} - \mathbf{A})$. Obtain the law of cosines relating C to A, B, and the angle between $\mathbf{A}$ and $\mathbf{B}$.

1.4.   The tips of three vectors $\mathbf{A}$, $\mathbf{B}$, and $\mathbf{C}$ drawn from a point determine a plane.

(a) Show that $(\mathbf{A} \times \mathbf{B} + \mathbf{B} \times \mathbf{C} + \mathbf{C} \times \mathbf{A})$ is normal to the plane.

(b) Show that the minimum distance from the point to the plane is

$$\frac{|\mathbf{A} \cdot [(\mathbf{A} - \mathbf{B}) \times (\mathbf{A} - \mathbf{C})]|}{|(\mathbf{A} - \mathbf{B}) \times (\mathbf{A} - \mathbf{C})|}$$

(c) Obtain an equation for the plane in terms of $\mathbf{A}$, $\mathbf{B}$, and $\mathbf{C}$.

1.5.   (a) In the expression for the differential displacement vector $d\mathbf{l}$ in the cartesian coordinate system given by (1-19a), substitute for x, y, z in terms of the cylindrical coordinates r, ϕ, z and obtain the expression for $d\mathbf{l}$ in the cylindrical coordinate system.

(b) Repeat (a) by substituting for x, y, z in terms of the spherical coordinates r, θ, ϕ to obtain the expression for $d\mathbf{l}$ in the spherical coordinate system.

(c) The parabolic cylindrical coordinates u, v, z are related to x, y, z as

$$x = \tfrac{1}{2}(u^2 - v^2) \qquad y = uv \qquad z = z$$

Obtain the expression for $d\mathbf{l}$ in the parabolic cylindrical coordinate system.

(d) What is the expression for the differential volume dv in the parabolic cylindrical coordinate system?

1.6.   Derive the relationships listed in Table 1.2 between the different sets of coordinates.

1.7.   In Fig. 1.33, a point of observation T on the surface of the earth is defined by a spherical coordinate system with the origin at the center of the earth. The spherical coordinates of T are its distance r_0 from the center of the earth, its colatitude θ_T and its east longitude ϕ_T. The colatitude is 90° minus the latitude, with south latitudes being negative. N is the north pole. A point P in space is now defined by a coordinate system centered at the point of observation T. The coordinates of P in this new coordinate system are the azimuthal angle α, which is the angle between the great circle path TN and the great circle path TR, where R is the projection of P onto the earth's surface, the elevation angle Δ in the plane TPR and the range S. The colatitude and east longitude of R are θ_R and ϕ_R, respectively.

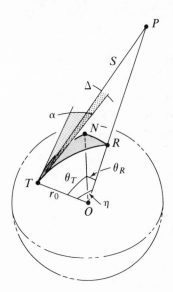

Fig. 1.33. For Problem 1.7.

(a) Show that

$$\cos \eta = \sin \theta_T \sin \theta_R \cos (\phi_T - \phi_R) + \cos \theta_T \cos \theta_R$$

(b) Show that

$$\cos \alpha = \frac{\cos \theta_R - \cos \eta \cos \theta_T}{\sin \eta \sin \theta_T}$$

(c) Find α, Δ, and S if T is at Urbana, Illinois (40.069° N latitude, 88.225° W longitude) and P represents a geostationary satellite parked above the equator at 50° W longitude. The earth radius r_0 is equal to 6370 km and the height h of the geostationary satellite above the earth's surface is equal to 35800 km.

(d) Find α if T represents Bondville, Illinois (40.1° N latitude, 88.4° W longitude) and R is located at Houston, Texas (29.4° N latitude, 95.0° W longitude). Repeat for the locations of T and R reversed.

1.8.   Derive the expressions listed in Tables 1.3 and 1.4 for the dot products and cross products of unit vectors in the different coordinate systems.

1.9. Derive the relationships listed in Table 1.5 between the components of a vector in the different coordinate systems.

1.10. Which of the following pairs of vectors are equal?

(a) $\mathbf{i}_x + 2\mathbf{i}_y + 3\mathbf{i}_z$ at $(1, 2, 3)$ and $\mathbf{i}_x + 2\mathbf{i}_y + 3\mathbf{i}_z$ at $(5.6, 9.8, 3.7)$ in cartesian coordinates.

(b) $\mathbf{i}_r + \mathbf{i}_\phi + 3\mathbf{i}_z$ at $(2, \pi/2, 3)$ and $\mathbf{i}_r + \mathbf{i}_\phi + 3\mathbf{i}_z$ at $(3.6, 3\pi/4, 9.4)$ in cylindrical coordinates.

(c) $\mathbf{i}_r + \mathbf{i}_\phi + 3\mathbf{i}_z$ at $(2, \pi/2, 3)$ and $\sqrt{2}\,\mathbf{i}_r + 3\mathbf{i}_z$ at $(3.6, 3\pi/4, 9.4)$ in cylindrical coordinates.

(d) $3\mathbf{i}_r + \sqrt{3}\,\mathbf{i}_\theta - 2\mathbf{i}_\phi$ at $(1, \pi/3, \pi/6)$ and $3\mathbf{i}_r + \sqrt{3}\,\mathbf{i}_\theta - 2\mathbf{i}_\phi$ at $(5.4, \pi/6, \pi/3)$ in spherical coordinates.

(e) $3\mathbf{i}_r + \sqrt{3}\,\mathbf{i}_\theta - 2\mathbf{i}_\phi$ at $(1, \pi/3, \pi/6)$ and $\mathbf{i}_r + \sqrt{3}\,\mathbf{i}_\theta - 2\sqrt{3}\,\mathbf{i}_\phi$ at $(5.4, \pi/6, \pi/3)$ in spherical coordinates.

1.11. Show that

(a) $(\mathbf{A} \times \mathbf{B}) \cdot (\mathbf{C} \times \mathbf{D}) = (\mathbf{A} \cdot \mathbf{C})(\mathbf{B} \cdot \mathbf{D}) - (\mathbf{B} \cdot \mathbf{C})(\mathbf{A} \cdot \mathbf{D})$.

(b) $(\mathbf{A} \times \mathbf{B}) \times (\mathbf{C} \times \mathbf{D}) = (\mathbf{A} \times \mathbf{B} \cdot \mathbf{D})\mathbf{C} - (\mathbf{A} \times \mathbf{B} \cdot \mathbf{C})\mathbf{D}$.

(c) $(\mathbf{A} \times \mathbf{B}) \cdot (\mathbf{B} \times \mathbf{C}) \times (\mathbf{C} \times \mathbf{A}) = (\mathbf{A} \times \mathbf{B} \cdot \mathbf{C})^2$.

(d) $\mathbf{A} \times (\mathbf{B} \times \mathbf{C}) + \mathbf{B} \times (\mathbf{C} \times \mathbf{A}) + \mathbf{C} \times (\mathbf{A} \times \mathbf{B}) = 0$.

1.12. Four vectors are given by

$$\mathbf{A} = \mathbf{i}_x + 2\mathbf{i}_y + 3\mathbf{i}_z$$
$$\mathbf{B} = 3\mathbf{i}_x + 2\mathbf{i}_y + \mathbf{i}_z$$
$$\mathbf{C} = \mathbf{i}_x - 2\mathbf{i}_y + \mathbf{i}_z$$
$$\mathbf{D} = -2\mathbf{i}_x + \mathbf{i}_y$$

Find

(a) $\mathbf{A} + \mathbf{B} - \mathbf{C}$, $\mathbf{C} - \mathbf{D} - \mathbf{A}$, $\mathbf{A} + \mathbf{B} + \mathbf{C} + \mathbf{D}$.

(b) $2\mathbf{A} + 3\mathbf{C}$, $\mathbf{A} - 3\mathbf{C} + 2\mathbf{D}$.

(c) $|\mathbf{C} - \mathbf{D}|$, $|\mathbf{C} - \mathbf{D} - \mathbf{A}|$.

(d) The unit vector along $(\mathbf{C} - \mathbf{D} - \mathbf{A})$.

(e) $\mathbf{A} \cdot \mathbf{B}$, $\mathbf{A} \cdot (\mathbf{C} - \mathbf{D})$, $\mathbf{B} \cdot (\mathbf{C} - \mathbf{D} - \mathbf{A})$.

(f) The cosines of the angles and the angles between $\mathbf{A}$ and $\mathbf{B}$, $\mathbf{A}$ and $(\mathbf{C} - \mathbf{D})$, $\mathbf{B}$ and $(\mathbf{C} - \mathbf{D} - \mathbf{A})$.

(g) $\mathbf{A} \times \mathbf{B}$, $\mathbf{B} \times \mathbf{C}$, $\mathbf{C} \times \mathbf{A}$, $\mathbf{A} \times (\mathbf{B} \times \mathbf{C})$, $\mathbf{B} \times (\mathbf{C} \times \mathbf{A})$, $\mathbf{C} \times (\mathbf{A} \times \mathbf{B})$.

(h) The sines of the angles and the angles between $\mathbf{A}$ and $(\mathbf{B} \times \mathbf{C})$, $\mathbf{B}$ and $(\mathbf{C} \times \mathbf{A})$, $\mathbf{C}$ and $(\mathbf{A} \times \mathbf{B})$.

(i) $(\mathbf{A} \times \mathbf{B}) \cdot (\mathbf{C} \times \mathbf{D})$; verify by using the identity of Problem 1.11(a).

(j) $\mathbf{A} \times \mathbf{B} \cdot \mathbf{D}$, $\mathbf{A} \times \mathbf{B} \cdot \mathbf{C}$, $\mathbf{B} \times \mathbf{C} \cdot \mathbf{A}$, $\mathbf{C} \times \mathbf{B} \cdot \mathbf{A}$.

(k) $(\mathbf{A} \times \mathbf{B}) \times (\mathbf{C} \times \mathbf{D})$; verify by using the identity of Problem 1.11(b).

(l) $(\mathbf{A} \times \mathbf{B}) \cdot (\mathbf{B} \times \mathbf{C}) \times (\mathbf{C} \times \mathbf{A})$; verify by using the identity of Problem 1.11(c).

(m) $\mathbf{A} \times (\mathbf{B} \times \mathbf{C}) + \mathbf{B} \times (\mathbf{C} \times \mathbf{A}) + \mathbf{C} \times (\mathbf{A} \times \mathbf{B})$.

(n) The components of $\mathbf{C}$ in cylindrical and spherical coordinates.

(o) A vector perpendicular to $(\mathbf{A} + \mathbf{B})$ by using a vector product; verify by using a dot product.

1.13. Let **A** and **B** be vectors in the xy plane making angles α and β with the x axis. With the aid of dot and cross products, prove the following trigonometric identities:

(a) $\cos(\alpha - \beta) = \cos\alpha\cos\beta + \sin\alpha\sin\beta$.

(b) $\sin(\alpha - \beta) = \sin\alpha\cos\beta - \cos\alpha\sin\beta$.

(c) $\cos(\alpha + \beta) = \cos\alpha\cos\beta - \sin\alpha\sin\beta$.

(d) $\sin(\alpha + \beta) = \sin\alpha\cos\beta + \cos\alpha\sin\beta$.

1.14. Write an expression for the component of a vector **A** along the direction of another vector **B** without the use of a coordinate system. Then find the component of $\mathbf{A} = 2\mathbf{i}_x - 3\mathbf{i}_y + \mathbf{i}_z$ along the direction of $\mathbf{B} = 3\mathbf{i}_x - \mathbf{i}_y - 2\mathbf{i}_z$.

1.15. Using two vectors in the plane $x + 2y + 3z = 3$, find the unit vector normal to that plane.

1.16. Show that the equation of the plane passing through the point (x_0, y_0, z_0) and normal to the vector $a\mathbf{i}_x + b\mathbf{i}_y + c\mathbf{i}_z$ is

$$a(x - x_0) + b(y - y_0) + c(z - z_0) = 0$$

1.17. For the following scalar functions, describe the shapes of the constant-magnitude surfaces:

(a) $T(x, y, z) = x^2 + 4y^2 + 9z^2$.

(b) $U(r, \phi, z) = (\cos\phi)/r$.

(c) $V(r, \theta, \phi) = (\sin\theta)/r$.

1.18. Using a spherical coordinate system with the origin at the center of the earth, write a vector function for the linear velocity of points inside the earth due to its spin motion. Describe the constant-magnitude surfaces and direction lines.

1.19. Using a spherical coordinate system with the origin at the center of the earth, write a vector function for the force experienced by a mass m in the gravitational field of the earth. Describe the constant-magnitude surfaces and direction lines.

1.20. Discuss the following vector fields with the aid of sketches:

(a) $\mathbf{A}(x, y, z) = (x - 2)\mathbf{i}_x$.

(b) $\mathbf{B}(r, \phi, z) = r(r - 1)\mathbf{i}_\phi$.

(c) $\mathbf{C}(r, \theta, \phi) = (1/r)\mathbf{i}_\theta$.

(d) $\mathbf{D}(r, \theta, \phi) = r\,\mathbf{i}_r$.

1.21. Derive the expressions listed in Table 1.6 for the partial derivatives of unit vectors with respect to the coordinates.

1.22. Let $\mathbf{r} = x\mathbf{i}_x + y\mathbf{i}_y + z\mathbf{i}_z = r_c\mathbf{i}_{rc} + z\mathbf{i}_z = r_s\mathbf{i}_{rs}$ be the position vector of a point P moving in three dimensions. Obtain the expressions for the velocity **v** and acceleration **a** of the point in all three coordinate systems.

1.23. (a) A point P moves along a curve in two dimensions such that its coordinates are given by $r = at$ and $\phi = bt$, where a and b are constants. Find the velocity and acceleration of the point.

(b) A point P moves along a curve in three dimensions such that its coordinates are given by $x = a\cos\omega t$, $y = b\sin\omega t$, and $z = ct$, where $a, b, c,$ and ω are constants. Find the velocity and acceleration of the point.

1.24. Verify Eqs. (1-63) and (1-64) by expansion in cartesian coordinates.

1.25. Find a unit vector normal to the surface $r^2 \cos 2\phi = 1$ at the point $(\sqrt{2}, \pi/6, 0)$ in the cylindrical coordinate system in two ways: (a) by using two vectors which are tangential to the surface at that point; and (b) by using the concept of the gradient of a scalar function.

1.26. Find the scalar functions whose gradients are given by the following vector functions:

(a) $\nabla T(x, y, z) = yz\mathbf{i}_x + zx\mathbf{i}_y + xy\mathbf{i}_z$.

(b) $\nabla U(x, y, z) = 3x^2yz^2\mathbf{i}_x + x^3z^2\mathbf{i}_y + 2x^3yz\mathbf{i}_z$.

(c) $\nabla V(r, \phi, z) = (1/r^2)(\cos \phi \, \mathbf{i}_r + \sin \phi \, \mathbf{i}_\phi)$.

(d) $\nabla W(r, \theta, \phi) = -n\mathbf{r}/r^{n+2}$, where $\mathbf{r}$ is the position vector.

1.27. Make up a table of gradients of the scalar functions defining the orthogonal surfaces in the three different coordinate systems.

1.28. Find the component of the unit vector normal to the surface $x^2 - y^2 = 3$ at the point $(2, 1, 1)$ in the direction of the vector joining the point $(1, -2, 0)$ to the point $(0, 0, 2)$.

1.29. Find the rate of change of $V = x^2y + yz^2 + zy^2$ in the direction normal to the surface $x^2y - yz + xz^2 = 5$ at the point $(1, 2, 3)$.

1.30. Find the equation of the plane tangential to the surface $xyz = 1$ at the point $(\frac{1}{2}, \frac{1}{4}, 8)$.

1.31. Evaluate the following volume integrals:

(a) $\int_V xyz \, dv$, where V is the volume enclosed by the planes $x = 0$, $y = 0$, $z = 0$, and $x + y + z = 1$.

(b) $\int_V \frac{1}{r} \, dv$, where V is the volume of a cylinder of radius a with the z axis as its axis and of length l.

(c) $\int_V x \, dv$, where V is that part of the volume of a sphere of radius unity lying in the first octant.

1.32. Given $\mathbf{A} = x^2yz\mathbf{i}_x + y^2zx\mathbf{i}_y + z^2xy\mathbf{i}_z$, evaluate $\oint \mathbf{A} \cdot d\mathbf{S}$ over the following closed surfaces:

(a) The surface of the cubical box bounded by the planes

$$x = 0, \, x = 1$$
$$y = 0, \, y = 1$$
$$z = 0, \, z = 1$$

(b) The surface of the box bounded by the planes

$$x = 0, \, y = 0, \, z = 0$$
$$x + 2y + 3z = 3$$

1.33. Given $\mathbf{A} = r \cos \phi \, \mathbf{i}_r - r \sin \phi \, \mathbf{i}_\phi$ in cylindrical coordinates, evaluate $\oint \mathbf{A} \cdot d\mathbf{S}$ over the following surfaces:

(a) The surface of the box bounded by the planes $z = 0$, $z = l$, and the cylinder $r = a$.

(b) The surface of the box bounded by the planes $x = 0$, $y = 0$, $z = 0$, $z = l$, and the cylinder $r = a$.

1.34. Given $\mathbf{A} = r^2\mathbf{i}_r + r \sin \theta\, \mathbf{i}_\theta$ in spherical coordinates, evaluate $\oint \mathbf{A} \cdot d\mathbf{S}$ over the following:

(a) The surface of that part of the spherical volume of radius unity lying in the first octant.

(b) The surface of a solid spherical shell lying between $r = a$ and $r = b$, where $b > a$ (note that this surface consists of two disconnected surfaces; the normal vectors to the surfaces must both be chosen to be away from or into the volume bounded by the surfaces).

1.35. For the force vector $\mathbf{F} = y\mathbf{i}_x + x\mathbf{i}_y$, find the work done by the force vector from the origin to the point $(\pi/2, 1, 0)$ along the following paths:

(a) $y = \sin^2 x$, $z = 0$.

(b) $y = (4/\pi^2)x^2$, $z = 0$.

(c) $x = (\pi/2)y^2$, $z = 0$.

(d) Any other path of your choice not necessarily in the $z = 0$ plane.

1.36. A certain vector field is given by

$$\mathbf{A} = a^2 y\mathbf{i}_x - b^2 x\mathbf{i}_y$$

where a and b are constants. Evaluable $\int \mathbf{A} \cdot d\mathbf{l}$ from the origin to the point $(1, 1, 1)$ along the following paths:

(a) $y = x = z^2$.

(b) The path given by $y = 0$, $z = 0$, then $x = 1$, $z = 0$, and then $x = y = 1$.

(c) The path given by $y = x$, $z = 0$, and then $x = y = 1$.

(d) The path given by $x = 0$, $z = 0$, then $y = 1$, $z = 0$, and then $x = y = 1$.

(e) $x = y = z$.

1.37. Given $\mathbf{A} = xy\mathbf{i}_x + yz\mathbf{i}_y + zx\mathbf{i}_z$, evaluate the circulation $\oint \mathbf{A} \cdot d\mathbf{l}$ around the contour *abcda* shown in Fig. 1.34.

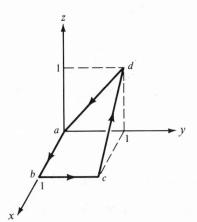

Fig. 1.34. For Problem 1.37.

1.38 Given $\mathbf{A} = 2r \cos \phi \, \mathbf{i}_r + r \mathbf{i}_\phi$ in cylindrical coordinates, find:

(a) $\oint_C \mathbf{A} \cdot d\mathbf{l}$, where C is the contour shown in Fig. 1.35(a).

(b) $\oint_{C_1} \mathbf{A} \cdot d\mathbf{l} + \oint_{C_2} \mathbf{A} \cdot d\mathbf{l}$, where C_1 and C_2 are the contours shown in Fig. 1.35(b).

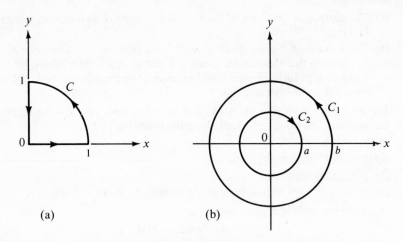

(a)   (b)

Fig. 1.35. For Problem 1.38.

1.39. Given $\mathbf{A} = (e^{-r}/r)\mathbf{i}_\theta$ in spherical coordinates, evaluate $\oint \mathbf{A} \cdot d\mathbf{l}$ around the contour *abca* shown in Fig. 1.36.

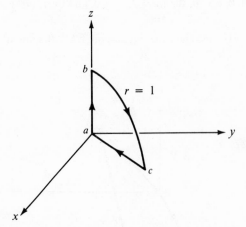

Fig. 1.36. For Problem 1.39.

1.40. Evaluate the following vector integrals:

(a) $\oint_C d\mathbf{l}$, where C is any closed path of your choice.

(b) $\oint_S d\mathbf{S}$, where S is the surface of the hemispherical volume of radius a above the xy plane and with center at the origin.

(c) $\int_V \mathbf{i}_\theta \, dv$, where V is the volume of the sphere of radius a centered at the origin.

1.41. Derive the expression for the divergence of a vector in cartesian coordinates given by (1-96).

1.42. Derive the expression for the divergence of a vector in spherical coordinates given by (1-97).

1.43. Make up a table of divergences of the unit vectors in the three coordinate systems.

1.44. Find the divergences of the following vectors:

(a) $\mathbf{A} = x^2yz\mathbf{i}_x + y^2zx\mathbf{i}_y + z^2xy\mathbf{i}_z$.

(b) $\mathbf{B} = 3x\mathbf{i}_x + (y - 3)\mathbf{i}_y + (2 + z)\mathbf{i}_z$.

(c) $\mathbf{C} = r \cos \phi \, \mathbf{i}_r - r \sin \phi \, \mathbf{i}_\phi$, cylindrical coordinates.

(d) $\mathbf{D} = (1/r^2)\mathbf{i}_r$, spherical coordinates.

(e) $\mathbf{E} = r^2\mathbf{i}_r + r \sin \theta \, \mathbf{i}_\theta$.

1.45. Using the position vector $\mathbf{r} = r\mathbf{i}_r$ in three dimensions, verify the divergence theorem by considering a sphere of radius a, and centered at the origin.

1.46. Verify your answers to Problem 1.32 by evaluating the appropriate volume integrals and using the divergence theorem.

1.47. Verify your answers to Problem 1.33 by evaluating the appropriate volume integrals and using the divergence theorem.

1.48. Verify your answers to Problem 1.34 by evaluating the appropriate volume integrals and using the divergence theorem.

1.49. For the vector $\mathbf{A} = yz\mathbf{i}_x + zx\mathbf{i}_y + xy\mathbf{i}_z$, use the divergence theorem to show that $\oint_S \mathbf{A} \cdot d\mathbf{S}$ is zero, where S is any closed surface. Then evaluate $\int \mathbf{A} \cdot d\mathbf{S}$ over the following surfaces:

(a) That part of the plane $x + 2y + 3z = 3$ lying in the first octant.

(b) That part of the cylindrical surface $r = 1$ lying in the first octant and between the planes $z = 0$ and $z = 1$.

(c) The upper half of the spherical surface $r = 1$.

(d) That part of the conical surface $\theta = \pi/4$ lying below the plane $z = 1$.

1.50. Derive the expression for the curl of a vector in cartesian coordinates given by (1-121).

1.51. Derive the expression for the curl of a vector in cylindrical coordinates given by (1-122).

1.52. Make up a table of curls of the unit vectors in the three coordinate systems.

1.53. Find the curls of the following vectors:

(a) $\mathbf{A} = xy\mathbf{i}_x + yz\mathbf{i}_y + zx\mathbf{i}_z$.

(b) $\mathbf{B} = y\mathbf{i}_x - x\mathbf{i}_y$.

(c) $\mathbf{C} = 2r \cos \phi\, \mathbf{i}_r + r\, \mathbf{i}_\phi$, cylindrical coordinates.

(d) $\mathbf{D} = (1/r)\mathbf{i}_\phi$, cylindrical coordinates.

(e) $\mathbf{E} = (e^{-r}/r)\mathbf{i}_\theta$.

1.54. Discuss the curls of the following vector fields by using the "paddle-wheel" device and also by expansion in the appropriate coordinate system:

(a) The velocity vector field associated with points inside the earth due to its spin motion.

(b) The position vector field associated with points in three-dimensional space.

(c) The velocity vector field associated with the flow of water in the stream of Fig. 1-30(a) such that the velocity varies uniformly from zero at the bottom of the stream to a maximum at the top surface.

(d) the vector field $\mathbf{F} = \mathbf{i}_\phi$.

1.55. By expansion in cartesian coordinates, verify

$$\nabla \cdot \nabla \times \mathbf{F} \equiv 0$$

$$\nabla \times \nabla V \equiv 0$$

1.56. Determine which of the following vectors can be expressed as the curl of another vector and which of them can be expressed as the gradient of a scalar:

(a) $\mathbf{A} = yz\mathbf{i}_x + zx\mathbf{i}_y + xy\mathbf{i}_z$.

(b) $\mathbf{B} = xy\mathbf{i}_x + yz\mathbf{i}_y + zx\mathbf{i}_z$.

(c) $\mathbf{C} = (x^2 - y^2)\mathbf{i}_x - 2xy\mathbf{i}_y + 4\mathbf{i}_z$.

(d) $\mathbf{D} = (e^{-r}/r)\mathbf{i}_\phi$, cylindrical coordinates.

(e) $\mathbf{E} = (1/r^2)(\cos \phi\, \mathbf{i}_r + \sin \phi\, \mathbf{i}_\phi)$, cylindrical coordinates.

(f) $\mathbf{F} = (1/r^3)(2 \cos \theta\, \mathbf{i}_r + \sin \theta\, \mathbf{i}_\theta)$, spherical coordinates.

1.57. Verify your answer to Problem 1.37 by evaluating the appropriate surface integral and using Stokes' theorem.

1.58. Verify your answers to Problem 1.38 by evaluating the appropriate surface integrals and using Stokes' theorem.

1.59. Verify your answer to Problem 1.39 by evaluating the appropriate surface integral and using Stoke's theorem.

1.60. For the vector $\mathbf{A} = yz\mathbf{i}_x + zx\mathbf{i}_y + xy\mathbf{i}_z$, use Stokes' theorem to show that $\oint_C \mathbf{A} \cdot d\mathbf{l}$ is zero, where C is any closed path. Then evaluate $\int \mathbf{A} \cdot d\mathbf{l}$ along the following paths:

(a) From the origin to the point $(1, \pi/2, 0)$ along the curve $r = t$, $\phi = (\pi/2)t$, $z = \sin \pi\, t$, in cylindrical coordinates.

(b) From the origin to the point $(1, 1, 1)$ along the curve $x = \sqrt{2} \sin t$, $y = \sqrt{2} \sin t$, $z = (4/\pi)t$.

(c) From the origin to the point $(22.34, 5.68, -6.93)$ in cartesian coordinates along any path of your choice.

1.61. Use Stokes' theorem and the divergence theorem to prove that $\nabla \cdot \nabla \times \mathbf{A} \equiv 0$, without the implication of a coordinate system.

1.62. From the definition of ∇V, show that $\oint_C \nabla V \cdot d\mathbf{l} \equiv 0$, where C is any closed path. Then use this result and Stoke's theorem to prove that $\nabla \times \nabla V \equiv 0$, without the implication of a coordinate system.

1.63. Find the Laplacians of the following scalar and vector functions:

(a) $T(x, y, z) = x^3yz^2$.

(b) $U(r, \phi, z) = (\cos \phi)/r$.

(c) $V(r, \theta, \phi) = e^{-r}/r$.

(d) $\mathbf{A}(x, y, z) = x^2yz\mathbf{i}_x + xy^2z\mathbf{i}_y + xyz^2\mathbf{i}_z$.

1.64. Derive the expansion for the Laplacian of a vector in cartesian coordinates given by (1-139).

1.65. Derive the expansion for the Laplacian of a vector in cylindrical coordinates given by (1-140).

1.66. Derive the expansion for the Laplacian of a vector in spherical coordinates given by (1-141).

1.67. Verify the general expressions for ∇V, $\nabla \cdot \mathbf{J}$, $\nabla \times \mathbf{F}$ and $\nabla^2 V$ given by (1-144), (1-145), (1-146), and (1-147), respectively.

1.68. By expansion in cartesian coordinates, show that

(a) $\nabla \cdot U\mathbf{A} = \mathbf{A} \cdot \nabla U + U \nabla \cdot \mathbf{A}$.

(b) $\nabla \times U\mathbf{A} = \nabla U \times \mathbf{A} + U \nabla \times \mathbf{A}$.

(c) $\nabla \cdot (\mathbf{A} \times \mathbf{B}) = \mathbf{B} \cdot \nabla \times \mathbf{A} - \mathbf{A} \cdot \nabla \times \mathbf{B}$.

(d) $\nabla \times (\mathbf{A} \times \mathbf{B}) = \mathbf{A} \nabla \cdot \mathbf{B} - \mathbf{B} \nabla \cdot \mathbf{A} + (\mathbf{B} \cdot \nabla)\mathbf{A} - (\mathbf{A} \cdot \nabla)\mathbf{B}$.

2

THE STATIC ELECTRIC FIELD

In Chapter 1 we learned the mathematical language of vector analysis so that we are now ready to use it for the study of electromagnetic field theory. Electromagnetic field theory is built upon four equations known as Maxwell's equations and an associated set of relations known as the constitutive relations. It is our goal to learn how to interpret these equations and to use them for various applications, important among them being electromagnetic waves. Maxwell's equations, in their general form, relate the time-varying or dynamic electric and magnetic fields with one another and with the electric charges and currents present in the medium. It is possible to study electromagnetic theory by starting with Maxwell's equations and another equation known as the Lorentz force equation as postulates. The Lorentz force equation is the defining equation for the electric and magnetic fields in terms of the forces experienced by the charges. Alternatively, it is possible to develop Maxwell's equations gradually from the electric and magnetic field concepts based on forces experienced by charges and currents and from a few experimental facts. We will take this latter approach. The electromagnetic field is one in which the electric and magnetic effects are coupled. Before we venture to discuss the electromagnetic field, we will study the electric and magnetic fields separately. This is best done by considering static or time-independent fields in free space. With this approach in mind, the present chapter is devoted to the static electric field in free space.

2.1 The Electric Field Concept

In the study of mechanics, we are familiar with the gravitational field as a force field associated with the mutual attraction of material bodies in space. For example, a small test mass m placed in the gravitational field of the earth experiences a force equal to mMG/r^2 directed towards the center of mass of the earth, where M is the mass of the earth, G is the constant of universal gravitation, and r is the distance of the test mass from the center of mass of the earth. We associate with every point in the vicinity of the earth a vector quantity $\mathbf{g}$, known as the gravitational field intensity, having a magnitude MG/r^2 and directed towards the center of the earth as shown in Fig. 2.1. In terms of the value of the test mass and the force experienced by the test mass, the gravitational field intensity is given by

$$\mathbf{g} = \frac{\mathbf{F}}{m} \tag{2-1}$$

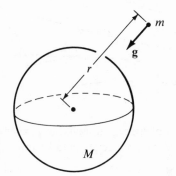

Fig. 2.1. Gravitational attraction of a test mass m towards the center of mass of the earth.

Just as the gravitational field is associated with the physical property known as "mass," a force field is associated with the physical property known as "charge" merely by virtue of its existence. This force field is known as the electric field. We will learn in the next chapter that a second kind of force field known as the magnetic field exists when charges are set in motion. A few words about charge are now in order. Matter can be regarded as composed of three types of elementary particles, known as protons, neutrons, and electrons. These particles are charged positive, zero, and negative, respectively. Table 2.1 gives the charge and mass for each of these particles.

TABLE 2.1. Charges and Masses of Elementary Particles

Particle	Charge, C	Mass, kg
Proton	1.6021×10^{-19}	1.6724×10^{-27}
Neutron	0	1.6747×10^{-27}
Electron	-1.6021×10^{-19}	9.1083×10^{-31}

Charges are conserved; that is, they can neither be created nor destroyed. They can only be transferred from one body to another. A material body is uncharged if it has no net charge. If the body acquires excess negative charge by some means, it is said to be negatively charged. On the other hand, if it loses some negative charge, it is said to be positively charged. The unit of charge is the coulomb (abbreviated C).

A small test charge q placed in the "electric field" of a larger charge Q experiences a force $\mathbf{F}$ given by

$$\mathbf{F} = q\mathbf{E} \tag{2-2}$$

as shown in Fig. 2.2, where $\mathbf{E}$ is the intensity of the electric field, analogous to the gravitational field intensity $\mathbf{g}$. Alternatively, we can say that if, in a

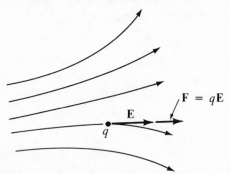

Fig. 2.2. Force experienced by a test charge in an electric field.

region of space, a test charge q experiences a force $\mathbf{F}$, then the region is characterized by an electric field of intensity $\mathbf{E}$ given by

$$\mathbf{E} = \frac{\mathbf{F}}{q} \tag{2-3}$$

Here we are assuming that the test charge q is so small that it does not alter the electric field in which it is placed. From a practical point of view, the test charge does influence the electric field irrespective of how small it is. However, theoretically, we can define $\mathbf{E}$ as the ratio of the force experienced by the test charge divided by the test charge in the limit that the test charge tends to zero; that is,

$$\mathbf{E} = \operatorname*{Lim}_{q \to 0} \frac{\mathbf{F}}{q} \tag{2-4}$$

The unit of electric field intensity is newton per coulomb (N/C).

EXAMPLE 2-1.   An electron placed at a point in an electric field experiences an acceleration of 10^5 m/sec² along the positive x axis. (a) What is the electric field intensity $\mathbf{E}$ at that point? (b) What acceleration does a proton placed at that point experience?

The force experienced by the electron is equal to $-1.6 \times 10^{-19}\,\mathbf{E}$. This is equal to the mass of the electron times the acceleration experienced by the electron. Hence

$$-1.6 \times 10^{-19}\,\mathbf{E} = 9.11 \times 10^{-31} \times 10^5\,\mathbf{i}_x$$

$$\mathbf{E} = \frac{9.11 \times 10^{-31} \times 10^5}{-1.6 \times 10^{-19}}\,\mathbf{i}_x = -5.7 \times 10^{-7}\,\mathbf{i}_x\ \text{N/C}$$

Thus the electric field intensity has a magnitude of about 5.7×10^{-7} N/C and it is directed along the negative x axis.

Now, if a proton is placed at the same point, the acceleration **a** experienced by it is given by

$$\mathbf{a} = \frac{\text{charge of proton} \times \mathbf{E}}{\text{mass of proton}}$$

$$= \frac{1.6 \times 10^{-19} \times (-5.7 \times 10^{-7})\mathbf{i}_x}{1.67 \times 10^{-27}} = -54.6\,\mathbf{i}_x\ \text{m/sec}^2$$

Thus the proton experiences an acceleration of about 54.6 m/sec² along the negative x axis. ▌

2.2   Coulomb's Law

In the previous section we introduced the concept of the electric field from an analogy with the gravitational field. It was mentioned that a small test charge placed in the electric field of a larger charge experiences a force. Actually, the larger charge also experiences a force just as two masses attract each other. This fact was proved experimentally by Coulomb. As a result of his experiments we have Coulomb's law, which relates the force between two charged bodies which are very small in size compared to their separation. Ideally, the charged bodies must be so small that they can be considered as "point charges." From Coulomb's experiments, the following conclusions were reached:

1. Like charges repel whereas unlike charges attract.
2. The magnitude of the force is proportional to the product of the magnitudes of the charges.
3. The magnitude of the force is inversely proportional to the square of the distance between the charges.
4. The direction of the force is along the line joining the charges.
5. The force depends upon the medium in which the charges are placed.

If we consider two point charges Q_1 and Q_2 C situated at points A and B separated by a distance R m, as shown in Fig. 2.3, we can express the foregoing five statements in equation form as

$$\mathbf{F}_A = k\frac{Q_1 Q_2}{R^2}\,\mathbf{i}_{BA} \tag{2-5}$$

$$\mathbf{F}_B = k\frac{Q_1 Q_2}{R^2}\,\mathbf{i}_{AB} \tag{2-6}$$

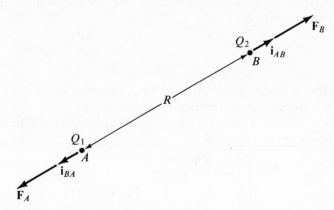

Fig. 2.3. Forces of repulsion between two charges Q_1 and Q_2 at points A and B.

where $\mathbf{F}_A$ and $\mathbf{F}_B$ are the forces experienced by Q_1 and Q_2, respectively, $\mathbf{i}_{BA}$ and $\mathbf{i}_{AB}$ are unit vectors along the line joining A and B (Fig. 2.3), and k is the constant of proportionality. Statement 1 is included in (2-5) and (2-6) since Q_1 and Q_2 represent the magnitudes as well as signs of the charges. If Q_1 and Q_2 are both positive charges or both negative charges, their product will be positive and hence positive forces act along $\mathbf{i}_{BA}$ and $\mathbf{i}_{AB}$. If one of the two charges is negative, then the product $Q_1 Q_2$ will be negative; hence negative forces act along $\mathbf{i}_{BA}$ and $\mathbf{i}_{AB}$ or positive forces act along directions opposite to $\mathbf{i}_{BA}$ and $\mathbf{i}_{AB}$, respectively. The constant of proportionality k is equal to $1/4\pi\epsilon_0$ for free space and in MKS rationalized units. The quantity ϵ_0 is known as the permittivity of free space and its value is 8.854×10^{-12} or approximately equal to $10^{-9}/36\pi$. Substituting for k in (2-5) and (2-6), we have

$$\mathbf{F}_A = \frac{Q_1 Q_2}{4\pi\epsilon_0 R^2} \mathbf{i}_{BA} \qquad (2\text{-}7)$$

$$\mathbf{F}_B = \frac{Q_1 Q_2}{4\pi\epsilon_0 R^2} \mathbf{i}_{AB} \qquad (2\text{-}8)$$

Equations (2-7) and (2-8) represent Coulomb's law. From these equations, we note that ϵ_0 has the units (coulombs)^2 per [(newton)(meter)^2]. These are commonly known as farads per meter (F/m).

2.3 The Electric Field of Point Charges

Let one of the two charges considered in the preceding section, say Q_2, be a small test charge q. Then, from a knowledge of the force experienced by this test charge due to the presence of the charge Q_1, we can obtain the expression for the electric field intensity due to the charge Q_1 using (2-3).

According to Coulomb's law, the force experienced by the test charge is given by

$$\mathbf{F}_B = \frac{Q_1 q}{4\pi\epsilon_0 R^2}\mathbf{i}_{AB} \tag{2-9}$$

From (2-3) we then have the electric field intensity $\mathbf{E}_B$ at point B due to the charge Q_1 as

$$\mathbf{E}_B = \frac{\mathbf{F}_B}{q} = \frac{Q_1}{4\pi\epsilon_0 R^2}\mathbf{i}_{AB} \tag{2-10}$$

We can generalize this result by making R variable, that is, by moving the test charge around in the medium, writing the expression for the force experienced by it, and dividing the force by the test charge. The result is the same as (2-10) except that R is now a variable since point B is a variable. Thus, omitting the subscripts in (2-10), we write the electric field intensity $\mathbf{E}$ of a point charge Q as

$$\mathbf{E} = \frac{Q}{4\pi\epsilon_0 R^2}\mathbf{i}_R \tag{2-11}$$

where R is the distance from the point charge to the point at which the field intensity is to be computed and $\mathbf{i}_R$ is the unit vector along the line joining the two points under consideration and directed away from the point charge. The electric field intensity of a point charge is thus directed everywhere radially away from the point charge, and on any spherical surface centered at the point charge its magnitude is constant. The situation is illustrated in Fig. 2.4. If the point charge is at the origin of a coordinate system, then we replace R by r and $\mathbf{i}_R$ by $\mathbf{i}_r$. The field represented by (2-11) is also known as the Coulomb field of a point charge.

If we now have several point charges $Q_1, Q_2, Q_3, \ldots, Q_n$ located at different points as shown in Fig. 2.5, we can invoke superposition and state that the force $\mathbf{F}$ experienced by a test charge situated at a point P is the vector

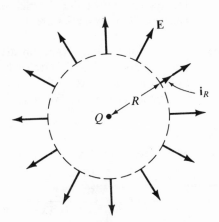

Fig. 2.4. The electric field of a point charge.

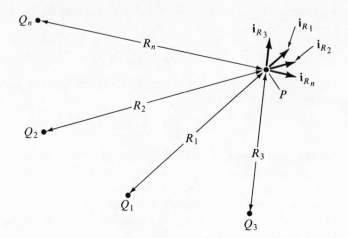

Fig. 2.5. Assembly of point charges and unit vectors along the direction of their electric field intensities at point P, due to the individual point charges.

sum of the forces experienced by the test charge due to the individual charges; that is,

$$\mathbf{F} = \frac{Q_1 q}{4\pi\epsilon_0 R_1^2}\mathbf{i}_{R_1} + \frac{Q_2 q}{4\pi\epsilon_0 R_2^2}\mathbf{i}_{R_2} + \frac{Q_3 q}{4\pi\epsilon_0 R_3^2}\mathbf{i}_{R_3} + \cdots + \frac{Q_n q}{4\pi\epsilon_0 R_n^2}\mathbf{i}_{R_n} \quad (2\text{-}12)$$

From (2-3) the electric field intensity $\mathbf{E}$ at the point P is

$$\mathbf{E} = \frac{\mathbf{F}}{q} = \frac{Q_1}{4\pi\epsilon_0 R_1^2}\mathbf{i}_{R_1} + \frac{Q_2}{4\pi\epsilon_0 R_2^2}\mathbf{i}_{R_2} + \cdots + \frac{Q_n}{4\pi\epsilon_0 R_n^2}\mathbf{i}_{R_n}$$

$$= \sum_{j=1}^{n} \frac{Q_j}{4\pi\epsilon_0 R_j^2}\mathbf{i}_{R_j} \quad (2\text{-}13)$$

The electric field intensity due to the assembly of the point charges is thus the vector sum of the electric field intensities due to the individual point charges. Some examples are now in order.

EXAMPLE 2-2. For a charge Q at an arbitrary point $A(x', y', z')$, obtain the x, y, and z components of the electric field intensity at an arbitrary point $B(x, y, z)$, as shown in Fig. 2.6.

From Coulomb's law, the electric field intensity at point B is given by

$$\mathbf{E} = \frac{Q}{4\pi\epsilon_0 (AB)^2}\mathbf{i}_{AB} \quad (2\text{-}14)$$

where from Fig. 2.6,

$$AB = \sqrt{(x - x')^2 + (y - y')^2 + (z - z')^2} \quad (2\text{-}15)$$

$$\mathbf{i}_{AB} = \frac{(x - x')\mathbf{i}_x + (y - y')\mathbf{i}_y + (z - z')\mathbf{i}_z}{\sqrt{(x - x')^2 + (y - y')^2 + (z - z')^2}} \quad (2\text{-}16)$$

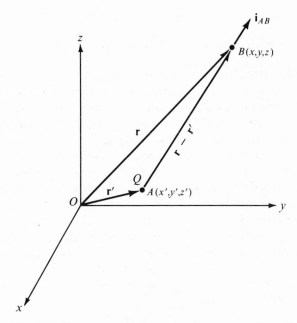

Fig. 2.6. Geometry pertinent to the computation of the electric field of a point charge located at an arbitrary point.

Substituting (2-15) and (2-16) into (2-14), we have

$$\mathbf{E} = \frac{Q}{4\pi\epsilon_0} \frac{(x - x')\mathbf{i}_x + (y - y')\mathbf{i}_y + (z - z')\mathbf{i}_z}{[(x - x')^2 + (y - y')^2 + (z - z')^2]^{3/2}} \qquad (2\text{-}17)$$

The x, y, and z components of $\mathbf{E}$ are therefore given by

$$E_x = \mathbf{E} \cdot \mathbf{i}_x = \frac{Q}{4\pi\epsilon_0} \frac{(x - x')}{[(x - x')^2 + (y - y')^2 + (z - z')^2]^{3/2}} \qquad (2\text{-}18a)$$

$$E_y = \mathbf{E} \cdot \mathbf{i}_y = \frac{Q}{4\pi\epsilon_0} \frac{(y - y')}{[(x - x')^2 + (y - y')^2 + (z - z')^2]^{3/2}} \qquad (2\text{-}18b)$$

$$E_z = \mathbf{E} \cdot \mathbf{i}_z = \frac{Q}{4\pi\epsilon_0} \frac{(z - z')}{[(x - x')^2 + (y - y')^2 + (z - z')]^{3/2}} \qquad (2\text{-}18c)$$

In vector notation, if we denote $\mathbf{r}'$ as the position vector for the source point A and $\mathbf{r}$ as the position vector for the point B at which the field is desired, then $AB = |\mathbf{r} - \mathbf{r}'|$ and $\mathbf{i}_{AB} = (\mathbf{r} - \mathbf{r}')/|\mathbf{r} - \mathbf{r}'|$ so that

$$\mathbf{E}(\mathbf{r}) = \frac{Q}{4\pi\epsilon_0 |\mathbf{r} - \mathbf{r}'|^2} \frac{\mathbf{r} - \mathbf{r}'}{|\mathbf{r} - \mathbf{r}'|} = \frac{Q}{4\pi\epsilon_0 |\mathbf{r} - \mathbf{r}'|^3}(\mathbf{r} - \mathbf{r}') \qquad (2\text{-}19)$$

If a number of charges Q_1, Q_2, $Q_3, \ldots, Q_n$ are located at points defined by position vectors $\mathbf{r}'_1, \mathbf{r}'_2, \mathbf{r}'_3, \ldots, \mathbf{r}'_n$, respectively, then

$$\mathbf{E}(\mathbf{r}) = \sum_{j=1}^{n} \frac{Q_j}{4\pi\epsilon_0 |\mathbf{r} - \mathbf{r}'_j|^3} (\mathbf{r} - \mathbf{r}'_j) \qquad (2\text{-}20)$$

where we have made use of superposition. ▮

EXAMPLE 2-3.   Two equal and opposite point charges Q and $-Q$ are situated on the
z axis at $d/2$ and $-d/2$, respectively, as shown in Fig. 2.7. Such an arrangement
is known as an electric dipole. It is desired to obtain the expression for the
electric field intensity due to the electric dipole at distances very large from
the origin compared to the spacing d.

   With reference to the geometry shown in Fig. 2.7, we note that the electric field intensity at any point P has only r and θ components if we use
the spherical coordinate system, whereas it has all three components if we
use the cartesian coordinate system. For fixed values of r and θ the field
intensity is independent of ϕ; that is, it has circular symmetry about the
z axis. Furthermore, we are interested only in the field at large distances
from the dipole, that is, for $r \gg d$. Hence we use the spherical coordinate
system. The electric field intensity $\mathbf{E}$ at P is the superposition of the electric
field intensities due to the two charges. Thus, with reference to the notation
in Fig. 2.7 we have

$$\mathbf{E} = \frac{Q}{4\pi\epsilon_0 r_+^2} \mathbf{i}_{r_+} - \frac{Q}{4\pi\epsilon_0 r_-^2} \mathbf{i}_{r_-} \qquad (2\text{-}21)$$

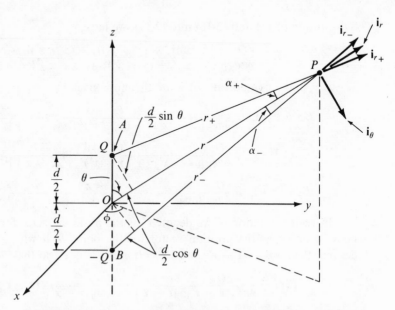

Fig. 2.7. Geometry pertinent to the computation of the electric
field due to a dipole.

Now, the r component of $\mathbf{E}$ is given by

$$E_r = \mathbf{E} \cdot \mathbf{i}_r$$

$$= \frac{Q}{4\pi\epsilon_0 r_+^2} \mathbf{i}_{r_+} \cdot \mathbf{i}_r - \frac{Q}{4\pi\epsilon_0 r_-^2} \mathbf{i}_{r_-} \cdot \mathbf{i}_r$$

$$= \frac{Q}{4\pi\epsilon_0 r_+^2} \cos \alpha_+ - \frac{Q}{4\pi\epsilon_0 r_-^2} \cos \alpha_- \qquad (2\text{-}22)$$

From the geometries of the triangles OAP and OBP, we have

$$\cos \alpha_+ = \frac{r_+^2 + r^2 - (d/2)^2}{2r_+ r} \quad \text{why?} \qquad (2\text{-}23)$$

$$\cos \alpha_- = \frac{r_-^2 + r^2 - (d/2)^2}{2r_- r} \qquad (2\text{-}24)$$

Substituting (2-23) and (2-24) into (2-22), we obtain

$$\begin{aligned}
E_r &= \frac{Q}{4\pi\epsilon_0} \left[\frac{r_+^2 + r^2 - (d/2)^2}{2r_+^3 r} - \frac{r_-^2 + r^2 - (d/2)^2}{2r_-^3 r} \right] \\
&= \frac{Q}{8\pi\epsilon_0 r_+^3 r_-^3 r} (r_- - r_+) \left\{ r_+^2 r_-^2 + \left[r^2 - \left(\frac{d}{2}\right)^2 \right] (r_-^2 + r_- r_+ + r_+^2) \right\} \\
&\approx \frac{Q}{8\pi\epsilon_0 r^7} (r_- - r_+)(r_+^2 r_-^2 + r^2 r_-^2 + r^2 r_- r_+ + r^2 r_+^2) \\
&\approx \frac{Q}{2\pi\epsilon_0 r^3} (r_- - r_+) \approx \frac{Q}{2\pi\epsilon_0 r^3} d \cos \theta
\end{aligned} \qquad (2\text{-}25)$$

where we have used the approximations that, for $r \gg d$,

$$r_+ \approx r - \frac{d}{2} \cos \theta$$

$$r_- \approx r + \frac{d}{2} \cos \theta$$

The θ component of $\mathbf{E}$ is given by

$$E_\theta = \mathbf{E} \cdot \mathbf{i}_\theta$$

$$= \frac{Q}{4\pi\epsilon_0 r_+^2} \mathbf{i}_{r_+} \cdot \mathbf{i}_\theta - \frac{Q}{4\pi\epsilon_0 r_-^2} \mathbf{i}_{r_-} \cdot \mathbf{i}_\theta$$

$$= \frac{Q}{4\pi\epsilon_0 r_+^2} \sin \alpha_+ + \frac{Q}{4\pi\epsilon_0 r_-^2} \sin \alpha_- \qquad (2\text{-}26)$$

$$\approx \frac{Q}{2\pi\epsilon_0 r^2} \sin \alpha_+$$

$$\approx \frac{Q}{4\pi\epsilon_0 r^3} d \sin \theta$$

Thus

$$\mathbf{E} = \frac{Qd}{4\pi\epsilon_0 r^3} (2 \cos \theta \, \mathbf{i}_r + \sin \theta \, \mathbf{i}_\theta) \qquad (2\text{-}27)$$

Equation (2-27) can be considered as a solution for the electric field intensity at very large distances compared to a fixed spacing d between the two point charges, or it can be considered as the solution for the electric field intensity at any point (r, θ, ϕ) in the limit that $d \to 0$, keeping Qd constant. It should be noted that to keep Qd constant as $d \to 0$ requires that $Q \to \infty$. The product Qd is known as the electric dipole moment p. The dipole moment also has an orientation associated with it which is from the negative charge to the positive charge. Substituting p for Qd in (2-27), we note that the electric field intensity due to an electric dipole moment p oriented along the positive z axis is given by

$$\mathbf{E} = \frac{p}{4\pi\epsilon_0 r^3} (2 \cos \theta \, \mathbf{i}_r + \sin \theta \, \mathbf{i}_\theta) \tag{2-28}$$

We note that, as compared to the inverse square distance dependence of the electric field intensity of a point charge, the dipole field drops off inversely proportional to the cube of the distance. Likewise, by an arrangement of two dipoles, a "quadrupole" can be created for which the field varies as inversely proportional to r^4. The process can be extended to "multipoles" step by step, with the power of r increasing by one for each step.   ∎

2.4   The Electric Field of Continuous Charge Distributions

In the previous section we considered collections of point charges at discrete points for which the field computation consists of finding the vector sums of the field intensities due to the individual point charges. In this section we will extend the computation to continuous charge distributions. Continuous charge distributions can be of three types:

(a) Line charge for which charge is distributed along a line (straight or curved).

(b) Surface charge for which charge is distributed on a surface (planar or nonplanar).

(c) Volume charge for which charge is distributed in a volume.

When a charge is distributed along a line or on a surface or in a volume, we have to deal with charge densities. The line charge density is the charge per unit length, the surface charge density is the charge per unit surface area, and the volume charge density is the charge per unit volume. We will use the symbols ρ_L, ρ_s, and ρ, respectively, for these charge densities. Obviously, the units of ρ_L, ρ_s, and ρ are coulombs per meter, coulombs per meter², coulombs per meter³, respectively. In each case we can divide the total charge into several infinitesimal parts, each of which can be considered as a point charge. We thus represent the total charge as a continuous collection of point charges and obtain the field intensity at any point due to the total charge as the vector superposition of the field intensities due to the individual point charges. However, we now have to evaluate integrals instead of summations of a few terms since the distribution of charges is continuous instead of being discrete. We will illustrate this process by considering three examples: (a) infinitely long line charge, (b) infinite sheet charge, and (c) spherical volume charge.

EXAMPLE 2-4.   An infinitely long line charge of uniform density ρ_{L0} C/m is situated along the z axis as shown in Fig. 2.8. We wish to obtain the electric field intensity due to this line charge.

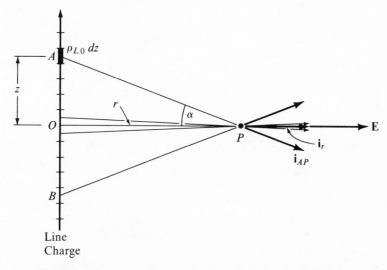

Fig. 2.8. Geometry for computing the electric field of an infinitely long line charge of uniform density ρ_{L0} C/m.

First, we divide the line into a number of infinitesimal segments each of length dz, as shown in Fig. 2.8, such that the charge $\rho_{L0}\,dz$ in each segment can be considered as a point charge. The electric field intensity due to each point charge is directed radially away from that point charge and varies inversely as the square of the distance from that charge. Now let us consider a point P at a distance r from the z axis, with the projection of the point P onto the z axis being the point O. The electric field intensity vectors at point P due to the infinitesimal segment immediately above O and the infinitesimal segment immediately below O have equal magnitudes and make equal angles with the line OP as shown in Fig. 2.8. The components of these two vectors perpendicular to OP (parallel to the z axis) therefore cancel, whereas the components along OP add to each other. Thus the resultant electric field intensity at P due to the two segments, one directly above O and another directly below O, is entirely directed along OP, that is, normal to the axis of the line charge. A similar argument can be made for the resultant electric field intensity vector at point P due to any other two segments which are equidistant from O with one above it and the other below it. Now, since there are as many (semiinfinite) segments above O as there are below it, the resultant field intensity at point P due to the entire line charge is directed radially away from it. The situation remains unchanged if we move P up or down, keeping r constant, since there are always a semiinfinite number of segments above the projection of P onto the line charge as well as below it. Thus the electric field intensity of an infinite line charge of uniform density at any arbitrary point is directed radially away from the line charge and is independent of the position of P parallel to the z axis. It is dependent only on the distance of P from the z axis. We have thus simplified the problem to one of finding the magnitude of the field intensity.

To determine the magnitude of **E**, let us once again refer to Fig. 2.8, and consider the segment at the point A at a distance z above O. The electric field intensity at point P due to this segment is equal to

$$\frac{\rho_{L0}\, dz}{4\pi\epsilon_0(r^2 + z^2)}\, \mathbf{i}_{AP}$$

The component of this electric field intensity along OP is

$$\frac{\rho_{L0}\, dz}{4\pi\epsilon_0(r^2 + z^2)}\, \mathbf{i}_{AP} \cdot \mathbf{i}_r = \frac{\rho_{L0}\, dz}{4\pi\epsilon_0(r^2 + z^2)}\cos\alpha = \frac{\rho_{L0}r\, dz}{4\pi\epsilon_0(r^2 + z^2)^{3/2}}$$

We need not consider the component normal to OP since it gets cancelled from the contribution due to another segment at the point B at a distance z below O. The component along OP is, on the other hand, doubled from the contribution due to this second segment. Thus the magnitude of the resultant electric field intensity at P due to the two segments at A and B is given by

$$dE = \frac{2\rho_{L0}r\, dz}{4\pi\epsilon_0(r^2 + z^2)^{3/2}} \tag{2-29}$$

The magnitude of the electric field intensity at P due to the entire line charge is now given by the integral of dE where the integration is to be performed between the limits $z = 0$ and $z = \infty$. Thus

$$E = \int_{z=0}^{\infty} dE = \frac{2\rho_{L0}r}{4\pi\epsilon_0} \int_{z=0}^{\infty} \frac{dz}{(r^2 + z^2)^{3/2}} \tag{2-30}$$

Introducing $z = r\tan\alpha$ in (2-30), we obtain

$$E = \frac{\rho_{L0}}{2\pi\epsilon_0 r} \int_{\alpha=0}^{\pi/2} \cos\alpha\, d\alpha = \frac{\rho_{L0}}{2\pi\epsilon_0 r} \tag{2-31}$$

Recalling that **E** is directed radially away from the line charge, we have

$$\mathbf{E} = \frac{\rho_{L0}}{2\pi\epsilon_0 r}\, \mathbf{i}_r \tag{2-32}$$

Equation (2-32) indicates that the electric field intensity of an infinite line charge of uniform density falls off only as the inverse of the distance from the line charge compared to the inverse square distance dependence in the case of the point charge. ∎

EXAMPLE 2-5. A sheet charge of uniform density ρ_{s0} C/m² extends over the entire xy plane as shown in Fig. 2.9. We wish to obtain the electric field intensity due to this infinite sheet charge.

Let us consider a point P at a distance z from the xy plane, with the projection of the point P on the xy plane being O, as shown in Fig. 2.9. The electric field intensities at point P due to two point charges situated at the diametrically opposite points A and B as shown in Fig. 2.9 have equal magnitudes but their directions are such that the resultant electric field intensity is directed along the line OP and away from the sheet charge. In fact, for any point charge on the ring of radius r, there is a diametrically opposite point charge which results in a resultant electric field intensity entirely along OP. Thus the field intensity at point P due to the charge on the entire ring of radius r and width dr is directed normally away from the sheet charge. This suggests that we divide the area of the xy plane into several

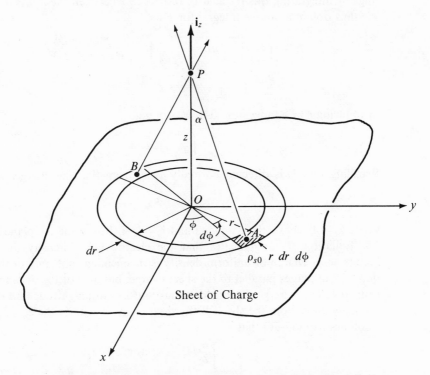

Fig. 2.9. Geometry for computing the electric field of an infinite sheet charge of uniform density ρ_{s0} C/m².

rings, each of width dr, and divide each ring into angular increments of $d\phi$, thus creating infinitesimal areas $r \, dr \, d\phi$ having charges $\rho_{so} r \, dr \, d\phi$ as shown in Fig. 2.9.

Now, since each ring results in an electric field intensity at point P, only along OP, the field intensity due to the entire sheet charge will also be along the same direction. If we move P sideways while keeping z constant, the situation remains unchanged so that the field intensity is independent of the position of P in planes parallel to the sheet charge. Once again, we have reduced the problem to one of finding the magnitude of **E**.

To find the magnitude of **E**, we note that the component along OP of the field intensity at P, due to the infinitesimal charge $\rho_{so} r \, dr \, d\phi$ at point A, is given by

$$dE = \frac{\rho_{so} r \, dr \, d\phi}{4\pi\epsilon_0 (r^2 + z^2)} \cos \alpha = \frac{\rho_{so} r z \, dr \, d\phi}{4\pi\epsilon_0 (r^2 + z^2)^{3/2}} \tag{2-33}$$

The resultant electric field intensity due to the ring of charge passing through A and B is obtained by adding up all the contributions due to the infinitesimal areas on the ring, that is, by integrating (2-33) with respect to ϕ between the limits 0 and 2π. We then add up all the contributions due to the several rings by integrating this result with respect to r between the limits 0 and ∞. We thus obtain a double integral for E as

$$E = \int_{r=0}^{\infty} \int_{\phi=0}^{2\pi} dE = \int_{r=0}^{\infty} \int_{\phi=0}^{2\pi} \frac{\rho_{so} r z \, dr \, d\phi}{4\pi\epsilon_0 (r^2 + z^2)^{3/2}}$$

$$= \frac{\rho_{so} z}{2\epsilon_0} \int_{r=0}^{\infty} \frac{r \, dr}{(r^2 + z^2)^{3/2}} \tag{2-34}$$

Introducing $r = z \tan \alpha$ in (2-34), we obtain

$$E = \frac{\rho_{so}}{2\epsilon_0} \int_{\alpha=0}^{\pi/2} \sin \alpha \, d\alpha = \frac{\rho_{so}}{2\epsilon_0} \tag{2-35}$$

Recalling that **E** is directed normally away from the line charge, we have

$$\mathbf{E} = \frac{\rho_{so}}{2\epsilon_0} \mathbf{i}_n \tag{2-36}$$

where $\mathbf{i}_n = \mathbf{i}_z$ above the xy plane and $\mathbf{i}_n = -\mathbf{i}_z$ below the xy plane in Fig. 2.9. Equation (2-36) indicates that the electric field intensity due to an infinite sheet charge of uniform density is independent not only of the position of P in planes parallel to the sheet charge, but also of the distance away from the sheet charge. The field is thus uniform in magnitude and directed normally away from the sheet. If the sheet charge occupies the $z = z_0$ plane, it follows from (2-36) that

$$\mathbf{E} = \begin{cases} \dfrac{\rho_{so}}{2\epsilon_0} \mathbf{i}_z & \text{for } z > z_0 \\[4mm] -\dfrac{\rho_{so}}{2\epsilon_0} \mathbf{i}_z & \text{for } z < z_0 \end{cases} \qquad \blacksquare$$

EXAMPLE 2-6.   A volume charge is distributed throughout a sphere of radius a, and centered at the origin, with uniform density ρ_0 C/m³. We wish to obtain the electric field intensity due to this volume charge.

With the experience gained in Examples 2-4 and 2-5, we will shorten the discussion concerning the direction of **E** by stating that, for every infinitesimal charge $\rho_0 r^2 \sin \theta \, dr \, d\theta \, d\phi$ in the infinitesimal volume $r^2 \sin \theta \, dr \, d\theta \, d\phi$ at point A inside the sphere as shown in Fig. 2.10, there is another infinitesimal charge such that the resultant electric field intensity at point P due to these two charges is directed entirely along OP, that is, radially away from the center of the sphere. Also, moving P on the surface of a sphere of radius z does not change the situation so that the field intensity is a function only of the distance from the center of the sphere. Thus it is sufficient if we evaluate the component of the electric field intensity at P along OP due to the infinitesimal charge $\rho_0 r^2 \sin \theta \, dr \, d\theta \, d\phi$ and perform a volume integration to obtain the electric

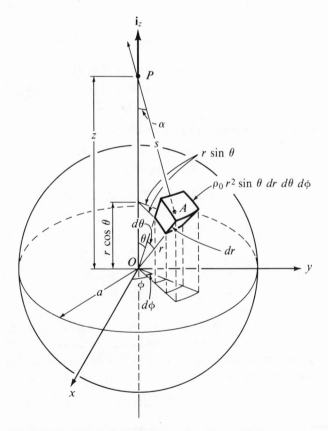

Fig. 2.10. Geometry for computing the electric field of a spherical volume charge of uniform density ρ_0 C/m³.

field intensity due to the entire spherical volume charge of radius a. The component, along OP, of the electric field intensity at P due to the infinitesimal charge at A is given by

$$dE = \frac{\rho_0 r^2 \sin \theta \, dr \, d\theta \, d\phi}{4\pi\epsilon_0 (r^2 + z^2 - 2rz \cos \theta)} \cos \alpha = \frac{\rho_0 (z - r \cos \theta) r^2 \sin \theta \, dr \, d\theta \, d\phi}{4\pi\epsilon_0 (r^2 + z^2 - 2rz \cos \theta)^{3/2}}$$

(2-37)

The electric field intensity due to the entire spherical charge is then given by

$$E = \int_{r=0}^{a} \int_{\theta=0}^{\pi} \int_{\phi=0}^{2\pi} dE = \int_{r=0}^{a} \int_{\theta=0}^{\pi} \int_{\phi=0}^{2\pi} \frac{\rho_0 (z - r \cos \theta) r^2 \sin \theta \, dr \, d\theta \, d\phi}{4\pi\epsilon_0 (r^2 + z^2 - 2rz \cos \theta)^{3/2}}$$

$$= \frac{\rho_0}{2\epsilon_0} \int_{r=0}^{a} \int_{\theta=0}^{\pi} \frac{(z - r \cos \theta) r^2 \sin \theta \, dr \, d\theta}{(r^2 + z^2 - 2rz \cos \theta)^{3/2}}$$

(2-38)

Introducing $s^2 = r^2 + z^2 - 2rz \cos \theta$, for integration with respect to θ, we have

$$\sin \theta \, d\theta = \frac{s \, ds}{rz}$$

(2-39a)

$$z - r \cos \theta = \frac{s^2 - r^2 + z^2}{2z}$$

(2-39b)

$$s = \begin{cases} z - r & \text{for} \quad \theta = 0, z > r \\ r - z & \text{for} \quad \theta = 0, 0 < z < r \\ z + r & \text{for} \quad \theta = \pi \end{cases}$$

(2-39c)

Substituting these into (2-38), we obtain, for $z > a$,

$$E = \frac{\rho_0}{2\epsilon_0} \int_{r=0}^{a} \frac{r \, dr}{2z^2} \int_{s=z-r}^{z+r} \frac{s^2 - r^2 + z^2}{s^2} \, ds$$

$$= \frac{\rho_0}{2\epsilon_0} \int_{r=0}^{a} \frac{4r^2 \, dr}{2z^2} = \frac{(4\pi a^3/3)\rho_0}{4\pi\epsilon_0 z^2}$$

(2-40)

For $0 < z < a$, we have

$$E = \frac{\rho_0}{2\epsilon_0} \int_{r=0}^{z} \frac{r \, dr}{2z^2} \int_{s=z-r}^{z+r} \frac{s^2 - r^2 + z^2}{s^2} \, ds + \frac{\rho_0}{2\epsilon_0} \int_{r=z}^{a} \frac{r \, dr}{2z^2} \int_{s=r-z}^{z+r} \frac{s^2 - r^2 + z^2}{s^2} \, ds$$

$$= \frac{\rho_0}{2\epsilon_0} \int_{r=0}^{z} \frac{4r^2 \, dr}{2z^2} + 0 = \frac{(4\pi z^3/3)\rho_0}{4\pi\epsilon_0 z^2}$$

(2-41)

Equations (2-40) and (2-41) give the magnitude of **E** at any radial distance z greater than a and less than a, respectively, from the center of the charge. Recalling that the direction of **E** is radially away from the center of the charge distribution and substituting r for z, we have

$$\mathbf{E} = \begin{cases} \dfrac{(4\pi a^3/3)\rho_0}{4\pi\epsilon_0 r^2} \mathbf{i}_r & \text{for} \quad r > a \\ \dfrac{(4\pi r^3/3)\rho_0}{4\pi\epsilon_0 r^2} \mathbf{i}_r & \text{for} \quad r < a \end{cases}$$

(2-42)

Noting that $4\pi r^3/3$ is the volume of a sphere of radius r and that there is no charge in the region $r > a$, we can combine the two results on the right side of (2-42) as

$$\mathbf{E}(r) = \frac{\text{charge enclosed by the spherical surface of radius } r}{4\pi\epsilon_0 r^2}\mathbf{i}_r \quad (2\text{-}43)$$

Viewed from any distance r from the center of the volume charge, the volume charge is equivalent to a point charge of value equal to the charge enclosed by the spherical surface of radius r. ∎

In the examples we have considered in this section, it was possible to determine the electric field intensity by evaluating a single scalar integral in each case because of the symmetries involved. In the general case, it would be necessary to evaluate three scalar integrals. Furthermore, in order not to get confused between the field points (i.e., points at which the field is desired) and the source points (i.e., points in the volume, surface, or contour occupied by the charge distribution), we must use a notation which distinguishes the two sets of points. Usually, the coordinates of the source points are denoted by primes, whereas the coordinates of the field points are unprimed. The integration is then to be performed with respect to the primed coordinates. This notation is known as the source point-field point notation. Thus, in general, if a line charge of density $\rho_L(\mathbf{r}')$ occupies a contour C', where $\mathbf{r}'$ is the position vector in the source point coordinate system, then the electric field intensity $\mathbf{E}(\mathbf{r})$ at a field point defined by the position vector $\mathbf{r}$ is given by

$$\mathbf{E}(\mathbf{r}) = \frac{1}{4\pi\epsilon_0} \int_{C'} \frac{[\rho_L(\mathbf{r}')\,dl'](\mathbf{r} - \mathbf{r}')}{|\mathbf{r} - \mathbf{r}'|^3} \quad (2\text{-}44\text{a})$$

The right side of Eq. (2-44a) is a vector integral and, in general, it requires the evaluation of three separate scalar integrals. Expressions similar to (2-44a) can be written for surface and volume charge distributions. Thus, for a surface charge of density $\rho_s(\mathbf{r}')$ occupying a surface S', we have

$$\mathbf{E}(\mathbf{r}) = \frac{1}{4\pi\epsilon_0} \int_{S'} \frac{[\rho_s(\mathbf{r}')\,dS'](\mathbf{r} - \mathbf{r}')}{|\mathbf{r} - \mathbf{r}'|^3} \quad (2\text{-}44\text{b})$$

For a volume charge of density $\rho(\mathbf{r}')$ occupying a volume V', we have

$$\mathbf{E}(\mathbf{r}) = \frac{1}{4\pi\epsilon_0} \int_{V'} \frac{[\rho(\mathbf{r}')\,dv'](\mathbf{r} - \mathbf{r}')}{|\mathbf{r} - \mathbf{r}'|^3} \quad (2\text{-}44\text{c})$$

We will use the source point-field point notation only wherever the same coordinate or coordinates for the source and field points appear in the integral. For example, if we wish to evaluate the electric field intensity due to a finitely long line charge along the z axis at a point (r, ϕ, z), then we will have to define the points occupied by the line charge using a z' coordinate so that no confusion arises with the z coordinate of the field point.

2.5 Direction Lines

In the previous two sections we obtained the expressions for the electric field intensities due to certain charge distributions both discrete and continuous. In simple cases, such as for the point charge and for the three examples of the previous section, it is easy to visualize, from a glance at the field expression, the direction of the electric field intensity vector everywhere in space. However, in a case such as the electric dipole (Example 2-3), it is not easy to visualize the direction of the electric field intensity vector by a glance at the field expression [Eq. (2-28)]. If we want to attack the problem directly in such a case, we can assign numerical values for the coordinates in the field expression and compute the direction of the field intensity vector at several points in the medium and then draw arrows along the computed directions. Alternatively and more elegantly, we ask the question: Suppose we place a test charge at a point in the electric field, what is the direction along which it experiences acceleration? Obviously, the test charge experiences acceleration along the direction of the electric field intensity vector at that point. If we stop the test charge after each infinitesimal distance and trace its path in the limit that the infinitesimal distance tends to zero, we get a line along which the electric field is everywhere tangential to it. Such lines, called "direction lines," are of great help in understanding the behavior of a given field, as suggested in Chapter 1. They are also known as "stream lines" and "flux lines."

To develop the technique of sketching the direction lines for a given field, let us consider a small test charge placed at a point $P(x, y, z)$ in the field as shown in Fig. 2.11. At the point P the force on the test charge is

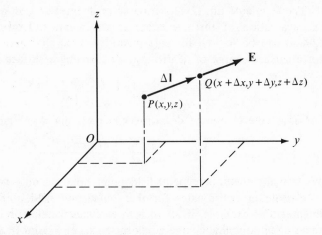

Fig. 2.11. Illustrating the proportionality of the electric field intensity vector **E** and the infinitesimal vector displacement $\Delta \mathbf{l}$ of a charge placed in the field.

directed along **E**. The test charge will travel for an infinitesimal distance Δl in the direction of **E** to point $Q(x + \Delta x, y + \Delta y, z + \Delta z)$. The vector displacement of the test charge is then equal to $\Delta x\, \mathbf{i}_x + \Delta y\, \mathbf{i}_y + \Delta z\, \mathbf{i}_z$. But this infinitesimal vector displacement is proportional to the force experienced by the charge which in turn is proportional to $\mathbf{E} = E_x\mathbf{i}_x + E_y\mathbf{i}_y + E_z\mathbf{i}_z$. Thus

$$\Delta x\, \mathbf{i}_x + \Delta y\, \mathbf{i}_y + \Delta z\, \mathbf{i}_z \propto E_x\mathbf{i}_x + E_y\mathbf{i}_y + E_z\mathbf{i}_z \qquad (2\text{-}45)$$

Two vectors are proportional if and only if their respective components are proportional by the same amount. Hence we have, from (2-45),

$$\frac{\Delta x}{E_x} = \frac{\Delta y}{E_y} = \frac{\Delta z}{E_z} \qquad (2\text{-}46)$$

But Eq. (2-46) is approximate since, in general, E varies continuously from point to point in magnitude and direction. However, it will be exact in the limit Δx, Δy, and Δz all tend to zero. It then reduces to

$$\frac{dx}{E_x} = \frac{dy}{E_y} = \frac{dz}{E_z} \qquad (2\text{-}47\text{a})$$

Knowing E_x, E_y, and E_z for a particular field, we can substitute in (2-47a) and solve the resulting differential equations to obtain the algebraic equations for the direction lines. We can obtain equations similar to (2-47a) for the cylindrical and spherical coordinate systems following similar arguments. These equations are

$$\frac{dr}{E_r} = \frac{r\, d\phi}{E_\phi} = \frac{dz}{E_z} \qquad \text{cylindrical} \qquad (2\text{-}47\text{b})$$

$$\frac{dr}{E_r} = \frac{r\, d\theta}{E_\theta} = \frac{r \sin \theta\, d\phi}{E_\phi} \qquad \text{spherical} \qquad (2\text{-}47\text{c})$$

We will now illustrate the use of these equations by considering an example.

EXAMPLE 2-7. In Example 2-3 we obtained the expression for **E** for an electric dipole of moment p oriented along the positive z axis as

$$\mathbf{E} = \frac{p}{4\pi\epsilon_0 r^3}(2 \cos \theta\, \mathbf{i}_r + \sin \theta\, \mathbf{i}_\theta)$$

It is desired to obtain the equation for the direction lines for this field. Noting that

$$E_r = \frac{2p \cos \theta}{4\pi\epsilon_0 r^3} \qquad E_\theta = \frac{p \sin \theta}{4\pi\epsilon_0 r^3} \qquad E_\phi = 0$$

we have, from (2-47c),

$$\frac{dr}{(2p \cos \theta)/4\pi\epsilon_0 r^3} = \frac{r\, d\theta}{(p \sin \theta)/4\pi\epsilon_0 r^3} = \frac{r \sin \theta\, d\phi}{0} \qquad (2\text{-}48)$$

or

$$\frac{dr}{r} = 2 \cot \theta \, d\theta \qquad\qquad d\phi = 0$$

$$\ln r = -2 \ln \operatorname{cosec} \theta + \text{constant} \qquad \phi = \text{constant}$$

$$r \operatorname{cosec}^2 \theta = \text{constant} \qquad\qquad \phi = \text{constant} \qquad (2\text{-}49)$$

The direction lines are thus intersections of the surfaces $r \operatorname{cosec}^2 \theta = \text{constant}$ and the planes $\phi = \text{constant}$. A few direction lines in constant ϕ plane are sketched in Fig. 2.12. The small arrow at the center indicates the dipole moment **p** with the direction of the arrow as the direction of orientation of the dipole. ∎

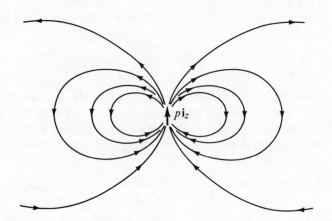

Fig. 2.12. Direction lines of **E** for electric dipole of moment $p\mathbf{i}_z$.

2.6   Gauss' Law in Integral Form

Let us consider the surface of a sphere of radius r and centered at a point charge Q at the origin. The electric field intensity due to the point charge is directed everywhere radially away from the point charge and hence is normal to the surface of the sphere as shown in Fig. 2.13. Its magnitude on the surface of the sphere is a constant equal to $Q/4\pi\epsilon_0 r^2$. If we now consider an infinitesimal area dS on the surface of the sphere, we have

$$\mathbf{E} \cdot d\mathbf{S} = \frac{Q}{4\pi\epsilon_0 r^2} \mathbf{i}_r \cdot dS \, \mathbf{i}_n = \frac{Q}{4\pi\epsilon_0 r^2} \mathbf{i}_r \cdot dS \, \mathbf{i}_r = \frac{Q \, dS}{4\pi\epsilon_0 r^2} \qquad (2\text{-}50)$$

The integral of $\mathbf{E} \cdot d\mathbf{S}$ over the surface S of the sphere is given by

$$\oint_s \mathbf{E} \cdot d\mathbf{S} = \oint_s \frac{Q}{4\pi\epsilon_0 r^2} dS = \frac{Q}{4\pi\epsilon_0 r^2} \oint_s dS \qquad (2\text{-}51)$$

since r is constant on the surface of the sphere. Proceeding further, we have

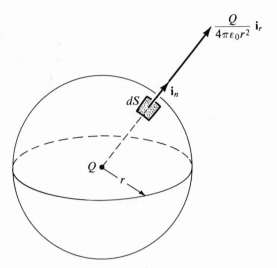

Fig. 2.13. For evaluating $\oint \mathbf{E} \cdot d\mathbf{S}$ on the surface of sphere centered at a point charge Q.

$$\oint_S \mathbf{E} \cdot d\mathbf{S} = \frac{Q}{4\pi\epsilon_0 r^2} \text{ (surface area of the sphere)}$$

$$= \frac{Q}{4\pi\epsilon_0 r^2}(4\pi r^2) = \frac{Q}{\epsilon_0} \tag{2-52}$$

The physical significance of (2-52) is obvious if we compare the electric field lines emanating from the point charge with the flow of a fluid away from the location of the point charge. The surface integral of the fluid flow density vector is the net amount of fluid flowing out of the surface. Similarly, the surface integral of the electric field intensity vector can be interpreted as the net flux of electric field emanating from the surface, although the electric field is not a fluid in the sense that it does not flow like a fluid.

Thus Eq. (2-52) states that the net electric field flux emanating from the surface of a sphere of radius r centered at a point charge Q is equal to Q/ϵ_0. It is independent of the radius r of the spherical surface. Whether $r = 1$ micron or 1000 km, the electric field flux is the same (provided, of course, that there is no other electric field in the medium). This is not surprising if we once again compare the flux of the electric field with the flow of the fluid. If the fluid is flowing radially away from a point source of the fluid, then the amount of fluid crossing a spherical surface of one radius must be the same as the amount crossing a spherical surface of another radius or, for that matter, any arbitrary closed surface enclosing the point source (provided, of course, there is no other source or sink of the fluid). Likewise, if we choose an arbitrary surface enclosing the point charge, the net electric field flux emanating from this surface must be equal to Q/ϵ_0. To prove this

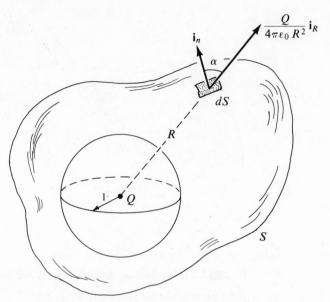

Fig. 2.14. For evaluating $\oint \mathbf{E} \cdot d\mathbf{S}$ over an arbitrary surface S enclosing a point charge Q.

mathematically, we refer to Fig. 2.14. Considering an infinitesimal area dS on the arbitrary surface, we find that the infinitesimal amount of electric field flux emanating from this area is given by

$$\mathbf{E} \cdot d\mathbf{S} = \frac{Q}{4\pi\epsilon_0 R^2} \mathbf{i}_R \cdot dS\,\mathbf{i}_n = \frac{Q\,dS}{4\pi\epsilon_0 R^2} \cos\alpha \qquad (2\text{-}53)$$

where α is the angle between the radial vector away from the point charge and the normal vector to the area dS. The total flux emanating from the entire closed surface S is then given by

$$\oint_S \mathbf{E} \cdot d\mathbf{S} = \oint_S \frac{Q\,dS}{4\pi\epsilon_0 R^2} \cos\alpha = \frac{Q}{4\pi\epsilon_0} \oint_S \frac{dS \cos\alpha}{R^2} \qquad (2\text{-}54)$$

In (2-54), $dS \cos\alpha$ is the projection of the area dS on the arbitrary surface S onto a spherical surface of radius R and centered at the point charge. Hence $(dS \cos\alpha)/R^2$ is the projection of dS onto a spherical surface of radius unity and centered at the point charge. It is known as the solid angle subtended at the point charge by the area dS. The unit of solid angle is steradian. The quantity $\oint (dS \cos\alpha)/R^2$ is the total solid angle subtended at the point charge by the closed surface S. It is the sum of the projections of all infinitesimal areas comprising the arbitrary surface S onto the spherical surface of radius unity and centered at the point charge. Thus it is equal to the surface area of the sphere of unit radius, that is, 4π. Substituting this result in (2-54), we have

$$\oint_{\substack{\text{surface} \\ \text{enclosing } Q}} \mathbf{E} \cdot d\mathbf{S} = \frac{Q}{4\pi\epsilon_0}(4\pi) = \frac{Q}{\epsilon_0} \qquad (2\text{-}55)$$

If an arbitrary surface does not enclose a point source of fluid, then the net amount of fluid emanating from the surface must be zero since there are equal amounts of fluid flowing in and out of the surface. Likewise, if the arbitrary surface does not enclose the point charge, the net electric field flux emanating from the surface must be zero. Thus

$$\oint_{\substack{\text{surface not} \\ \text{enclosing } Q}} \mathbf{E} \cdot d\mathbf{S} = 0 \qquad (2\text{-}56)$$

It will be left as an exercise for the student to provide a mathematical proof of (2-56).

If, instead of one point charge, we have five point charges Q_1, Q_2, Q_3, Q_4, Q_5 as shown in Fig. 2.15, then for an arbitrary surface S enclosing point

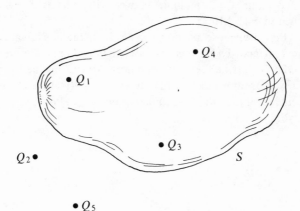

Fig. 2.15. An arbitrary surface enclosing three point charges.

charges Q_1, Q_3, and Q_4 but not Q_2 and Q_5, we can obtain the net electric field flux emanating from the surface using superposition. Thus, if $\mathbf{E}_1$, $\mathbf{E}_2$, $\mathbf{E}_3$, $\mathbf{E}_4$, and $\mathbf{E}_5$ are the electric field intensity vectors due to Q_1, Q_2, Q_3, Q_4, and Q_5, respectively, we have

$$\oint_S \mathbf{E} \cdot d\mathbf{S} = \oint_S \mathbf{E}_1 \cdot d\mathbf{S} + \oint_S \mathbf{E}_2 \cdot d\mathbf{S} + \oint_S \mathbf{E}_3 \cdot d\mathbf{S} + \oint_S \mathbf{E}_4 \cdot d\mathbf{S}$$
$$+ \oint_S \mathbf{E}_5 \cdot d\mathbf{S}$$
$$= \frac{1}{\epsilon_0}\left(Q_1 + 0 + Q_3 + Q_4 + 0\right) = \frac{1}{\epsilon_0}\left(Q_1 + Q_3 + Q_4\right) \qquad (2\text{-}57)$$
$$= \frac{1}{\epsilon_0}\left(\text{charge enclosed by the surface } S\right)$$

The discussion can be extended to a continuous charge distribution if we note that a continuous charge distribution can be represented as a continuous collection of charges occupying infinitesimal volumes, each of which can be considered as a point charge. Those charges enclosed by the arbitrary surface result in a net electric field flux in accordance with (2-55), whereas those which are not enclosed by the surface result in zero flux in accordance with (2-56). We can summarize these conclusions in a single statement that "the net electric field flux emanating from a closed surface is equal to the net charge enclosed by the surface divided by ϵ_0." This statement is Gauss' law—one of the important laws in electromagnetic field theory. In equation form, Gauss' law is written as

$$\oint_S \mathbf{E} \cdot d\mathbf{S} = \frac{1}{\epsilon_0}(\text{charge enclosed by the surface } S) \qquad (2\text{-}58)$$

EXAMPLE 2-8. An infinitely long line charge of uniform density ρ_{L0} C/m is situated along the z axis. It is desired to find the electric field flux cutting the portion of the plane $x = 1$ m lying between the planes $z = 0$ m and $z = 1$ m as shown in Fig. 2.16.

First we will solve this problem by actually evaluating $\int \mathbf{E} \cdot d\mathbf{S}$ over the given surface. To do this, we note that $\mathbf{E}$ due to the line charge is given by $(\rho_{L0}/2\pi\epsilon_0 r)\mathbf{i}_r$, where r is the radial distance from the line charge and $\mathbf{i}_r$ is the unit vector directed radially away from the line charge. Considering an infinitesimal area $dy\,dz$ at the location $(1, y, z)$ on the given plane, the infini-

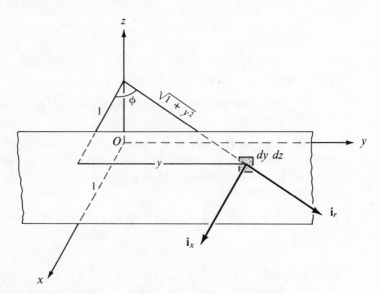

Fig. 2.16. For evaluation of electric field flux emanating from an infinite line charge and cutting a portion of the $x = 1$ plane.

tesimal amount of flux cutting this area is given by

$$\mathbf{E} \cdot d\mathbf{S} = \frac{\rho_{L0}}{2\pi\epsilon_0\sqrt{1+y^2}}\, \mathbf{i}_r \cdot dy\, dz\, \mathbf{i}_x = \frac{\rho_{L0}\, dy\, dz}{2\pi\epsilon_0(1+y^2)} \qquad (2\text{-}59)$$

The total flux cutting the portion of the plane $x = 1$ m lying between the planes $z = 0$ m and $z = 1$ m is then given by

$$\int_{y=-\infty}^{\infty}\int_{z=0}^{1} \mathbf{E} \cdot d\mathbf{S} = \int_{y=-\infty}^{\infty}\int_{z=0}^{1} \frac{\rho_{L0}\, dy\, dz}{2\pi\epsilon_0(1+y^2)}$$

$$= \frac{\rho_{L0}}{2\pi\epsilon_0}\int_{\phi=-\pi/2}^{\pi/2} d\phi = \frac{\rho_{L0}}{2\epsilon_0} \qquad (2\text{-}60)$$

This result can, however, be obtained without performing the integration if we note that the electric field intensity due to the line charge is independent of ϕ and hence the electric field flux from the line charge emanates from it uniformly in ϕ. Thus half of the electric field flux emanating from that portion of the line charge lying between $z = 0$ m and $z = 1$ m cuts the given surface. Since the total flux emanating from this portion of the line charge is $\rho_{L0}(1)/\epsilon_0 = \rho_{L0}/\epsilon_0$, according to Gauss' law, the flux cutting the specified surface is $\rho_{L0}/2\epsilon_0$. ■

Given $\mathbf{E}$ and a closed surface S, it is always possible to compute the charge enclosed by the surface by evaluating $\oint_S \mathbf{E} \cdot \mathbf{S}$ analytically or numerically and then multiplying the result by ϵ_0 in accordance with Gauss' law as given by (2-58). The inverse problem of finding $\mathbf{E}$ for a given charge distribution by using (2-58) is possible only for certain simple cases involving a high degree of symmetry, since the unknown quantity $\mathbf{E}$ appears in the integrand. As a first step, the symmetry of the electric field must be determined by making use of the fact that the electric field due to a point charge is directed radially away from it. We have illustrated this in Examples 2-4, 2-5, and 2-6. Next, we should be able to choose a closed surface S such that $\oint_S \mathbf{E} \cdot d\mathbf{S}$ can be reduced to an algebraic quantity involving the magnitude of $\mathbf{E}$. Such a surface is known as a Gaussian surface. Obviously, the Gaussian surface must be such that the magnitude of $\mathbf{E}$ is uniform and the direction of $\mathbf{E}$ is normal to the surface over the whole or part of the surface, while the magnitude of $\mathbf{E}$ is zero or the direction of $\mathbf{E}$ is tangential to the surface over the rest of the surface in the latter case. We will illustrate this method of obtaining $\mathbf{E}$ by reconsidering Examples 2-4, 2-5, and 2-6.

EXAMPLE 2-9.   An infinitely long line charge of uniform density ρ_{L0} C/m is situated along the z axis as shown in Fig 2.17. We wish to obtain the electric field intensity due to this line charge using Gauss' law.

In Example 2-4, we established from purely qualitative arguments that $\mathbf{E}$ due to the infinite line charge of uniform density is directed radially away from the line charge and its magnitude is dependent only on its distance

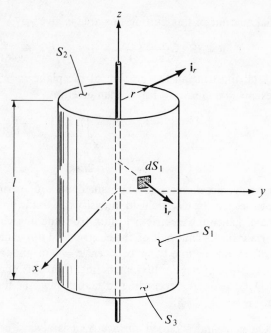

Fig. 2.17. Gaussian surface for computing the electric field of an infinitely long line charge of uniform density.

from the line charge. Thus

$$\mathbf{E} = E_r(r)\mathbf{i}_r \qquad (2\text{-}61)$$

Choosing the Gaussian surface S as the surface of a cylinder of radius r with the line charge as its axis and of length l, as shown in Fig. 2.17, we have

$$\underset{\substack{\text{surface of} \\ \text{cylinder, } S}}{\oint} \mathbf{E} \cdot d\mathbf{S} = \underset{\substack{\text{curved} \\ \text{surface } S_1}}{\int} \mathbf{E} \cdot d\mathbf{S} + \underset{\substack{\text{plane sur-} \\ \text{faces } S_2, S_3}}{\int} \mathbf{E} \cdot d\mathbf{S} \qquad (2\text{-}62)$$

The second integral on the right side of (2-62) is zero since $\mathbf{E}$ is tangential to the surfaces; that is, $\mathbf{E} \cdot d\mathbf{S}$ is zero throughout the surfaces. Noting that E_r is constant on the curved surface S_1, we find that the first integral can be written as

$$\underset{\substack{\text{curved} \\ \text{surface } S_1}}{\int} \mathbf{E} \cdot d\mathbf{S} = \int_{S_1} E_r \mathbf{i}_r \cdot dS_1\, \mathbf{i}_r = E_r \int_{S_1} dS_1$$

$$= E_r \text{ (surface area of } S_1) = E_r(2\pi r l) \qquad (2\text{-}63)$$

Thus

$$\oint_S \mathbf{E} \cdot d\mathbf{S} = 2\pi r l E_r \qquad (2\text{-}64)$$

But, from Gauss' law,

$$\oint_S \mathbf{E} \cdot d\mathbf{S} = \frac{\text{charge enclosed by } S}{\epsilon_0} = \frac{\rho_{L0} l}{\epsilon_0} \qquad (2\text{-}65)$$

Comparing (2-64) and (2-65), we have

$$E_r = \frac{\rho_{L0}}{2\pi\epsilon_0 r} \qquad (2\text{-}66)$$

$$\mathbf{E} = \frac{\rho_{L0}}{2\pi\epsilon_0 r} \mathbf{i}_r \qquad (2\text{-}67)$$

which agrees with the result obtained in Example 2-4.   ∎

EXAMPLE 2-10.   A sheet charge of uniform density ρ_{s0} C/m² extends over the entire *xy* plane as shown in Fig. 2.18. We wish to obtain the electric field intensity due to this infinite sheet charge using Gauss' law.

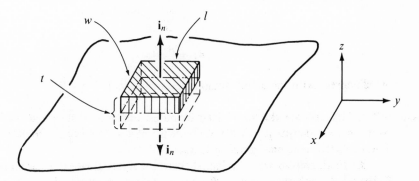

Fig. 2.18. Gaussian surface for computing the electric field of an infinite sheet charge of uniform density.

In Example 2-5 we established from purely qualitative arguments that **E** due to the infinite sheet charge of uniform density is directed normally away from the sheet charge and that it is uniform in planes parallel to the sheet charge. Thus

$$\mathbf{E} = E_n \mathbf{i}_n \qquad (2\text{-}68)$$

Choosing the Gaussian surface S as the surface of a rectangular pill box of sides l, w, and t as shown in Fig. 2.18, such that half of the box is above the sheet charge and the other half below it, we have

$$\oint_S \mathbf{E} \cdot d\mathbf{S} = \underbrace{\int \mathbf{E} \cdot d\mathbf{S}}_{\substack{\text{top} \\ \text{surface}}} + \underbrace{\int \mathbf{E} \cdot d\mathbf{S}}_{\substack{\text{bottom} \\ \text{surface}}} + \underbrace{\int \mathbf{E} \cdot d\mathbf{S}}_{\substack{\text{side} \\ \text{surfaces}}} \qquad (2\text{-}69)$$

But the last integral on the right side of (2-69) is equal to zero since **E** is parallel to the side surfaces and hence **E** · *d***S** is zero throughout these surfaces. Because E_n is constant on both the top and bottom surfaces and E_n is the same on both these surfaces, since they are equidistant from the sheet charge, Eq. (2-69) then reduces to

$$\oint_S \mathbf{E} \cdot d\mathbf{S} = 2 \int_{\substack{\text{top} \\ \text{surface}}} \mathbf{E} \cdot d\mathbf{S} = 2 \int_{\substack{\text{top} \\ \text{surface}}} E_n \mathbf{i}_n \cdot d\mathbf{S}\, \mathbf{i}_n$$

$$= 2E_n \int_{\substack{\text{top} \\ \text{surface}}} dS = 2E_n \text{ (surface area of top surface)} \qquad (2\text{-}70)$$

$$= 2E_n lw$$

But, from Gauss' law,

$$\oint_S \mathbf{E} \cdot d\mathbf{S} = \frac{\text{charge enclosed by } S}{\epsilon_0} = \frac{\rho_{s0} lw}{\epsilon_0} \qquad (2\text{-}71)$$

Comparing (2-70) and (2-71), we have

$$E_n = \frac{\rho_{s0}}{2\epsilon_0} \qquad (2\text{-}72)$$

$$\mathbf{E} = \frac{\rho_{s0}}{2\epsilon_0} \mathbf{i}_n \qquad (2\text{-}73)$$

which agrees with the result obtained in Example 2-5. ∎

EXAMPLE 2-11.   A volume charge is distributed throughout a sphere of radius a with uniform density ρ_0 C/m³. We wish to obtain the electric field intensity due to this volume charge using Gauss' law.

In Example 2-6 we established from purely qualitative arguments that **E** due to the spherical volume charge of uniform density is directed radially away from the center of the charge and is a function only of the distance from the center of the sphere. Thus

$$\mathbf{E} = E_r(r)\mathbf{i}_r \qquad (2\text{-}74)$$

Choosing the Gaussian surface S as the surface of a sphere of radius $r \geqq a$, concentric with the spherical charge, as shown in Fig. 2.19, we have

$$\oint_S \mathbf{E} \cdot d\mathbf{S} = \oint_S E_r \mathbf{i}_r \cdot d\mathbf{S}\, \mathbf{i}_r = E_r \oint_S dS$$

$$= E_r \text{ (surface area of the sphere of radius } r)$$

$$= E_r(4\pi r^2) \qquad (2\text{-}75)$$

But, from Gauss' law,

$$\oint_S \mathbf{E} \cdot d\mathbf{S} = \frac{\text{charge enclosed by } S}{\epsilon_0}$$

$$= \frac{\text{charge enclosed by spherical surface of radius } r}{\epsilon_0} \qquad (2\text{-}76)$$

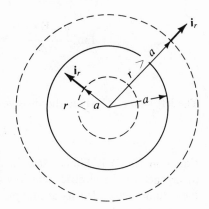

Fig. 2.19. Gaussian surfaces for computing the electric field of a spherical volume charge of uniform density.

Comparing (2-75) and (2-76), we have

$$E_r = \frac{\text{charge enclosed by spherical surface of radius } r}{4\pi\epsilon_0 r^2} \qquad (2\text{-}77)$$

$$\mathbf{E} = \frac{\text{charge enclosed by spherical surface of radius } r}{4\pi\epsilon_0 r^2}\mathbf{i}_r \qquad (2\text{-}78)$$

which agrees with the result of Example 2-6. ∎

2.7   Gauss' Law in Differential Form (Maxwell's Divergence Equation for the Electric Field)

Let us consider a volume charge distribution with the charge density ρ as a given function of the coordinate system. The charge enclosed by an arbitrary closed surface S is given by the volume integral of the charge density throughout the volume V enclosed by the surface S; that is, $\int_V \rho \, dv$. According to Gauss' law (2-58), we have

$$\oint_S \mathbf{E} \cdot d\mathbf{S} = \frac{1}{\epsilon_0} \int_V \rho \, dv \qquad (2\text{-}79)$$

If we now shrink the volume to a very small value ΔV, so that the surface area becomes very small ΔS, we can write (2-79) for this infinitesimal surface as

$$\oint_{\Delta S} \mathbf{E} \cdot d\mathbf{S} = \frac{1}{\epsilon_0} \int_{\Delta v} \rho \, dv \qquad (2\text{-}80)$$

Since the volume is very small, we can consider the charge density ρ to be uniform inside that volume so that $\int_{\Delta v} \rho \, dv \approx \rho \, \Delta v$. This is exact in the limit that $\Delta v \to 0$. Dividing both sides of (2-80) by Δv and letting $\Delta v \to 0$, we have

$$\lim_{\Delta v \to 0} \frac{\oint_{\Delta S} \mathbf{E} \cdot d\mathbf{S}}{\Delta v} = \lim_{\Delta v \to 0} \frac{(1/\epsilon_0) \int_{\Delta v} \rho \, dv}{\Delta v}$$

$$= \frac{1}{\epsilon_0} \lim_{\Delta v \to 0} \frac{\rho \, \Delta v}{\Delta v} = \frac{1}{\epsilon_0} \rho \qquad (2\text{-}81)$$

The left side of (2-81) is the divergence of $\mathbf{E}$ so that we have

$$\boldsymbol{\nabla} \cdot \mathbf{E} = \frac{1}{\epsilon_0} \rho \qquad (2\text{-}82)$$

Equation (2-82) is Gauss' law in differential form, which states that the divergence of the electric field intensity at any point is equal to $1/\epsilon_0$ times the volume charge density at that point. This is Maxwell's divergence equation for the electric field.

The right side of (2-82) represents a volume charge density. Suppose we are considering problems involving point charges, line charges, and surface charges. The question then arises as to how we should represent the right side of (2-82) since, for such charges, the volume charge density is infinity. We can resolve this problem by resorting to the Dirac delta function or the impulse function. We will illustrate this for the case of a surface charge in the following example.

EXAMPLE 2-12.   A sheet charge of uniform density ρ_{s0} C/m² extends over the entire *xy* plane. It is desired to write Gauss' law in differential form for this sheet charge.

Let us consider a slab of charge lying between the planes $z = -a$ and $z = +a$ and of uniform density ρ_0 C/m³ as shown in Fig. 2.20(a). The volume charge density as a function of z for such a charge distribution is sketched in

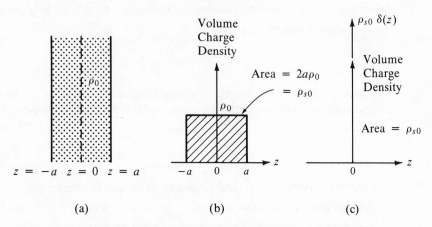

Fig. 2.20. For deriving the volume charge density corresponding to a surface charge.

Fig. 2.20(b). The charge per unit surface area of the slab charge is given by $\int_{z=-a}^{a} \rho_0 \, dz = \rho_0 2a =$ area under the curve of Fig. 2.20(b). Let this quantity be ρ_{s0}. Suppose we now shrink a to zero, increasing ρ_0 such that ρ_{s0} remains constant. We then obtain a sheet charge of density ρ_{s0} C/m². What happens to the sketch of Fig. 2.20(b)? The width of the pulse-shaped sketch decreases to zero and the height increases to infinity but maintaining the area under it equal to ρ_{s0}. The resulting function is sketched in Fig. 2.20(c). This function is known as the Dirac delta function of strength ρ_{s0} and is represented as $\rho_{s0} \, \delta(z)$, where $\delta(z)$ satisfies the properties

$$\delta(z) = \begin{cases} 0 & \text{for } z \neq 0 \\ \infty & \text{for } z = 0 \end{cases} \tag{2-83}$$

$$\int_{z=-\infty}^{\infty} \delta(z) \, dz = \int_{z=0-}^{0+} \delta(z) \, dz = \lim_{a \to 0} \int_{z=-a}^{a} \frac{1}{2a} \, dz = 1 \tag{2-84}$$

$$\int_{z=-\infty}^{\infty} f(z) \, \delta(z) \, dz = f(0) \tag{2-85}$$

Thus the volume charge density corresponding to the sheet charge of density ρ_{s0} lying in the $z = 0$ plane is $\rho_{s0} \, \delta(z)$. Gauss' law in differential form for the sheet charge is then given by

$$\mathbf{V} \cdot \mathbf{E} = \frac{1}{\epsilon_0} \rho_{s0} \, \delta(z) \tag{2-86}$$

If the sheet charge lies in the $z = z_0$ plane, then the Dirac delta function is shifted to $z = z_0$ and is written as $\delta(z - z_0)$, having the properties

$$\delta(z - z_0) = \begin{cases} 0 & \text{for } z \neq z_0 \\ \infty & \text{for } z = z_0 \end{cases} \tag{2-87}$$

$$\int_{z=-\infty}^{\infty} \delta(z - z_0) \, dz = 1 \tag{2-88}$$

$$\int_{z=-\infty}^{\infty} f(z) \, \delta(z - z_0) \, dz = f(z_0) \tag{2-89}$$

Gauss' law in differential form is modified to read

$$\mathbf{V} \cdot \mathbf{E} = \frac{1}{\epsilon_0} \rho_{s0} \, \delta(z - z_0) \tag{2-90}$$

It is left to the student to derive equations similar to (2-90) for line and point charges, involving two-dimensional and three-dimensional Dirac delta functions, respectively (See Problems 2.33 and 2.34). ∎

2.8 Potential Difference

In the study of mechanics, we are familiar with potential energy associated with the movement of a mass in the gravitational field of the earth. If the movement of the mass is along the direction of the gravitational field, that is, from a higher elevation to a lower elevation, the gravitational field does the work. If the movement is opposite to the direction of the gravitational field, that is, from a lower elevation to a higher elevation, certain work has to be performed by an external source to overcome the gravitational force. Likewise, since the electric field is a force field in so far as charges are concerned, there is work associated with the movement of charges in an electric field. If a test charge is moved along the direction of the field, work is done by the field since the force exerted by the field on the charge is in the direction of its movement and hence it accelerates the test charge. If the charge is moved against the direction of the field, an external agent has to supply the energy to overcome the force exerted on the charge by the field, since this force is opposite to the direction of movement of the charge.

Let us consider the displacement of a test charge q by an infinitesimal distance $d\mathbf{l}$ from A to B at an angle α with the electric field $\mathbf{E}$ at the point A as shown in Fig. 2.21(a). The force exerted on the test charge by the field

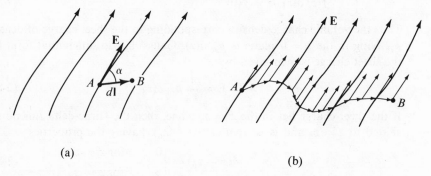

(a)

(b)

Fig. 2.21. Movement of a test charge in an electric field.

has magnitude qE and is directed along $\mathbf{E}$. Its component along the line from A to B is $qE \cos \alpha$. If the charge is moved from A to B, the amount of work dW done by the field is the product of the force and the displacement; that is,

$$dW = qE \cos \alpha \, dl = q\mathbf{E} \cdot d\mathbf{l} \tag{2-91}$$

where $d\mathbf{l}$ is the vector from A to B. Note that dW is positive if $0 < \alpha < 90°$ so that work is done by the field; dW is negative if $90° < \alpha < 180°$ so that negative work is done by the field, which amounts to stating that work is done against the field by an external agent. For $\alpha = 90°$, dW is zero, which is analogous to the movement of a mass on a frictionless surface at right angles to the gravitational field. Now let us consider two points A and B which are widely separated as shown in Fig. 2.21(b). The work W_{AB} done by the field in moving a test charge q from A to B along a given path can be obtained by dividing the path into several segments of infinitesimal length dl, then applying (2-91) to each segment, and adding up all the contributions.

The result is a line integral expression given by

$$W_{AB} = q \int_A^B \mathbf{E} \cdot d\mathbf{l} \qquad (2\text{-}92)$$

where the integration is performed along the given path from A to B. The evaluation of line integrals was discussed in Section 1.7.

In the gravitational field, when a mass moves from a higher elevation to a lower elevation, it loses some potential energy and vice versa. Likewise, in the electric field, we can state that the test charge has certain potential energy associated with it by virtue of its location in the electric field. W_{AB} as given by (2-92) is then the loss of potential energy associated with the movement of the charge from A to B. If we divide W_{AB} by q, we obtain the loss of potential energy per unit charge. This quantity denoted by V_{AB} is known as the potential difference between the points A and B. Thus

$$V_{AB} = \frac{W_{AB}}{q} = \int_A^B \mathbf{E} \cdot d\mathbf{l} \qquad (2\text{-}93)$$

If V_{AB} is positive, there is a loss in potential energy associated with the movement of the charge from A to B; that is, the field does the work. If V_{AB} is negative, there is a gain in potential energy associated with the movement of the charge from A to B; that is, an external agent has to do the work. The units of potential difference are newton-meters per coulomb or joules per coulomb, commonly known as volts. This gives the units of volts per meter to the electric field intensity.

EXAMPLE 2-13. In cartesian coordinates, the electric field intensity is given by

$$\mathbf{E} = yz\mathbf{i}_x + zx\mathbf{i}_y + xy\mathbf{i}_z$$

Find the potential difference between the points $A(0, 22.7, 99)$ and $B(1, 1, 1)$. Is it necessary to specify a path for line integration between the two points?

In cartesian coordinates, $d\mathbf{l} = dx\,\mathbf{i}_x + dy\,\mathbf{i}_y + dz\,\mathbf{i}_z$ so that

$$V_{AB} = \int_A^B \mathbf{E} \cdot d\mathbf{l} = \int_A^B (yz\mathbf{i}_x + zx\mathbf{i}_y + xy\mathbf{i}_z) \cdot (dx\,\mathbf{i}_x + dy\,\mathbf{i}_y + dz\,\mathbf{i}_z)$$

$$= \int_A^B (yz\,dx + zx\,dy + xy\,dz)$$

$$= \int_A^B d(xyz) = [xyz]_A^B$$

Since $\mathbf{E} \cdot d\mathbf{l}$ is the total derivative of a function of x, y, z, it is not necessary to specify a path for the line integration between the two points. V_{AB} is dependent only on the coordinates of the end points A and B. We will find in Section 2.11 that this is a general characteristic of the static electric field. Here, we have

$$V_{AB} = [xyz]_A^B = [xyz]_{0,\,22.7,\,99}^{1,\,1,\,1} = 1. \quad \blacksquare$$

2.9   The Potential Field of Point Charges

Let us now consider two points A and B in the electric field of a point charge Q situated at distances r_A and r_B, respectively, from the point charge as shown in Fig. 2.22. Using (2-93), the potential difference between A and B can be computed for any specified path from A to B. Noting that $\mathbf{E} = (Q/4\pi\epsilon_0 r^2)\mathbf{i}_r$ for a point charge and that the differential length vector $d\mathbf{l}$ is given in spherical coordinates as

$$d\mathbf{l} = dr\,\mathbf{i}_r + r\,d\theta\,\mathbf{i}_\theta + r\sin\theta\,d\phi\,\mathbf{i}_\phi \qquad (2\text{-}94)$$

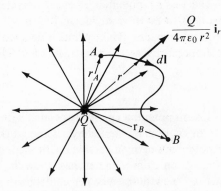

Fig. 2.22. Computation of the potential difference between two points in the electric field of a point charge.

we have, from (2-93),

$$V_{AB} = \int_A^B \mathbf{E} \cdot d\mathbf{l} = \int_A^B \left(\frac{Q}{4\pi\epsilon_0 r^2}\,\mathbf{i}_r\right) \cdot (dr\,\mathbf{i}_r + r\,d\theta\,\mathbf{i}_\theta + r\sin\theta\,d\phi\,\mathbf{i}_\phi)$$

$$= \int_{r=r_A}^{r_B} \frac{Q}{4\pi\epsilon_0 r^2}\,dr = \frac{Q}{4\pi\epsilon_0 r_A} - \frac{Q}{4\pi\epsilon_0 r_B} \qquad (2\text{-}95)$$

Equation (2-95) indicates that, for a given charge Q, the potential difference between the two points is dependent only upon their distances from the point charge and not on the path from A to B chosen for its evaluation. Furthermore, the potential difference is the difference between two terms, one of which is dependent on r_A only and the other dependent on r_B only. We can call these terms the potentials at r_A and r_B, respectively. If we denote these potentials as V_A and V_B, respectively, we have, from (2-95),

$$V_A = \frac{Q}{4\pi\epsilon_0 r_A} \qquad (2\text{-}96)$$

$$V_B = \frac{Q}{4\pi\epsilon_0 r_B} \qquad (2\text{-}97)$$

The right sides of Eqs. (2-96) and (2-97) are, however, not unique expressions for V_A and V_B since, on the right side of (2-95), we can add and subtract any arbitrary constant C without altering its value; that is,

$$V_{AB} = \left(\frac{Q}{4\pi\epsilon_0 r_A} + C\right) - \left(\frac{Q}{4\pi\epsilon_0 r_B} + C\right) \tag{2-98}$$

which then leads to

$$V_A = \frac{Q}{4\pi\epsilon_0 r_A} + C \tag{2-99}$$

$$V_B = \frac{Q}{4\pi\epsilon_0 r_B} + C \tag{2-100}$$

If we let $C = Q/4\pi\epsilon_0 r_0$, where r_0 is a constant, we have

$$V_A = \frac{Q}{4\pi\epsilon_0 r_A} - \frac{Q}{4\pi\epsilon_0 r_0} \tag{2-101a}$$

$$V_B = \frac{Q}{4\pi\epsilon_0 r_B} - \frac{Q}{4\pi\epsilon_0 r_0} \tag{2-101b}$$

Comparing (2-101a) with (2-95), we note that V_A is the potential difference between point A and another point situated at a distance r_0 from the point charge, which we will call the reference point. Similarly, V_B is the potential difference between the point B and the same reference point. Thus the potential at any point is simply the potential difference between that point and an arbitrary reference point. But then, what is the potential at the reference point? The answer to this question is obtained by substituting $r_A = r_0$ in (2-101a) or $r_B = r_0$ in (2-101b), both of which result in zero. The potential at the reference point is therefore zero. To complete the definition, we state that the potential at any point is the potential difference between that point and an arbitrary reference point at which the potential is zero. In the case of a point charge, a convenient reference point is $r_0 = \infty$. We then have

$$V(r) = \frac{Q}{4\pi\epsilon_0 r} \tag{2-102}$$

The potential at a distance r from the point charge is thus the work done per unit charge by the field in the movement of a test charge from that point to infinity or, it is the work done per unit charge by an external agent in bringing a test charge from infinity to that point; that is,

$$V(r) = \int_r^\infty \mathbf{E} \cdot d\mathbf{l} = -\int_\infty^r \mathbf{E} \cdot d\mathbf{l} \tag{2-103}$$

The right side of (2-102) represents the potential field of a point charge. It is also known as the Coulomb potential of a point charge. In contrast to the vector nature of the electric field intensity, the potential field is a scalar field.

Surfaces on which potential is a constant are known as equipotential surfaces. If a test charge is moved on such a surface from one point to another, no work is involved since the potential difference between any two points is zero. For the point charge, the equipotential surfaces are, according to (2-102), $r = $ constant, that is, surfaces of spheres centered at the point charge. The equipotential surfaces are thus orthogonal to the direction lines of **E** which are radial, as shown in Fig. 2.23. This result is to be expected not only for a point charge but for any charge distribution, since if we move a test charge along a path everywhere normal to the direction lines, there is no component of force acting on the charge along the direction of the path and hence the work involved is zero.

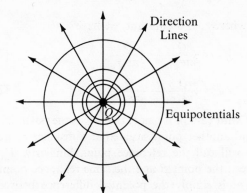

Fig. 2.23. Cross sections of equipotential surfaces and direction lines of **E** for a point charge.

For several point charges located at different points as shown in Fig. 2.5, the potential at any point P is the work done per unit charge by an external agent in bringing a test charge from infinity to that point in the combined electric field **E** of all the charges; that is,

$$V(P) = -\int_{\infty}^{P} \mathbf{E} \cdot d\mathbf{l}$$

$$= -\int_{\infty}^{P} (\mathbf{E}_1 + \mathbf{E}_2 + \mathbf{E}_3 + \cdots + \mathbf{E}_n) \cdot d\mathbf{l} \qquad (2\text{-}104)$$

$$= -\int_{\infty}^{P} \mathbf{E}_1 \cdot d\mathbf{l} - \int_{\infty}^{P} \mathbf{E}_2 \cdot d\mathbf{l} - \cdots - \int_{\infty}^{P} \mathbf{E}_n \cdot d\mathbf{l}$$

where $\mathbf{E}_1, \mathbf{E}_2, \mathbf{E}_3, \ldots, \mathbf{E}_n$ are the electric field intensities due to the individual point charges $Q_1, Q_2, Q_3, \ldots, Q_n$, respectively. But each term on the right side of (2-104) is equal to the potential at the point P due to the corresponding charge. Thus

$$V(P) = \frac{Q_1}{4\pi\epsilon_0 R_1} + \frac{Q_2}{4\pi\epsilon_0 R_2} + \cdots + \frac{Q_n}{4\pi\epsilon_0 R_n}$$
$$= \sum_{j=1}^{n} \frac{Q_j}{4\pi\epsilon_0 R_j} \qquad (2\text{-}105)$$

The potential at P due to the collection of point charges is the sum of the potentials at P due to the individual charges. In the vector notation defined in connection with Eq. (2-20), we write

$$V(\mathbf{r}) = \sum_{j=1}^{n} \frac{Q_j}{4\pi\epsilon_0 |\mathbf{r} - \mathbf{r}'_j|} \qquad (2\text{-}106)$$

EXAMPLE 2-14. For the electric dipole arrangement of Fig. 2.7, it is desired to find the potential at distances very far from the dipole compared to the spacing d.

With reference to the notation of Fig. 2.7, the electric potential at point P is given by

$$V(r) = \frac{Q}{4\pi\epsilon_0 r_+} - \frac{Q}{4\pi\epsilon_0 r_-} \qquad (2\text{-}107)$$

For $r \gg d$, (2-107) can be approximated as

$$V(r) \approx \frac{Q}{4\pi\epsilon_0[r - (d/2)\cos\theta]} - \frac{Q}{4\pi\epsilon_0[r + (d/2)\cos\theta]}$$
$$= \frac{Qd\cos\theta}{4\pi\epsilon_0[r^2 - (d^2/4)\cos^2\theta]} \approx \frac{Qd\cos\theta}{4\pi\epsilon_0 r^2} \qquad (2\text{-}108)$$

Equation (2-108) becomes exact in the limit $d \to 0$, keeping the dipole moment $p = Qd$ constant. We then have the potential field of dipole moment $\mathbf{p} = p\mathbf{i}_z$ given by

$$V(r) = \frac{p\cos\theta}{4\pi\epsilon_0 r^2} = \frac{\mathbf{p}\cdot\mathbf{i}_r}{4\pi\epsilon_0 r^2} = \frac{\mathbf{p}\cdot\mathbf{r}}{4\pi\epsilon_0 r^3} \qquad (2\text{-}109)$$

The potential field of a dipole drops off inversely as the square of the distance, as compared to the inverse distance dependence of the potential field of a point charge. Likewise, the potential field of a quadrupole can be shown to vary inversely as r^3. The potential fields of successive higher-order multipoles vary inversely as $r^4, r^5, \ldots$. From (2-109), we note that the equipotential surfaces for the dipole field are $(\cos\theta)/r^2 = $ constant, or

$$r^2 \sec\theta = \text{constant} \qquad (2\text{-}110)$$

Cross sections of these surfaces are sketched in Fig. 2.24, in which the direction lines of $\mathbf{E}$ taken from Fig. 2.12 are also shown. It is left as an exercise for the student to show that the equipotential surfaces given by (2-110) and the direction lines given by (2-49) are orthogonal. ▉

EXAMPLE 2-15. A point charge Q is situated at a vector distance $\mathbf{r}'$ from the origin of a coordinate system as shown in Fig. 2.25. It is desired to find the potential due to this point charge at distances $\mathbf{r}$ from the origin large in magnitude compared to $\mathbf{r}'$ in the form of a power series in r.

Let P be the point at which the potential is desired. Then, from (2-106), the potential at P due to Q is given by

$$V(\mathbf{r}) = \frac{Q}{4\pi\epsilon_0|\mathbf{r} - \mathbf{r}'|}$$
$$= \frac{Q}{4\pi\epsilon_0(r^2 + r'^2 - 2rr'\cos\alpha)^{1/2}} \qquad (2\text{-}111)$$
$$= \frac{Q}{4\pi\epsilon_0 r}\left(1 + \frac{r'^2}{r^2} - \frac{2\mathbf{r}'\cdot\mathbf{r}}{r^2}\right)^{-1/2}$$

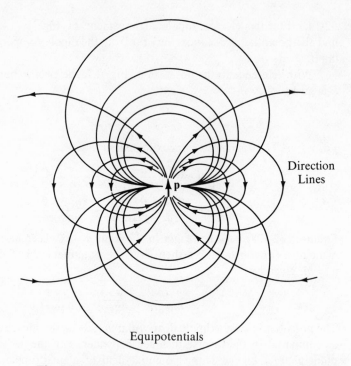

Fig. 2.24. Cross sections of equipotential surfaces and direction lines of **E** for an electric dipole.

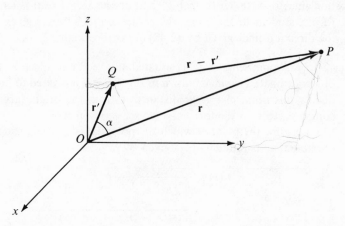

Fig. 2.25. For the computation of potential due to a point charge at distances large compared to its distance from the origin.

Using the binomial theorem,

$$(1 + x)^n = 1 + nx + \frac{n(n-1)}{2!}x^2 + \cdots$$

we have

$$V(\mathbf{r}) = \frac{Q}{4\pi\epsilon_0 r}\left[1 + \left(-\frac{1}{2}\right)\left(\frac{r'^2}{r^2} - \frac{2\mathbf{r}'\cdot\mathbf{r}}{r^2}\right) + \frac{1}{2}\left(-\frac{1}{2}\right)\left(-\frac{3}{2}\right)\left(\frac{r'^2}{r^2} - \frac{2\mathbf{r}'\cdot\mathbf{r}}{r^2}\right)^2\right.$$

$$\left. + \cdots \text{higher-order terms}\right]$$

$$= \frac{Q}{4\pi\epsilon_0 r}\left\{1 + \frac{\mathbf{r}'\cdot\mathbf{r}}{r^2} + \frac{1}{2r^4}[3(\mathbf{r}'\cdot\mathbf{r})^2 - r^2 r'^2] + \cdots \text{higher-order terms}\right\}$$

$$(2\text{-}112)$$

For $r'/r \ll 1$, the magnitudes of the successive terms on the right side of (2-112) decrease rapidly as can be seen by writing (2-112) as

$$V(\mathbf{r}) = \frac{Q}{4\pi\epsilon_0 r}\left[1 + \left(\frac{r'}{r}\right)\cos\alpha + \left(\frac{r'}{r}\right)^2\left(\frac{3\cos^2\alpha - 1}{2}\right)\right.$$

$$\left. + \cdots \text{higher-order terms}\right]$$

$$(2\text{-}113)$$

Hence, for $r' \ll r$, only the first few terms are significant. Furthermore, writing

$$V(\mathbf{r}) = \frac{Q}{4\pi\epsilon_0 r} + \frac{Q\mathbf{r}'\cdot\mathbf{r}}{4\pi\epsilon_0 r^3} + \frac{Q}{8\pi\epsilon_0 r^5}[3(\mathbf{r}'\cdot\mathbf{r})^2 - r^2 r'^2] + \cdots$$

$$= \frac{Q}{4\pi\epsilon_0 r} + \frac{Qr'\cos\alpha}{4\pi\epsilon_0 r^2} + \frac{Qr'^2}{4\pi\epsilon_0 r^3}\left(\frac{3\cos^2\alpha - 1}{2}\right) + \cdots$$

$$(2\text{-}114)$$

we observe that, on the right side of (2-114), the first term is the potential at P due to a point charge Q at the origin; the second term is the potential at P due a dipole moment $\mathbf{p} = Q\mathbf{r}'$ at the origin; the third term seems like the potential at P due to a quadrupole at the origin since it varies as $1/r^3$, and so on.

If we have several point charges $Q_1, Q_2, Q_3, \ldots, Q_n$ situated at $\mathbf{r}'_1, \mathbf{r}'_2, \mathbf{r}'_3, \ldots, \mathbf{r}'_n$, the potential at $\mathbf{r}$ due to this collection of point charges can be written by applying superposition to (2-114) as

$$V(\mathbf{r}) = \sum_{j=1}^{n}\left\{\frac{Q_j}{4\pi\epsilon_0 r} + \frac{Q_j\mathbf{r}'_j\cdot\mathbf{r}}{4\pi\epsilon_0 r^3} + \frac{Q_j}{8\pi\epsilon_0 r^5}[3(\mathbf{r}'_j\cdot\mathbf{r})^2 - r^2 r_j'^2] + \cdots\right\}$$

$$= \frac{\sum_{j=1}^{n}Q_j}{4\pi\epsilon_0 r} + \frac{\sum_{j=1}^{n}Q_j\mathbf{r}'_j\cdot\mathbf{r}}{4\pi\epsilon_0 r^3} + \cdots$$

$$(2\text{-}115)$$

The potential due to the collection of point charges at large distances from the collection is thus a superposition of the potentials due to a point charge of value $\sum\limits_{j=1}^{n} Q_j$, a dipole moment $\sum\limits_{j=1}^{n} Q_j \mathbf{r}'_j$, and so on, all situated at the origin. We note that if the sum of the charges is zero, the first significant term is that of the dipole moment. Likewise, if the sum of the charges as well as the dipole moment are zero, the first significant term is the quadrupole term, and so on. Usually, two significant terms will suffice. ∎

EXAMPLE 2-16.   Point charges are located at the corners of a cube of sides 1 m², with one corner placed at the origin and three edges coinciding with the coordinate axes as shown in Fig. 2.26. Values of the point charges in coulombs are indicated at the respective corners. Find the first two significant terms in the potential of this collection of charges at large distances from it.

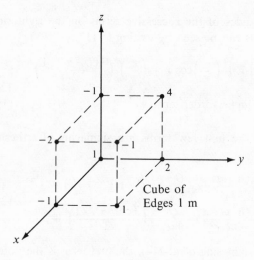

Fig. 2.26. Point charges located at the corners of a cube. Values of the point charges indicated at the respective corners are in coulombs.

The solution to this problem consists of evaluating $\sum Q$ and $\sum Q \mathbf{r}'$ for the collection of point charges and substituting the results in (2-115). These quantities are evaluated with the aid of Table 2.2.

The potential for large r correct to the first two significant terms is then given by

$$V = \frac{\sum Q}{4\pi\epsilon_0 r} + \frac{\sum Q \mathbf{r}' \cdot \mathbf{r}}{4\pi\epsilon_0 r^3}$$

$$= \frac{3}{4\pi\epsilon_0 r} + \frac{(-3\mathbf{i}_x + 6\mathbf{i}_y) \cdot \mathbf{i}_r}{4\pi\epsilon_0 r^2} \qquad (2\text{-}116)$$

$$= \frac{3}{4\pi\epsilon_0 r} + \frac{-3\sin\theta\cos\phi + 6\sin\theta\sin\phi}{4\pi\epsilon_0 r^2}$$

If, in Table 2.2, $\sum Q$ is zero, then we have to evaluate the third term if the result is to be correct to the first two significant terms, and so on. ∎

TABLE 2.2. Computation of $\sum Q$ and $\sum Q\mathbf{r}'$ for the Arrangement of Point Charges in Fig. 2.26

Location (x, y, z)	Charge, Q	r'	Qr'
0, 0, 0	1	0	0
1, 0, 0	−1	$\mathbf{i}_x$	$-\mathbf{i}_x$
0, 1, 0	2	$\mathbf{i}_y$	$2\mathbf{i}_y$
0, 0, 1	−1	$\mathbf{i}_z$	$-\mathbf{i}_z$
1, 1, 0	1	$\mathbf{i}_x + \mathbf{i}_y$	$\mathbf{i}_x + \mathbf{i}_y$
0, 1, 1	4	$\mathbf{i}_y + \mathbf{i}_z$	$4\mathbf{i}_y + 4\mathbf{i}_z$
1, 0, 1	−2	$\mathbf{i}_x + \mathbf{i}_z$	$-2\mathbf{i}_x - 2\mathbf{i}_z$
1, 1, 1	−1	$\mathbf{i}_x + \mathbf{i}_y + \mathbf{i}_z$	$-\mathbf{i}_x - \mathbf{i}_y - \mathbf{i}_z$
	$\sum Q = 3$		$\sum Q\mathbf{r}' = -3\mathbf{i}_x + 6\mathbf{i}_y$

2.10 The Potential Field of Continuous Charge Distributions

In the previous section we considered the potential field of collections of point charges at discrete points. In this section we will extend the discussion to continuous charge distributions. As in Section 2.4, we divide the continuous charge distribution into several infinitesimal parts, each of which can be considered as a point charge, and obtain the potential at any point due to the total charge as the superposition of the potentials due to the individual point charges. To do this, we again have to evaluate integrals as in Section 2.4. However, the integrals involve the scalar quantity potential instead of the vector quantity electric field intensity. Hence, for a particular charge distribution, the potential at any point is given by a single integral, whereas for the determination of the electric field intensity as in Section 2.4, it is necessary to evaluate three integrals for the three components in the general case. We will illustrate the determination of the potential for continuous charge distributions through some examples.

EXAMPLE 2-17.   An infinitely long line charge of uniform density ρ_{L0} C/m is situated along the z axis. It is desired to obtain the potential field due to this charge.

First we divide the line into a number of infinitesimal segments each of length dz as shown in Fig. 2.27, such that the charge $\rho_{L0}\, dz$ in each segment can be considered as a point charge. Let us consider a point P at a distance r from the z axis, with the projection of P onto the z axis being O. For the sake of generality, we consider the point P_0 at a distance r_0 from O along OP as the reference point for zero potential and write the potential dV at P due to the infinitesimal charge $\rho_{L0}\, dz$ at A as

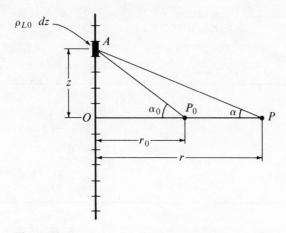

Fig. 2.27. Geometry for the computation of the potential field of an infinitely long line charge of uniform density ρ_{L0} C/m.

$$dV = \frac{\rho_{L0} \, dz}{4\pi\epsilon_0 (AP)} - \frac{\rho_{L0} \, dz}{4\pi\epsilon_0 (AP_0)}$$

$$= \frac{\rho_{L0} \, dz}{4\pi\epsilon_0 \sqrt{r^2 + z^2}} - \frac{\rho_{L0} \, dz}{4\pi\epsilon_0 \sqrt{r_0^2 + z^2}} \tag{2-117}$$

We will, however, find later that we have to choose the reference point for zero potential at a finite value of r, in contrast to the case of the point charge for which the reference point can be chosen to be infinity. The potential V at P due to the entire line charge is now given by the integral of (2-117), where the integration is to be performed between the limits $z = -\infty$ and $z = \infty$. Thus

$$V = \int_{z=-\infty}^{\infty} dV = \int_{z=-\infty}^{\infty} \left(\frac{\rho_{L0} \, dz}{4\pi\epsilon_0 \sqrt{r^2 + z^2}} - \frac{\rho_{L0} \, dz}{4\pi\epsilon_0 \sqrt{r_0^2 + z^2}} \right)$$

$$= \frac{\rho_{L0}}{2\pi\epsilon_0} \int_{z=0}^{\infty} \left(\frac{dz}{\sqrt{r^2 + z^2}} - \frac{dz}{\sqrt{r_0^2 + z^2}} \right) \tag{2-118}$$

Introducing $z = r \tan \alpha$ and $z = r_0 \tan \alpha_0$ in the first and second terms, respectively, in the integrand on the right side of (2-118), we have

$$V = \frac{\rho_{L0}}{2\pi\epsilon_0} \left(\int_{\alpha=0}^{\pi/2} \sec \alpha \, d\alpha - \int_{\alpha_0=0}^{\pi/2} \sec \alpha_0 \, d\alpha_0 \right)$$

$$= \frac{\rho_{L0}}{2\pi\epsilon_0} \left\{ [\ln (\sec \alpha + \tan \alpha)]_{\alpha=0}^{\pi/2} - [\ln (\sec \alpha_0 + \tan \alpha_0)]_{\alpha_0=0}^{\pi/2} \right\}$$

$$= \frac{\rho_{L0}}{2\pi\epsilon_0} \left[\ln \frac{(\sqrt{r^2 + z^2} + z) r_0}{(\sqrt{r_0^2 + z^2} + z) r} \right]_{z=0}^{\infty} \tag{2-119}$$

$$= -\frac{\rho_{L0}}{2\pi\epsilon_0} \ln \frac{r}{r_0}$$

In view of the cylindrical symmetry about the line charge, (2-119) is the general expression in cylindrical coordinates for the potential field of the infinitely long line charge of uniform density. It can be seen from (2-119) that a choice of $r_0 = \infty$ is not a good choice, since then the potential would be infinity at all points. The difficulty lies in the fact that infinity plus a finite number is still infinity. We also note from (2-119) that the equipotential surfaces are $\ln r/r_0 = $ constant or $r = $ constant, that is, surfaces of cylinders with the line charge as their axis. In Ex. 2-4, we found that the electric field intensity due to the line charge is directed radially away from the line charge. Thus the direction lines of **E** and the equipotential surfaces are indeed orthogonal to each other. ▌

Generalizing the expression for the computation of potential for a line charge distribution of density $\rho_L(\mathbf{r}')$ occupying a contour C', we have

$$V(\mathbf{r}) = \int_{C'} \frac{\rho_L(\mathbf{r}')}{4\pi\epsilon_0 |\mathbf{r} - \mathbf{r}'|} \, dl' \qquad (2\text{-}120a)$$

This is the Coulomb potential of line charge distribution $\rho_L(\mathbf{r}')$. Similarly, for a surface charge distribution of density $\rho_s(\mathbf{r}')$ occupying a surface S', we have

$$V(\mathbf{r}) = \int_{S'} \frac{\rho_s(\mathbf{r}')}{4\pi\epsilon_0 |\mathbf{r} - \mathbf{r}'|} \, dS' \qquad (2\text{-}120b)$$

and for a volume charge distribution of density $\rho(\mathbf{r}')$ occupying a volume V', we have

$$V(\mathbf{r}) = \int_{V'} \frac{\rho(\mathbf{r}')}{4\pi\epsilon_0 |\mathbf{r} - \mathbf{r}'|} \, dv' \qquad (2\text{-}120c)$$

EXAMPLE 2-18.   A cube charge of sides 1 m² is situated with one corner at the origin and three edges coinciding with the coordinate axes. The charge density ρ within the cube is given by

$$\rho = (x + y + z) \, C$$

Find the potential field of the cube charge at large distances from it correct to the first two significant terms.

This problem is an extension of Example 2-16 to a continuous charge distribution. For a continuous charge distribution, the summations of (2-115) have to be replaced by integrals so that we have

$$V = \frac{\int_{\text{vol}} dQ}{4\pi\epsilon_0 r} + \frac{\left(\int_{\text{vol}} dQ \, \mathbf{r}' \cdot \mathbf{r}\right)}{4\pi\epsilon_0 r^3} + \cdots \qquad (2\text{-}121)$$

For the specified charge distribution

$$\int_{vol} dQ = \int_{vol} \rho \, dv$$
$$= \int_{x=0}^{1} \int_{y=0}^{1} \int_{z=0}^{1} (x + y + z) \, dx \, dy \, dz \qquad (2\text{-}122)$$
$$= \tfrac{3}{2} \, \text{C}$$

$$\int_{vol} (dQ \, \mathbf{r}') = \int_{vol} \rho \, dv \, \mathbf{r}'$$
$$= \int_{x=0}^{1} \int_{y=0}^{1} \int_{z=0}^{1} (x + y + z) \, dx \, dy \, dz \, (x\mathbf{i}_x + y\mathbf{i}_y + z\mathbf{i}_z)$$
$$= \tfrac{5}{6}(\mathbf{i}_x + \mathbf{i}_y + \mathbf{i}_z)\text{C-m} \qquad (2\text{-}123)$$

Substituting (2-122) and (2-123) into (2-121), we obtain

$$V = \frac{3}{8\pi\epsilon_0 r} + \frac{5(\mathbf{i}_x + \mathbf{i}_y + \mathbf{i}_z) \cdot \mathbf{r}}{24\pi\epsilon_0 r^3}$$
$$= \frac{3}{8\pi\epsilon_0 r} + \frac{5}{24\pi\epsilon_0 r^2} (\sin\theta\cos\phi + \sin\theta\sin\phi + \cos\theta) \qquad (2\text{-}124)$$

correct to the first two significant terms.   ∎

2.11   Maxwell's Curl Equation for the Static Electric Field

In Section 2.9 we showed that the potential difference between two points A and B, that is, the quantity $\int_A^B \mathbf{E} \cdot d\mathbf{l}$, in the field of a point charge is independent of the path followed from A to B to evaluate it. Suppose we now consider two different paths ACB and ADB in the field of a point charge as shown in Fig. 2.28. We then have

$$\int_{ACB} \mathbf{E} \cdot d\mathbf{l} = \int_{ADB} \mathbf{E} \cdot d\mathbf{l} \qquad (2\text{-}125)$$

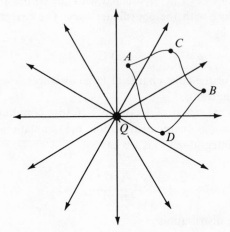

Fig. 2.28. Two different paths between points A and B in the electric field of a point charge.

where

$$\mathbf{E} = \frac{Q}{4\pi\epsilon_0 r^2}\,\mathbf{i}_r$$

Equation (2-125) can be rearranged to read

$$\int_{ACB} \mathbf{E} \cdot d\mathbf{l} - \int_{ADB} \mathbf{E} \cdot d\mathbf{l} = 0$$

or

$$\int_{ACB} \mathbf{E} \cdot d\mathbf{l} + \int_{BDA} \mathbf{E} \cdot d\mathbf{l} = 0 \qquad (2\text{-}126)$$

or

$$\oint_{ACBDA} \mathbf{E} \cdot d\mathbf{l} = 0 \qquad (2\text{-}127)$$

where we have introduced a circle in the integral sign to indicate that the integral is evaluated around a closed path.

If we now have a collection of point charges discrete or continuous, we can apply superposition in the usual manner and arrive at the result that, for any static electric field **E**,

$$\oint \mathbf{E} \cdot d\mathbf{l} = 0 \qquad (2\text{-}128)$$

Equation (2-128) states that the line integral of the electric field intensity vector of any static charge distribution evaluated around a closed path, or the circulation of the static electric field, is equal to zero. Multiplying both sides of (2-128) by a test charge q, we obtain

$$\oint q\mathbf{E} \cdot d\mathbf{l} = 0 \qquad (2\text{-}129)$$

which states that the work involved in moving a test charge around a closed path in a static electric field is equal to zero. If a certain amount of work is done by an external agent during a portion of the closed path, the same amount of work must be done by the field during the remainder of the closed path. It is now evident that (2-129) is simply a statement of conservation of energy, which is so familiar in the case of the gravitational field as the work done in moving a mass around a closed path is zero. Fields which satisfy this property are known as conservative fields. The static electric field is thus a conservative field. In Chapter 4 we will learn that a time-varying electric field does not satisfy this property.

From the definition of the curl of a vector, we have

$$\nabla \times \mathbf{E} = \lim_{\Delta S \to 0} \frac{\oint_{\Delta C} \mathbf{E} \cdot d\mathbf{l}}{\Delta S}\,\mathbf{i}_n \qquad (2\text{-}130)$$

where ΔS is the area bounded by the closed path ΔC and $\mathbf{i}_n$ is the normal vector to the area which should be oriented such that $\oint_{\Delta C} \mathbf{E} \cdot d\mathbf{l}$ is a maximum. However, for the static electric field, $\oint \mathbf{E} \cdot d\mathbf{l} = 0$ for any closed path and hence the right side of (2-130) is identically zero, thus giving

$$\mathbf{V} \times \mathbf{E} = 0 \qquad (2\text{-}131)$$

Equation (2-131) is Maxwell's curl equation for the static electric field. It states that the curl of the static electric field intensity vector is everywhere equal to zero. Fields which satisfy the property of zero curl are known as irrotational fields; that is, such fields cannot rotate the paddle wheel discussed in Section 1.9. Together with Maxwell's divergence equation for the electric field given by (2-82), (2-131) completely defines the properties of the static electric field. Equation (2-131) determines whether or not a given vector field is realizable as a static electric field whereas Eq. (2-82) relates the field to the charge distribution responsible for producing the field. As an alternative approach to that which we followed in this chapter, it is possible to accept these two equations as a starting point and obtain the electric field intensity of a point charge and other charge distributions.

EXAMPLE 2-19. Determine if the following fields are realizable as static electric fields.

(a) $\mathbf{F}_a = -y\mathbf{i}_x + x\mathbf{i}_y$ cartesian coordinates

(b) $\mathbf{F}_b = (p_L/2\pi\epsilon_0 r^2)(\cos\phi\,\mathbf{i}_r + \sin\phi\,\mathbf{i}_\phi)$ cylindrical coordinates

(c) $\mathbf{F}_c = \sin\theta\,\mathbf{i}_r + \cos\theta\,\mathbf{i}_\theta$ spherical coordinates

(a)

$$\mathbf{V} \times \mathbf{F}_a = \begin{vmatrix} \mathbf{i}_x & \mathbf{i}_y & \mathbf{i}_z \\ \dfrac{\partial}{\partial x} & \dfrac{\partial}{\partial y} & \dfrac{\partial}{\partial z} \\ -y & x & 0 \end{vmatrix} \neq 0$$

Hence $\mathbf{F}_a$ cannot be realized as a static electric field.

(b)

$$\mathbf{V} \times \mathbf{F}_b = \begin{vmatrix} \dfrac{\mathbf{i}_r}{r} & \mathbf{i}_\phi & \dfrac{\mathbf{i}_z}{r} \\ \dfrac{\partial}{\partial r} & \dfrac{\partial}{\partial \phi} & \dfrac{\partial}{\partial z} \\ \dfrac{p_L \cos\phi}{2\pi\epsilon_0 r^2} & \dfrac{p_L \sin\phi}{2\pi\epsilon_0 r} & 0 \end{vmatrix} = 0$$

Hence $\mathbf{F}_b$ is realizable as a static electric field. It is left as an exercise (Problem 2.15) for the student to show that $\mathbf{F}_b$ is the field of a two-dimensional electric dipole of moment p_L.

(c)

$$\mathbf{V} \times \mathbf{F}_c = \begin{vmatrix} \dfrac{\mathbf{i}_r}{r^2 \sin\theta} & \dfrac{\mathbf{i}_\theta}{r \sin\theta} & \dfrac{\mathbf{i}_\phi}{r} \\ \dfrac{\partial}{\partial r} & \dfrac{\partial}{\partial \theta} & \dfrac{\partial}{\partial \phi} \\ \sin\theta & r\cos\theta & 0 \end{vmatrix} = 0$$

Hence $\mathbf{F}_c$ can be realized as a static electric field. In fact, if we note that, in cylindrical coordinates, $\mathbf{F}_c = \mathbf{i}_r$, the irrotational nature of $\mathbf{F}_c$ becomes obvious. ∎

2.12   The Relationship Between Electric Field Intensity and Potential

In Section 1.9, we learned that the curl of any vector which can be expressed as the gradient of a scalar is zero. Conversely, if the curl of a vector is equal to zero, the vector can be expressed as the gradient of a scalar. From (2-131), we can say therefore, that the static electric field vector **E** can be expressed as the gradient of a scalar, say, Φ. The question that arises now is: What is this scalar function Φ? For a hint, let us compare the direction of the gradient of the potential V with the direction of **E**. The direction of the gradient of a scalar function at any point is the normal to the surface passing through that point and on which the scalar function has a constant value. Hence the direction of ∇V is normal to the equipotential surfaces. But we found in Section 2.9 that **E** is normal to the equipotential surfaces. Thus the directions of ∇V and **E** at a point have to be either the same or opposite.

To determine which of these is correct and to probe the relationship between **E** and V further, let us consider two equipotential surfaces in a static electric field as shown in Fig. 2.29. Let the potentials on these surfaces

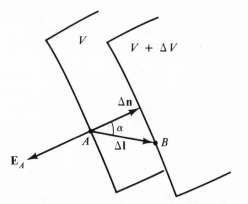

Fig. 2.29. For the determination of the relationship between **E** and V.

be V and $V + \Delta V$, where ΔV is infinitesimal. Since ΔV is infinitesimal, the two surfaces are infinitesimally close so that we can assume that the electric field intensity between the two surfaces in the neighborhood of point A is uniform and equal to the electric field intensity $\mathbf{E}_A$ at point A. We know from previous discussion that $\mathbf{E}_A$ is normal to the equipotential surface V at A. To decide whether $\mathbf{E}_A$ is directed towards the equipotential surface $V + \Delta V$ or away from it, we note that, if a test charge is moved along the direction of **E**, the field does the work; that is, the charge accelerates and hence loses potential energy. This is the same as stating that the charge moves from a higher potential to a lower potential. Thus $\mathbf{E}_A$ is directed away from the equipotential surface $V + \Delta V$ as shown in Fig. 2.29. Now, the potential difference between point A and another point B on the equipotential surface $V + \Delta V$ can be written, using (2-93), as

$$V_{AB} = \int_A^B \mathbf{E} \cdot d\mathbf{l} = \mathbf{E}_A \cdot \Delta\mathbf{l} \tag{2-132}$$

But

$$V_{AB} = V - (V + \Delta V) = -\Delta V \tag{2-133}$$

Also, if $\Delta\mathbf{n}$ is the normal vector from the surface V up to the surface $V + \Delta V$, we have

$$\mathbf{E}_A \cdot \Delta\mathbf{l} = -E_A \,\Delta l \cos\alpha = -E_A \,\Delta n \tag{2-134}$$

Substituting (2-133) and (2-134) into (2-132), we obtain

$$-\Delta V = -E_A \,\Delta n \tag{2-135}$$

or

$$E_A = \frac{\Delta V}{\Delta n} \tag{2-136}$$

and

$$\mathbf{E}_A = -\frac{\Delta V}{\Delta n}\,\mathbf{i}_n \tag{2-137}$$

where $\mathbf{i}_n$ is the unit vector along $\Delta\mathbf{n}$. If we now let Δn tend to zero, $(\Delta V/\Delta n)\mathbf{i}_n$ becomes ∇V. Dropping the subscript A in (2-137), since the same arguments can be applied to any other point in the field, we obtain a relationship between the static electric field intensity vector and the potential at a point as

$$\mathbf{E} = -\nabla V \tag{2-138}$$

Equation (2-138) permits us to compute $\mathbf{E}$ from a knowledge of V using differentiation.

Substituting (2-138) into Maxwell's divergence equation for the electric field, $\nabla \cdot \mathbf{E} = \rho/\epsilon_0$, we have

$$\nabla \cdot (-\nabla V) = \frac{\rho}{\epsilon_0} \tag{2-139}$$

Recalling that $\nabla \cdot \nabla V$ is the Laplacian of V, denoted as $\nabla^2 V$, we see that Eq. (2-139) becomes

$$\nabla^2 V = -\frac{\rho}{\epsilon_0} \tag{2-140}$$

This is known as Poisson's equation. It is a differential equation which relates the potential at a point to the volume charge density at that point. If the volume charge density in a region is zero, then the right side of (2-140) is zero for that region so that (2-140) reduces to

$$\nabla^2 V = 0 \tag{2-141}$$

This is known as Laplace's equation. It states that the Laplacian of the electrostatic potential in a region devoid of charge is equal to zero. We will discuss the solutions of Poisson's and Laplace's equations in Chapter 6.

PROBLEMS

2.1. Find the electric field intensity required to counteract the earth's gravitational force on a charge of q C having a mass m kg. Compute the value of this electric field intensity if the charge is an electron.

2.2. A radial electric field given by

$$\mathbf{E} = \frac{E_0}{r}\mathbf{i}_r$$

where E_0 is a constant exists between two cylindrical surfaces $r = a$ and $r = b$. A test charge q having a mass m enters the electric field region at a radius r_0 with a velocity $\mathbf{v} = v_0\,\mathbf{i}_\phi$. Find the value of E_0 for which the test charge follows a circular orbit of radius r_0.

2.3. An electric field given by

$$\mathbf{E} = E_0\mathbf{i}_y$$

where E_0 is a constant exists in the space between two parallel metallic plates of length L as shown in Fig. 2.30. A small test charge q having a mass m enters the region between the plates at $t = 0$ with a velocity $\mathbf{v} = v_0\,\mathbf{i}_x$ as shown in the figure.

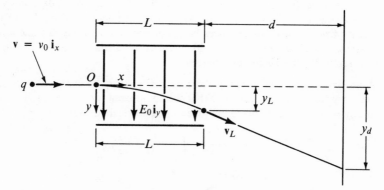

Fig. 2.30. For Problem 2.3.

(a) Show that the path of the test charge between the plates is parabolic.

(b) Find the position y_L along the y direction and velocity $\mathbf{v}_L$ of the test charge just after it emerges from the field region.

(c) Find the deflection y_d undergone by the test charge along the y direction at a distance d from the plates along the x direction.

2.4. Three point charges Q, kQ, and kQ are arranged as shown in Fig. 2.31. Find k in terms of x if a test charge placed at a point on $y = x$ in the plane of the charges is to experience no force. Compute k for $x = 1$ and $x = \frac{1}{4}$.

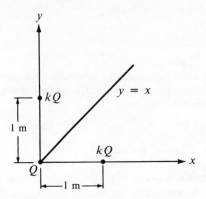

Fig. 2.31. For Problem 2.4.

2.5. Three point charges, each of mass m and charge Q, are suspended by strings of length L from a common point. It is found that the common point and the points occupied by the three charges form the corners of a tetrahedron. Find the relationship between Q, m, L, and the acceleration due to gravity, g.

2.6. Eight point charges, each of value 1 C, are situated at the corners of a cube of edges 2 m with one corner placed at the origin and three edges lying along the coordinate axes. (a) Find the force experienced by each charge. (b) Find the electric field intensity at the point $(2, 2, 2)$. (c) Find the electric field intensity at the point $(0, 0, 2)$.

2.7. Point charges Q, $-2Q$, and Q are located at $(0, 0, d)$, $(0, 0, 0)$, and $(0, 0, -d)$, respectively. Such an arrangement is known as a linear quadrupole. (a) Find the electric field intensity at distances large compared to d along the line joining the charges. (b) Find the electric field intensity at distances large compared to d normal to the line joining the charges.

2.8. A line charge is situated along the z axis. Consider the charge density ρ_L to be arbitrary function of z and show that the components of the electric field intensity at any point in the xy plane are given in cylindrical coordinates by

$$E_r = \frac{r}{4\pi\epsilon_0} \int_{z=-\infty}^{\infty} \frac{\rho_L \, dz}{(r^2 + z^2)^{3/2}}$$

$$E_\phi = 0$$

$$E_z = -\frac{1}{4\pi\epsilon_0} \int_{z=-\infty}^{\infty} \frac{\rho_L z \, dz}{(r^2 + z^2)^{3/2}}$$

Evaluate the field components for the following charge distributions:

(a) $\rho_L = \rho_{L0}$ $-\infty < z < \infty$
(b) $\rho_L = \rho_{L0}$ $-z_0 < z < z_0$
(c) $\rho_L = |z|$ $-z_0 < z < z_0$
(d) $\rho_L = z$ $-z_0 < z < z_0$

where ρ_{L0} is a constant. Discuss your results from considerations of symmetry. Verify your results by considering limiting cases wherever appropriate.

2.9. A ring charge of radius a is situated in the xy plane with its center at the origin. Consider the charge density ρ_L to be an arbitrary function of ϕ and show that the components of the electric field intensity at a point $(0, 0, z)$ are given in cartesian coordinates by

$$E_x = \frac{-a^2}{4\pi\epsilon_0(a^2 + z^2)^{3/2}} \int_{\phi=0}^{2\pi} \rho_L \cos\phi \, d\phi$$

$$E_y = \frac{-a^2}{4\pi\epsilon_0(a^2 + z^2)^{3/2}} \int_{\phi=0}^{2\pi} \rho_L \sin\phi \, d\phi$$

$$E_z = \frac{az}{4\pi\epsilon_0(a^2 + z^2)^{3/2}} \int_{\phi=0}^{2\pi} \rho_L \, d\phi$$

Evaluate the field components for the following charge distributions:

(a) $\rho_L = \rho_{L0}$ $0 < \phi < 2\pi$

(b) $\rho_L = \begin{cases} \rho_{L0} & 0 < \phi < \pi \\ -\rho_{L0} & \pi < \phi < 2\pi \end{cases}$

(c) $\rho_L = \rho_{L0} \cos\phi$ $0 < \phi < 2\pi$

(d) $\rho_L = \rho_{L0} \sin\phi$ $0 < \phi < 2\pi$

where ρ_{L0} is a constant. Discuss your results from considerations of symmetry. Verify your results by considering limiting cases wherever appropriate.

2.10. A sheet charge is situated in the xy plane. Consider the charge density ρ_s to be an arbitrary function of r and ϕ and show that the components of the electric field intensity at a point $(0, 0, z)$ are given by

$$E_x = -\frac{1}{4\pi\epsilon_0} \int_{r=0}^{\infty} \int_{\phi=0}^{2\pi} \frac{\rho_s r^2 \cos\phi \, dr \, d\phi}{(r^2 + z^2)^{3/2}}$$

$$E_y = -\frac{1}{4\pi\epsilon_0} \int_{r=0}^{\infty} \int_{\phi=0}^{2\pi} \frac{\rho_s r^2 \sin\phi \, dr \, d\phi}{(r^2 + z^2)^{3/2}}$$

$$E_z = \frac{z}{4\pi\epsilon_0} \int_{r=0}^{\infty} \int_{\phi=0}^{2\pi} \frac{\rho_s r \, dr \, d\phi}{(r^2 + z^2)^{3/2}}$$

Evaluate the field components for the following charge distributions:

(a) $\rho_s = \rho_{s0}$ $0 < r < \infty, 0 < \phi < 2\pi$

(b) $\rho_s = \begin{cases} \rho_{s0} & 0 < r < r_0, 0 < \phi < 2\pi \\ 0 & r_0 < r < \infty, 0 < \phi < 2\pi \end{cases}$

(c) $\rho_s = \begin{cases} 0 & 0 < r < r_0, 0 < \phi < 2\pi \\ \rho_{s0} & r_0 < r < \infty, 0 < \phi < 2\pi \end{cases}$

(d) $\rho_s = \dfrac{\rho_{s0} \cos\phi}{r}$ $0 < r < \infty, 0 < \phi < 2\pi$

(e) $\rho_s = \dfrac{\rho_{s0} \sin\phi}{r}$ $0 < r < \infty, 0 < \phi < 2\pi$

where ρ_{s0} is a constant. Discuss your results from considerations of symmetry. Verify your results by considering limiting cases wherever appropriate.

2.11.   A surface charge is distributed over a spherical surface of radius a and centered at the origin. Consider the charge density ρ_s to be uniform in ϕ but not necessarily in θ and show that the electric field intensity at a point $(0, 0, z)$ has only a z-component given by

$$E_z = \frac{a^2}{2\epsilon_0} \int_{\theta=0}^{\pi} \frac{\rho_s(z - a\cos\theta)\sin\theta \, d\theta}{(a^2 + z^2 - 2az\cos\theta)^{3/2}}$$

Evaluate E_z both for $|z| < a$ and for $|z| > a$ for the following charge distributions:

(a) $\rho_s = \rho_{s0}$ $\qquad\qquad 0 < \theta < \pi$

(b) $\rho_s = \rho_{s0}\cos\theta$ $\qquad 0 < \theta < \pi$

where ρ_{s0} is a constant.

2.12.   A volume charge is distributed throughout an infinite slab of thickness $2a$ symmetrically placed about the xy plane. Consider the charge density ρ to be uniform in x and y but not necessarily in z and show that the electric field intensity at any point (x, y, z) has only a z component given by

$$E_z = \begin{cases} \dfrac{1}{2\epsilon_0} \displaystyle\int_{z=-a}^{a} \rho \, dz & z > a \\[3mm] \dfrac{1}{2\epsilon_0}\left(\displaystyle\int_{z=-a}^{a} \rho \, dz - \int_{z=z}^{a} \rho \, dz \right) & -a < z < a \\[3mm] -\dfrac{1}{2\epsilon_0} \displaystyle\int_{z=-a}^{a} \rho \, dz & z < a \end{cases}$$

Evaluate E_z as a function of z for $-\infty < z < \infty$ for the following charge distributions:

(a) $\rho = \rho_0$ $\qquad\qquad -a < z < a$

(b) $\rho = \begin{cases} \rho_0 & 0 < z < a \\ -\rho_0 & -a < z < 0 \end{cases}$

(c) $\rho = |z|$ $\qquad\qquad -a < z < a$

(d) $\rho = z$ $\qquad\qquad -a < z < a$

where ρ_0 is a constant. Discuss your results from considerations of symmetry.

2.13.   A volume charge is distributed with uniform density ρ_0 C/m³ throughout an infinitely long cylinder of radius a m. Obtain the electric field intensity at points both inside and outside the cylinder by dividing the cylindrical charge into several infinitesimal parts each of which can be considered as a point charge.

2.14.   A small hole is drilled through the center of the spherical volume charge of Example 2-6., as shown in Fig. 2.32. The size of the hole is negligible compared to the size of the sphere. A point charge $q(< 0)$ is placed at one end of the hole and released from rest at $t = 0$. Assume that the magnitude of q is very small compared to the total charge Q (> 0) contained in the sphere. (a) Derive the equation of motion of the point charge. (b) Solve the equation for the position and velocity of the point charge as functions of time. (c) What is the frequency of oscillation of the point charge?

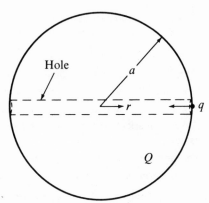

Fig. 2.32. For Problem 2.14.

2.15. Two infinitely long line charges of uniform but opposite densities ρ_{L0} and $-\rho_{L0}$ are situated parallel to the z axis and passing through $(d/2, 0, 0)$ and $(-d/2, 0, 0)$, respectively. The arrangement is known as a two-dimensional electric dipole, in contrast to the three-dimensional electric dipole made up of two equal but opposite point charges. (a) Obtain the electric field intensity due to the two-dimensional electric dipole in the limit that $d \longrightarrow 0$, keeping the dipole moment $\rho_{L0}d$ constant. (b) Find and sketch the direction lines.

2.16. Two infinitely long line charges of uniform densities ρ_{L1} and ρ_{L2}, respectively, are situated parallel to each other at a distance d apart. Show that the equation for the direction lines of **E** is

$$\alpha_1 \rho_{L1} + \alpha_2 \rho_{L2} = \text{constant}$$

in the plane normal to the line charges, where α_1 and α_2 are the angles made by the lines drawn from any point P to the line charges with the line joining the charges as shown in Fig. 2.33. Obtain and sketch the direction lines for the following cases:

(a) $\rho_{L1} = \rho_{L2} = \rho_{L0}$
(b) $\rho_{L1} = \rho_{L0}, \rho_{L2} = -\rho_{L0}$ (two-dimensional dipole)

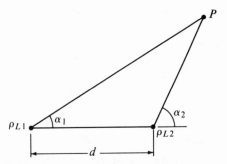

Fig. 2.33. For Problem 2.16.

2.17. Obtain the electric field intensity of a finitely long line charge of uniform density ρ_{L0} and length $2a$ at an arbitrary point. Show that the direction lines are hyperbolas with the ends of the line charge as their foci.

2.18. Carry out the mathematical proof to show that the net electric field flux emanating from an arbitrary surface not enclosing a point charge is zero.

2.19. Find the solid angle subtended by

(a) One face of a regular tetrahedron at the center of the tetrahedron

(b) One face of a cube at the center of the cube

(c) One face of a cube at one of the corners of the opposite face

(d) A hemispherical surface at a point on the base of the hemisphere other than its center

(e) The first quadrant of the xy plane at a point on the z axis

(f) The portion of any plane in the first octant at the origin.

2.20. An infinitely long line charge of uniform density ρ_{L0} C/m is situated along the z axis. Find the electric field flux cutting the portion of the plane $x + y = 1$ m lying in the first octant and bounded by the planes $z = 0$ and $z = 1$ m by evaluating $\int \mathbf{E} \cdot d\mathbf{S}$. Check your answer from considerations of symmetry of electric field flux emanating from the line charge.

2.21. A point charge Q C is located at the origin. Find the electric field flux cutting the portion of the plane $x + y = 1$ m lying in the first octant by evaluating $\int \mathbf{E} \cdot d\mathbf{S}$. Check your answer from considerations of symmetry of electric field flux emanating from the line charge.

2.22. Charges are located, in cartesian coordinates, as follows: (a) point charge, 1 C, at $(0.23, 0.73, 0)$; (b) infinitely long line charge of uniform density 1 C/m parallel to the z axis and passing through $(0.6, 0, 0)$; and (c) an infinite sheet charge of uniform density 1 C/m^2 in the $z = 0.5$ plane. Determine the total electric field flux cutting the upper half of the spherical surface of radius unity and centered at the origin.

2.23. Using Gauss' law in integral form, obtain the electric fields due to the following volume charge distributions, in cartesian coordinates:

(a) $\rho = \begin{cases} \rho_0 & |z| < a \\ 0 & |z| > a \end{cases}$

(b) $\rho = \begin{cases} \rho_0 & 0 < z < a \\ -\rho_0 & -a < z < 0 \end{cases}$

(c) $\rho = \begin{cases} |z| & |z| < a \\ 0 & |z| > a \end{cases}$

(d) $\rho = \begin{cases} z & |z| < a \\ 0 & |z| > a \end{cases}$

(e) $\rho = \begin{cases} a - |z| & |z| < a \\ 0 & |z| > a \end{cases}$

where ρ_0 is a constant.

2.24. Using Gauss' law in integral form, obtain the electric fields due to the following volume charge distributions, in cylindrical coordinates:

(a) $\rho = \begin{cases} \rho_0 & 0 < r < a \\ 0 & a < r < \infty \end{cases}$

(b) $\rho = \begin{cases} 0 & 0 < r < a \\ \rho_0 & a < r < b \\ 0 & b < r < \infty \end{cases}$

(c) $\rho = \begin{cases} \rho_0 \dfrac{r}{a} & 0 < r < a \\ 0 & a < r < \infty \end{cases}$

where ρ_0 is a constant.

2.25. Using Gauss' law in integral form, obtain the electric fields due to the following volume charge distributions, in spherical coordinates:

(a) $\rho = \begin{cases} 0 & 0 < r < a \\ \rho_0 & a < r < b \\ 0 & b < r < \infty \end{cases}$

(b) $\rho = \begin{cases} \rho_0 \dfrac{r}{a} & 0 < r < a \\ 0 & a < r < \infty \end{cases}$

(c) $\rho = \begin{cases} \rho_0\left(1 - \dfrac{r^2}{a^2}\right) & 0 < r < a \\ 0 & a < r < \infty \end{cases}$

where ρ_0 is a constant.

2.26. Using Gauss' law in integral form, obtain the electric fields due to the following surface charge distributions:

(a) $\rho_s = \begin{cases} \rho_{s0} & z = a \\ -\rho_{s0} & z = -a \end{cases}$ cartesian coordinates

(b) $\rho_s = \rho_{s0} \qquad r = a$ cylindrical coordinates

(c) $\rho_s = \begin{cases} \rho_{s0} & r = a \\ -\rho_{s0}\dfrac{a}{b} & r = b \end{cases}$ cylindrical coordinates

(d) $\rho_s = \rho_{s0} \qquad r = a$ spherical coordinates

(e) $\rho_s = \begin{cases} \rho_{s0} & r = a \\ -\rho_{s0}\dfrac{a^2}{b^2} & r = b \end{cases}$ spherical coordinates

where ρ_{s0} in a constant.

2.27. Volume charge of uniform density ρ_0 C/m³ is distributed in the region between two infinitely long, parallel cylindrical surfaces of radii a and b $(< a)$ and with their axes separated by distance c $(< a - b)$ as shown in Fig. 2.34. Find the electric field intensity in the charge-free region inside the cylindrical surface of radius b.

2.28. Verify your answers to Problem 2.23 by using Gauss' law in differential form.

2.29. Verify your answers to Problem 2.24 by using Gauss' law in differential form.

2.30. Verify your answers to Problem 2.25 by using Gauss' law in differential form.

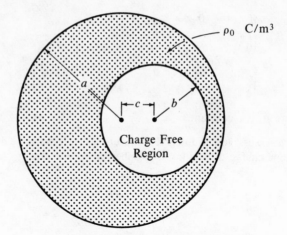

Fig. 2.34. For Problem 2.27.

2.31. For each of the following electric fields, find the charge distribution which produces the field, using Gauss' law in differential form:

(a) $\mathbf{E} = \begin{cases} -\dfrac{\rho_{s0}}{\epsilon_0}\mathbf{i}_z & -\infty < z < 0 \\[2mm] -\dfrac{\rho_{s0}}{3\epsilon_0}\mathbf{i}_z & 0 < z < a \\[2mm] \dfrac{\rho_{s0}}{\epsilon_0}\mathbf{i}_z & a < z < \infty \end{cases}$ cartesian coordinates

(b) $\mathbf{E} = \dfrac{1 - e^r}{\epsilon_0 r}\mathbf{i}_r \qquad 0 < r < \infty$ cylindrical coordinates

(c) $\mathbf{E} = \begin{cases} 0 & 0 < r < a \\[2mm] \dfrac{Q}{4\pi\epsilon_0 r^2}\mathbf{i}_r & a < r < b \\[2mm] 0 & b < r < \infty \end{cases}$ spherical coordinates

where ρ_{s0} and Q are constants.

2.32. A surface charge of density ρ_s C/m² occupies the spherical surface of radius r_0 and centered at the origin. Show that

$$\nabla \cdot \mathbf{E} = \frac{1}{\epsilon_0}\,\rho_s\,\delta(r - r_0)$$

2.33. An infinitely long line charge of density ρ_{L0} C/m is situated parallel to the z axis and passes through the point (r_0, ϕ_0) in the $z = 0$ plane. Show that

$$\nabla \cdot \mathbf{E} = \frac{1}{\epsilon_0}\,\rho_{L0}\,\frac{\delta(r - r_0)\,\delta(\phi - \phi_0)}{r_0}$$

where

$$\int_{r=0}^{\infty}\int_{\phi=0}^{2\pi} f(r, \phi)\,\frac{\delta(r - r_0)\,\delta(\phi - \phi_0)}{r_0}\,r\,dr\,d\phi = f(r_0, \phi_0)$$

2.34. A point charge Q C is located at the point (r_0, θ_0, ϕ_0). Show that

$$\mathbf{V} \cdot \mathbf{E} = \frac{1}{\epsilon_0} Q \frac{\delta(r - r_0)\, \delta(\theta - \theta_0)\, \delta(\phi - \phi_0)}{r_0^2 \sin \theta_0}$$

where

$$\int_{r=0}^{\infty} \int_{\theta=0}^{\pi} \int_{\phi=0}^{2\pi} f(r, \theta, \phi) \frac{\delta(r - r_0)\, \delta(\theta - \theta_0)\, \delta(\phi - \phi_0)}{r_0^2 \sin \theta_0} r^2 \sin \theta\, dr\, d\theta\, d\phi$$
$$= f(r_0, \theta_0, \phi_0)$$

2.35. The electric field intensity is given in cylindrical coordinates by

$$\mathbf{E} = \frac{\cos \phi}{r^2} \mathbf{i}_r + \frac{\sin \phi}{r^2} \mathbf{i}_\phi$$

Find the work associated with the movement of a test charge from the point $(1, 0, -22.7)$ to the point $(0.5, \pi/2, 43.8)$. Is this work done by the field or by an external agent?

2.36. The electric field intensity is given, in cartesian coordinates, by

$$\mathbf{E} = \begin{cases} 0 & -\infty < x < 0 \\ 2x\mathbf{i}_x & 0 < x < 1 \\ \dfrac{2}{x^2}\mathbf{i}_x & 1 < x < \infty \end{cases}$$

Obtain and draw a graph of the potential difference between $x = 1$ and an arbitrary value of x.

2.37. For the three-dimensional electric dipole, show that the equipotential surfaces, $r^2 \sec \theta = $ constant, are orthogonal to the direction lines of $\mathbf{E}$ given by the intersections of $r \operatorname{cosec}^2 \theta = $ constant and $\phi = $ constant.

2.38. For the linear quadrupole consisting of an arrangement of point charges Q, $-2Q$, and Q at $(0, 0, d)$, $(0, 0, 0)$, and $(0, 0, -d)$, respectively, obtain the expression for the potential at distances large compared to d.

2.39. For the rectangular quadrupole consisting of an arrangement of four point charges as shown in Fig. 2.35, obtain the potential at distances large compared to the dimensions of the quadrupole.

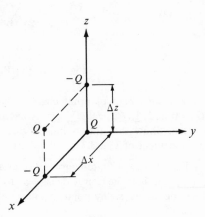

Fig. 2.35. For Problem 2.39.

2.40. For the arrangement of point charges shown in Fig. 2.36, obtain the expression for the potential at distances large compared to d.

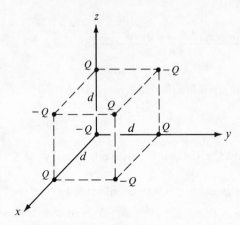

Fig. 2.36. For Problem 2.40.

2.41. For each of the arrangements of point charges shown in Fig. 2.37, find the first two significant terms in the potential at large distances from the origin.

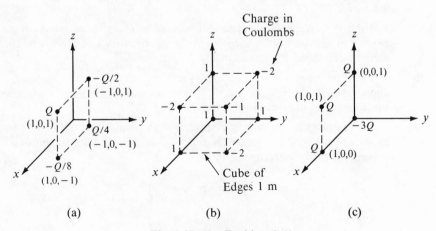

(a)   (b)   (c)

Fig. 2.37. For Problem 2.41.

2.42. For the arrangement of point charges shown in Fig. 2.37(c), $\sum Q = 0$. When $\sum Q = 0$ and $\sum Q\mathbf{r}' \neq 0$, $\sum Q\mathbf{r}'$ is independent of the point about which it is computed. Show that this is indeed true by computing the dipole moment for the arrangement of Fig. 2.37(c) about an arbitrary point (x, y, z).

2.43. For a line charge of finite length situated along the z axis between $z = -z_0$ and $z = +z_0$, consider the charge density ρ_L to be an arbitrary function of z and show that the potential at any point in the xy plane at a distance r from the origin is given by

$$V = \frac{1}{4\pi\epsilon_0} \int_{z=-z_0}^{z_0} \frac{\rho_L \, dz}{\sqrt{r^2 + z^2}}$$

Evaluate the integral for the following charge distributions:

(a) $\rho_L = \rho_{L0}$, a constant $-z_0 < z < z_0$

(b) $\rho_L = |z|$ $-z_0 < z < z_0$

(c) $\rho_L = z$ $-z_0 < z < z_0$

Discuss your results from considerations of symmetry. Verify your results by considering limiting cases wherever appropriate.

2.44. For the ring charge of Problem 2.9, show that the potential at a point $(0, 0, z)$ is given by

$$V = \frac{a}{4\pi\epsilon_0 (a^2 + z^2)^{1/2}} \int_{\phi=0}^{2\pi} \rho_L \, d\phi$$

Evaluate V for the charge distributions specified in Problem 2.9, and discuss the results from considerations of symmetry. Verify your results by considering limiting cases wherever appropriate.

2.45. For the sheet charge of Problem 2.10, show that the potential at a point $(0, 0, z)$ is given by

$$V = \frac{1}{4\pi\epsilon_0} \int_{r=0}^{\infty} \int_{\phi=0}^{2\pi} \left[\frac{\rho_s r \, dr \, d\phi}{(r^2 + z^2)^{1/2}} - \frac{\rho_s r \, dr \, d\phi}{(r^2 + z_0^2)^{1/2}} \right]$$

where $(0, 0, z_0)$ is the reference point for zero potential. Evaluate V for the charge distributions specified in Problem 2.10, and discuss the results from considerations of symmetry. Verify your results by considering limiting cases wherever appropriate.

2.46. For the surface charge of Problem 2.11, show that the potential at a point $(0, 0, z)$ is given by

$$V = \frac{a^2}{2\epsilon_0} \int_{\theta=0}^{\pi} \frac{\rho_s \sin \theta \, d\theta}{(a^2 + z^2 - 2az \cos \theta)^{1/2}}$$

Evaluate V both for $|z| < a$ and for $|z| > a$ for the charge distributions specified in Problem 2.11 and discuss your results from considerations of symmetry.

2.47. Obtain the potential field of a finitely long line charge of uniform density ρ_{L0} and length $2a$ at an arbitrary point. Show that the equipotential surfaces are ellipsoids with the ends of the line as their foci. Establish their orthogonality with the direction lines deduced in Problem 2.17.

2.48. For the two-dimensional electric dipole of Problem 2.15, (a) obtain the potential field and (b) show that the equipotential surfaces are orthogonal to the direction lines deduced in Problem 2.15.

2.49. A volume charge is distributed throughout a sphere of radius a, and centered at the origin, with uniform density ρ_0 C/m³. Find the potential field of the volume charge distribution.

2.50. For the volume charge distributions specified in Problem 2.23, obtain the potential fields by evaluating $\int \mathbf{E} \cdot d\mathbf{l}$.

2.51. For the volume charge distributions specified in Problem 2.24, obtain the potential fields by evaluating $\int \mathbf{E} \cdot d\mathbf{l}$.

2.52. For the volume charge distributions specified in Problem 2.25, obtain the potential fields by evaluating $\int \mathbf{E} \cdot d\mathbf{l}$.

2.53. For the following surface charge distributions, obtain the potential fields:

(a) $\rho_s = \begin{cases} \rho_{s0} & z = a \\ -\rho_{s0} & z = -a \end{cases}$ cartesian coordinates

(b) $\rho_s = \begin{cases} \rho_{s0} & r = a \\ -\rho_{s0}\dfrac{a}{b} & r = b \end{cases}$ cylindrical coordinates

(c) $\rho_s = \begin{cases} \rho_{s0} & r = a \\ -\rho_{s0}\dfrac{a^2}{b^2} & r = b \end{cases}$ spherical coordinates

where ρ_{s0} is a constant.

2.54. A volume charge is distributed with uniform density ρ_0 C/m³ in the portion of a sphere of radius a centered at the origin and lying in the first octant. Find the potential field at large distances from the charge distribution correct to the first two significant terms.

2.55. A ring charge of radius a is situated in the xy plane with its center at the origin. Find the dipole moments about the origin for the following charge densities, where ρ_{L0} is constant:

(a) $\rho_L = \rho_{L0} \cos \phi$ $0 < \phi < 2\pi$
(b) $\rho_L = \rho_{L0} \sin 2\phi$ $0 < \phi < 2\pi$
(c) $\rho_L = \rho_{L0} \phi \sin \phi$ $0 < \phi < 2\pi$

What are the dipole moments for cases (a), and (b) about any point other than the origin? Explain.

2.56. Determine if the following fields are realizable as static electric fields:

(a) $\mathbf{A} = \dfrac{1}{y^2}(y\mathbf{i}_x - x\mathbf{i}_y)$ cartesian coordinates

(b) $\mathbf{B} = \dfrac{1}{r}\mathbf{i}_\phi$ cylindrical coordinates

(c) $\mathbf{C} = \left(1 + \dfrac{1}{r^2}\right)\cos\phi\,\mathbf{i}_r - \left(1 - \dfrac{1}{r^2}\right)\sin\phi\,\mathbf{i}_\phi$ cylindrical coordinates

(d) $\mathbf{D} = \left(1 + \dfrac{2}{r^3}\right)\cos\theta\,\mathbf{i}_r - \left(1 - \dfrac{1}{r^3}\right)\sin\theta\,\mathbf{i}_\theta$ spherical coordinates

2.57. Check that $\mathbf{E} = -\nabla V$ by substituting independently obtained expressions for $\mathbf{E}$ and V for the following charge distributions:

(a) An infinitely long line charge of uniform density.
(b) A finitely long line charge of uniform density.
(c) A three-dimensional dipole.
(d) A two-dimensional dipole.
(e) A spherical surface charge of radius a and uniform density ρ_{s0}.
(f) A spherical volume charge of radius a and uniform density ρ_0.

2.58. In shorthand notation, the three-dimensional Dirac delta function situated at the origin is written as $\delta(\mathbf{r})$, and is defined as

$$\delta(\mathbf{r}) = \lim_{\substack{r_0 \to 0 \\ \theta_0 \to 0 \\ \phi_0 \to 0}} \frac{\delta(r - r_0)\,\delta(\theta - \theta_0)\,\delta(\phi - \phi_0)}{r_0^2 \sin^2 \theta_0}$$

$$\int_V \delta(\mathbf{r})\,dv = \begin{cases} 1 & \text{if the volume } V \text{ contains the origin} \\ 0 & \text{if the volume } V \text{ does not contain the origin} \end{cases}$$

By performing volume integration of $\nabla^2 (1/r) = \nabla \cdot \nabla(1/r)$ throughout a sphere of radius a and centered at the origin and then letting $a \to 0$, show that

$$\nabla^2 \left(\frac{1}{r} \right) = -4\pi\,\delta(\mathbf{r})$$

Hence, show that the potential field of a point charge Q located at the origin is $Q/4\pi\epsilon_0 r$.

3

THE STATIC MAGNETIC FIELD

In Chapter 2 we introduced the electric field as a force field associated with a region of space in which charges at rest experience forces. In this chapter we introduce a second kind of force field, known as the magnetic field and associated with a region in which charges in motion experience forces. These forces experienced by moving charges are in addition to any electric forces experienced by them by virtue of an electric field in the region. Just as we were concerned only with the static electric field in free space in Chapter 2, we are in this chapter concerned only with the static magnetic field in free space. We know that the motion of charges constitutes a current. Currents are, however, classified into different categories according to how they are produced. Currents arising from movement of charges such as space charges in vacuum tubes and electron beams in cathode-ray tubes are called convection currents. Two other types of current known as conduction and polarization currents result from different effects on charges in material media under the influence of electric fields, as we will learn in Chapter 5. Yet another type of current is the magnetization current which results from magnetic effects in materials, as we will learn also in Chapter 5. For the purposes of this chapter, it is not necessary to distinguish between them because they are all basically equivalent to rate of flow of charges with time in free space. Thus the laws which we will learn in this chapter can be applied equally well to all of these currents.

134

3.1 The Magnetic Field Concept

In Section 2.1 we learned that if, in a region of space, a fixed test charge q experiences a force $\mathbf{F}$, then the region is characterized by an electric field of intensity $\mathbf{E}$ given by

$$\mathbf{E} = \frac{\mathbf{F}}{q} \qquad (2\text{-}3)$$

Here we introduce the concept of magnetic field by considering a test charge moving in a region of space. If the test charge q moving with a velocity $\mathbf{v}$ experiences a force $\mathbf{F}$, then the region is said to be characterized by a magnetic field, which we will represent by the symbol $\mathbf{B}$. This force $\mathbf{F}$ is related to q, $\mathbf{v}$, and $\mathbf{B}$ as given by

$$\mathbf{F} = q\mathbf{v} \times \mathbf{B} \qquad (3\text{-}1)$$

According to (3-1), the force experienced by the moving charge due to the magnetic field is directed normal to both $\mathbf{v}$ and $\mathbf{B}$, as shown in Fig. 3.1, in

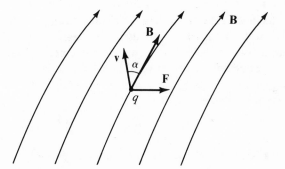

Fig. 3.1. Force experienced by a test charge moving with a velocity $\mathbf{v}$ in a magnetic field $\mathbf{B}$.

contrast to the same directions of electric force and electric field intensity. The magnitude of the force is equal to $qvB \sin \alpha$, where α is the angle between $\mathbf{v}$ and $\mathbf{B}$. Since the force is always normal to $\mathbf{v}$, there is no acceleration along the direction of motion. Thus, the magnetic field changes only the direction of motion of the charge and does not alter the kinetic energy associated with it.

From Eq. (3-1), we note that if the test charge moves in, or opposite to, the direction of $\mathbf{B}$, it does not experience a force. Also, rewriting Eq. (3-1) as

$$\mathbf{F} = qvB\,\mathbf{i}_v \times \mathbf{i}_B = qvB \sin \alpha\, \mathbf{i}_F \qquad (3\text{-}2)$$

where $\mathbf{i}_v$, $\mathbf{i}_B$, and $\mathbf{i}_F$ are unit vectors along $\mathbf{v}$, $\mathbf{B}$, and $\mathbf{F}$, respectively, we observe that it is only possible to deduce $B \sin \alpha$ by knowing the force for only one direction of motion of the test charge. On the other hand, if we know two nonzero forces $\mathbf{F}_1$ and $\mathbf{F}_2$ for two velocities $\mathbf{v}_1$ and $\mathbf{v}_2$ in different directions, then we have

$$\begin{aligned}
\mathbf{F}_1 \times \mathbf{F}_2 &= (q\mathbf{v}_1 \times \mathbf{B}) \times (q\mathbf{v}_2 \times \mathbf{B}) \\
&= q^2[(\mathbf{v}_1 \times \mathbf{B} \cdot \mathbf{B})\mathbf{v}_2 - (\mathbf{v}_1 \times \mathbf{B} \cdot \mathbf{v}_2)\mathbf{B}] \\
&= -q(\mathbf{F}_1 \cdot \mathbf{v}_2)\mathbf{B}
\end{aligned} \tag{3-3}$$

or

$$\mathbf{B} = \frac{\mathbf{F}_2 \times \mathbf{F}_1}{q(\mathbf{F}_1 \cdot \mathbf{v}_2)} \tag{3-4}$$

Alternatively, we note from (3-1) or (3-2) that the force is maximum for $\mathbf{v}$ normal to $\mathbf{B}$ so that if we find a maximum force $\mathbf{F}_m$ by trying several directions of $\mathbf{v}$, keeping its magnitude constant, then

$$\mathbf{B} = \frac{\mathbf{F}_m \times \mathbf{i}_m}{qv} \tag{3-5}$$

where $\mathbf{i}_m$ is the direction of $\mathbf{v}$ for which the force is $\mathbf{F}_m$.

As in the case of defining the electric field, we assume that the movement of the test charge does not alter the magnetic field in which it is placed. From a practical point of view, the movement of the charge does influence the magnetic field irrespective of how small it is and how slowly it is moved. However, theoretically, we can define $\mathbf{B}$ as the right side of (3-5) in the limit that qv tends to zero; that is,

$$\mathbf{B} = \lim_{qv \to 0} \frac{\mathbf{F}_m \times \mathbf{i}_m}{qv} \tag{3-6}$$

From (3-5), we observe that the units of $\mathbf{B}$ are

$$\frac{\text{newtons per coulomb}}{\text{meters per second}} = \frac{\text{newton-seconds}}{\text{coulomb-meter}} = \frac{\text{newton-meter}}{\text{coulomb}} \times \frac{\text{seconds}}{(\text{meter})^2}$$

Recalling that newton-meter per coulomb is a volt, we can write these units as volt-seconds per square meter, commonly known as webers per square meter, and abbreviated Wb/m², giving the character of a flux density for $\mathbf{B}$. Accordingly, $\mathbf{B}$ is known as the magnetic flux density vector.

EXAMPLE 3-1.   An electron moving with a velocity $\mathbf{v}_1 = \mathbf{i}_x$ m/sec at a point in a magnetic field experiences a force $\mathbf{F}_1 = e(-\mathbf{i}_y + \mathbf{i}_z)$ N, where e is the charge of the electron. If the electron is moving with a velocity $\mathbf{v}_2 = \mathbf{i}_y$ m/sec at the same point, it experiences a force $\mathbf{F}_2 = e(\mathbf{i}_x - \mathbf{i}_z)$ N. Find $\mathbf{B}$ at that point.

Using (3-4), we have

$$\begin{aligned}
\mathbf{B} = \frac{\mathbf{F}_2 \times \mathbf{F}_1}{q(\mathbf{F}_1 \cdot \mathbf{v}_2)} &= \frac{e(\mathbf{i}_x - \mathbf{i}_z) \times e(-\mathbf{i}_y + \mathbf{i}_z)}{e[e(-\mathbf{i}_y + \mathbf{i}_z) \cdot \mathbf{i}_y]} \\
&= \frac{e^2(-\mathbf{i}_x - \mathbf{i}_y - \mathbf{i}_z)}{-e^2} \\
&= (\mathbf{i}_x + \mathbf{i}_y + \mathbf{i}_z) \text{ Wb/m}^2 \quad \blacksquare
\end{aligned}$$

3.2 Force on a Current Element

In Section 3.1 we defined the magnetic field in terms of force experienced by a moving test charge, involving explicitly the charge and its velocity. This form of the definition, given by Eq. (3-1) is, however, not convenient for use with currents. Hence it is necessary to formulate Eq. (3-1) in terms of current. The current crossing a surface is defined as the rate at which charge flows across the surface; that is,

$$I = \frac{dQ}{dt} \tag{3-7}$$

Let us now consider a region in which charges distributed with a density ρ are moving with a velocity $\mathbf{v}$, where ρ and $\mathbf{v}$ can, in general, be nonuniform. At a point P in this region, let us consider an infinitesimal area dS normal to the direction of flow of charges as shown in Fig. 3.2. In a time dt, the

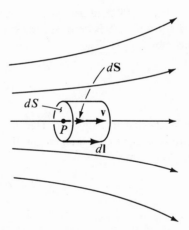

Fig. 3.2. Volume occupied by charge crossing a surface dS normal to it with a velocity $\mathbf{v}$, in time $dt = dl/v$.

distance traveled by the charge crossing this surface is equal to $\mathbf{v}\,dt$. Let $\mathbf{v}\,dt$ be equal to $d\mathbf{l}$, as shown in Fig. 3.2, so that the charge dQ crossing the surface dS in time dt is that contained in the infinitesimal volume $(d\mathbf{l} \cdot d\mathbf{S})$. The current crossing the surface dS is then given by

$$I = \frac{dQ}{dt} = \frac{dQ}{dl}v \tag{3-8}$$

But the current crossing the surface is also equal to $\mathbf{J} \cdot d\mathbf{S}$, where $\mathbf{J}$ is the current density at P. Since $\mathbf{J}$ and $d\mathbf{S}$ are in the same direction, $\mathbf{J} \cdot d\mathbf{S} = J\,dS$. Thus

$$J = \frac{I}{dS} = \frac{dQ}{(dl)(dS)}v = \rho v \tag{3-9}$$

and

$$\mathbf{J} = \rho \mathbf{v} \tag{3-10}$$

Now, the force experienced by the charge dQ moving with the velocity $\mathbf{v}$ is given by

$$\begin{aligned}
d\mathbf{F} &= dQ\,\mathbf{v} \times \mathbf{B} \\
&= \rho(dl)(dS)\mathbf{v} \times \mathbf{B} \\
&= (d\mathbf{l} \cdot d\mathbf{S})\rho\mathbf{v} \times \mathbf{B} \\
&= \mathbf{J} \times \mathbf{B}\,d(\text{vol})
\end{aligned} \tag{3-11}$$

where $d(\text{vol})$ is the differential volume $(d\mathbf{l} \cdot d\mathbf{S})$. Thus the magnetic force experienced by the charges in a differential volume in a region of current is given by (3-11). To obtain the total force experienced in a large volume, we need to integrate the right side of (3-11) throughout the volume under consideration; that is,

$$\mathbf{F} = \int_{\text{vol}} \mathbf{J} \times \mathbf{B}\,d(\text{vol}) \tag{3-12}$$

For a filamentary wire carrying current I, the current density J is infinity since dS is zero but the product $\mathbf{J} \cdot d\mathbf{S}$ is equal to I so that (3-11) becomes

$$\begin{aligned}
d\mathbf{F} &= (dl)(dS)\,\mathbf{J} \times \mathbf{B} \\
&= (\mathbf{J} \cdot d\mathbf{S})d\mathbf{l} \times \mathbf{B} \\
&= I\,d\mathbf{l} \times \mathbf{B}
\end{aligned} \tag{3-13}$$

as illustrated in Fig. 3.3. The total force experienced by the filamentary wire is obtained by integrating the right side of (3-13) along the length of the wire. Thus

$$\mathbf{F} = \int_{\text{wire}} (I\,d\mathbf{l} \times \mathbf{B}) = I \int_{\text{wire}} (d\mathbf{l} \times \mathbf{B}) \tag{3-14}$$

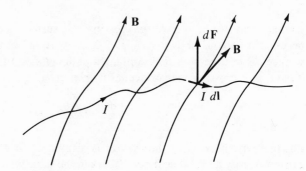

Fig. 3.3. Illustrating the force experienced by an infinitesimal segment of a filamentary wire carrying current I in a magnetic field $\mathbf{B}$.

EXAMPLE 3-2.   Show that the total magnetic force experienced by a closed loop of wire carrying a current I in a uniform magnetic field $\mathbf{B}$ is equal to zero.

Applying (3-14) for the contour C of the wire, we have

$$\mathbf{F} = I \oint_C (d\mathbf{l} \times \mathbf{B}) = I \left(\oint_C d\mathbf{l} \right) \times \mathbf{B} \qquad (3\text{-}15)$$

where, since $\mathbf{B}$ is uniform, we have taken it outside the integral on the right side of (3-15). Now,

$$\oint_C d\mathbf{l} = \oint_C (dx\,\mathbf{i}_x + dy\,\mathbf{i}_y + dz\,\mathbf{i}_z)$$
$$= \left(\oint_C dx \right) \mathbf{i}_x + \left(\oint_C dy \right) \mathbf{i}_y + \left(\oint_C dz \right) \mathbf{i}_z = 0 \qquad (3\text{-}16)$$

Hence $\mathbf{F} = 0$. ∎

3.3 Ampere's Law of Force

In Chapter 2 the concept of electric field was introduced in terms of force experienced by a small test charge placed in the presence of a larger charge in analogy with the gravitational force associated with two masses. We then presented an experimental law known as Coulomb's law and obtained from it the expression for the electric field intensity of a point charge. Just as static charges which are influenced by electric fields are themselves sources of electric fields, moving charges or currents which are influenced by magnetic fields are themselves sources of magnetic fields. To demonstrate this, we will in this section present an experimental law known as Ampere's law of force, analogous to Coulomb's law, and use it in the next section to obtain the expression for the magnetic field due to a current element.

Ampere's law of force is concerned with the forces experienced by two loops of wire carrying currents I_1 and I_2, as shown in Fig. 3.4. As a result of

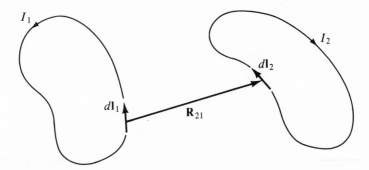

Fig. 3.4. Two loops of wire carrying currents I_1 and I_2.

experimental findings by Ampere, the force experienced by current loop 2 is given by

$$\mathbf{F}_{21} = k \oint_{C_1} \oint_{C_2} \frac{I_2 \, d\mathbf{l}_2 \times (I_1 \, d\mathbf{l}_1 \times \mathbf{R}_{21})}{R_{21}^3} \tag{3-17}$$

where C_1 and C_2 are the contours of loops 1 and 2, respectively, k is a constant of proportionality, and $\mathbf{R}_{21}$ is the vector drawn from a differential length element $d\mathbf{l}_1$ in loop 1 to a differential length element $d\mathbf{l}_2$ in loop 2. The constant of proportionality k is equal to $\mu_0/4\pi$ for free space and in the MKS system of units. The quanity μ_0 is known as the permeability of free space and is equal to $4\pi \times 10^{-7}$. Since (3-17) is valid for any orientation of the current loops, it follows that the differential force $d\mathbf{F}_{21}$ experienced by the differential current element $I_2 \, d\mathbf{l}_2$ due to the differential current element $I_1 \, d\mathbf{l}_1$ is

$$d\mathbf{F}_{21} = \frac{\mu_0}{4\pi} \frac{I_2 \, d\mathbf{l}_2 \times (I_1 \, d\mathbf{l}_1 \times \mathbf{R}_{21})}{R_{21}^3} \tag{3-18}$$

where we have substituted $\mu_0/4\pi$ for k. From (3-18), we note that μ_0 has the units newtons per ampere squared. These are commonly known as henrys per meter. Recalling that the permittivity of free space, ϵ_0, is equal to $10^{-9}/36\pi$ C^2/N-m², we note that

$$\frac{1}{\sqrt{\mu_0 \epsilon_0}} = \frac{1}{\sqrt{4\pi \times 10^{-7} \times (10^{-9}/36\pi)}} \text{ amp-m/C} \tag{3-19}$$
$$= 3 \times 10^8 \text{ m/sec}$$

which is the velocity of light in free space.

Some of the features evident from Eq. (3-18) are as follows:

(a) The magnitude of the force is proportional to the product of the magnitudes of the currents.

(b) The magnitude of the force is inversely proportional to the square of the distance between the current elements.

(c) To determine the direction of the force, we first find the cross product $d\mathbf{l}_1 \times \mathbf{R}_{21}$ and then cross $d\mathbf{l}_2$ into the resulting vector. The parenthesis on the right side of (3-18) is very important since, for a triple cross product, $\mathbf{A} \times (\mathbf{B} \times \mathbf{C}) \neq (\mathbf{A} \times \mathbf{B}) \times \mathbf{C}$.

By interchanging $I_1 \, d\mathbf{l}_1$ and $I_2 \, d\mathbf{l}_2$ and replacing $\mathbf{R}_{21}$ by $\mathbf{R}_{12}$ in (3-18), we obtain the expression for the force experienced by $I_1 \, d\mathbf{l}_1$ due to $I_2 \, d\mathbf{l}_2$ as

$$d\mathbf{F}_{12} = \frac{\mu_0}{4\pi} \frac{I_1 \, d\mathbf{l}_1 \times (I_2 \, d\mathbf{l}_2 \times \mathbf{R}_{12})}{R_{12}^3} \tag{3-20}$$

It may be noted that $d\mathbf{F}_{12}$ is not necessarily equal to $-d\mathbf{F}_{21}$. This can be illustrated by considering a simple case in which $d\mathbf{l}_1$ and $d\mathbf{l}_2$ are normal, as shown in Fig. 3.5. The construction of Fig. 3.5(a) shows that $d\mathbf{F}_{21}$ is nonzero and directed parallel to $d\mathbf{l}_1$ whereas the construction of Fig. 3.5(b) shows that

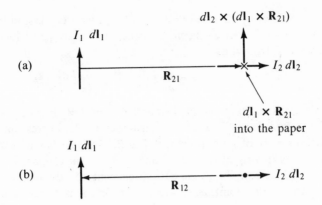

Fig. 3.5. For showing that the force experienced by $I_1\,d\mathbf{l}_1$ due to $I_2\,d\mathbf{l}_2$ is not necessarily equal and opposite to the force experienced by $I_2\,d\mathbf{l}_2$ due to $I_1\,d\mathbf{l}_1$.

$(d\mathbf{l}_2 \times \mathbf{R}_{12})$ is zero and hence $d\mathbf{F}_{12}$ is zero. The fact that $d\mathbf{F}_{21}$ and $d\mathbf{F}_{12}$ are not equal and opposite is not a violation of Newton's third law since isolated current elements do not exist without sources and sinks of charges at their ends. On the other hand, Newton's third law must and does hold for current loops. It is left as an exercise for the student to prove this (Problem 3.8).

3.4   The Magnetic Field of Filamentary Currents

In Section 3.2 we derived Eq. (3-13) for the differential force experienced by a filamentary current element located in a magnetic field. In Section 3.3 we introduced Ampere's law of force, which expresses the forces experienced by two current-carrying loops of wires, and from it obtained the expression (3-18) for the differential force experienced by a current element in the presence of another current element. Now, comparing the forms of the right sides of (3-13) and (3-18), we observe that the force experienced by $I_2\,d\mathbf{l}_2$ is due to the magnetic field of $I_1\,d\mathbf{l}_1$. If we denote this magnetic field by $d\mathbf{B}_1$, we can then write

$$I_2\,d\mathbf{l}_2 \times d\mathbf{B}_1 = \frac{\mu_0}{4\pi}\frac{I_2\,d\mathbf{l}_2 \times (I_1\,d\mathbf{l}_1 \times \mathbf{R}_{21})}{R_{21}^3} \qquad (3\text{-}21)$$

where we have introduced the appropriate subscripts on the left side of (3-21). Similarly, by comparing the right sides of (3-13) and (3-20), we note that the force experienced by $I_1\,d\mathbf{l}_1$ is due to the magnetic field of $I_2\,d\mathbf{l}_2$. If we denote this magnetic field by $d\mathbf{B}_2$, we can then write

$$I_1\,d\mathbf{l}_1 \times d\mathbf{B}_2 = \frac{\mu_0}{4\pi}\frac{I_1\,d\mathbf{l}_1 \times (I_2\,d\mathbf{l}_2 \times \mathbf{R}_{12})}{R_{12}^3} \qquad (3\text{-}22)$$

where we have introduced the appropriate subscripts on the left side of (3-22).

Equations (3-21) and (3-22) yield a general expression for the magnetic flux density due to a current element $I\,d\mathbf{l}$ at any point located at a vector distance $\mathbf{R}$ from it as

$$d\mathbf{B} = \frac{\mu_0}{4\pi}\frac{I\,d\mathbf{l}\times\mathbf{R}}{R^3} = \frac{\mu_0}{4\pi}\frac{I\,d\mathbf{l}\times\mathbf{i}_R}{R^2} \qquad (3\text{-}23)$$

where $\mathbf{i}_R$ is the unit vector in the direction of $\mathbf{R}$. Equation (3-23) is known as the Biot–Savart law and is analogous to the expression for the electric field intensity of a point charge. The Biot–Savart law tells us that the magnetic flux density at a point P due to a current element is directed normal to the plane containing the current element and the line joining the current element to the point, as shown in Fig. 3.6. It is therefore directed circular

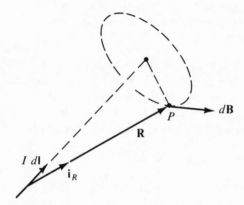

Fig. 3.6. The magnetic field $d\mathbf{B}$ due to a current element $I\,d\mathbf{l}$, at a distance $\mathbf{R}$ from the current element.

to the straight-line axis along the current element. In particular, the sense of the normal is that towards which the fingers are curled when the filamentary wire is grabbed with the right hand and with the thumb pointing in the direction of the current; it is the same as the sense of turning of a right-hand screw as it advances in the direction of $I\,d\mathbf{l}$. The magnitude of the magnetic flux density is proportional to the current I, the element length dl, and the sine of the angle between the current element and the line from it to the point P, and inversely proportional to the square of the distance from the current element to the point P. Hence the magnetic field is zero along the straight line in the direction of the current element. The magnetic flux density $\mathbf{B}$ due to a filamentary wire of any length can now be obtained by integrating the right side of (3-23) along the contour C of the wire. Thus

$$\mathbf{B} = \frac{\mu_0}{4\pi}\int_C \frac{I\,d\mathbf{l}\times\mathbf{i}_R}{R^2} \qquad (3\text{-}24)$$

In evaluating the integral in (3-24), we note that $\mathbf{i}_R$ and $\mathbf{R}$ are functions of the location of $d\mathbf{l}$. In terms of source point-field point notation, (3-24) is written as

$$\mathbf{B}(\mathbf{r}) = \frac{\mu_0}{4\pi} \int_{C'} \frac{I \, d\mathbf{l'} \times (\mathbf{r} - \mathbf{r'})}{|\mathbf{r} - \mathbf{r'}|^3} \qquad (3\text{-}25)$$

where C' is the contour occupied by the wire.

EXAMPLE 3-3.   A straight wire carrying current I amp lies along the z axis as shown in Fig. 3.7. Find the magnetic flux density vector due to the portion of the wire lying between $z = -a$ and $z = +a$ and then extend the result to that of an infinitely long wire.

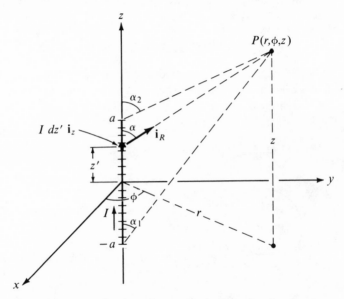

Fig. 3.7. For evaluating the magnetic flux density due to a straight wire carrying current I amp and lying along the z axis between $z = -a$ and $z = +a$.

First we divide the wire into a number of infinitesimal segments, each of which can be considered as a current element. The magnetic flux density due to a current element is given by (3-23). For a current element oriented along the z axis, $\mathbf{i}_z \times \mathbf{i}_R$ is in the $\mathbf{i}_\phi$ direction and hence the magnetic field is in the $\mathbf{i}_\phi$ direction. Also, its magnitude is independent of ϕ. Since all current elements making up the wire are along the z axis, the contributions due to them are all in the $\mathbf{i}_\phi$ direction and independent of ϕ. Thus the magnetic field has circular symmetry about the axis of the wire.

Let us now consider a point $P(r, \phi, z)$. The magnetic flux density at P due to a current element $I \, dz' \, \mathbf{i}_z$ at distance z' from the origin is given by

$$d\mathbf{B} = \frac{\mu_0}{4\pi} \frac{I \, dz' \, \mathbf{i}_z \times \mathbf{i}_R}{R^2} = \frac{\mu_0}{4\pi} \frac{I \, dz' \sin \alpha}{[r^2 + (z - z')^2]} \mathbf{i}_\phi$$

The magnetic flux density at P due to the segment of the wire lying between $z = -a$ and $z = +a$ is then given by

$$\mathbf{B} = \int_{z'=-a}^{a} d\mathbf{B} = \frac{\mu_0}{4\pi} \int_{z'=-a}^{a} \frac{I\,dz'\sin\alpha}{[r^2 + (z-z')^2]}\mathbf{i}_\phi \qquad (3\text{-}26)$$

Introducing $(z - z') = r\cot\alpha$ in (3-26), we obtain

$$\mathbf{B} = \frac{\mu_0 I}{4\pi r} \int_{\alpha=\alpha_1}^{\alpha_2} \sin\alpha\,d\alpha\,\mathbf{i}_\phi = \frac{\mu_0 I}{4\pi r}(\cos\alpha_1 - \cos\alpha_2)\mathbf{i}_\phi \qquad (3\text{-}27)$$

where α_1 and α_2 are the angles which the lines joining the ends of the wire segment to the point P make with the z axis. Now, for an infinitely long wire, $\alpha_1 = 0$ and $\alpha_2 = \pi$. Hence

$$\mathbf{B} = \frac{\mu_0 I}{4\pi r}(\cos 0 - \cos\pi)\mathbf{i}_\phi = \frac{\mu_0 I}{2\pi r}\mathbf{i}_\phi \qquad (3\text{-}28)$$

Thus the magnetic flux density due to an infinitely long straight wire is dependent only on the distance away from the wire, analogous to the electric field intensity due to an infinitely long line charge of uniform density. The field is sketched in Fig. 3.8. ∎

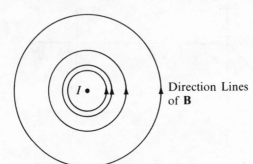

Direction Lines of **B**

Fig. 3.8. The direction lines of magnetic field due to an infinitely long straight wire carrying current out of the plane of the paper.

EXAMPLE 3-4. A circular loop of wire of radius a m and carrying a current I amp lies in the xy plane with its center at the origin. Such an arrangement is known as a magnetic dipole. Obtain the expression for the magnetic flux density due to the magnetic dipole at distances very large from the origin compared to the radius a.

   With reference to the geometry shown in Fig. 3.9, we note that, at any point P, the ϕ component of the magnetic field due to a current element in the ring is cancelled by the ϕ component of the magnetic field due to another current element situated symmetrically about P so that the ϕ component due to the entire ring is zero. Thus the magnetic field has only r and θ components. Furthermore, since the ring is circular about the origin and is in the xy plane, the magnetic field has circular symmetry about the z axis. Hence we consider, for simplicity, a point P having the spherical coordinates $(r, \theta, \pi/2)$. The magnetic field at P due to the current element 1 situated at

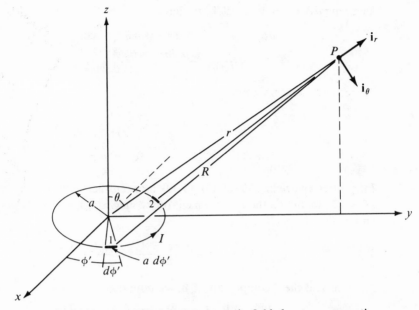

Fig. 3.9. For evaluating the magnetic field due to a magnetic dipole at distances very large from it compared to its radius.

$\phi = \phi'$ is then given by

$$dB_1 = \frac{\mu_0}{4\pi} \frac{\begin{Bmatrix} Ia\, d\phi'\, (-\sin \phi'\, \mathbf{i}_x + \cos \phi'\, \mathbf{i}_y) \\ \times\, [-a \cos \phi'\, \mathbf{i}_x + (r \sin \theta - a \sin \phi')\mathbf{i}_y + r \cos \theta\, \mathbf{i}_z] \end{Bmatrix}}{(a^2 + r^2 - 2ar \sin \theta \sin \phi')^{3/2}}$$

$$= \frac{\mu_0 Ia\, d\phi'\, [r \cos \theta \cos \phi'\, \mathbf{i}_x + r \cos \theta \sin \phi'\, \mathbf{i}_y + (a - r \sin \theta \sin \phi')\mathbf{i}_z]}{4\pi(a^2 + r^2 - 2ar \sin \theta \sin \phi')^{3/2}}$$

The magnetic field at P due to the symmetrically situated current element

2 at $\phi = \pi - \phi'$ is given by

$$dB_2 = \frac{\mu_0}{4\pi} \frac{\begin{Bmatrix} Ia\, d\phi'\, (-\sin \phi'\, \mathbf{i}_x - \cos \phi'\, \mathbf{i}_y) \\ \times\, [a \cos \phi'\, \mathbf{i}_x + (r \sin \theta - a \sin \phi')\mathbf{i}_y + r \cos \theta\, \mathbf{i}_z] \end{Bmatrix}}{(a^2 + r^2 - 2ar \sin \theta \sin \phi')^{3/2}}$$

$$= \frac{\mu_0 Ia\, d\phi'[-r \cos \theta \cos \phi'\, \mathbf{i}_x + r \cos \theta \sin \phi'\, \mathbf{i}_y + (a - r \sin \theta \sin \phi')\, \mathbf{i}_z]}{4\pi(a^2 + r^2 - 2ar \sin \theta \sin \phi')^{3/2}}$$

The contribution to the magnetic field at P due to the pair of current elements 1 and 2 is then given by

$$dB = dB_1 + dB_2$$

$$= \frac{\mu_0 Ia\, d\phi'\, [r \cos \theta \sin \phi'\, \mathbf{i}_y + (a - r \sin \theta \sin \phi')\mathbf{i}_z]}{2\pi(a^2 + r^2 - 2ar \sin \theta \sin \phi')^{3/2}}$$

Denoting $d\mathbf{B} = dB_r\,\mathbf{i}_r + dB_\theta\,\mathbf{i}_\theta$, we have

$$dB_r = d\mathbf{B} \cdot \mathbf{i}_r = d\mathbf{B} \cdot (\sin\theta\,\mathbf{i}_y + \cos\theta\,\mathbf{i}_z)$$

$$= \frac{\mu_0 I a^2 \cos\theta\, d\phi'}{2\pi(a^2 + r^2 - 2ar\sin\theta\sin\phi')^{3/2}} \tag{3-29}$$

Proceeding further, we obtain

$$dB_r = \frac{\mu_0 I a^2 \cos\theta\, d\phi'}{2\pi r^3[(a/r)^2 + 1 - 2(a/r)\sin\theta\sin\phi']^{3/2}}$$

$$\approx \frac{\mu_0 I a^2 \cos\theta\, d\phi'}{2\pi r^3} \qquad \text{for } r \gg a \tag{3-30}$$

Integrating the right side of Eq. (3-30) between the limits $\phi' = -\pi/2$ and $\phi' = \pi/2$, we obtain the r component of the magnetic flux density due to the entire ring as

$$B_r = \int_{\phi'=-\pi/2}^{\pi/2} \frac{\mu_0 I a^2 \cos\theta\, d\phi'}{2\pi r^3} = \frac{\mu_0 I \pi a^2 \cos\theta}{2\pi r^3} \tag{3-31}$$

Now, to find the θ component of $\mathbf{B}$, we note that

$$dB_\theta = d\mathbf{B} \cdot \mathbf{i}_\theta = d\mathbf{B} \cdot (\cos\theta\,\mathbf{i}_y - \sin\theta\,\mathbf{i}_z)$$

$$= \frac{\mu_0 I a\, d\phi'\,(-a\sin\theta + r\sin\phi')}{2\pi(a^2 + r^2 - 2ar\sin\theta\sin\phi')^{3/2}} \tag{3-32}$$

Proceeding further, we obtain

$$dB_\theta = \frac{\mu_0 I a\, d\phi'}{2\pi r^2}\left[-\left(\frac{a}{r}\right)\sin\theta + \sin\phi'\right]\left[1 - 2\left(\frac{a}{r}\right)\sin\theta\sin\phi' + \left(\frac{a}{r}\right)^2\right]^{-3/2}$$

$$= \frac{\mu_0 I a\, d\phi'}{2\pi r^2}\left[-\left(\frac{a}{r}\right)\sin\theta + \sin\phi'\right]\left[1 + 3\left(\frac{a}{r}\right)\sin\theta\sin\phi' + \cdots\right]$$

$$= \frac{\mu_0 I a\, d\phi'}{2\pi r^2}\left[-\left(\frac{a}{r}\right)\sin\theta + \sin\phi' + 3\left(\frac{a}{r}\right)\sin\theta\sin^2\phi'\right.$$

$$\left. + \ldots \text{terms involving higher powers of } \left(\frac{a}{r}\right)\right]$$

$$\approx \frac{\mu_0 I a}{2\pi r^2}\left[-\left(\frac{a}{r}\right)\sin\theta + \sin\phi' + 3\left(\frac{a}{r}\right)\sin\theta\sin^2\phi'\right]d\phi' \qquad \text{for } r \gg a \tag{3-33}$$

where we have retained the (a/r) term since the $\sin\phi'$ term yields zero when integrated between $\phi' = -\pi/2$ and $\phi' = \pi/2$. Integrating the right side of Eq.(3-33) between these limits, we obtain the θ component of the magnetic flux density due to the entire ring as

$$B_\theta = \int_{\phi'=-\pi/2}^{\pi/2} \frac{\mu_0 I a}{2\pi r^2}\left[-\left(\frac{a}{r}\right)\sin\theta + \sin\phi' + 3\left(\frac{a}{r}\right)\sin\theta\sin^2\phi'\right]d\phi'$$

$$= \frac{\mu_0 I \pi a^2 \sin\theta}{4\pi r^3} \tag{3-34}$$

Thus

$$\mathbf{B} = \frac{\mu_0 I \pi a^2}{4\pi r^3}(2\cos\theta\,\mathbf{i}_r + \sin\theta\,\mathbf{i}_\theta) \tag{3-35}$$

We can consider Eq. (3-25) as the solution for the magnetic flux density at very large distances compared to the radius a or as the solution for the magnetic flux density at any point (r, θ, ϕ) in the limit that $a \to 0$, keeping $I\pi a^2$ constant. It should be noted that to keep $I\pi a^2$ constant as $a \to 0$ requires that $I \to \infty$. The product $I\pi a^2$ is known as the magnetic dipole moment m. The magnetic dipole moment has also an orientation associated with it which is normal to the surface of the loop. In particular, the sense of the normal is that towards which the fingers pierce through the area of the ring when the loop is grabbed with the right hand and with the thumb pointing in the direction of the current. It is the same as the direction of advance of a right-hand screw as it is turned in the sense of the loop current. Substituting m for $I\pi a^2$ in (3-35), the magnetic flux density due to a magnetic dipole of moment m oriented along the positive z axis is given by

$$\mathbf{B} = \frac{\mu_0 m}{4\pi r^3}(2\cos\theta\,\mathbf{i}_r + \sin\theta\,\mathbf{i}_\theta) \tag{3-36}$$

The magnetic field given by (3-36) is analogous to the electric field due to an electric dipole of moment p oriented along the z-axis and given by (2-28). ∎

EXAMPLE 3-5.   A solenoid consists of continuously wound, circular current loops. Let us consider an infinitely long, uniformly wound solenoid of radius a and n turns per unit length, each carrying the same current I and with the z axis as its axis. It is desired to find the magnetic flux density due to the infinitely long solenoid.

Since the solenoid is uniformly wound and infinitely long, and since it possesses cylindrical symmetry about the z axis, the magnetic flux density must be independent of z and must possess cylindrical symmetry about the z axis. Hence it is sufficient if we compute the magnetic flux density at a point P on the y axis. To do this, let us consider two sections of the solenoid symmetrically placed about the xy plane at distances z' from it and having infinitesimal lengths dz' as shown in Fig. 3.10. Since the lengths are infinitesimal, these sections can be considered as current loops carrying currents $nI\,dz'$.

In each of these current loops, let us consider two differential elements of lengths $a\,d\phi'$ symmetrically situated about the yz plane, as shown in Fig. 3.10. Applying the notation of Fig. 3.10 to (3-29) and (3-32), we obtain the magnetic field at P due to the pair of current elements 1 and 2 as

$$d\mathbf{B}_1 = \mu_0 nI\,dz' \frac{a^2\cos\alpha\,d\phi'\,\mathbf{i}_1 + a\,d\phi'\,[-a\sin\alpha + (y/\sin\alpha)\sin\phi']\mathbf{i}_2}{2\pi(a^2 + y^2 + z'^2 - 2ay\sin\phi')^{3/2}} \tag{3-37}$$

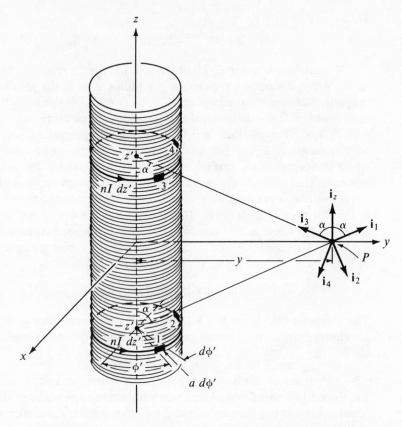

Fig. 3.10. For evaluating the magnetic field due to an infinitely long, uniformly wound solenoid of radius a and n turns per unit length.

and the magnetic field at P due to the pair of current elements 3 and 4 as

$$d\mathbf{B}_2 = \mu_0 nI\, dz'\, \frac{a^2 \cos\alpha\, d\phi'\, \mathbf{i}_3 + a\, d\phi'\, [-a\sin\alpha + (y/\sin\alpha)\sin\phi']\mathbf{i}_4}{2\pi(a^2 + y^2 + z'^2 - 2ay\sin\phi')^{3/2}}$$

(3-38)

where α, $\mathbf{i}_1$, $\mathbf{i}_2$, $\mathbf{i}_3$, and $\mathbf{i}_4$ are defined in Fig. 3.10. Adding (3-37) and (3-38) and simplifying, we obtain the magnetic field at P due to the four current elements 1, 2, 3, and 4 as

$$d\mathbf{B} = d\mathbf{B}_1 + d\mathbf{B}_2 = \mu_0 nI\, dz'\, \frac{(a^2 - ay\sin\phi')\, d\phi'}{\pi(a^2 + y^2 + z'^2 - 2ay\sin\phi')^{3/2}}\mathbf{i}_z$$

(3-39)

Performing double integration of the right side of (3-39) between the appropriate limits, we obtain the magnetic flux density at P due to the entire solenoid as

$$
\begin{aligned}
\mathbf{B} &= \int_{\phi'=-\pi/2}^{\pi/2}\int_{z'=0}^{\infty}\frac{\mu_0 nI(a^2 - ay\sin\phi')\,d\phi'\,dz'}{\pi(a^2 + y^2 + z'^2 - 2ay\sin\phi')^{3/2}}\mathbf{i}_z \\
&= \frac{\mu_0 nIa}{\pi}\int_{\phi'=-\pi/2}^{\pi/2}\frac{(a - y\sin\phi')d\phi'}{(a^2 + y^2 - 2ay\sin\phi')}\mathbf{i}_z \\
&= \begin{cases} 0 & \text{for } y > a \\ \mu_0 nI\mathbf{i}_z & \text{for } y < a \end{cases}
\end{aligned}
\tag{3-40}
$$

Thus the magnetic field due to the infinitely long solenoid is zero outside the solenoid and uniform inside the solenoid, having a value $\mu_0 nI$ and directed along the axis of the solenoid. ∎

3.5   The Magnetic Field of Current Distributions

In the previous section we considered the magnetic field computation for filamentary wires carrying current. In this section we will extend the computation to current distributions. Current distributions can be of two types:

(a) Surface current for which current is distributed on a surface (planar or nonplanar).

(b) Volume current for which current is distributed in a volume.

As in the case of continuous charge distributions, introduced in Section 2.4, we have to work with current densities when a current is distributed on a surface or in a volume. We have already introduced the current density for volume currents in Sections 1.7 and 3.2. The magnitude of the volume current density $\mathbf{J}$ at a point is the current per unit area crossing an infinitesimal area at that point with the orientation of the area adjusted so as to maximize the current, in the limit that the area tends to zero. The direction of $\mathbf{J}$ at that point is the direction to which the normal to the area approaches in the limit. Similarly, the magnitude of the surface current density at a point is the current per unit width crossing an infinitesimal line segment at that point with the orientation of the segment adjusted so as to maximize the current, in the limit that the width of the line segment tends to zero. The direction of the surface current density at that point is the direction to which the normal to the line and tangent to the surface approaches in the limit. We will use the symbol $\mathbf{J}_s$ for the surface current density, in contrast to $\mathbf{J}$ for the volume current density. In each case, we represent the total current as a continuous collection of appropriate filamentary currents and evaluate the magnetic field as the vector superposition of the contributions due to the individual filamentary currents.

EXAMPLE 3-6.   A sheet of current with the surface current density given by

$$
\mathbf{J}_s = J_{s0}\mathbf{i}_z \text{ amp/m}
$$

where J_{s0} is a constant, occupies the entire xz plane. Find the magnetic flux density vector due to the portion of the current sheet lying between $x = -a$ and $x = +a$ as shown in Fig. 3.11(a) and then extend the result to that of the infinite sheet.

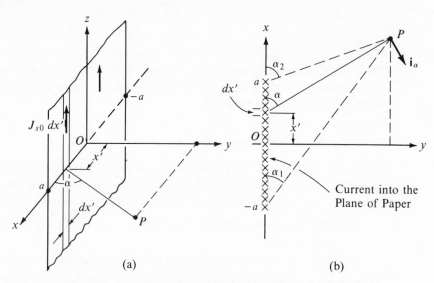

(a)   **(b)**

Fig. 3.11. For evaluating the magnetic field due to a sheet of current flowing in the z direction and lying in the xz plane between $x = -a$ and $x = +a$.

We divide the current sheet into a number of filaments of infinitesimal width in the x direction, each of which can be considered as an infinitely long wire parallel to the z axis. Let us consider a filament of width dx' located at $x = x'$ in the plane of the sheet, as shown in Fig. 3.11(a). From Example 3-3, we know that the magnetic flux density due to an infinitely long wire is dependent only on the distance away from the wire and is oriented circular to the wire. Hence the magnetic field due to the current sheet will not be dependent on the z coordinate and also will have only x and y components, so that it is sufficient if we consider the two-dimensional geometry shown in Fig. 3.11(b). Since the current density is $J_{s0}\mathbf{i}_z$, the current flowing in the filament of width dx' is $J_{s0}\, dx'$. Applying (3-28) to the geometry associated with this filament, we obtain the magnetic flux density due to it at any point $P(x, y, z)$ as

$$dB = \frac{\mu_0 J_{s0}\, dx'}{2\pi\sqrt{(x - x')^2 + y^2}}\mathbf{i}_\alpha \qquad (3\text{-}41)$$

where $\mathbf{i}_\alpha$ is the unit vector normal to the line drawn from the filament to the point P as shown in Fig. 3.11(b). Expressing $d\mathbf{B}$ in terms of its components along the coordinate axes, we have

$$dB = \frac{\mu_0 J_{s0}\, dx'}{2\pi\sqrt{(x-x')^2 + y^2}}(-\sin\alpha\, \mathbf{i}_x + \cos\alpha\, \mathbf{i}_y) \qquad (3\text{-}42)$$

The magnetic flux density at P due to the portion of the infinite current sheet between $x = -a$ and $x = +a$ is then given by

$$\mathbf{B} = \int_{x'=-a}^{a} d\mathbf{B}$$

$$= \int_{x'=-a}^{a} \left[-\frac{\mu_0 J_{s0}\sin\alpha\, dx'}{2\pi\sqrt{(x-x')^2 + y^2}}\mathbf{i}_x + \frac{\mu_0 J_{s0}\cos\alpha\, dx'}{2\pi\sqrt{(x-x')^2 + y^2}}\mathbf{i}_y \right] \qquad (3\text{-}43)$$

$$= \frac{\mu_0 J_{s0}}{2\pi}\left[(\alpha_1 - \alpha_2)\mathbf{i}_x + \ln\left(\frac{\sin\alpha_2}{\sin\alpha_1}\right)\mathbf{i}_y \right]$$

where we have used the transformation $(x - x') = y\cot\alpha$ for evaluating the integrals in (3-43), and the angles α_1 and α_2 are as shown in Fig. 3.11(b). Now, for the infinite sheet of current, $\alpha_1 = 0$ and $\alpha_2 = \pi$ for $y > 0$, and $\alpha_1 = 2\pi$ and $\alpha_2 = \pi$ for $y < 0$. However, to evaluate $\ln(\sin\alpha_2/\sin\alpha_1)$, we note that

$$\lim_{a\to\infty}\frac{\sin\alpha_2}{\sin\alpha_1} = \lim_{a\to\infty}\left[\frac{(x+a)^2 + y^2}{(x-a)^2 + y^2}\right]^{1/2} = 1$$

and hence

$$\lim_{a\to\infty}\ln\frac{\sin\alpha_2}{\sin\alpha_1} = 0 \qquad (3\text{-}44)$$

Substituting for α_1 and α_2 in (3-43), we then obtain the magnetic flux density due to the infinite sheet of current as

$$\mathbf{B} = \begin{cases} -\dfrac{\mu_0 J_{s0}}{2}\mathbf{i}_x & \text{for } y > 0 \\[2mm] \dfrac{\mu_0 J_{s0}}{2}\mathbf{i}_x & \text{for } y < 0 \end{cases} \qquad (3\text{-}45)$$

$$= \frac{\mu_0}{2}\mathbf{J}_s \times \mathbf{i}_n \qquad \text{where } \mathbf{i}_n = \begin{cases} \mathbf{i}_y \text{ for } y > 0 \\ -\mathbf{i}_y \text{ for } y < 0 \end{cases}$$

The field given by (3-45) is sketched in Fig. 3.12. If the sheet current occupies the $y = y_0$ plane, it follows from (3-45) that

$$\mathbf{B} = \begin{cases} -\dfrac{\mu_0 J_{s0}}{2}\mathbf{i}_x & \text{for } y > y_0 \\[2mm] \dfrac{\mu_0 J_{s0}}{2}\mathbf{i}_x & \text{for } y < y_0 \quad\blacksquare \end{cases}$$

EXAMPLE 3-7.   Current flows in the axial direction in an infinitely long cylinder of radius a with uniform density J_0 amp/m^2. Find the magnetic flux density both inside and outside the cylinder.

Choosing the z axis as the axis of the infinitely long cylinder as shown

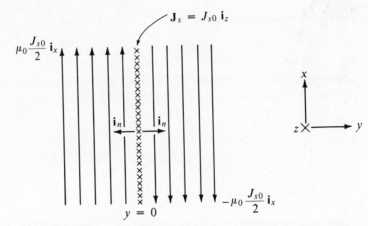

Fig. 3.12. The direction lines of magnetic field due to an infinite sheet of current flowing into the plane of the paper with uniform density.

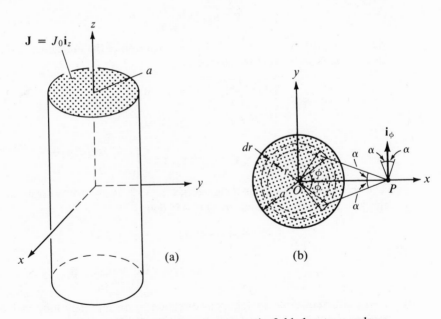

Fig. 3.13. For evaluating the magnetic field due to a volume current flowing along an infinitely long cylinder of radius *a* with uniform density.

in Fig. 3.13(a), we have the volume current density as

$$\mathbf{J} = J_0 \mathbf{i}_z$$

The cylindrical current distribution can be thought of as a superpositon of filamentary currents parallel to the z axis so that the magnetic field is independent of z. Hence it is sufficient if we consider the two-dimensional geometry shown in Fig. 3.13(b). Furthermore, for every filamentary current and for a given point P, there is another filamentary current so that the combined magnetic field due to these two filamentary currents is entirely in the ϕ direction. This is illustrated in Fig. 3.13(b) for a point P on the x axis. Thus the magnetic field due to the entire current distribution has only a ϕ component and possesses cylindrical symmetry about the z axis. Let us therefore consider two filamentary currents corresponding to the infinitesimal areas $r\,dr\,d\phi$ at (r, ϕ) and $(r, -\phi)$ as shown in Fig. 3.13(b). The magnetic field at P due to these two filamentary currents is given by

$$
\begin{aligned}
d\mathbf{B} &= \frac{\mu_0 J_0 r\,dr\,d\phi}{2\pi(r^2 + x^2 - 2rx\cos\phi)^{1/2}}\,2\cos\alpha\,\mathbf{i}_\phi \\
&= \frac{\mu_0 J_0 r\,dr\,d\phi(x - r\cos\phi)}{\pi(r^2 + x^2 - 2rx\cos\phi)}\mathbf{i}_\phi
\end{aligned}
\tag{3-46}
$$

The magnetic field at P due to the entire current distribution is then given by

$$
\begin{aligned}
\mathbf{B} &= \int_{r=0}^{a}\int_{\phi=0}^{\pi} d\mathbf{B} \\
&= \frac{\mu_0 J_0}{\pi}\int_{r=0}^{a} r\,dr \int_{\phi=0}^{\pi} \frac{(x - r\cos\phi)\,d\phi}{(r^2 + x^2 - 2rx\cos\phi)}\mathbf{i}_\phi \\
&= \frac{\mu_0 J_0}{\pi}\int_{r=0}^{a} r\,dr \begin{pmatrix} 0 & \text{for } x < r \\ \dfrac{\pi}{x} & \text{for } x > r \end{pmatrix}\mathbf{i}_\phi \\
&= \begin{cases} \dfrac{\mu_0 J_0}{\pi}\displaystyle\int_{r=0}^{a} \dfrac{\pi}{x} r\,dr\,\mathbf{i}_\phi & \text{for } x > a \\[3mm] \dfrac{\mu_0 J_0}{\pi}\displaystyle\int_{r=0}^{x} \dfrac{\pi}{x} r\,dr\,\mathbf{i}_\phi & \text{for } x < a \end{cases} \\
&= \begin{cases} \dfrac{\mu_0 J_0}{\pi x}\dfrac{\pi a^2}{2}\mathbf{i}_\phi & \text{for } x > a \\[3mm] \dfrac{\mu_0 J_0}{\pi x}\dfrac{\pi x^2}{2}\mathbf{i}_\phi & \text{for } x < a \end{cases}
\end{aligned}
\tag{3-47}
$$

Recalling that $\mathbf{B}$ has cylindrical symmetry about the z axis, we substitute r for x in (3-47) and obtain

$$
\mathbf{B} = \begin{cases} \mu_0 \dfrac{J_0(\pi a^2/2)}{2\pi r}\mathbf{i}_\phi & \text{for } r > a \\[3mm] \mu_0 \dfrac{J_0(\pi r^2/2)}{2\pi r}\mathbf{i}_\phi & \text{for } r < a \end{cases}
\tag{3-48}
$$

Noting that $\pi r^2/2$ is the area of cross section of a wire of radius r, and that there is no current for $r > a$, we can combine the two results on the right side of (3-48) as

$$\mathbf{B}(r) = \mu_0 \frac{\text{current enclosed by the circular path of radius } r}{2\pi r}\mathbf{i}_\phi \qquad (3\text{-}49)$$

Viewed from any distance r from the axis of the infinitely long cylinder carrying current, the current distribution is equivalent to an infinitely long filamentary current of value equal to the current enclosed by the circular path of radius r. ▌

3.6   Ampere's Circuital Law in Integral Form

In Section 2.6 we started with the electric field intensity of a point charge and derived Gauss' law, which was later found to be very convenient for computing the electric field due to certain symmetrical charge distributions. Similarly, in this section we will start with the magnetic flux density due to an infinitely long wire carrying current and derive Ampere's circuital law. We will later find Ampere's circuital law to be very useful compared to the Biot–Savart law for computing the magnetic field due to certain symmetrical current distributions.

   Let us consider an infinitely long filamentary wire along the z axis carrying current I amp. The magnetic flux density due to this wire is directed everywhere circular to the wire and its magnitude is dependent only on the distance from the wire. Let us consider a circular path C of radius r in the plane normal to the wire and centered at the wire as shown in Fig. 3.14. For an infinitesimal length $d\mathbf{l} = dl\,\mathbf{i}_\phi$ on this contour C, we have

$$\mathbf{B} \cdot d\mathbf{l} = \frac{\mu_0 I}{2\pi r}\mathbf{i}_\phi \cdot dl\,\mathbf{i}_\phi = \frac{\mu_0 I\,dl}{2\pi r} \qquad (3\text{-}50)$$

The integral of $\mathbf{B} \cdot d\mathbf{l}$ along the entire path C is then given by

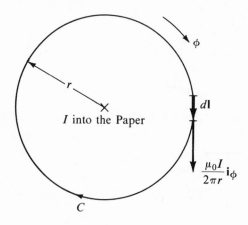

I into the Paper

$\dfrac{\mu_0 I}{2\pi r}\mathbf{i}_\phi$

C

Fig. 3.14. For evaluating $\oint_C \mathbf{B} \cdot d\mathbf{l}$, where C is a circular path of radius r in the plane normal to a straight, infinitely long wire carrying current I and centered at the wire.

$$\oint_C \mathbf{B} \cdot d\mathbf{l} = \oint_C \frac{\mu_0 I \, dl}{2\pi r} = \frac{\mu_0 I}{2\pi r} \oint_C dl \qquad (3\text{-}51)$$

where we have taken $\mu_0 I / 2\pi r$ outside the integral since r is constant for the contour C. Proceeding further, we have

$$\oint_C \mathbf{B} \cdot d\mathbf{l} = \frac{\mu_0 I}{2\pi r} \, (\text{circumference of } C)$$

$$= \frac{\mu_0 I}{2\pi r} (2\pi r) = \mu_0 I \qquad (3\text{-}52)$$

Equation (3-52) states that the line integral of $\mathbf{B}$ around a circular path in the plane normal to an infinitely long wire carrying current I and centered at the wire is equal to $\mu_0 I$. It is independent of the radius r of the circular path. Whether $r = 1$ micron or 1000 km, the value of the line integral is the same (provided, of course, that there is no other magnetic field in the medium). It should be noted that the current I in (3-52) is the current which flows in the direction of advance of a right-hand screw as it is turned in the sense in which the line integral around C is evaluated.

Before we proceed further, a few words about the line integral of $\mathbf{B}$ are in order. In Chapter 2 we learned that $\int_a^b \mathbf{E} \cdot d\mathbf{l}$ has the meaning of work or change in potential energy per unit charge associated with the movement of a test charge from point a to point b in the electric field $\mathbf{E}$. This is because the force experienced by a charge due to an electric field is in the same direction as the electric field. On the other hand, in a magnetic field $\mathbf{B}$, the force experienced by a test charge moving in the direction of $d\mathbf{l}$ (or by a current element $I \, d\mathbf{l}$) is perpendicular to both $\mathbf{B}$ and $d\mathbf{l}$. Hence the work associated with the movement of the test charge is zero. Thus $\int \mathbf{B} \cdot d\mathbf{l}$ does not have the meaning of work. Just as $\oint_S \mathbf{E} \cdot d\mathbf{S}$ provides us information about charges enclosed by S, $\oint_C \mathbf{B} \cdot d\mathbf{l}$ tells us about the current enclosed by C. Therefore, in this respect $\oint_C \mathbf{B} \cdot d\mathbf{l}$ is analogous to $\oint_S \mathbf{E} \cdot d\mathbf{S}$. We will simply call it the circulation of $\mathbf{B}$.

Let us now consider an arbitrary path C (not necessarily in a plane) enclosing the current as shown in Fig. 3.15. For an infinitesimal segment $d\mathbf{l}$ at P along this path,

$$\mathbf{B} \cdot d\mathbf{l} = \frac{\mu_0 I}{2\pi R} \mathbf{i}_\phi \cdot d\mathbf{l} = \frac{\mu_0 I \, dl \cos \alpha}{2\pi R} \qquad (3\text{-}53)$$

where R is the distance of P from the wire, $\mathbf{i}_\phi$ is the unit vector at P directed circular to the wire, and α is the angle between $d\mathbf{l}$ and $\mathbf{i}_\phi$. The circulation of $\mathbf{B}$ around the arbitrary path C is

$$\oint_C \mathbf{B} \cdot d\mathbf{l} = \oint_C \frac{\mu_0 I \, dl \cos \alpha}{2\pi R} = \frac{\mu_0 I}{2\pi} \oint_C \frac{dl \cos \alpha}{R} \qquad (3\text{-}54)$$

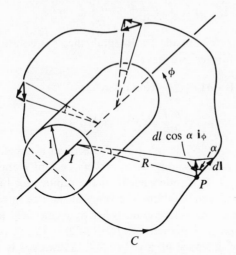

Fig. 3.15. For evaluating $\oint_C \mathbf{B} \cdot d\mathbf{l}$, where C is an arbitrary closed path enclosing a straight, infinitely long wire carrying current I.

In (3-54), $dl \cos \alpha$ is the projection of $d\mathbf{l}$ onto the circle of radius R centered at the wire and passing through P. Hence $(dl \cos \alpha)/R$ is the projection of $d\mathbf{l}$ on to the circle of radius unity in the plane normal to the wire and centered at the wire, and $\oint_C (dl \cos \alpha)/R$ is the sum of the projections of all infinitesimal segments comprising the contour C onto the circle of radius unity. Thus it is equal to the circumference of the circle of unit radius, that is, 2π. Substituting this result in (3-54), we have

$$\oint_{\substack{\text{contour}\\\text{enclosing } I}} \mathbf{B} \cdot d\mathbf{l} = \frac{\mu_0 I}{2\pi}(2\pi) = \mu_0 I \qquad (3\text{-}55)$$

If the arbitrary contour does not enclose the current, then, in evaluating $\oint_C (dl \cos \alpha)/R$, we start at one point on the circle of unit radius, traverse to another point on it and return to the starting point along the same path in the opposite direction, obtaining a result of zero in this process. Hence

$$\oint_{\substack{\text{contour not}\\\text{enclosing } I}} \mathbf{B} \cdot d\mathbf{l} = 0 \qquad (3\text{-}56)$$

Equations (3-55) and (3-56) may be combined into a single statement which reads as

$$\oint_C \mathbf{B} \cdot d\mathbf{l} = \mu_0(\text{current enclosed by the contour } C) \qquad (3\text{-}57)$$

This is Ampere's circuital law. Although we have derived it here for an infinitely long straight wire, it can be proved for a current loop of arbitrary shape. Also, if we have a number of current loops or infinitely long wires

carrying currents or continuous current distributions in the form of surface
or volume current, we can invoke superposition and conclude that Ampere's
circuital law as given by (3-57) holds for any closed path C provided the
current enclosed by C is uniquely defined.

Let us now discuss the uniqueness of a closed path enclosing or not
enclosing a current. To do this, let us consider the case of a straight filamentary wire of finite length in the plane of the paper carrying current I,
as shown in Fig. 3.16. This can be achieved by having a source of point

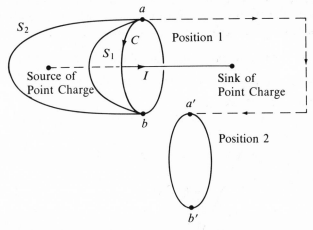

Fig. 3.16. For illustrating that the current enclosed by closed
path C surrounding a finitely long filamentary wire is not
uniquely defined.

charge at one end of the wire and a sink of point charge at the other end.
Let a closed path C be in the plane normal to the paper, emerging out of
the paper at a and going into it at b. Let us denote this position of the closed
path as position 1. Imagining the closed path to be rigid, we can bring it to
position 2 by sliding it parallel to the wire for some distance, pulling it down,
and then sliding it back parallel to the wire as shown by the dashed lines.
We are able to achieve this without cutting through the wire. We then say
that the current enclosed by the closed path C is not uniquely defined. Alternatively, we can define the current enclosed by a path as that which pierces
through (passes from one side to the other side of) a surface whose perimeter
is the closed path. For the closed path C in Fig. 3.16, let us consider two
bowl-shaped surfaces S_1 and S_2. It can be seen that the wire pierces through
S_1 but not through S_2. This suggests that we cannot uniquely define the
current enclosed by C in Fig. 3.16. It is clear that Ampere's circuital law
(3-57) cannot be used for the case of Fig. 3.16. In fact, if we evaluate $\oint_C \mathbf{B} \cdot d\mathbf{l}$

around the contour C in Fig. 3.16, we will not obtain $\mu_0 I$ for the answer. On the other hand, if the wire is infinitely long, we cannot bring the closed path from position 1 to position 2 without cutting through the wire and there can be no surface whose perimeter is C and through which the wire does not pierce. The current enclosed by C is then uniquely defined. Similarly, for surfaces whose perimeter is position 2 of the closed path in Fig. 3.16, the infinitely long wire does not pierce at all or it pierces through an even number of times, entering from one side and emerging out on the same side so that the net current enclosed by the path is always zero. Thus we can summarize the discussion in this paragraph by stating that the current enclosed by a path is uniquely defined if the net current which passes through each possible surface whose perimeter is the closed path is the same.

EXAMPLE 3-8. An infinitely long filamentary wire along the z axis carries current I amp. Find $\int_{P}^{Q} \mathbf{B} \cdot d\mathbf{l}$ along the straight line joining P to Q, where P and Q are $(1, -1, 0)$ and $(1, 1, 0)$, respectively, in cartesian coordinates.

The geometry of the problem in the xy plane is shown in Fig. 3.17.

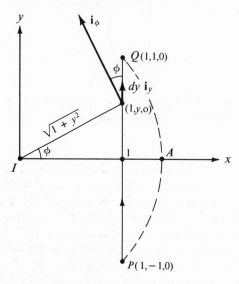

Fig. 3.17. For evaluating $\int_{P}^{Q} \mathbf{B} \cdot d\mathbf{l}$ along the straight line from P to Q in the field of an infinitely long wire carrying current I.

First we will solve this problem by actually evaluating $\int_{P}^{Q} \mathbf{B} \cdot d\mathbf{l}$ along the given path. To do this, let us consider an infinitesimal segment $d\mathbf{l} = dy\, \mathbf{i}_y$ at $(1, y, 0)$. Since $\mathbf{B}$ at this point due to the line current is $[\mu_0 I/(2\pi\sqrt{1 + y^2})]\mathbf{i}_\phi$, we have

$$\mathbf{B} \cdot d\mathbf{l} = \frac{\mu_0 I}{2\pi\sqrt{1 + y^2}}\mathbf{i}_\phi \cdot dy\, \mathbf{i}_y$$

$$= \frac{\mu_0 I\, dy}{2\pi\sqrt{1 + y^2}} \cos \phi = \frac{\mu_0 I\, dy}{2\pi(1 + y^2)}$$

Thus

$$\int_P^Q \mathbf{B} \cdot d\mathbf{l} = \int_{y=-1}^1 \frac{\mu_0 I \, dy}{2\pi(1+y^2)}$$

$$= \frac{\mu_0 I}{2\pi} \int_{\phi=-\pi/4}^{\pi/4} d\phi = \frac{\mu_0 I}{4} \tag{3-58}$$

This result can, however, be obtained without performing the integration if we note that, according to Ampere's circuital law,

$$\oint_{PQAP} \mathbf{B} \cdot d\mathbf{l} = 0 \tag{3-59}$$

where QAP is part of a circle centered at the line current. Equation (3-59) may be written as

$$\int_P^Q \mathbf{B} \cdot d\mathbf{l} + \int_{QAP} \mathbf{B} \cdot d\mathbf{l} = 0$$

which yields

$$\int_P^Q \mathbf{B} \cdot d\mathbf{l} = -\int_{QAP} \mathbf{B} \cdot d\mathbf{l} \tag{3-60}$$

However, from symmetry considerations, $\int_{QAP} \mathbf{B} \cdot d\mathbf{l}$ is equal to $-\mu_0 I(QAP)$ divided by the circumference of the circle, or $-\mu_0 I(\pi/2)/2\pi = -\mu_0 I/4$. From (3-60), we then obtain a value $\mu_0 I/4$ for $\int_P^Q \mathbf{B} \cdot d\mathbf{l}$, which agrees with (3-58).   ∎

Given $\mathbf{B}$ and a closed path C, it is always possible to compute the current enclosed by the path by evaluating $\oint_C \mathbf{B} \cdot d\mathbf{l}$ analytically or numerically and then dividing the result by μ_0 in accordance with Ampere's circuital law given by (3-57). The inverse problem of finding $\mathbf{B}$ for a given current distribution by using (3-57) is possible only for certain simple cases involving a high degree of symmetry, just as in the case of the application of Gauss' law for finding $\mathbf{E}$ for a given charge distribution. First, the symmetry of the magnetic field must be determined from the Biot–Savart law and second, we should be able to choose a closed path C such that $\oint_C \mathbf{B} \cdot d\mathbf{l}$ can be reduced to an algebraic quantity involving the magnitude of $\mathbf{B}$. Obviously, the closed path must be chosen such that the magnitude of $\mathbf{B}$ is uniform and the direction of $\mathbf{B}$ is tangential to the path along all or part of the path, while the magnitude of $\mathbf{B}$ is zero or the direction of $\mathbf{B}$ is normal to the path along the rest of the path in the latter case. We will illustrate this method of obtaining $\mathbf{B}$ by reconsidering Examples 3-6 and 3-7.

EXAMPLE 3-9.   A sheet of current with the surface current density given by

$$\mathbf{J}_s = J_{s0}\mathbf{i}_z$$

where J_{s0} is a constant, occupies the entire xz plane as shown in Fig. 3.18.

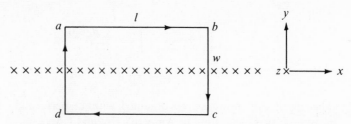

Fig. 3.18. For evaluating the magnetic flux density due to an infinite plane sheet of current.

The magnetic field due to such a current sheet was found in Example 3-6 by using the Biot–Savart law. It is here desired to find the magnetic flux density due to this infinite sheet of current using Ampere's circuital law.

From purely qualitative reasoning based upon the magnetic flux density due to an infinitely long, straight filamentary wire of current, we can conclude that the magnetic flux density due to the infinite sheet of current of uniform density is (a) entirely in the $+x$ direction for $y > 0$ and in the $-x$ direction for $y < 0$, (b) uniform in planes parallel to the current sheet, and (c) symmetrical about $y = 0$. Thus

$$\mathbf{B} = B_t \mathbf{i}_t \qquad (3\text{-}61)$$

where $\mathbf{i}_t$ is the unit tangential vector to the current sheet given by

$$\mathbf{i}_t = \mathbf{i}_z \times \mathbf{i}_n \qquad (3\text{-}62)$$

in which $\mathbf{i}_n$ is the unit normal vector to the current sheet. We can therefore choose a rectangular path *abcda* having length l parallel to the current sheet and width w normal to the current sheet and symmetrical about the current sheet as shown in Fig. 3.18. Then

$$\oint_{abcda} \mathbf{B} \cdot d\mathbf{l} = \int_a^b \mathbf{B} \cdot d\mathbf{l} + \int_b^c \mathbf{B} \cdot d\mathbf{l} + \int_c^d \mathbf{B} \cdot d\mathbf{l} + \int_d^a \mathbf{B} \cdot d\mathbf{l} \qquad (3\text{-}63)$$

But $\int_b^c \mathbf{B} \cdot d\mathbf{l}$ and $\int_d^a \mathbf{B} \cdot d\mathbf{l}$ are equal to zero since $\mathbf{B}$ is normal to the paths *bc* and *da*. For paths *ab* and *cd*, $\mathbf{B}$ is parallel and directed along these paths. Furthermore, the magnitudes of $\mathbf{B}$ are the same for these paths since they are equidistant from the current sheet. Thus (3-63) reduces to

$$\oint_{abcda} \mathbf{B} \cdot d\mathbf{l} = 2\int_a^b \mathbf{B} \cdot d\mathbf{l} = 2\int_a^b B_t \mathbf{i}_t \cdot d\mathbf{l}\, \mathbf{i}_t$$
$$= 2B_t \int_a^b dl = 2B_t l \qquad (3\text{-}64)$$

But, from Ampere's circuital law,

$$\oint_{abcda} \mathbf{B} \cdot d\mathbf{l} = \mu_0 \,(\text{current enclosed by } abcda) = \mu_0 J_{s0} l \qquad (3\text{-}65)$$

Comparing (3-64) and (3-65), we have

$$B_t = \frac{\mu_0 J_{s0}}{2} \tag{3-66}$$

$$\mathbf{B} = \frac{\mu_0 J_{s0}}{2} \mathbf{i}_z \times \mathbf{i}_n = \frac{\mu_0}{2} \mathbf{J}_s \times \mathbf{i}_n \tag{3-67}$$

which agrees with the result obtained in Example 3-6. ▮

EXAMPLE 3-10.   Current flows in the axial (z) direction in an infinitely long cylinder of radius a with uniform density J_0 amp/m² as shown in Fig. 3.19. The magnetic field due to such a current distribution was found in Example 3-7 by using the Biot–Savart law. It is here desired to find the magnetic flux density both inside and outside the cylinder using Ampere's circuital law.

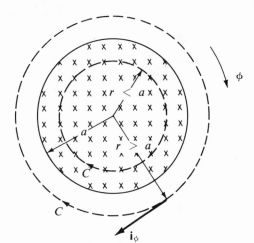

Fig. 3.19. For evaluating the magnetic flux density due to a volume current flowing with uniform density along an infinitely long cylinder.

In Example 3-7 we established from purely qualitative arguments that **B**, due to the given current distribution, has only a ϕ component and possesses cylindrical symmetry so that it is a function only of the distance from the axis of the cylinder. Thus

$$\mathbf{B} = B_\phi(r)\mathbf{i}_\phi \tag{3-68}$$

Choosing, therefore, a circular path C of radius $r \gtrless a$ centered at the axis of the cylinder and in the plane normal to the axis, as shown in Fig. 3.19, we have

$$\oint_C \mathbf{B} \cdot dl = \oint_C B_\phi \mathbf{i}_\phi \cdot dl\,\mathbf{i}_\phi = B_\phi \oint_C dl$$
$$= B_\phi(\text{circumference of the circle of radius } r) \tag{3-69}$$
$$= B_\phi(2\pi r)$$

But, from Ampere's circuital law,

$$\oint_C \mathbf{B} \cdot d\mathbf{l} = \mu_0(\text{current enclosed by } C)$$

$$= \mu_0(\text{current enclosed by circular path of radius } r) \tag{3-70}$$

Comparing (3-69) and (3-70), we have

$$B_\phi = \mu_0 \frac{\text{current enclosed by circular path of radius } r}{2\pi r}$$

$$\mathbf{B} = \mu_0 \frac{\text{current enclosed by circular path of radius } r}{2\pi r} \mathbf{i}_\phi \tag{3-71}$$

which agrees with the result of Example 3-7. ∎

3.7   Ampere's Circuital Law in Differential Form (Maxwell's Curl Equation for the Static Magnetic Field)

Let us consider a volume current distribution with the current density vector **J** as a given function of the coordinates. The current enclosed by an arbitrary closed path C is given by the surface integral of the current density over any surface S bounded by the closed path C; that is, $\int_S \mathbf{J} \cdot d\mathbf{S}$. According-ing to Ampere's circuital law (3-57), we then have

$$\oint_C \mathbf{B} \cdot d\mathbf{l} = \mu_0 \int_S \mathbf{J} \cdot d\mathbf{S} \tag{3-72}$$

where C is traversed in the sense in which a right-hand screw needs to be turned if it is to advance to the side of S towards which the current on the right side of (3-72) is evaluated. If we now shrink the path C to a very small size ΔC so that the surface area bounded by it becomes very small, ΔS, we can write (3-72) as

$$\oint_{\Delta C} \mathbf{B} \cdot d\mathbf{l} = \mu_0 \int_{\Delta S} \mathbf{J} \cdot d\mathbf{S} \tag{3-73}$$

Since the surface area ΔS is very small, we can consider the current density to be uniform over the surface so that $\int_{\Delta S} \mathbf{J} \cdot d\mathbf{S} \approx \mathbf{J} \cdot \mathbf{i}_n \Delta S$, where $\mathbf{i}_n$ is the normal vector to ΔS pointed to the side towards which a right-hand screw advances as it is turned in the sense of the closed path. This relation becomes exact in the limit $\Delta S \longrightarrow 0$. Dividing both sides of (3-73) by ΔS and letting $\Delta S \longrightarrow 0$, we have

$$\lim_{\Delta S \to 0} \frac{\oint_{\Delta C} \mathbf{B} \cdot d\mathbf{l}}{\Delta S} = \lim_{\Delta S \to 0} \frac{\mu_0 \int_{\Delta S} \mathbf{J} \cdot d\mathbf{S}}{\Delta S}$$

$$= \mu_0 \lim_{\Delta S \to 0} \frac{\mathbf{J} \cdot \mathbf{i}_n \Delta S}{\Delta S} \tag{3-74}$$

$$= \mu_0 \mathbf{J} \cdot \mathbf{i}_n$$

Now, the curl of **B** is defined as the vector having the magnitude given by the maximum value of the quantity on the left side of (3-74) and the direction given by the normal to the ΔS for which the quantity is maximized. Looking at the right side of (3-74), we note that this maximum value occurs for an orientation of ΔS for which the direction of $\mathbf{i}_n$ coincides with the direction of **J** and it is equal to μ_0 times the magnitude of *J*. Thus

$$|\nabla \times \mathbf{B}| = \text{maximum value of}\left(\lim_{\Delta S \to 0} \frac{\oint_{\Delta C} \mathbf{B} \cdot d\mathbf{l}}{\Delta S}\right) = \mu_0 |\mathbf{J}| \qquad (3\text{-}75a)$$

$$\text{direction of } \nabla \times \mathbf{B} = \text{direction of } \mathbf{J} \qquad (3\text{-}75b)$$

so that

$$\nabla \times \mathbf{B} = \mu_0 \mathbf{J} \qquad (3\text{-}76)$$

Equation (3-76) is Ampere's circuital law in differential form. It states that the curl of the magnetic flux density at any point is equal to μ_0 times the volume current density at that point. This is Maxwell's curl equation for the static magnetic field.

The right side of (3-76) represents a volume current density. For problems involving line and surface currents, we make use of Dirac delta functions just as in the case of Gauss' law in differential form for point charges, line charges, and surface charges. For example, following the method employed in Example 2-12, we obtain for a surface current of density $\mathbf{J}_s$ occupying the $y = y_0$ plane,

$$\nabla \times \mathbf{B} = \mu_0 \mathbf{J}_s \, \delta(y - y_0) \qquad (3\text{-}77)$$

3.8   Magnetic Vector Potential

Thus far we have discussed the determination of the magnetic field due to a current distribution directly from the current distribution using initially the Biot–Savart law and then Ampere's circuital law. In Chapter 2, we first discussed the determination of the electric field due to a charge distribution directly from the charge distribution using initially an integral formulation based on the electric field intensity due to a point charge and then Gauss' law. Later we introduced the electric potential field from energy considerations and discovered the relationship of the electric field intensity to the scalar potential through the gradient operation as an alternative approach to the determination of the electric field. In this section we introduce a similar alternative method for the computation of the magnetic field due to a given current distribution.

To do this, we note from (3-25) that, for a filamentary wire carrying current *I*, the magnetic flux density is given by

$$\mathbf{B}(\mathbf{r}) = \frac{\mu_0}{4\pi} \int_{C'} \frac{I \, d\mathbf{l}' \times \mathbf{i}_R(\mathbf{r}, \mathbf{r}')}{R^2(\mathbf{r}, \mathbf{r}')} \qquad (3\text{-}78)$$

where C' is the contour of the wire, $\mathbf{r}'$ is the position vector defining the infinitesimal length element $d\mathbf{l}'$ on C', $\mathbf{r}$ is the position vector of the field point, $\mathbf{i}_R(\mathbf{r}, \mathbf{r}')$ is the unit vector along $\mathbf{r} - \mathbf{r}'$, and $R(\mathbf{r}, \mathbf{r}')$ is equal to $|\mathbf{r} - \mathbf{r}'|$. Substituting

$$\mathbf{V}\left(\frac{1}{R}\right) = -\frac{1}{R^2}\mathbf{i}_R$$

in (3-78), we have

$$\mathbf{B}(\mathbf{r}) = -\frac{\mu_0 I}{4\pi}\int_{C'} d\mathbf{l}' \times \mathbf{V}\left[\frac{1}{R(\mathbf{r}, \mathbf{r}')}\right] \tag{3-79}$$

Using the vector identity

$$\mathbf{A} \times \mathbf{V}V = V\mathbf{V} \times \mathbf{A} - \mathbf{V} \times (V\mathbf{A})$$

we can write (3-79) as

$$\mathbf{B}(\mathbf{r}) = -\frac{\mu_0 I}{4\pi}\int_{C'}\left[\frac{1}{R(\mathbf{r}, \mathbf{r}')}\mathbf{V} \times d\mathbf{l}' - \mathbf{V} \times \frac{d\mathbf{l}'}{R(\mathbf{r}, \mathbf{r}')}\right] \tag{3-80}$$

In (3-80), the integration is with respect to the points on the filamentary wire, whereas the curl operation has to do with differentiation with respect to the coordinates of the field point. Hence $\mathbf{V} \times d\mathbf{l}' = 0$ and also, the two operations can be interchanged to give us

$$\mathbf{B} = \frac{\mu_0 I}{4\pi}\int_{C'}\mathbf{V} \times \frac{d\mathbf{l}'}{R(\mathbf{r}, \mathbf{r}')} = \mathbf{V} \times \left(\frac{\mu_0}{4\pi}\int_{C'}\frac{I\,d\mathbf{l}'}{R}\right) \tag{3-81a}$$

If, instead of a filamentary wire, we have a surface current of density $\mathbf{J}_s$ on a surface S', or a volume current of density $\mathbf{J}$ in a volume V', we obtain similar relationships as follows, respectively:

$$\mathbf{B} = \mathbf{V} \times \left(\frac{\mu_0}{4\pi}\int_{S'}\frac{\mathbf{J}_s\,dS'}{R}\right) \tag{3-81b}$$

$$\mathbf{B} = \mathbf{V} \times \left(\frac{\mu_0}{4\pi}\int_{V'}\frac{\mathbf{J}\,dv'}{R}\right) \tag{3-81c}$$

In (3-81a)–(3-81c), we have expressions which permit us to compute $\mathbf{B}$ by finding the curl of a vector quantity. Denoting this vector quantity as $\mathbf{A}$, we have

$$\mathbf{B} = \mathbf{V} \times \mathbf{A} \tag{3-82}$$

where

$$\mathbf{A} = \frac{\mu_0}{4\pi}\int_{C'}\frac{I\,d\mathbf{l}'}{R} \qquad \text{for line current} \tag{3-83a}$$

$$\mathbf{A} = \frac{\mu_0}{4\pi}\int_{S'}\frac{\mathbf{J}_s\,dS'}{R} \qquad \text{for surface current} \tag{3-83b}$$

$$\mathbf{A} = \frac{\mu_0}{4\pi}\int_{V'}\frac{\mathbf{J}\,dv'}{R} \qquad \text{for volume current} \tag{3-83c}$$

We note the similarity of the right sides of (3-83a)–(3-83c) with the expressions for the electrostatic potential V due to line, surface, and volume charges given, respectively, by

$$V = \frac{1}{4\pi\epsilon_0} \int_{C'} \frac{\rho_L \, dl'}{R} \qquad \text{for line charge}$$

$$V = \frac{1}{4\pi\epsilon_0} \int_{S'} \frac{\rho_S \, dS'}{R} \qquad \text{for surface charge}$$

$$V = \frac{1}{4\pi\epsilon_0} \int_{V'} \frac{\rho \, dv'}{R} \qquad \text{for volume charge}$$

In view of this similarity, and since $\mathbf{A}$ is a vector in contrast to the scalar nature of V, $\mathbf{A}$ is called the magnetic vector potential. Unlike V, $\mathbf{A}$ does not have a physical significance. It serves as a convenient intermediate step for the computation of $\mathbf{B}$. This is especially so because of the similarity of the expressions for V and the expressions for $\mathbf{A}$. The components of $\mathbf{A}$ due to a particular current distribution can be written without actually evaluating the integrals if the analogous integrals for the electrostatic potential have already been evaluated in the corresponding electrostatic problem.

EXAMPLE 3-11.   An infinitely long straight wire carrying current I amp lies along the z axis. Obtain the magnetic vector potential due to this wire and then find the magnetic flux density by performing the curl operation on the vector potential.

Applying (3-83a) to the infinitely long wire, we have the vector potential given by

$$\mathbf{A} = \frac{\mu_0}{4\pi} \int_{z'=-\infty}^{\infty} \left(\frac{I \, dz' \, \mathbf{i}_z}{R} \right)$$

or

$$\mathbf{A} = \left(\frac{\mu_0}{4\pi} \int_{z'=-\infty}^{\infty} \frac{I \, dz'}{R} \right) \mathbf{i}_z \qquad (3\text{-}84)$$

where R is the distance of the point P at which $\mathbf{A}$ is to be computed from an infinitesimal current element $I \, dz' \, \mathbf{i}_z$, as shown in Fig. 3.20. Let us now consider the quantity

$$\left(\frac{1}{4\pi\epsilon_0} \int_{z'=-\infty}^{\infty} \frac{\rho_{L0} \, dz'}{R} \right)$$

This is the integral for computing the electrostatic potential due to an infinitely long line charge of uniform density ρ_{L0} lying along the z axis. This expression is analogous to the expression inside the parentheses on the right side of (3-84). Thus, finding the vector potential due to the infinitely long wire is analogous to determining the electrostatic potential due to the infinitely long line charge of uniform density. However, we already know the

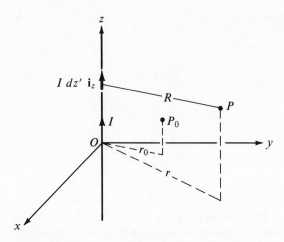

Fig. 3.20. For evaluating the magnetic vector potential due to an infinitely long, straight wire carrying current *I*.

solution for this electrostatic potential from Example 2-17. This is given by

$$V = -\frac{\rho_{L0}}{2\pi\epsilon_0} \ln \frac{r}{r_0} \qquad (2\text{-}119)$$

where r is the distance of the point P, at which V is desired, from the line charge and r_0 is the distance from the line charge to the point at which the potential is zero, as explained in Example 2-17. Thus

$$\frac{1}{4\pi\epsilon_0} \int_{z'=-\infty}^{\infty} \frac{\rho_{L0}\,dz'}{R} = -\frac{\rho_{L0}}{2\pi\epsilon_0} \ln \frac{r}{r_0} \qquad (3\text{-}85)$$

We can immediately write down by analogy that

$$\frac{\mu_0}{4\pi} \int_{z'=-\infty}^{\infty} \frac{I\,dz'}{R} = -\frac{\mu_0 I}{2\pi} \ln \frac{r}{r_0} \qquad (3\text{-}86)$$

Substituting this result into (3-84), we obtain the vector potential due to the infinitely long wire as

$$\mathbf{A} = -\frac{\mu_0 I}{2\pi} \ln \frac{r}{r_0} \mathbf{i}_z \qquad (3\text{-}87)$$

Using the expression for the curl in cylindrical coordinates, we then have

$$\mathbf{B} = \nabla \times \mathbf{A} = \begin{vmatrix} \dfrac{\mathbf{i}_r}{r} & \mathbf{i}_\phi & \dfrac{\mathbf{i}_z}{r} \\[2mm] \dfrac{\partial}{\partial r} & \dfrac{\partial}{\partial \phi} & \dfrac{\partial}{\partial z} \\[2mm] 0 & 0 & A_z \end{vmatrix}$$

$$= \frac{1}{r}\frac{\partial A_z}{\partial \phi}\mathbf{i}_r - \frac{\partial A_z}{\partial r}\mathbf{i}_\phi = \frac{\mu_0 I}{2\pi r}\mathbf{i}_\phi$$

which is the same as the result obtained in Example 3-3. ▮

EXAMPLE 3-12. A loop of wire carrying current I amp occupies an arbitrary contour C' as shown in Fig. 3.21. Find the vector potential due to this current loop at distances $\mathbf{r}$ from the origin large in magnitude compared to the distances of the points on the loop from the origin.

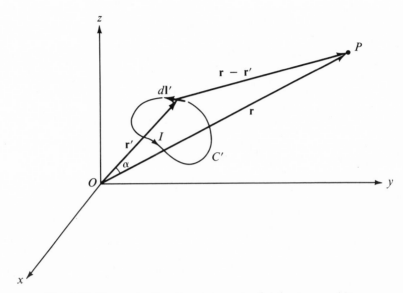

Fig. 3.21. For evaluating the vector potential due to an arbitrary loop of current I at large distances from the origin compared to the distances of the points on the loop from the origin.

Let P be the point at which the vector potential is desired. Then, from (3-83a), the vector potential at P due to the current loop is given by

$$
\begin{aligned}
\mathbf{A}(\mathbf{r}) &= \frac{\mu_0}{4\pi} \oint_{C'} \frac{I\, d\mathbf{l'}}{|\mathbf{r} - \mathbf{r'}|} \\
&= \frac{\mu_0 I}{4\pi} \oint_{C'} \frac{d\mathbf{l'}}{(r^2 + r'^2 - 2rr' \cos \alpha)^{1/2}} \\
&= \frac{\mu_0 I}{4\pi r} \oint_{C'} \left(1 + \frac{r'^2}{r^2} - \frac{2\mathbf{r'} \cdot \mathbf{r}}{r^2}\right)^{-1/2} d\mathbf{l'}
\end{aligned}
\tag{3-88}
$$

Using the binomial expansion employed in Example 2-15, we have

$$
\begin{aligned}
\mathbf{A} &= \frac{\mu_0 I}{4\pi r} \oint_{C'} \left\{ 1 + \frac{\mathbf{r'} \cdot \mathbf{r}}{r^2} + \frac{1}{2r^4}[3(\mathbf{r'} \cdot \mathbf{r})^2 - r^2 r'^2] \right. \\
&\quad \left. + \ldots \text{higher-order terms} \right\} d\mathbf{l'}
\end{aligned}
\tag{3-89}
$$

$$
= \frac{\mu_0 I}{4\pi r} \left[\oint_{C'} d\mathbf{l'} + \oint_{C'} \frac{\mathbf{r'} \cdot \mathbf{r}}{r^2} d\mathbf{l'} + \oint_{C'} \frac{3(\mathbf{r'} \cdot \mathbf{r})^2 - r^2 r'^2}{2r^4} d\mathbf{l'} + \ldots \right]
$$

In (3-89), $\oint_{C'} d\mathbf{l}' = 0$ for any C' so that the second term is the first significant term. Furthermore, for $r' \ll r$, it is sufficient if we consider the first
significant term. Thus, for $r \gg r'$,

$$\mathbf{A} = \frac{\mu_0 I}{4\pi r^3} \oint_{C'} (\mathbf{r}' \cdot \mathbf{r}) \, d\mathbf{l}' \tag{3-90}$$

Now, using the vector identity

$$\mathbf{A} \times (\mathbf{B} \times \mathbf{C}) = (\mathbf{A} \cdot \mathbf{C})\mathbf{B} - (\mathbf{A} \cdot \mathbf{B})\mathbf{C}$$

we have

$$\mathbf{r} \times (d\mathbf{l}' \times \mathbf{r}') = (\mathbf{r} \cdot \mathbf{r}') \, d\mathbf{l}' - (\mathbf{r} \cdot d\mathbf{l}')\mathbf{r}'$$

or

$$
\begin{aligned}
(\mathbf{r}' \cdot \mathbf{r}) \, d\mathbf{l}' &= \mathbf{r} \times (d\mathbf{l}' \times \mathbf{r}') + (\mathbf{r} \cdot d\mathbf{l}')\mathbf{r}' \\
&= \tfrac{1}{2}\mathbf{r} \times (d\mathbf{l}' \times \mathbf{r}') + \tfrac{1}{2}(\mathbf{r} \cdot d\mathbf{l}')\mathbf{r}' + \tfrac{1}{2}(\mathbf{r}' \cdot \mathbf{r}) \, d\mathbf{l}'
\end{aligned}
\tag{3-91}
$$

We further note that

$$
\begin{aligned}
(\mathbf{r} \cdot d\mathbf{l}')&\mathbf{r}' + (\mathbf{r}' \cdot \mathbf{r}) \, d\mathbf{l}' \\
&= (x \, dx' + y \, dy' + z \, dz')(x'\mathbf{i}_x + y'\mathbf{i}_y + z'\mathbf{i}_z) \\
&\quad + (x'x + y'y + z'z)(dx' \, \mathbf{i}_x + dy' \, \mathbf{i}_y + dz' \, \mathbf{i}_z) \\
&= (2xx' \, dx' + yy' \, dx' + zz' \, dx' + yx' \, dy' + zx' \, dz')\mathbf{i}_x \\
&\quad + (xx' \, dy' + 2yy' \, dy' + zz' \, dy' + xy' \, dx' + zy' \, dz')\mathbf{i}_y \\
&\quad + (xx' \, dz' + yy' \, dz' + 2zz' \, dz' + xz' \, dx' + yz' \, dy')\mathbf{i}_z \\
&= d[(xx' + yy' + zz')(x'\mathbf{i}_x + y'\mathbf{i}_y + z'\mathbf{i}_z)] \\
&= d[(\mathbf{r} \cdot \mathbf{r}')\mathbf{r}']
\end{aligned}
\tag{3-92}
$$

Substituting (3-92) into (3-91), we obtain

$$(\mathbf{r}' \cdot \mathbf{r}) \, d\mathbf{l}' = \tfrac{1}{2}\mathbf{r} \times (d\mathbf{l}' \times \mathbf{r}') + \tfrac{1}{2} \, d[(\mathbf{r} \cdot \mathbf{r}')\mathbf{r}'] \tag{3-93}$$

Substituting (3-93) into (3-90), we have

$$
\begin{aligned}
\mathbf{A} &= \frac{\mu_0 I}{4\pi r^3} \oint_{C'} \frac{1}{2}\mathbf{r} \times (d\mathbf{l}' \times \mathbf{r}') + \frac{\mu_0 I}{8\pi r^3} \oint_{C'} d[(\mathbf{r} \cdot \mathbf{r}')\mathbf{r}'] \\
&= \frac{\mu_0 I}{4\pi r^3} \oint_{C'} \frac{1}{2}\mathbf{r} \times (d\mathbf{l}' \times \mathbf{r}')
\end{aligned}
\tag{3-94}
$$

since the second integral, being an integral of a total differential around a
closed contour, is equal to zero. Finally, defining

$$\mathbf{m} = \frac{1}{2} \oint_{C'} \mathbf{r}' \times I \, d\mathbf{l}' \tag{3-95}$$

we obtain

$$
\begin{aligned}
\mathbf{A} &= \frac{\mu_0}{4\pi r^3} \left[\oint_{C'} \frac{1}{2}(\mathbf{r}' \times I \, d\mathbf{l}') \right] \times \mathbf{r} \\
&= \frac{\mu_0}{4\pi r^3} \mathbf{m} \times \mathbf{r}
\end{aligned}
\tag{3-96}
$$

Thus, at large distances from the current loop, the vector potential falls off inversely as r^2 in contrast to the inverse distance dependence of the electrostatic potential at large distances from an arbitrary charge distribution, provided the total charge is not equal to zero. The quantity **m** is the dipole moment of the current loop about the origin. ∎

EXAMPLE 3-13.   Show that, for a plane loop of wire carrying current *I*, the dipole moment **m** given by

$$\mathbf{m} = \frac{1}{2} \oint_{C'} \mathbf{r}' \times I \, d\mathbf{l}'$$

has a magnitude equal to the area of the loop and a direction normal to the plane of the loop drawn towards the direction of advance of a right-hand screw as it is turned in the sense of the contour *C'* of the loop.

First we show that the dipole moment about the origin is the same as the dipole moment about any other point. Letting the position vector of this arbitrary point be $\mathbf{r}_0$, we have

$$
\begin{aligned}
\mathbf{m} &= \frac{1}{2} \oint_{C'} \mathbf{r}' \times I \, d\mathbf{l}' \\
&= \frac{1}{2} \oint_{C'} (\mathbf{r}' - \mathbf{r}_0) \times I \, d\mathbf{l}' + \frac{1}{2} \oint_{C'} \mathbf{r}_0 \times I \, d\mathbf{l}' \\
&= \frac{1}{2} \oint_{C'} (\mathbf{r}' - \mathbf{r}_0) \times I \, d\mathbf{l}' + \frac{1}{2} I \mathbf{r}_0 \times \oint_{C'} d\mathbf{l}' \\
&= \frac{1}{2} \oint_{C'} (\mathbf{r}' - \mathbf{r}_0) \times I \, d\mathbf{l}'
\end{aligned}
\tag{3-97}
$$

since $\oint_{C'} d\mathbf{l}' = 0$. Thus the dipole moment of the current loop is independent of the point about which it is computed. Let us therefore choose this point to be in the plane of the loop and inside the loop as shown in Fig. 3.22. Then

$$\tfrac{1}{2}(\mathbf{r}' - \mathbf{r}_0) \times d\mathbf{l}' = \tfrac{1}{2} |\mathbf{r}' - \mathbf{r}_0| \, dl' \sin \alpha \, \mathbf{i}_n \tag{3-98}$$

where α is the angle between $(\mathbf{r}' - \mathbf{r}_0)$ and $d\mathbf{l}'$ and $\mathbf{i}_n$ is the normal vector to the plane of the loop, drawn towards the direction of advance of a right-hand screw as it is turned in the sense of *C'* as shown in Fig. 3.22. But the magnitude on the right side of (3-98) is the area of the triangle formed by $(\mathbf{r}' - \mathbf{r}_0)$ and $d\mathbf{l}'$. Thus, since $(\mathbf{r}' - \mathbf{r}_0) \times d\mathbf{l}'$ is along $\mathbf{i}_n$ for all $d\mathbf{l}'$ on *C'*, we have

$$\frac{1}{2} \oint_{C'} (\mathbf{r}' - \mathbf{r}_0) \times d\mathbf{l}' = \begin{pmatrix} \text{sum of areas of triangles formed by all} \\ d\mathbf{l}' \text{ with the corresponding } \mathbf{r}' - \mathbf{r}_0 \end{pmatrix} \mathbf{i}_n \tag{3-99}$$

$$= (\text{area of the loop})\mathbf{i}_n$$

This result is consistent with the dipole moment defined in Example 3-4 for the plane circular loop of radius *a* lying in the *xy* plane. ∎

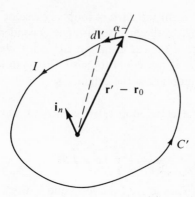

Fig. 3.22. For evaluating the dipole moment of a plane loop of wire carrying current I.

Returning to Eq. (3-82) and taking the curl of both sides, we obtain

$$\mathbf{V} \times \mathbf{B} = \mathbf{V} \times \mathbf{V} \times \mathbf{A} = \mathbf{V}(\mathbf{V} \cdot \mathbf{A}) - \mathbf{V}^2\mathbf{A} \qquad (3\text{-}100)$$

where we have used the vector identity for $\mathbf{V} \times \mathbf{V} \times \mathbf{A}$. But, from Ampere's circuital law in differential form, we have

$$\mathbf{V} \times \mathbf{B} = \mu_0\mathbf{J} \qquad (3\text{-}76)$$

Thus, from (3-100) and (3-76), we get

$$\mathbf{V}(\mathbf{V} \cdot \mathbf{A}) - \mathbf{V}^2\mathbf{A} = \mu_0\mathbf{J} \qquad (3\text{-}101)$$

However, considering a current loop, we have

$$\begin{aligned}
\mathbf{V} \cdot \mathbf{A} &= \mathbf{V} \cdot \oint_{C'} \frac{\mu_0 I \, d\mathbf{l}'}{4\pi R} \\
&= \frac{\mu_0 I}{4\pi} \oint_{C'} \mathbf{V} \cdot \frac{d\mathbf{l}'}{R}
\end{aligned} \qquad (3\text{-}102)$$

where C' is the contour of the current loop and $d\mathbf{l}'$ is an infinitesimal length element on C'. Using the vector identity

$$\mathbf{V} \cdot V\mathbf{A} = \mathbf{A} \cdot \mathbf{V}V + V\mathbf{V} \cdot \mathbf{A}$$

we write (3-102) as

$$\mathbf{V} \cdot \mathbf{A} = \frac{\mu_0 I}{4\pi}\left(\oint_{C'} d\mathbf{l}' \cdot \mathbf{V}\frac{1}{R} + \oint_{C'} \frac{1}{R}\mathbf{V} \cdot d\mathbf{l}' \right) \qquad (3\text{-}103)$$

On the right side of (3-103), the second integral is zero since $\mathbf{V} \cdot d\mathbf{l}' = 0$. Using $\mathbf{V}(1/R) = -\mathbf{V}'(1/R)$ where the prime denotes differentiation with respect to the primed variables, and then using Stoke's theorem, the first integral can be written as

$$\oint_{C'} d\mathbf{l}' \cdot \mathbf{V}\frac{1}{R} = -\oint_{C'} \mathbf{V}'\frac{1}{R} \cdot d\mathbf{l}' = -\int_{S'} \mathbf{V}' \times \mathbf{V}'\frac{1}{R} \cdot d\mathbf{S}' \qquad (3\text{-}104)$$

where S' is any surface whose perimeter is C'. But the curl of the gradient of a scalar is identically equal to zero. Hence, the right side of (3-104) is zero. Thus, for a current loop, $\mathbf{V} \cdot \mathbf{A} = 0$. If we now consider a region of volume current in which there is no accumulation of charge, we can represent the volume current as a superposition of a number of current loops for each of which $\mathbf{V} \cdot \mathbf{A} = 0$ so that, for the entire volume current, $\mathbf{V} \cdot \mathbf{A} = 0$. Substituting this result in (3-101), we obtain

$$\nabla^2 \mathbf{A} = -\mu_0 \mathbf{J} \tag{3-105}$$

In analogy with

$$\nabla^2 V = -\frac{\rho}{\epsilon_0} \tag{2-140}$$

Equation (3-105) is known as the Poisson's equation for the vector potential. It is a differential equation which relates the magnetic vector potential at a point to the volume current density at that point, just as (2-140) is a differential equation which relates the electrostatic potential at a point to the volume charge density at that point. Equation (3-105) is a vector equation and hence it is equivalent to three scalar equations. For example, in rectangular coordinates,

$$\nabla^2 \mathbf{A} = (\nabla^2 A_x)\mathbf{i}_x + (\nabla^2 A_y)\mathbf{i}_y + (\nabla^2 A_z)\mathbf{i}_z$$

so that we have

$$\nabla^2 A_x = -\mu_0 J_x \tag{3-106a}$$

$$\nabla^2 A_y = -\mu_0 J_y \tag{3-106b}$$

$$\nabla^2 A_z = -\mu_0 J_z \tag{3-106c}$$

If the volume current density is zero in a region, then the right side of (3-105) is zero for that region so that (3-105) reduces to

$$\nabla^2 \mathbf{A} = 0 \quad \text{for } \mathbf{J} = 0 \tag{3-107}$$

which is Laplace's equation for the magnetic vector potential, in analogy with Laplace's equation for the electrostatic potential given by

$$\nabla^2 V = 0 \quad \text{for } \rho = 0 \tag{2-141}$$

It states that the Laplacian of the magnetic vector potential in a region devoid of current is zero, just as (2-141) states that the Laplacian of the electrostatic potential in a region devoid of charges is zero. Again, using the expansion for $\nabla^2 \mathbf{A}$ in rectangular coordinates, we obtain the three component equations for (3-107) as

$$\nabla^2 A_x = 0 \tag{3-108a}$$

$$\nabla^2 A_y = 0 \tag{3-108b}$$

$$\nabla^2 A_z = 0 \tag{3-108c}$$

For a given current distribution, the solution to Poisson's equation (3-105) is obtained by solving the three component equations (3-106a)–(3-106c). Again, we can take advantage of the similarity of (3-106a)–(3-106c) with (2-140) and in many cases simply write down the solution from previous knowledge of electrostatics, without the necessity of solving the differential equations.

3.9 Maxwell's Divergence Equation for the Magnetic Field

The divergence of the curl of a vector is identically zero. Since

$$\mathbf{B} = \mathbf{\nabla} \times \mathbf{A} \tag{3-82}$$

it then follows that

$$\mathbf{\nabla} \cdot \mathbf{B} = 0 \tag{3-109}$$

Equation (3-109) is Maxwell's divergence equation for the magnetic field. Together with Maxwell's curl equation for the static magnetic field given by (3-76), (3-109) completely defines the properties of the static magnetic field. Equation (3-109) determines whether or not a given vector field is realizable as a magnetic field, whereas Eq. (3-76) relates the field to the current distribution responsible for producing the field. When compared with Maxwell's divergence equation for the electric field intensity,

$$\mathbf{\nabla} \cdot \mathbf{E} = \frac{\rho}{\epsilon_0} \tag{2-82}$$

Eq. (3-109) reveals the fact that isolated magnetic charges do not exist.

Taking the volume integral of both sides of (3-109) in a volume V, we have

$$\int_V (\mathbf{\nabla} \cdot \mathbf{B}) \, dv = 0 \tag{3-110}$$

But, according to the divergence theorem,

$$\oint_S \mathbf{B} \cdot d\mathbf{S} = \int_V (\mathbf{\nabla} \cdot \mathbf{B}) \, dv$$

where S is the surface bounding the volume V. Since (3-110) is true for any volume, we obtain the result that

$$\oint_S \mathbf{B} \cdot d\mathbf{S} = 0 \tag{3-111}$$

for any closed surface S. Equation (3-111) is the integral form of the divergence equation (3-109). Since $\mathbf{B}$ is the magnetic flux density, $\oint_S \mathbf{B} \cdot d\mathbf{S}$ is the total magnetic flux emanating from the surface S. Thus Eq. (3-111) states that the total magnetic flux emanating from any closed surface is equal to zero. Whatever flux goes into the volume bounded by the surface must come out of it. The magnetic field lines form closed paths, unlike electric field lines which begin from positive charges and terminate on negative charges. Since

$$\oint_C \mathbf{B} \cdot d\mathbf{l} = \mu_0 (\text{current enclosed by } C) \tag{3-57}$$

the closed paths must form around the current producing the magnetic field. Vectors which, in this manner, are characterized by zero net flux over all possible closed surfaces are said to be solenoidal. The current density vector $\mathbf{J}$ for static fields is another example of a solenoidal vector since, from

$$\mathbf{\nabla} \cdot \mathbf{\nabla} \times \mathbf{B} = 0 \qquad (3\text{-}112)$$

we have

$$\mathbf{\nabla} \cdot \mu_0 \mathbf{J} = 0$$

or

$$\mathbf{\nabla} \cdot \mathbf{J} = 0 \qquad (3\text{-}113)$$

The solenoidal nature of $\mathbf{J}$ follows from the fact that, in the absence of accumulation of charge at a point with time, current must flow in closed paths. Since we are here considering static phenomena, there cannot be any accumulation of charge and hence $\mathbf{\nabla} \cdot \mathbf{J} = 0$. On the other hand, when we consider time-varying or dynamic fields, we can allow for the accumulation of charge, in which case we will find that (3-113) does not necessarily hold everywhere.

EXAMPLE 3-14.   Determine if the following vector fields are realizable as magnetic fields:

(a) $\mathbf{F}_a = (-y\mathbf{i}_x + x\mathbf{i}_y)$ cartesian coordinates

(b) $\mathbf{F}_b = \dfrac{\mu_0 m_L}{2\pi r^2} (-\sin \phi \, \mathbf{i}_r + \cos \phi \, \mathbf{i}_\phi)$ cylindrical coordinates

(c) $\mathbf{F}_c = (\sin \theta \, \mathbf{i}_r + \cos \theta \, \mathbf{i}_\theta)$ spherical coordinates

(a) $\mathbf{\nabla} \cdot \mathbf{F}_a = \dfrac{\partial}{\partial x}(-y) + \dfrac{\partial}{\partial y}(x) = 0$

Hence $\mathbf{F}_a$ can be realized as a magnetic field. In fact, if we note that, in cylindrical coordinates, $\mathbf{F}_a = r\mathbf{i}_\phi$, the solenoidal nature of $\mathbf{F}_a$ becomes obvious.

(b) $\mathbf{\nabla} \cdot \mathbf{F}_b = \dfrac{1}{r} \dfrac{\partial}{\partial r}\left(-\dfrac{\mu_0 m_L}{2\pi r} \sin \phi\right) + \dfrac{1}{r} \dfrac{\partial}{\partial \phi}\left(\dfrac{\mu_0 m_L}{2\pi r^2} \cos \phi\right) = 0$

Hence $\mathbf{F}_b$ can be realized as a magnetic field. It is left as an exercise (Problem 3.21) for the student to show that $\mathbf{F}_b$ is the magnetic field due to a two-dimensional magnetic dipole of moment m_L.

(c) $\mathbf{\nabla} \cdot \mathbf{F}_c = \dfrac{1}{r^2} \dfrac{\partial}{\partial r}(r^2 \sin \theta) + \dfrac{1}{r \sin \theta} \dfrac{\partial}{\partial \theta}(\sin \theta \cos \theta) \neq 0$

Hence $\mathbf{F}_c$ cannot be realized as a magnetic field.   ∎

EXAMPLE 3-15.   In Example 3-5, the magnetic field due to an infinitely long, uniformly wound solenold of radius a and n turns per unit length carrying current I was found by using the Biot–Savart law. It is here desired to find the magnetic field due to the solenoid from Ampere's circuital law and the solenoidal character of the magnetic field.

Employing a cylindrical coordinate system with the z axis as the axis of the solenoid, let us assume that the magnetic field due to the solenoid has all three components B_r, B_ϕ, and B_z. Because of the cylindrical symmetry and infinite length of the solenoid, all three components must be independent of ϕ and z. Thus B_r, B_ϕ, and B_z can be functions of r only. Now, applying (3-111) to a cylindrical box of radius b, length l and coaxial with the solenoid, as shown in Fig. 3.23(a), we have

$$\oint_{\substack{\text{surface of the}\\\text{cylindrical box}}} \mathbf{B} \cdot d\mathbf{S} = 0 \tag{3-114}$$

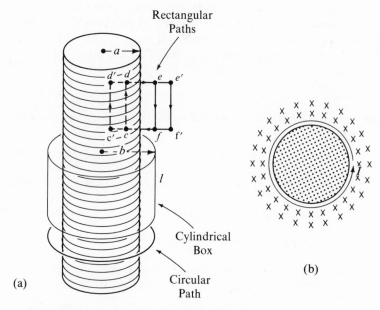

Fig. 3.23. For evaluating the magnetic field due to an infinitely long, uniformly wound solenoid using Ampere's circuital law and the solenoidal character of the magnetic field.

But

$$\oint_{\substack{\text{surface of the}\\\text{cylindrical box}}} \mathbf{B} \cdot d\mathbf{S} = \int_{\substack{\text{cylindrical}\\\text{surface}}} \mathbf{B} \cdot d\mathbf{S} + \int_{\substack{\text{upper plane}\\\text{surface}}} \mathbf{B} \cdot d\mathbf{S} + \int_{\substack{\text{lower plane}\\\text{surface}}} \mathbf{B} \cdot d\mathbf{S}$$

$$\tag{3-115}$$

On the cylindrical surface,

$$\mathbf{B} \cdot d\mathbf{S} = [B_r \mathbf{i}_r + B_\phi \mathbf{i}_\phi + B_z \mathbf{i}_z]_{r=b} \cdot b \, d\phi \, dz \, \mathbf{i}_r = [B_r]_{r=b} b \, d\phi \, dz$$

$$\int \mathbf{B} \cdot d\mathbf{S} = \int_{z=z}^{z+l} \int_{\phi=0}^{2\pi} [B_r]_{r=b} b \, d\phi \, dz = 2\pi b l [B_r]_{r=b} \tag{3-116}$$

since $[B_r]_{r=b}$ is a constant.

On the upper plane surface,

$$\mathbf{B} \cdot d\mathbf{S} = (B_r \mathbf{i}_r + B_\phi \mathbf{i}_\phi + \mathbf{B}_z \mathbf{i}_z) \cdot r \, dr \, d\phi \, \mathbf{i}_z = B_z(r) r \, dr \, d\phi \qquad (3\text{-}117a)$$

On the lower plane surface

$$\mathbf{B} \cdot d\mathbf{S} = (B_r \mathbf{i}_r + B_\phi \mathbf{i}_\phi + B_z \mathbf{i}_z) \cdot (-r \, dr \, d\phi \, \mathbf{i}_z) = -B_z(r) r \, dr \, d\phi \qquad (3\text{-}117b)$$

We see from (3-117a) and (3-117b) that $\int \mathbf{B} \cdot d\mathbf{S}$ on the upper plane surface cancels exactly with $\int \mathbf{B} \cdot d\mathbf{S}$ on the lower plane surface since the integrands are equal and opposite and the limits of integration are the same. Thus

$$\oint_{\substack{\text{surface of the} \\ \text{cylindrical box}}} \mathbf{B} \cdot d\mathbf{S} = 2\pi b l [B_r]_{r=b} \qquad (3\text{-}118)$$

Comparing (3-118) and (3-114), we obtain the result that $[B_r]_{r=b} = 0$. Since the radius b can be chosen to be any value, it follows that

$$B_r = 0 \qquad \text{for all } r$$

Applying Ampere's circuital law to a circular path of radius b, as shown in Fig. 3.23(a) in the plane normal to the axis of the solenoid and centered at the axis of the solenoid, we have

$$\oint_{\substack{\text{circular} \\ \text{path}}} \mathbf{B} \cdot d\mathbf{l} = 0 \qquad (3\text{-}119)$$

since the path does not enclose any current. But, along the circular path,

$$\mathbf{B} \cdot d\mathbf{l} = [B_r \mathbf{i}_r + B_\phi \mathbf{i}_\phi + B_z \mathbf{i}_z]_{r=b} \cdot b \, d\phi \, \mathbf{i}_\phi = [B_\phi]_{r=b} b \, d\phi$$

$$\oint \mathbf{B} \cdot d\mathbf{l} = \int_{\phi=0}^{2\pi} [B_\phi]_{r=b} b \, d\phi = 2\pi b [B_\phi]_{r=b} \qquad (3\text{-}120)$$

since $[B_\phi]_{r=b}$ is a constant. Comparing (3-120) with (3-119), we obtain the result that $[B_\phi]_{r=b} = 0$. Since the radius b can be chosen to be any value, it follows that

$$B_\phi = 0 \qquad \text{for all } r$$

Thus the magnetic field due to the solenoid has only a z component and we are now left with the task of finding this component.

Applying Ampere's circuital law for two rectangular paths *cdefc* and *cde'f'c* in the plane containing the solenoid axis, as shown in Fig. 3.23(a), we have

$$\oint_{cdefc} \mathbf{B} \cdot d\mathbf{l} = \oint_{cde'f'c} \mathbf{B} \cdot d\mathbf{l} = \mu_0 n I(cd) \qquad (3\text{-}121)$$

Since the three sides *cd*, *de*, and *fc* are common to the two rectangular paths, (3-121) gives us

$$\int_e^f \mathbf{B} \cdot d\mathbf{l} = \int_{e'}^{f'} \mathbf{B} \cdot d\mathbf{l} \qquad (3\text{-}122)$$

Along paths *ef* and *e'f'*,

$$\mathbf{B} \cdot d\mathbf{l} = [B_r \mathbf{i}_r + B_\phi \mathbf{i}_\phi + B_z \mathbf{i}_z] \cdot dz\, \mathbf{i}_z = B_z\, dz$$

and since B_z is independent of z, (3-122) yields

$$[B_z]_{ef}(ef) = [B_z]_{e'f'}(e'f')$$

or

$$[B_z]_{ef} = [B_z]_{e'f'} \tag{3-123}$$

Thus B_z is independent of r (in addition to ϕ and z) outside the solenoid. Similarly, by applying Ampere's circuital law to the two rectangular paths *cdefc* and *c'd'efc'* in the plane containing the solenoid axis, we can show that B_z is independent of r (in addition to ϕ and z) inside the solenoid. Thus the values of B_z both inside and outside the solenoid are constants. This requires that B_z outside the solenoid be equal to zero since, if it is nonzero, the amount of magnetic flux outside the solenoid will be infinity and for this flux to return in the opposite direction inside the solenoid as shown in Fig. 3.23(b), the flux density inside the solenoid must be infinity. But then, if the flux density inside the solenoid is infinity and that outside the solenoid is finite, (3-121) cannot be satisfied. On the other hand, for a finite amount of flux inside the solenoid in one direction to return in the opposite direction outside the solenoid, it requires zero flux density outside the solenoid since the area of cross section outside the solenoid is infinity ($\infty \times 0 = $ nonzero). Thus we conclude that B_z is zero outside the solenoid. It remains to evaluate B_z inside the solenoid. To do this, we write (3-121) as

$$\int_c^d \mathbf{B} \cdot d\mathbf{l} + \int_d^e \mathbf{B} \cdot d\mathbf{l} + \int_e^f \mathbf{B} \cdot d\mathbf{l} + \int_f^c \mathbf{B} \cdot d\mathbf{l} = \mu_0 nI(cd) \tag{3-124}$$

In (3-124),

$$\int_c^d \mathbf{B} \cdot d\mathbf{l} = [B_z]_{cd}(cd) \tag{3-125a}$$

$$\int_d^e \mathbf{B} \cdot d\mathbf{l} = 0 \qquad \text{since } \mathbf{B} \text{ is normal to the path} \tag{2-125b}$$

$$\int_e^f \mathbf{B} \cdot d\mathbf{l} = 0 \qquad \text{since } \mathbf{B} \text{ is zero outside the solenoid} \tag{3-125c}$$

$$\int_f^c \mathbf{B} \cdot d\mathbf{l} = 0 \qquad \text{since } \mathbf{B} \text{ is normal to the path} \tag{3-125d}$$

Substituting (3-125a)–(3-125d) into (3-124), we obtain

$$[B_z]_{cd}(cd) = \mu_0 nI(cd)$$

or

$$[B_z]_{cd} = \mu_0 nI \tag{3-126}$$

The constant value of B_z inside the solenoid is equal to $\mu_0 nI$. Thus

$$\mathbf{B} = \begin{cases} \mu_0 nI\, \mathbf{i}_z & \text{inside the solenoid} \\ 0 & \text{outside the solenoid} \end{cases}$$

which agrees with the result obtained in Example 3-5 by using the Biot–
Savart law. However, compared with Example 3-5, we have here obtained
the solution in a conceptual manner, gaining in this process considerable
insight into the properties of the magnetic field.   ∎

3.10   Summary and Further Discussion of Static Electric and Magnetic Field Laws and Formulas

Now that we have gained familiarity with the static magnetic field as well
as the static electric field, it is worthwhile to list the basic laws governing
the two fields and important formulas derived from them and make a few
further comments. Accordingly, these laws and formulas are summarized
in Table 3.1. Note that we have repeated Maxwell's equations at the end of the
table. These equations pertain to the divergence and curl of the static electric
and magnetic fields. We note from these equations that static vector fields,
that is, vector fields independent of time, may be classified into four groups,
depending on the values of their divergence and curl in the region of interest. These groups are as follows:

(a) Divergence of the field is not zero but its curl is zero. This represents
    a static electric field.

(b) Divergence of the field is zero but its curl is not zero. This represents
    a static magnetic field.

(c) Both divergence and curl of the field are zero. This represents either
    a static electric field in a charge-free region or a static magnetic
    field in a current-free region.

(d) Both divergence and curl of the field are not zero. Obviously, this
    represents a combination of the fields belonging to groups (a) and
    (b) and hence cannot be realized solely as a static electric field or
    solely as a static magnetic field.

In Table 3.2 we list the expressions for the electric and magnetic fields
for two simple analogous pairs of source distributions: infinitely long line
charge of uniform density versus infinitely long filamentary wire of current
along the z axis, and infinite sheet charge of uniform density versus infinite
sheet current of uniform density. For each pair, the analogy between the two
fields is obvious from the expressions: The magnitudes of the fields are
proportional to each other whereas their directions are orthogonal. This
analogy is actually more general than is indicated by these two cases. To
illustrate this, let us consider a charge distribution of density $\rho(x, y)$ and a
current distribution of density $\mathbf{J} = J_z(x, y)\mathbf{i}_z$ such that

$$J_z(x, y) = k\rho(x, y) \tag{3-127}$$

where k is a proportionality constant. The electrostatic potential $V(x, y)$
corresponding to $\rho(x, y)$ and the magnetic vector potential $\mathbf{A} = A_z(x, y)\mathbf{i}_z$

TABLE 3.1. Summary of Basic Laws and Important Formulas Associated with the Static Electric and Magnetic Fields

	Static Electric Field	Static Magnetic Field
Definition	$\mathbf{F} = q\mathbf{E}$	$\mathbf{F} = q\mathbf{v} \times \mathbf{B} = I\,d\mathbf{l} \times \mathbf{B}$
Experimental laws	Coulomb's law: $$\mathbf{F}_{21} = \frac{Q_1 Q_2}{4\pi\epsilon_0 R_{21}^3}\mathbf{R}_{21}$$	Ampere's law of force: $$\mathbf{F}_{21} = \frac{\mu_0}{4\pi}\oint_{C_1}\oint_{C_2}\frac{I_2\,d\mathbf{l}_2 \times (I_1\,d\mathbf{l}_1 \times \mathbf{R}_{21})}{R_{21}^3}$$
Fields due to point sources	$$\mathbf{E} = \frac{Q}{4\pi\epsilon_0 R^3}\mathbf{R}$$	$$\mathbf{B} = \mu_0\frac{I\,d\mathbf{l} \times \mathbf{R}}{4\pi R^3}$$
Fields due to continuous source distributions:		
Line	$$\mathbf{E} = \int_{C'}\frac{[\rho_L(\mathbf{r}')](\mathbf{r} - \mathbf{r}')\,dl'}{4\pi\epsilon_0|\mathbf{r} - \mathbf{r}'|^3}$$	$$\mathbf{B} = \int_{C'}\frac{\mu_0 I\,d\mathbf{l}' \times (\mathbf{r} - \mathbf{r}')}{4\pi|\mathbf{r} - \mathbf{r}'|^3}$$
Surface	$$\mathbf{E} = \int_{S'}\frac{[\rho_s(\mathbf{r}')](\mathbf{r} - \mathbf{r}')\,dS'}{4\pi\epsilon_0|\mathbf{r} - \mathbf{r}'|^3}$$	$$\mathbf{B} = \int_{S'}\frac{\mu_0 \mathbf{J}_s(\mathbf{r}') \times (\mathbf{r} - \mathbf{r}')\,dS'}{4\pi|\mathbf{r} - \mathbf{r}'|^3}$$
Volume	$$\mathbf{E} = \int_{V'}\frac{[\rho(\mathbf{r}')](\mathbf{r} - \mathbf{r}')\,dv'}{4\pi\epsilon_0|\mathbf{r} - \mathbf{r}'|^3}$$	$$\mathbf{B} = \int_{V'}\frac{\mu_0 \mathbf{J}(\mathbf{r}') \times (\mathbf{r} - \mathbf{r}')\,dv'}{4\pi|\mathbf{r} - \mathbf{r}'|^3}$$
Integral laws involving sources	Gauss' law: $$\oint_S \mathbf{E} \cdot d\mathbf{S} = \frac{1}{\epsilon_0}\left(\begin{matrix}\text{charge} \\ \text{enclosed by } S\end{matrix}\right)$$	Ampere's circuital law: $$\oint_C \mathbf{B} \cdot d\mathbf{l} = \mu_0\left(\begin{matrix}\text{current enclosed} \\ \text{by } C\end{matrix}\right)$$
Differential laws involving sources	$$\mathbf{\nabla} \cdot \mathbf{E} = \frac{\rho}{\epsilon_0}$$	$$\mathbf{\nabla} \times \mathbf{B} = \mu_0\mathbf{J}$$
Integral laws independent of sources	Conservative property: $$\oint_C \mathbf{E} \cdot d\mathbf{l} = 0$$	Solenoidal property: $$\oint_S \mathbf{B} \cdot d\mathbf{S} = 0$$
Differential laws independent of sources	$$\mathbf{\nabla} \times \mathbf{E} = 0$$	$$\mathbf{\nabla} \cdot \mathbf{B} = 0$$
Potentials	Scalar potential: $$\mathbf{E} = -\mathbf{\nabla}V$$	Vector potential: $$\mathbf{B} = \mathbf{\nabla} \times \mathbf{A}$$
Potentials due to point sources	$$V = \frac{Q}{4\pi\epsilon_0 R}$$	$$\mathbf{A} = \frac{\mu_0 I\,d\mathbf{l}}{4\pi R}$$
Potentials due to continuous source distributions:		
Line	$$V = \int_{C'}\frac{\rho_L(\mathbf{r}')\,dl'}{4\pi\epsilon_0|\mathbf{r} - \mathbf{r}'|}$$	$$\mathbf{A} = \int_{C'}\frac{\mu_0 I\,d\mathbf{l}'}{4\pi|\mathbf{r} - \mathbf{r}'|}$$
Surface	$$V = \int_{S'}\frac{\rho_s(\mathbf{r}')\,dS'}{4\pi\epsilon_0|\mathbf{r} - \mathbf{r}'|}$$	$$\mathbf{A} = \int_{S'}\frac{\mu_0 \mathbf{J}_s(\mathbf{r}')\,dS'}{4\pi|\mathbf{r} - \mathbf{r}'|}$$
Volume	$$V = \int_{V'}\frac{\rho(\mathbf{r}')\,dv'}{4\pi\epsilon_0|\mathbf{r} - \mathbf{r}'|}$$	$$\mathbf{A} = \int_{V'}\frac{\mu_0 \mathbf{J}(\mathbf{r}')\,dv'}{4\pi|\mathbf{r} - \mathbf{r}'|}$$
Differential equations for potentials	$$\nabla^2 V = -\frac{\rho}{\epsilon_0}$$	$$\nabla^2\mathbf{A} = -\mu_0\mathbf{J}$$
Maxwell's equations:		
Divergence equation	$$\mathbf{\nabla} \cdot \mathbf{E} = \frac{\rho}{\epsilon_0}$$	$$\mathbf{\nabla} \cdot \mathbf{B} = 0$$
Curl equation	$$\mathbf{\nabla} \times \mathbf{E} = 0$$	$$\mathbf{\nabla} \times \mathbf{B} = \mu_0\mathbf{J}$$

TABLE 3.2. Electric and Magnetic Fields for Two Pairs of Analogous Source Distributions

Electric Field	*Magnetic Field*
Infinitely long, straight line charge of uniform density ρ_{L0}:	Infinitely long, straight wire of current I:
$\mathbf{E} = \dfrac{\rho_{L0}}{2\pi\epsilon_0 r}\mathbf{i}_r$	$\mathbf{B} = \dfrac{\mu_0 I}{2\pi r}\mathbf{i}_\phi$
Infinite sheet charge of uniform density ρ_{s0}:	Infinite sheet current of uniform density $\mathbf{J}_{s0}$:
$\mathbf{E} = \dfrac{\rho_{s0}}{2\epsilon_0}\mathbf{i}_n$	$\mathbf{B} = \dfrac{\mu_0}{2}\mathbf{J}_{s0} \times \mathbf{i}_n$

corresponding to $J_z(x, y)\mathbf{i}_z$ satisfy the equations

$$\nabla^2 V = -\frac{\rho}{\epsilon_0} \tag{3-128}$$

and

$$(\nabla^2 A_z)\mathbf{i}_z = -\mu_0 J_z \mathbf{i}_z = -\mu_0 k\rho\mathbf{i}_z \tag{3-129}$$

respectively. Comparing (3-128) and (3-129), we have

$$A_z = k\mu_0\epsilon_0 V \tag{3-130}$$

We then obtain

$$\begin{aligned}
\frac{E}{B} &= \frac{|\nabla V|}{|\nabla \times A_z\mathbf{i}_z|} = \frac{[(\partial V/\partial x)^2 + (\partial V/\partial y)^2]^{1/2}}{[(\partial A_z/\partial x)^2 + (\partial A_z/\partial y)^2]^{1/2}} \\
&= \frac{1}{k\mu_0\epsilon_0}\frac{[(\partial V/\partial x)^2 + (\partial V/\partial y)^2]^{1/2}}{[(\partial V/\partial x)^2 + (\partial V/\partial y)^2]^{1/2}} = \frac{1}{k\mu_0\epsilon_0}
\end{aligned} \tag{3-131}$$

and

$$\begin{aligned}
\mathbf{E} \cdot \mathbf{B} &= -\nabla V \cdot (\nabla \times A_z\mathbf{i}_z) \\
&= -\nabla V \cdot (\nabla \times k\mu_0\epsilon_0 V\mathbf{i}_z) \\
&= -k\mu_0\epsilon_0\, \nabla V \cdot (\nabla \times V\mathbf{i}_z) \\
&= -k\mu_0\epsilon_0\, \nabla V \cdot (\nabla V \times \mathbf{i}_z + V\nabla \times \mathbf{i}_z) \\
&= -k\mu_0\epsilon_0[\nabla V \cdot \nabla V \times \mathbf{i}_z] = 0
\end{aligned} \tag{3-132}$$

Thus, for analogous charge and current distributions which vary only in two dimensions x and y (or r and ϕ in cylindrical coordinates) and with the current flow along the z direction, the electric and magnetic fields are proportional in magnitude and orthogonal in direction. We will use this important result in chapter 6.

PROBLEMS

3.1. An electron moving with a velocity $\mathbf{v}_1 = \mathbf{i}_x$ m/sec at a point in a magnetic field experiences a force $\mathbf{F}_1 = -e\mathbf{i}_y$ N, where e is the charge of the electron. If the electron is moving with a velocity $\mathbf{v}_2 = (\mathbf{i}_y + \mathbf{i}_z)$ m/sec at the same point, it experiences a force $\mathbf{F}_2 = e\mathbf{i}_x$ N. Find the force the electron would experience if it were moving with a velocity $\mathbf{v}_3 = \mathbf{v}_1 \times \mathbf{v}_2$ at the same point.

3.2. A mass spectrograph is a device for separating charged particles having different masses. Consider two particles of the same charge q but different masses m_1 and m_2 injected into the region of a uniform magnetic field $\mathbf{B}$ with a known velocity $\mathbf{v}$ normal to the magnetic field as shown in Fig. 3.24. Show that the particles are separated by a distance $d = |2(m_2 - m_1)\mathbf{v}|/|q\mathbf{B}|$ in the plane normal to the incident velocity.

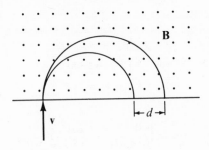

Fig. 3.24. For Problem 3.2.

3.3. A magnetic field given by

$$\mathbf{B} = B_0\mathbf{i}_z$$

where B_0 is a constant exists in the space between two parallel metallic plates of length L as shown in Fig. 3.25. A small test charge q having a mass m enters the

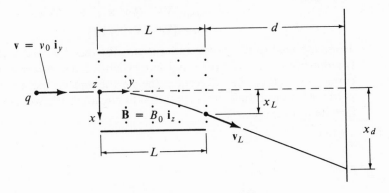

Fig. 3.25. For Problem 3.3

region between the plates at $t = 0$ with a velocity $\mathbf{v} = v_0\mathbf{i}_y$ as shown in the figure.

(a) Show that the path of the test charge between the plates is circular.

(b) Find the position x_L along the x direction and the velocity $\mathbf{v}_L$ of the test charge just after it emerges from the field region.

(c) Find the deflection x_d undergone by the test charge along the x direction at a distance d from the plates in the y direction.

3.4. In a region of magnetic field $\mathbf{B} = B_0\mathbf{i}_z$, where B_0 is a constant, an electron starts out at the origin with an initial velocity $\mathbf{v}_0 = v_{x0}\mathbf{i}_x + v_{y0}\mathbf{i}_y + v_{z0}\mathbf{i}_z$. Obtain the equations of motion of the electron and show that the path of the electron is a helix of radius $m\sqrt{v_{x0}^2 + v_{y0}^2}/|eB_0|$ and pitch $2\pi m|v_{z0}|/|eB_0|$, where e and m are the charge and mass of the electron.

3.5. Find the current required to counteract the earth's gravitational force on a horizontal filamentary wire of length l and mass m and oriented in the east-west direction in a uniform magnetic field B_0 directed northward. Compute the value of this current for a wire of length 1 meter and mass 30 grams situated in the earth's magnetic field at the magnetic equator assuming a value of 0.3×10^{-4} Wb/m² for B_0.

3.6. A rigid loop of wire in the form of a square of sides a m is hung by pivoting one side along the x axis as shown in Fig. 3.26. The loop is free to swing about the pivoted side without friction. The mass of the wire is m kg/m. If the wire is situated in a uniform magnetic field $\mathbf{B} = B_0\mathbf{i}_z$ and carries a current I amp, find the angle by which the loop swings from the vertical.

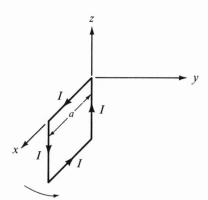

Fig. 3.26. For Example 3.6

3.7. A rigid rectangular loop of wire carrying current I amp and symmetrically situated about the z axis is in the yz plane as shown in Fig. 3.27. If the loop is situated in a uniform magnetic field $\mathbf{B}$ and is free to swing about the z axis, show that the torque acting on the loop is $IA(\mathbf{i}_y \cdot \mathbf{B})\mathbf{i}_z$ where A is the area of the loop.

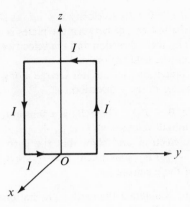

Fig. 3.27. For Problem 3.7.

3.8.   Show that the total force experienced by a current loop C_1 carrying current I_1 due to another current loop C_2 carrying current I_2 is equal and opposite to the total force experienced by the current loop C_2 due to the current loop C_1; that is, show that Newton's third law holds for current loops.

3.9.   Two circular loops of radii 1 m carrying currents I_1 and I_2 amp are situated in the $z = 0$ m and $z = 1$ m planes, respectively, and with their centers on the z axis, as shown in Fig. 3.28. Find the forces experienced by the current elements $I_1\,d\mathbf{l}_1$, $I_2\,d\mathbf{l}_2$ and $I_2\,d\mathbf{l}_3$ due to each other.

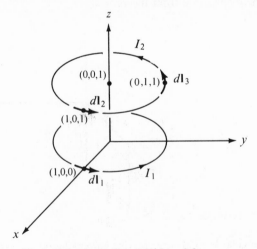

Fig. 3.28. For Problem 3.9.

3.10. Two square loops of sides a m are placed parallel to each other and separated by a distance d m as shown in Fig. 3.29. If the currents carried by the loops are I_1 and I_2 amp, respectively, as shown in Fig. 3.29, find the force acting on one loop due to the second loop.

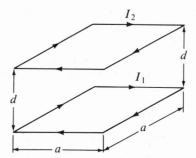

Fig. 3.29. For Problem 3.10.

3.11.   An infinitely long straight wire carrying current I_1 amp is situated in the plane of and parallel to one side of a rectangular loop of wire carrying current I_2 amp as shown in Fig. 3.30. Evaluate independently the force experienced by the infinitely long wire due to the magnetic field of the rectangular loop of wire and the force experienced by the rectangular loop of wire due to the magnetic field of the infinitely long wire.

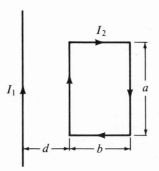

Fig. 3.30. For Problem 3.11.

3.12.   Four infinitely long, straight filamentary wires occupy the lines $x = 0$, $y = 0$; $x = 1$, $y = 0$; $x = 1$, $y = 1$ and $x = 0$, $y = 1$. Each wire carries a current of value 1 amp in the z direction.
(a) Find the force experienced per unit length of each wire.
(b) Find the magnetic flux density at the point $(2, 2, 0)$.
(c) Find the magnetic flux density at the point $(0, 2, 0)$.

3.13.   Two identical rigid filamentary wires, each of length l and weight W, are suspended horizontally from the ceiling by long weightless threads, each of length L. The wires are arranged to be parallel and separated by a distance d, where d is very small compared to l and L. A current I amp is passed through both wires through flexible connections so as to cause the wires to be attracted towards each other. If the current is gradually increased from zero, the wires will gradually approach each other. A condition may be reached at which any further increase of current will cause the wires to swing and touch each other. Determine the critical current at which this would happen.

3.14. A circular loop of wire of radius a lying in the xy plane with its center at the origin carries a current I in the ϕ direction. Find **B** at the point $(0, 0, z)$. Verify your answer by letting $z \longrightarrow 0$.

3.15. A loop of wire lying in the xy plane and carrying a current I is in the shape of a regular polygon of n sides inscribed in a circle of radius a with its center at the origin. If the current flow is in the sense of the ϕ direction, find **B** at the point $(0, 0, z)$. Verify your answer by letting $n \longrightarrow \infty$ and comparing the result with the answer to Problem 3.14.

3.16. A V-shaped filamentary wire with semi-infinitely long legs making an angle α at its vertex P and lying in the plane of the paper carries a current I amp as shown in Fig. 3.31. Find **B** at a point distance d directly above the vertex P. Verify your result by letting $\alpha \longrightarrow \pi$.

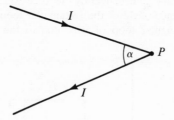

Fig. 3.31. For Problem 3.16.

3.17. Two circular loops of filamentary wire each of radius a and with their centers on the z axis are situated parallel to and symmetrically about the xy plane with the separation equal to $2b$ as shown in Fig. 3.32. The loops carry a current of I amp each in the ϕ direction. (a) Obtain the expression for **B** at a point on the z axis. (b) Show that if $b = a/2$, the first three derivatives of **B** evaluated at the origin are equal to zero.

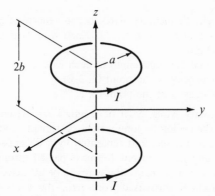

Fig. 3.32. For Problem 3.17.

3.18. A finitely long, uniformly wound solenoid of radius a, consisting of n turns per unit length and carrying a current I in the ϕ direction, lies between $z = -L_1$ and $z = L_2$ with the z axis as its axis. Find **B** at a point $(0, 0, z)$. Verify your answer by letting L_1 and $L_2 \longrightarrow \infty$.

3.19. A filamentary wire closely wound in the form of a spiral in the xy plane, starting at the origin and ending at radius a, carries a current I in the ϕ direction. Consider the turn density n to be an arbitrary function of r and show that the magnetic flux density at a point $(0, 0, z)$ is given by

$$\mathbf{B} = \frac{\mu_0 I}{2} \int_{r=0}^{a} \frac{nr^2 \, dr}{(r^2 + z^2)^{3/2}} \mathbf{i}_z$$

Evaluate $\mathbf{B}$ for the following turn density distributions:
(a) $n = n_0$

(b) $n = \dfrac{n_0}{r}$

(c) $n = \dfrac{n_0}{r^2}$

where n_0 is a constant.

3.20. A filamentary wire carrying a current I is closely wound on the surface of a sphere of radius a and centered at the origin. The winding starts at $(0, 0, a)$ and ends at $(0, 0, -a)$ with the turns in the planes normal to the z axis and carrying current in the ϕ direction. Consider the turn density to be an arbitrary function of θ and show that the magnetic flux density at a point $(0, 0, z)$ is given by

$$\mathbf{B} = \frac{\mu_0 I a^3}{2} \int_{\theta=0}^{\pi} \frac{n \sin^2 \theta \, d\theta}{[a^2 + z^2 - 2az \cos \theta]^{3/2}} \mathbf{i}_z$$

Evaluate $\mathbf{B}$ both for $|z| < a$ and for $|z| > a$ for the following turn density distributions:
(a) $n = n_0 \sin \theta$
(b) $n = n_0/\sin \theta$
where n_0 is a constant.

3.21. Two infinitely long, straight filamentary wires situated parallel to the z axis and passing through $(d/2, 0, 0)$ and $(-d/2, 0, 0)$, respectively, carry currents I in the $+z$ and $-z$ directions, respectively. The arrangement is known as a two-dimensional magnetic dipole in contrast to the three-dimensional magnetic dipole consisting of a circular loop of current. (a) Obtain the magnetic flux density due to the two-dimensional magnetic dipole in the limit that $d \longrightarrow 0$, keeping the dipole moment Id constant. (b) Find and sketch the direction lines of the magnetic flux density.

3.22. Two infinitely long, straight filamentary wires situated parallel to the z axis and passing through $(d/2, 0, 0)$ and $(-d/2, 0, 0)$ carry currents I_1 and I_2, respectively, in the z direction. Show that the equation for the direction lines of the magnetic flux density is

$$I_1 \ln \left[\left(x + \frac{d}{2} \right)^2 + y^2 \right] + I_2 \ln \left[\left(x - \frac{d}{2} \right)^2 + y^2 \right] = \text{constant}$$

Obtain and sketch the direction lines for the following cases:
(a) $I_1 = I_2 = I_0$
(b) $I_1 = I_0, I_2 = -I_0$

3.23. Two circular loops of filamentary wire, each of radius a and with their centers on the z axis, are situated parallel to and symmetrically about the xy plane with the separation equal to $2d$. The loops carry currents of I amp each in opposite directions. Such an arrangement is known as the magnetic quadrupole. Obtain the magnetic flux density due to the magnetic quadrupole at distances from the origin large compared to a and d, at points along (a) the z-axis and (b) in the xy plane.

3.24. A sheet of surface current flowing in the z direction occupies the portion of the $y = 0$ plane lying between $x = -a$ and $x = +a$. Consider the z-directed surface current density $\mathbf{J}_s$ to be an arbitrary function of x and show that the components of the magnetic flux density at a point $(0, 0, y)$ are given in cartesian coordinates by

$$B_x = -\frac{\mu_0 y}{2\pi} \int_{x=-a}^{a} \frac{J_s \, dx}{(x^2 + y^2)}$$

$$B_y = -\frac{\mu_0}{2\pi} \int_{x=-a}^{a} \frac{J_s x \, dx}{(x^2 + y^2)}$$

$$B_z = 0$$

Evaluate the field components for the following surface current density distributions:

(a) $\mathbf{J}_s = J_{s0}\mathbf{i}_z$

(b) $\mathbf{J}_s = J_{s0}\left(1 - \frac{|x|}{a}\right)\mathbf{i}_z$

(c) $\mathbf{J}_s = J_{s0}\frac{x}{a}\mathbf{i}_z$

where J_{s0} is a constant.

3.25. Current flows on the xy plane radially away from the origin with density given by

$$\mathbf{J}_s = \frac{I}{2\pi r}\mathbf{i}_r \text{ amps/m}$$

Show that the magnetic flux density at any point above the xy plane is the same as that which would be produced by a filamentary wire along the negative z axis carrying current I from the origin to $z = -\infty$. Show also that the magnetic flux density at any point below the xy plane is the same as that which would be produced by a filamentary wire along the positive z axis carrying current I from the origin to $z = \infty$.

3.26. Current flows in the z direction in an infinite slab of thickness $2a$ symmetrically placed about the xz plane. Consider the z-directed current density $\mathbf{J}$ to be uniform in x but not necessarily in y and show that the magnetic flux density at any point (x, y, z) has only an x component given by

$$B_x = \begin{cases} -\dfrac{\mu_0}{2}\displaystyle\int_{y=-a}^{a} J \, dy & y > a \\[3mm] \dfrac{\mu_0}{2}\left(\displaystyle\int_{y=y}^{a} J \, dy - \displaystyle\int_{y=-a}^{y} J \, dy\right) & -a < y < a \\[3mm] \dfrac{\mu_0}{2}\displaystyle\int_{y=-a}^{a} J \, dy & y < -a \end{cases}$$

Evaluate B_x as a function of y for $-\infty < y < \infty$ for the following current distributions:

(a) $\mathbf{J} = J_0 \mathbf{i}_z \qquad -a < y < a$

(b) $\mathbf{J} = \begin{cases} J_0 \mathbf{i}_z & -a < y < 0 \\ -J_0 \mathbf{i}_z & 0 < y < a \end{cases}$

(c) $\mathbf{J} = |y| \mathbf{i}_z \qquad -a < y < a$

(d) $\mathbf{J} = y \mathbf{i}_z \qquad -a < y < a$

where J_0 is a constant. Discuss your results from considerations of symmetry.

3.27. Current flows in the axial direction in an infinitely long cylinder of radius a having the z axis as its axis. Consider the z-directed current density $\mathbf{J}$ to be uniform in ϕ but an arbitrary function of r and show that the magnetic flux density is given by

$$\mathbf{B} = \frac{\mu_0}{r} \int_{r=0}^{r} Jr \, dr \, \mathbf{i}_\phi$$

Evaluate $\mathbf{B}$ for the following current density distributions:

(a) $\mathbf{J} = J_0 \mathbf{i}_z, \qquad 0 < r < a$

(b) $\mathbf{J} = \begin{cases} 0 & 0 < r < a \\ J_0 \mathbf{i}_z & a < r < b \\ 0 & b < r < \infty \end{cases}$

(c) $\mathbf{J} = J_0 \left(\dfrac{r}{a}\right)^n \mathbf{i}_z, n \geq 1 \quad 0 < r < a$

where J_0 is a constant.

3.28. An infinitely long straight filamentary wire occupying the z axis carries current I amp in the z direction. Evaluate $\int \mathbf{B} \cdot d\mathbf{l}$ for the following paths:

(a) From $(1, 0, 0)$ to $(0, \frac{1}{2}, 0)$ along the path $x + 2y = 1, z = 0$.

(b) From $(2, 0, 0)$ to $(1, 1, 1)$ along a straight line path.

Check your answers from considerations of symmetry and Ampere's circuital law in integral form.

3.29. Using Ampere's circuital law in integral form, obtain the magnetic flux densities due to the following volume current distributions in cartesian coordinates:

(a) $\mathbf{J} = \begin{cases} J_0 \mathbf{i}_z & |y| < a \\ 0 & |y| > a \end{cases}$

(b) $\mathbf{J} = \begin{cases} J_0 \mathbf{i}_z & -a < y < 0 \\ -J_0 \mathbf{i}_z & 0 < y < a \end{cases}$

(c) $\mathbf{J} = \begin{cases} |y| \mathbf{i}_z & |y| < a \\ 0 & |y| > a \end{cases}$

(d) $\mathbf{J} = \begin{cases} y \mathbf{i}_z & |y| < a \\ 0 & |y| > a \end{cases}$

(e) $\mathbf{J} = \begin{cases} (a - |y|) \mathbf{i}_z & |y| < a \\ 0 & |y| > a \end{cases}$

where J_0 is a constant.

3.30. Using Ampere's circuital law in integral form, obtain the magnetic flux densities due to the following volume current distributions in cylindrical coordinates:

(a) $\mathbf{J} = \begin{cases} 0 & 0 < r < a \\ J_0 \mathbf{i}_z & a < r < b \\ 0 & b < r < \infty \end{cases}$

(b) $\mathbf{J} = \begin{cases} J_0 \left(\dfrac{r}{a}\right)^n \mathbf{i}_z, n \geq 1 & 0 < r < a \\ 0 & a < r < \infty \end{cases}$

(c) $\mathbf{J} = \begin{cases} \dfrac{I}{\pi a^2}\mathbf{i}_z & 0 < r < a \\ 0 & a < r < b \\ -\dfrac{I}{\pi(c^2 - b^2)}\mathbf{i}_z & b < r < c \\ 0 & c < r < \infty \end{cases}$

where J_0 and I are constants.

3.31. Using Ampere's circuital law in integral form, obtain the magnetic flux densities due to the following surface current distributions:

(a) $\mathbf{J}_s = \begin{cases} J_{s0}\mathbf{i}_z & y = a \\ -J_{s0}\mathbf{i}_z & y = -a \end{cases}$ cartesian coordinates

(b) $\mathbf{J}_s = J_{s0}\mathbf{i}_z \qquad\quad\ r = a \qquad$ cylindrical coordinates

(c) $\mathbf{J}_s = \begin{cases} J_{s0}\mathbf{i}_z & r = a \\ -J_{s0}\dfrac{a}{b}\mathbf{i}_z & r = b \end{cases}$ cylindrical coordinates

where J_{s0} is a constant.

3.32. A toroid with a circular cross section is formed by rotating about the z axis the circle of radius $a\ (< b)$ in the xz plane and centered at $(b, 0, 0)$ as shown in Fig. 3.33. A filamentary wire carrying current I is closely wound around the toroid uniformly with n turns per unit length along the mean circumference. Using Ampere's circuital law in integral form, find the magnetic field both inside and outside the toroid.

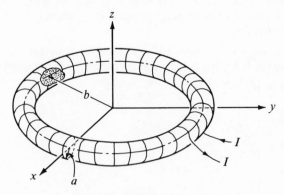

Fig. 3.33. For Problem 3.32.

3.33. Current I amp flows in a filamentary wire along the z axis from $z = \infty$ to $z = a$ and then to the point $z = -a$ via a spherical surface of radius a centered at the origin, continuing on to $z = -\infty$ along a filamentary wire from $z = -a$ to $z = -\infty$. The surface current density on the spherical surface is given by

$$\mathbf{J}_s = \frac{I}{2\pi a \sin\theta}\mathbf{i}_\theta \text{ amp/m}$$

Using Ampere's circuital law in integral form, find **B** both inside and outside the sphere of radius *a*.

3.34. Current flows axially with uniform density J_0 amp/m² in the region between two infinitely long, parallel cylindrical surfaces of radii *a* and *b* ($<$ a) and with their axes separated by a distance *c* ($< a - b$) as shown in Fig. 3.34. Find the magnetic flux density in the current-free region inside the cylindrical surface of radius *b*.

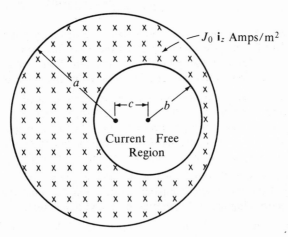

Fig. 3.34. For Problem 3.34.

3.35. Verify your answers to Problem 3.29 by using Ampere's circuital law in differential form.

3.36. Verify your answers to Problem 3.30 by using Ampere's circuital law in differential form.

3.37. For each of the following magnetic fields, find the current distribution which produces the field, using Ampere's circuital law in differential form:

$$\text{(a)} \quad \mathbf{B} = \begin{cases} \mu_0 J_{s0} \mathbf{i}_x & -\infty < y < 0 \\ \dfrac{\mu_0 J_{s0}}{3} \mathbf{i}_x & 0 < y < a \\ -\mu_0 J_{s0} \mathbf{i}_x & a < y < \infty \end{cases} \quad \begin{array}{l} \text{cartesian} \\ \text{coordinates} \end{array}$$

$$\text{(b)} \quad \mathbf{B} = \begin{cases} \mu_0 J_0 r^2 \mathbf{i}_\phi & 0 < r < a \\ \mu_0 J_0 \dfrac{a^3}{r} \mathbf{i}_\phi & a < r < b \\ 0 & b < r < \infty \end{cases} \quad \begin{array}{l} \text{cylindrical} \\ \text{coordinates} \end{array}$$

$$\text{(c)} \quad \mathbf{B} = \begin{cases} \mu_0 J_{s0}(\cos\theta\, \mathbf{i}_r - \sin\theta\, \mathbf{i}_\theta) & 0 < r < a \\ \dfrac{\mu_0 J_{s0}}{2}\left(\dfrac{a}{r}\right)^3 (2\cos\theta\, \mathbf{i}_r + \sin\theta\, \mathbf{i}_\theta) & a < r < \infty \end{cases} \quad \begin{array}{l} \text{spherical} \\ \text{coordinates} \end{array}$$

where J_{s0} and J_0 are constants.

3.38. A surface current of density $\mathbf{J}_s$ amp/m occupies the plane surface $y = y_0$. Show that

$$\nabla \times \mathbf{B} = \mu_0 \mathbf{J}_s \, \delta(y - y_0)$$

3.39.   A surface current of density $\mathbf{J}_s$ amp/m occupies the cylindrical surface $r = r_0$. Show that

$$\nabla \times \mathbf{B} = \mu_0 \mathbf{J}_s \, \delta(r - r_0)$$

3.40.   An infinitely long filamentary wire carrying current I amp in the z direction is situated parallel to the z axis and passes through the point (r_0, ϕ_0) in the $z = 0$ plane. Show that

$$\nabla \times \mathbf{B} = \mu_0 I \frac{\delta(r - r_0) \, \delta(\phi - \phi_0)}{r_0} \mathbf{i}_z$$

3.41.   Obtain the magnetic vector potential at an arbitrary point due to a finitely long, straight filamentary wire lying along the z axis between $z = -a$ and $z = +a$ and carrying a current I amp in the $+z$ direction. Then evaluate $\mathbf{B}$ by performing the curl operation on the magnetic vector potential and compare the result with (3-27).

3.42.   Two infinitely long, straight filamentary wires situated parallel to the z axis and passing through $(d/2, 0, 0)$ and $(-d/2, 0, 0)$, respectively, carry currents I in the $+z$ and $-z$ directions, respectively. (a) Obtain the magnetic vector potential $\mathbf{A}$. (b) Find $\mathbf{A}$ in the limit that $d \to 0$, keeping Id constant. (c) Evaluate the curl of $\mathbf{A}$ found in part (b) and compare with the result of Problem 3.21.

3.43.   For the magnetic dipole of Fig. 3-9, obtain the vector potential at distances very large from the dipole compared to the radius a. Find the magnetic flux density by performing the curl operation on the vector potential.

3.44.   For the magnetic quadrupole arrangement of Problem 3.23, obtain the magnetic vector potential at distances from it large compared to the dimensions of the quadrupole. Then find $\mathbf{B}$ by evaluating the curl of the magnetic vector potential and verify the results for the special cases of Problem 3.23.

3.45.   For the volume current distributions specified in Problem 3.29, obtain the magnetic vector potentials.

3.46.   For the volume current distributions specified in Problem 3.30, obtain the magnetic vector potentials.

3.47.   For the following surface current distributions, obtain the magnetic vector potentials:

(a) $\mathbf{J}_s = \begin{cases} J_{s0}\mathbf{i}_z & y = a \\ -J_{s0}\mathbf{i}_z & y = -a \end{cases}$ cartesian coordinates

(b) $\mathbf{J}_s = \begin{cases} J_{s0}\mathbf{i}_z & r = a \\ -J_{s0}\dfrac{a}{b}\mathbf{i}_z & r = b \end{cases}$ cylindrical coordinates

where J_{s0} is a constant.

3.48.   For each of the arrangements of current loops shown in Fig. 3.35, find the magnetic vector potential at distances very far from the loop.

3.49.   For the spirally wound filamentary wire of Problem 3.19, show that the magnetic dipole moment $\mathbf{m}$ is given by

$$\mathbf{m} = \pi I \left(\int_{r=0}^{a} nr^2 \, dr \right) \mathbf{i}_z$$

Evaluate $\mathbf{m}$ and hence $\mathbf{A}$ at large distances from the spiral for each of the three cases specified in Problem 3.19.

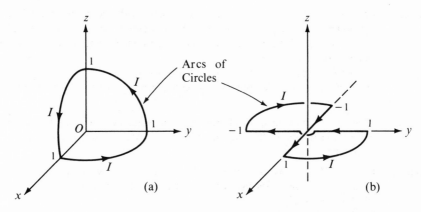

Fig. 3.35. For Problem 3.48.

3.50. For the filamentary wire wound on the surface of a sphere as specified in Problem 3.20, show that the magnetic dipole moment **m** is given by

$$\mathbf{m} = \pi a^3 I \left(\int_{\theta=0}^{\pi} n \sin^2 \theta \, d\theta \right) \mathbf{i}_z$$

Evaluate **m** and hence **A** at large distances from the sphere for each of the two cases listed in Problem 3.20.

3.51. A spherical volume charge of radius a m and having uniform density ρ_0 C/m³ and centered at the origin spins about the z axis with constant angular velocity ω_0 in the ϕ direction. Obtain the magnetic vector potential due to the spinning sphere of charge at distances from the origin large compared to a.

3.52. Show that the magnetic flux enclosed by a closed path C in a magnetic field **B** is equal to $\oint_C \mathbf{A} \cdot d\mathbf{l}$, where **A** is the magnetic vector potential corresponding to **B**. Use this result to find the magnetic flux enclosed by the rectangular loop of Fig. 3.30 due to the current flowing in the infinitely long wire. Check your answer by evaluating $\int_S \mathbf{B} \cdot d\mathbf{S}$, where S is the surface bounded by the rectangular loop.

3.53. Show that, if $\mathbf{A} = A_z \mathbf{i}_z$, where A_z is independent of z, the direction lines of $\mathbf{B} = \nabla \times \mathbf{A}$ are the cross sections of the constant $|\mathbf{A}|$ surfaces in the $z = $ constant plane. Use this result to find and sketch the direction lines of the magnetic flux density due to the infinitely long, filamentary wire-pair arrangement of Problem 3.42.

3.54. Determine if the following fields are realizable as magnetic fields:

(a) $\mathbf{A} = \dfrac{1}{y^2}(y\mathbf{i}_x - x\mathbf{i}_y)$ cartesian coordinates

(b) $\mathbf{B} = \dfrac{1}{r^n}\mathbf{i}_\phi$ cylindrical coordinates

(c) $\mathbf{C} = \left(1 + \dfrac{1}{r^2}\right) \cos\phi \, \mathbf{i}_r - \left(1 - \dfrac{1}{r^2}\right) \sin\phi \, \mathbf{i}_\phi$ cylindrical coordinates

(d) $\mathbf{D} = \left(1 + \dfrac{2}{r^3}\right) \cos\theta \, \mathbf{i}_r - \left(1 - \dfrac{1}{r^3}\right) \sin\theta \, \mathbf{i}_\theta$ spherical coordinates

3.55. For the following current distributions, start with the assumption that all three components of **B** exist and use Ampere's circuital law and the solenoidal nature of the magnetic field to eliminate some components and evaluate the remaining components:

(a) Infinite sheet of current with uniform density.

(b) Surface current flowing axially with uniform density along an infinitely long cylinder.

3.56. Make use of the solenoidal character of the magnetic field to find the radial derivative of the magnetic flux density due to a circular loop of current I at a point on its axis.

3.57. In Sec. 3-10, we classified static vector fields into four groups. Determine to which of the four groups does each of the following fields belong:

(a) $\mathbf{A} = x\mathbf{i}_x + y\mathbf{i}_y$

(b) $\mathbf{B} = xy\mathbf{i}_x + yz\mathbf{i}_y + zx\mathbf{i}_z$

(c) $\mathbf{C} = (x^2 - y^2)\mathbf{i}_x - 2xy\mathbf{i}_y + 4\mathbf{i}_z$

(d) $\mathbf{D} = \dfrac{e^{-r}}{r}\mathbf{i}_\phi$, cylindrical coordinates

(e) $\mathbf{E} = \dfrac{\cos\phi}{r^2}\mathbf{i}_r + \dfrac{\sin\phi}{r^2}\mathbf{i}_\phi$, cylindrical coordinates

3.58. From the examples and problems of Chapters 2 and 3, identify and prepare a table of analogous pairs of charge and current distributions which vary only in two dimensions x and y (or r and ϕ) and with the current flow in the z direction. List the expressions for the corresponding electric and magnetic fields and demonstrate that the fields are proportional in magnitude and orthogonal in direction.

4

THE ELECTROMAGNETIC FIELD

In Chapter 2 we studied the static or time-independent electric field in free space. We introduced Maxwell's equations for the static electric field gradually from the experimental law of Coulomb concerning the force between two charges. In Chapter 3 we studied the static or time-independent magnetic field in free space. We introduced Maxwell's equations for the static magnetic field gradually from the experimental law of Ampere concerning the force between two current loops. In this chapter we will study time-varying electric and magnetic fields. We will learn that Maxwell's curl equations for the static electric and magnetic fields have to be modified for time-varying fields in accordance with an experimental law of Faraday and a purely mathematical contribution of Maxwell. When these modifications are made, we will find that the time-varying electric and magnetic fields are coupled; that is, they are interdependent and hence the name "electromagnetic field." As in the case of Chapters 2 and 3, we will in this chapter be concerned with the electromagnetic field in free space only.

4.1 The Lorentz Force Equation

In Section 2.1 we introduced the electric field concept in terms of a force field acting upon charges, whereas in Section 3.1 we introduced the magnetic field concept, also in terms of a force field acting upon charges but only when they are in motion. If an electric field $\mathbf{E}$ as well as a magnetic field $\mathbf{B}$ exist in a region, then the force $\mathbf{F}$ experienced by a test charge q moving with velocity **v** is simply the sum of the electric and magnetic forces given by (2-2) and (3-1), respectively. Thus

$$\mathbf{F} = q\mathbf{E} + q\mathbf{v} \times \mathbf{B} = q(\mathbf{E} + \mathbf{v} \times \mathbf{B}) \tag{4-1}$$

Equation (4-1) is known as the Lorentz force equation, and the force given by it is known as the Lorentz force. For a continuous charge distribution of density ρ moving with a velocity **v**, we can define a force per unit volume, **f**. Considering an infinitesimal volume dv, we then have

$$\mathbf{f} \, dv = \rho \, dv \, (\mathbf{E} + \mathbf{v} \times \mathbf{B}) = (\rho\mathbf{E} + \mathbf{J} \times \mathbf{B}) \, dv$$

or

$$\mathbf{f} = \rho\mathbf{E} + \mathbf{J} \times \mathbf{B} \tag{4-2}$$

where $\mathbf{J} = \rho\mathbf{v}$ is the volume current density.

EXAMPLE 4-1.   A test charge q C, moving with a velocity $\mathbf{v} = (\mathbf{i}_x + \mathbf{i}_y)$ m/sec, experiences no force in a region of electric and magnetic fields. If the magnetic flux density $\mathbf{B} = (\mathbf{i}_x - 2\mathbf{i}_z)$ Wb/m², find **E**.

From (4-1), the electric field intensity **E** must be equal to $-\mathbf{v} \times \mathbf{B}$ for the charge to experience no force. Thus

$$\mathbf{E} = -(\mathbf{i}_x + \mathbf{i}_y) \times (\mathbf{i}_x - 2\mathbf{i}_z)$$
$$= (2\mathbf{i}_x - 2\mathbf{i}_y + \mathbf{i}_z) \text{ volts/m} \quad \blacksquare$$

EXAMPLE 4-2.   A region is characterized by crossed electric and magnetic fields, $\mathbf{E} = E_0\mathbf{i}_y$ and $\mathbf{B} = B_0\mathbf{i}_z$ as shown in Fig. 4.1, where E_0 and B_0 are constants. A small test charge q having a mass m starts from rest at the origin at $t = 0$. We wish to obtain the parametric equations of motion of the test charge.

The force exerted by the crossed electric and magnetic fields on the test charge is

$$\mathbf{F} = q(\mathbf{E} + \mathbf{v} \times \mathbf{B}) = q[E_0\mathbf{i}_y + (v_x\mathbf{i}_x + v_y\mathbf{i}_y + v_z\mathbf{i}_z) \times (B_0\mathbf{i}_z)] \tag{4-3}$$

The equations of motion of the test charge can therefore be written as

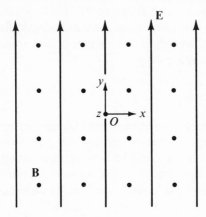

Fig. 4.1. A region of crossed electric and magnetic fields.

$$\frac{dv_x}{dt} = \frac{qB_0}{m}v_y \tag{4-4a}$$

$$\frac{dv_y}{dt} = -\frac{qB_0}{m}v_x + \frac{q}{m}E_0 \tag{4-4b}$$

$$\frac{dv_z}{dt} = 0 \tag{4-4c}$$

Eliminating v_y from (4-4a) and (4-4b), we have

$$\frac{d^2v_x}{dt^2} + \left(\frac{qB_0}{m}\right)^2 v_x = \left(\frac{q}{m}\right)^2 B_0 E_0 \tag{4-5}$$

The solution for (4-5) is

$$v_x = \frac{E_0}{B_0} + C_1 \cos \omega_c t + C_2 \sin \omega_c t \tag{4-6}$$

where C_1 and C_2 are arbitrary constants and $\omega_c = qB_0/m$. Substituting (4-6) into (4-4a), we obtain

$$v_y = -C_1 \sin \omega_c t + C_2 \cos \omega_c t \tag{4-7}$$

Using initial conditions given by

$$v_x = v_y = 0 \qquad \text{at } t = 0$$

to evaluate C_1 and C_2 in (4-6) and (4-7), we obtain

$$v_x = \frac{E_0}{B_0} - \frac{E_0}{B_0} \cos \omega_c t \tag{4-8}$$

$$v_y = \frac{E_0}{B_0} \sin \omega_c t \tag{4-9}$$

Integrating (4-8) and (4-9) with respect to t, we have

$$x = \frac{E_0}{B_0}t - \frac{E_0}{\omega_c B_0} \sin \omega_c t + C_3 \tag{4-10}$$

$$y = -\frac{E_0}{\omega_c B_0} \cos \omega_c t + C_4 \tag{4-11}$$

Using initial conditions given by

$$x = y = 0 \qquad \text{at } t = 0$$

to evaluate C_3 and C_4 in (4-10) and (4-11), we obtain

$$x = \frac{E_0}{B_0}t - \frac{E_0}{\omega_c B_0} \sin \omega_c t = \frac{E_0}{\omega_c B_0}(\omega_c t - \sin \omega_c t) \tag{4-12}$$

$$y = -\frac{E_0}{\omega_c B_0} \cos \omega_c t + \frac{E_0}{\omega_c B_0} = \frac{E_0}{\omega_c B_0}(1 - \cos \omega_c t) \tag{4-13}$$

Equation (4-4c), together with the initial conditions $v_z = 0$ and $z = 0$ at $t = 0$, yields a solution

$$z = 0 \tag{4-14}$$

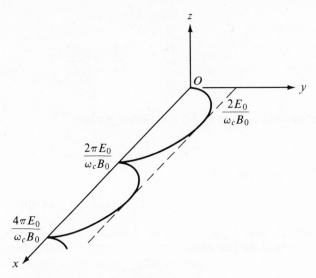

Fig. 4.2. Path of a test charge q in crossed electric and magnetic fields $\mathbf{E} = E_0 \mathbf{i}_y$ and $\mathbf{B} = B_0 \mathbf{i}_z$.

The equations of motion of the test charge in the crossed electric and magnetic field region are thus given by (4-12), (4-13), and (4-14). These equations represent a cycloid in the $z = 0$ plane, as shown in Fig. 4.2. ∎

4.2 Faraday's Law in Integral Form

We learned in Section 2.2 that Coulomb's experiments demonstrated that charges at rest experience forces as given by Coulomb's law, leading to the interpretation of an electric field set up by charges at rest. Similarly, we learned in Section 3.3 that Ampere's experiments showed that current loops experience forces as given by Ampere's law, leading to the interpretation of a magnetic field being set up by currents, that is, charges in motion. In this section we present the results of experiments by yet another scientist, Michael Faraday. Faraday demonstrated that a magnetic field changing with time results in a flow of current in a loop of wire placed in the magnetic field region. When the magnetic field does not change with time, there is no current flow in the wire. This implies that a time-varying magnetic field exerts electric-type forces on charges. Thus Faraday's experiments demonstrate that a time-varying magnetic field produces an electric field.

The electric field produced by the time-varying magnetic field is such that the work done by it around a closed path C per unit charge in the limit that the charge tends to zero, that is, its circulation around the closed path C, is equal to the negative of the time rate of change of the magnetic flux ψ enclosed by the path C. In equation form we have

$$\text{circulation of } \mathbf{E} \text{ around } C = -\frac{d\psi}{dt} \qquad (4\text{-}15)$$

The circulation of $\mathbf{E}$ around a closed path C is $\oint_C \mathbf{E} \cdot d\mathbf{l}$. The magnetic flux enclosed by C is given by the surface integral of the magnetic flux density evaluated over a surface S bounded by the contour C, that is, $\int_S \mathbf{B} \cdot d\mathbf{S}$. In evaluating $\int_S \mathbf{B} \cdot d\mathbf{S}$, we choose the normals to the infinitesimal surfaces comprising S to be pointing towards the side of advance of a right-hand screw as it is turned in the sense of C. Equation (4-15) is thus written as

$$\oint_C \mathbf{E} \cdot d\mathbf{l} = -\frac{d}{dt} \int_S \mathbf{B} \cdot d\mathbf{S} \qquad (4\text{-}16)$$

The statement represented by (4-15) or (4-16) is known as Faraday's law. Note that the time derivative on the right side of (4-16) operates on the entire integral so that the circulation of $\mathbf{E}$ can be due to a change in $\mathbf{B}$ or a change in the surface S or both. Classically, the quantity $\oint \mathbf{E} \cdot d\mathbf{l}$ on the left side of (4-16) is known under different names, for example, induced electromotive force, induced electromotance, induced voltage. Certainly the word force is not appropriate, since $\mathbf{E}$ is force per unit charge and $\int \mathbf{E} \cdot d\mathbf{l}$ is work per unit charge. We shall simply refer to $\mathbf{E}$ as the induced electric field and to $\oint \mathbf{E} \cdot d\mathbf{l}$ as the circulation of $\mathbf{E}$.

The minus sign on the right side of (4-16) needs an explanation. We know that the normal to a surface at a point on the surface can be directed towards either side of the surface. In formulating (4-16), we always direct the normal towards the side of advance of a right-hand screw as it is turned around C in the sense in which C is defined. For simplicity, let us consider the plane surface S bounded by a closed path C and let the magnetic flux density be uniform and directed normal to the surface, as shown in Fig. 4.3. If the flux density is increasing with time, $d\psi/dt$ is positive and $-d\psi/dt$ is negative so that $\oint_C \mathbf{E} \cdot d\mathbf{l}$ is negative. Hence the electric field produced by the increasing magnetic flux acts opposite to the sense of the contour C. If we place a test charge at a point on C, it will move opposite to C; if C is occupied by a wire, a current will flow in the sense opposite to

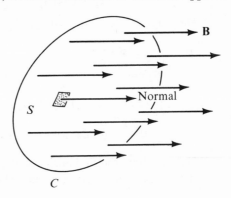

Fig. 4.3. Uniform magnetic field **B** directed normal to a plane surface S.

that of C. Such a current will produce a magnetic field directed to the side opposite to that of the normal since, if the wire is grabbed with the right hand and with the thumb pointing in the direction of the current, the fingers will be curled opposite to the normal as they penetrate the surface S. Thus the current will produce magnetic flux which opposes the increase in the original flux. Likewise, if the flux density is decreasing with time, $d\psi/dt$ is negative and $-d\psi/dt$ is positive so that $\oint_C \mathbf{E} \cdot d\mathbf{l}$ is positive. The electric field produced by the decreasing magnetic flux acts in the sense of the contour C so that, if C is occupied by a wire, a current will flow in the same sense as that of C. Such a current will produce a magnetic field directed to the side of the normal, thereby opposing the decrease in the flux. Thus the minus sign on the right side of (4-16) signifies that the induced electric field is such that it opposes the change in the magnetic flux producing it. This fact is known as Lenz' law. If the induced electric field is such that it aids the change in the magnetic flux instead of opposing it, any small change in the magnetic flux will set up a chain reaction by inducing an electric field, which will aid the change in the magnetic flux, which will increase the electric field, and so on, thereby violating the conservation of energy. Hence Lenz' law must be obeyed and the minus sign on the right side of (4-16) is very important.

EXAMPLE 4-3. The magnetic flux density is given by

$$\mathbf{B} = B_0 \cos \omega_1 t \, \mathbf{i}_x$$

where B_0 and ω_1 are constants. A rectangular loop of wire of area A is placed symmetrically with respect to the z axis and rotated about the z axis at a constant angular velocity ω_2 as shown in Fig. 4.4, such that the angle ϕ which the normal to the plane of the loop makes with the x axis is given by

$$\phi = \phi_0 + \omega_2 t$$

It is desired to find the circulation of the induced electric field around the contour C of the loop.

The unit vector normal to the plane of the loop is

$$\mathbf{i}_n = \cos(\phi_0 + \omega_2 t) \, \mathbf{i}_x + \sin(\phi_0 + \omega_2 t) \, \mathbf{i}_y \qquad (4\text{-}17)$$

The magnetic flux enclosed by the loop is

$$\psi = \int_{\substack{\text{plane surface} \\ S \text{ bounded} \\ \text{by } C}} \mathbf{B} \cdot d\mathbf{S}$$

$$= \int_S (B_0 \cos \omega_1 t \, \mathbf{i}_x) \cdot [\cos(\phi_0 + \omega_2 t) \, \mathbf{i}_x + \sin(\phi_0 + \omega_2 t) \, \mathbf{i}_y] \, dS$$

$$= \int_S B_0 \cos \omega_1 t \cos(\phi_0 + \omega_2 t) \, dS = B_0 A \cos \omega_1 t \cos(\phi_0 + \omega_2 t) \quad (4\text{-}18)$$

This is simply the flux enclosed at any time t by the projection of the loop at that time on to the yz plane, which is normal to the flux density. From

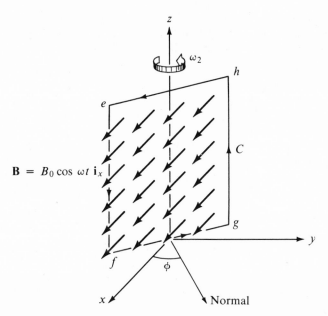

Fig. 4.4. A rectangular loop of wire rotating about the z axis with a constant angular velocity and situated in a time-varying magnetic field.

Faraday's law, we now have

$$\oint_C \mathbf{E}' \cdot d\mathbf{l} = -\frac{d\psi}{dt} = -\frac{d}{dt}[B_0 A \cos \omega_1 t \cos (\phi_0 + \omega_2 t)]$$
$$= B_0 A[\omega_1 \sin \omega_1 t \cos (\phi_0 + \omega_2 t) \qquad \text{(4-19)}$$
$$+ \omega_2 \cos \omega_1 t \sin (\phi_0 + \omega_2 t)]$$

where the prime in $\mathbf{E}'$ denotes that the electric field is associated with the contour of the moving loop. Note that the right side of (4-19) reduces to $B_0 A \omega_1 \cos \phi_0 \sin \omega_1 t$ for $\omega_2 = 0$, that is, for a stationary loop in a time-varying magnetic field and to $B_0 A \omega_2 \sin (\phi_0 + \omega_2 t)$ for $\omega_1 = 0$, that is, for a moving loop in a static magnetic field. ∎

EXAMPLE 4-4.   The magnetic flux density is given in cylindrical coordinates by

$$\mathbf{B} = \begin{cases} B_0 \sin \omega t \, \mathbf{i}_z & \text{for } r < a \\ 0 & \text{for } r > a \end{cases}$$

where B_0 and ω are constants. It is desired to find the induced electric field everywhere.

We note that the time-varying magnetic field has circular symmetry about the z axis and is independent of z. Hence the induced electric field must also possess circular symmetry about the z axis and must be independent of z; that is, **E** can be a function of r only. Choosing a circular contour C of radius r and centered at the origin, as shown in Fig. 4.5, we note that the magnetic flux enclosed by the contour C is

$$\psi = \int_S \mathbf{B} \cdot d\mathbf{S} \tag{4-20}$$

where S is the plane surface bounded by the contour C. Substituting for **B** and $d\mathbf{S}$ in (4-20), we get, for $r < a$,

$$\psi = \int_S \mathbf{B} \cdot d\mathbf{S} = \int_S B_0 \sin \omega t \, \mathbf{i}_z \cdot d\mathbf{S} \, \mathbf{i}_z$$

$$= B_0 \sin \omega t \int_S d\mathbf{S} = \pi r^2 B_0 \sin \omega t$$

For $r > a$,

$$\psi = \int_S \mathbf{B} \cdot d\mathbf{S} = \int_{S_1} \mathbf{B} \cdot d\mathbf{S} + \int_{S_2} \mathbf{B} \cdot d\mathbf{S} \tag{4-21}$$

where S_1 is the plane surface enclosed by the circular contour of radius a and S_2 is the remainder of the surface S. The magnetic field is zero, however, on the surface S_2 and hence the second integral on the right side of (4-21) is zero. Hence, for $r > a$,

$$\psi = \int_{S_1} \mathbf{B} \cdot d\mathbf{S} = \pi a^2 B_0 \sin \omega t$$

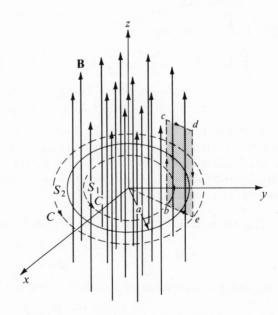

Fig. 4.5. For evaluating the induced electric field due a time-varying magnetic field possessing cylindrical symmetry.

Thus

$$\psi = \begin{cases} \pi r^2 B_0 \sin \omega t & \text{for } r < a \\ \pi a^2 B_0 \sin \omega t & \text{for } r > a \end{cases} \tag{4-22}$$

Now,

$$\oint_C \mathbf{E} \cdot d\mathbf{l} = \int_{\phi=0}^{2\pi} E_\phi r \, d\phi = 2\pi r E_\phi \tag{4-23}$$

From Faraday's law, we then have

$$2\pi r E_\phi = -\frac{d\psi}{dt} = \begin{cases} -\pi r^2 B_0 \omega \cos \omega t & \text{for } r < a \\ -\pi a^2 B_0 \omega \cos \omega t & \text{for } r > a \end{cases}$$

or

$$E_\phi = \begin{cases} -\dfrac{B_0 r \omega}{2} \cos \omega t & \text{for } r < a \\ -\dfrac{B_0 a^2 \omega}{2r} \cos \omega t & \text{for } r > a \end{cases} \tag{4-24}$$

Any r component of **E** independent of ϕ and z will have nonzero curl and hence can be attributed to sources appropriate for a static electric field, that is, an electric field originating from charges at rest. Any z component will have to be independent of r since the magnetic field has no ϕ component. This is because if we consider a rectangular contour *bcdeb* in a plane containing the z axis as shown in Fig. 4.5, the magnetic flux enclosed by this contour is zero. Hence $\oint_{bcdeb} \mathbf{E} \cdot d\mathbf{l}$ is zero or $\int_b^c E_z \, dz + \int_d^e E_z \, dz$ is equal to zero, leading to the conclusion that E_z along *bc* is the same as E_z along *ed*. Since the curl of a field which has a z component independent of r and ϕ is zero, it can also be attributed to sources appropriate for a static field. Thus the induced electric field due to the time-varying magnetic field has a ϕ component only, thereby surrounding the magnetic field, and it is given by

$$\mathbf{E} = \begin{cases} -\dfrac{B_0 r \omega}{2} \cos \omega t \, \mathbf{i}_\phi & \text{for } r < a \\ -\dfrac{B_0 a^2 \omega}{2r} \cos \omega t \, \mathbf{i}_\phi & \text{for } r > a \end{cases} \tag{4-25}$$

The fact that the induced electric field surrounds the time-varying magnetic field can also be seen if we recognize that Faraday's law is similar in form to Ampere's circuital law

$$\oint_C \mathbf{B} \cdot d\mathbf{l} = \mu_0 (\text{current } I \text{ enclosed by C})$$

The magnetic field due to the current I surrounds the current. Likewise, the electric field due to the changing magnetic flux should surround the flux. The induced electric field is thus solenoidal in character, as compared to the irrotational nature of the electric field due to charges at rest. ∎

One of the consequences of Faraday's law is that $\int \mathbf{E} \cdot d\mathbf{l}$ evaluated between two points a and b is, in general, dependent on the path followed from a to b to evaluate the integral, unlike in the case of the static electric field. To illustrate this, let us consider a region of uniform but time-varying magnetic field. Applying Faraday's law to two different closed paths *acbea* and *adbea* as shown in Fig. 4.6, we obtain two different results for $\oint \mathbf{E} \cdot d\mathbf{l}$

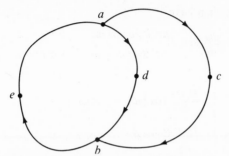

Fig. 4.6. Two different closed paths *acbea* and *adbea*.

since the paths enclose different areas. However, path *bea* is common to both the closed paths, and the contributions from the path *bea* to $\displaystyle\oint_{acbea} \mathbf{E} \cdot d\mathbf{l}$ and to $\displaystyle\oint_{adbea} \mathbf{E} \cdot d\mathbf{l}$ are the same. It then follows that $\displaystyle\int_{acb} \mathbf{E} \cdot d\mathbf{l}$ is not equal to $\displaystyle\int_{adb} \mathbf{E} \cdot d\mathbf{l}$. Thus the work done per unit charge in carrying a test charge from a to b in an electromagnetic field, that is, $\displaystyle\int_a^b \mathbf{E} \cdot d\mathbf{l}$ in an electromagnetic field, is not uniquely defined. It depends upon the path followed from a to b in evaluating $\displaystyle\int_a^b \mathbf{E} \cdot d\mathbf{l}$. The quantity $\displaystyle\int_a^b \mathbf{E} \cdot d\mathbf{l}$ is known as the voltage between the points a and b in the case of time-varying fields. The word "voltage" is interchangeable with "potential difference" for the case of static electric field only. For time-varying fields, the electric field cannot be expressed exclusively in terms of a time-varying electric scalar potential as we will learn in the following section. Hence, the two words are not interchangeable in the time-varying case.

Now, let us consider two different surfaces S_1 and S_2 bounded by a contour C with the normals defining the surfaces directed out of the volume bounded by $S_1 + S_2$ as shown in Fig. 4.7. Then, applying Faraday's law to C, we have

$$\oint_C \mathbf{E} \cdot d\mathbf{l} = -\frac{d}{dt}\int_{S_1} \mathbf{B} \cdot d\mathbf{S} = \frac{d}{dt}\int_{S_2} \mathbf{B} \cdot d\mathbf{S} \qquad (4\text{-}26)$$

It follows from (4-26) that

$$\frac{d}{dt}\left(\int_{S_1} \mathbf{B} \cdot d\mathbf{S} + \int_{S_2} \mathbf{B} \cdot d\mathbf{S}\right) = 0$$

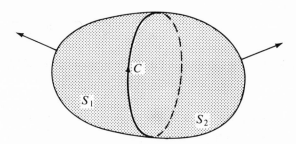

Fig. 4.7. Two surfaces S_1 and S_2 bounded by a contour C.

or

$$\oint_{S_1+S_2} \mathbf{B} \cdot d\mathbf{S} = \text{constant with time} \tag{4-27}$$

The constant on the right side of (4-27) must, however, be equal to zero since a nonzero value for any surface requires the existence forever of isolated magnetic charge within the volume bounded by that surface. There is no experimental evidence of the existence of such magnetic charge. Thus, it follows from Faraday's law in integral form that

$$\oint_{S} \mathbf{B} \cdot d\mathbf{S} = 0 \tag{3-111}$$

where S is any closed surface.

4.3 Faraday's Law in Differential Form (Maxwell's First Curl Equation for the Electromagnetic Field)

In the previous section we introduced Faraday's law in integral form, given by

$$\oint_{C} \mathbf{E} \cdot d\mathbf{l} = -\frac{d}{dt} \int_{S} \mathbf{B} \cdot d\mathbf{S} \tag{4-16}$$

where S is any surface bounded by the contour C. According to Stokes' theorem, we have

$$\oint_{C} \mathbf{E} \cdot d\mathbf{l} = \int_{S} (\mathbf{\nabla} \times \mathbf{E}) \cdot d\mathbf{S}$$

where S is any surface bounded by the contour C. In particular, choosing the same surface as for the integral on the right side of (4-16), we obtain

$$\int_{S} (\mathbf{\nabla} \times \mathbf{E}) \cdot d\mathbf{S} = -\frac{d}{dt} \int_{S} \mathbf{B} \cdot d\mathbf{S} \tag{4-28}$$

If the surface S is stationary, that is, independent of time, then

$$\frac{d}{dt} \int_{S} \mathbf{B} \cdot d\mathbf{S} = \int_{S} \frac{\partial \mathbf{B}}{\partial t} \cdot d\mathbf{S} \tag{4-29}$$

and

$$\int_S (\mathbf{V} \times \mathbf{E}) \cdot d\mathbf{S} = \int_S -\frac{\partial \mathbf{B}}{\partial t} \cdot d\mathbf{S} \tag{4-30}$$

Comparing the integrands on the two sides of (4-30), we have

$$\mathbf{V} \times \mathbf{E} = -\frac{\partial \mathbf{B}}{\partial t} \tag{4-31}$$

This is the differential form of Faraday's law and Maxwell's first curl equation for the electromagnetic field.

If, in addition to the variation of the magnetic field with time, the surface S is also changing with time due to a displacement of the contour as shown in Fig. 4.8, then we evaluate $\frac{d}{dt} \int \mathbf{B} \cdot d\mathbf{S}$ by considering two times t_1 and t_2, where $t_2 = t_1 + \Delta t$. If S_1 and S_2 are the surfaces bounded

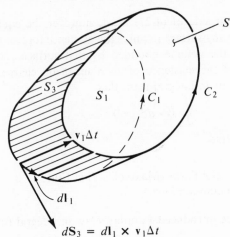

Fig. 4.8. Displacement of contour C_1 with time and the associated surfaces.

$$dS_3 = dl_1 \times v_1 \Delta t$$

by C_1 and C_2 at t_1 and t_2, respectively, we have, from the definition of differentiation,

$$\left[\frac{d}{dt} \int_S \mathbf{B} \cdot d\mathbf{S} \right]_{t_1} = \lim_{t_2 \to t_1} \frac{1}{t_2 - t_1} \left\{ \left[\int_S \mathbf{B} \cdot d\mathbf{S} \right]_{t_2} - \left[\int_S \mathbf{B} \cdot d\mathbf{S} \right]_{t_1} \right\}$$

$$= \lim_{\Delta t \to 0} \frac{1}{\Delta t} \left(\int_{S_2} \mathbf{B}_2 \cdot d\mathbf{S}_2 - \int_{S_1} \mathbf{B}_1 \cdot d\mathbf{S}_1 \right) \tag{4-32}$$

where $\mathbf{B}_2$ and $\mathbf{B}_1$ are $\mathbf{B}(t_2)$ and $\mathbf{B}(t_1)$, respectively. Applying the divergence theorem at time t_2 to the volume V bounded by the two surfaces S_1 and S_2 and the surface S_3 formed by the movement of the contour C, we have

$$\int_V \mathbf{V} \cdot \mathbf{B}_2 \, dv = \oint_{S_1 + S_2 + S_3} \mathbf{B}_2 \cdot d\mathbf{S}$$

$$= -\int_{S_1} \mathbf{B}_2 \cdot d\mathbf{S}_1 + \int_{S_2} \mathbf{B}_2 \cdot d\mathbf{S}_2 + \int_{S_3} \mathbf{B}_2 \cdot d\mathbf{S}_3 \tag{4-33}$$

where the minus sign associated with the first of the three integrals on the right side of (4-33) is due to the direction of $d\mathbf{S}_1$ pointing into the volume V. Also, in the third integral, we choose the direction of $d\mathbf{S}_3$ as pointing out of the volume V.

Since $\nabla \cdot \mathbf{B} = 0$, we have, from (4-33),

$$\int_{S_2} \mathbf{B}_2 \cdot d\mathbf{S}_2 - \int_{S_1} \mathbf{B}_2 \cdot d\mathbf{S}_1 = - \int_{S_3} \mathbf{B}_2 \cdot d\mathbf{S}_3 \qquad (4\text{-}34)$$

If the velocity with which an element $d\mathbf{l}_1$ in the contour C_1 is displaced is $\mathbf{v}_1$, the infinitesimal area $d\mathbf{S}_3$ swept by the element in the time Δt is $d\mathbf{l}_1 \times \mathbf{v}_1 \, \Delta t$ as shown in Fig. 4.8. Hence

$$\int_{S_3} \mathbf{B}_2 \cdot d\mathbf{S}_3 = \oint_{C_1} \mathbf{B}_2 \cdot d\mathbf{l}_1 \times \mathbf{v}_1 \, \Delta t \qquad (4\text{-}35)$$

Substituting (4-35) into (4-34), we have

$$\int_{S_2} \mathbf{B}_2 \cdot d\mathbf{S}_2 - \int_{S_1} \mathbf{B}_2 \cdot d\mathbf{S}_1 = - \oint_{C_1} \mathbf{B}_2 \cdot d\mathbf{l}_1 \times \mathbf{v}_1 \, \Delta t \qquad (4\text{-}36)$$

Now, expanding $\mathbf{B}(t)$ in a Taylor's series at time t_1, we have

$$\mathbf{B}_2 = \mathbf{B}_1 + \left[\frac{\partial \mathbf{B}}{\partial t}\right]_{t_1} \Delta t + \frac{1}{2}\left[\frac{\partial^2 \mathbf{B}}{\partial t^2}\right]_{t_1} (\Delta t)^2 + \cdots \qquad (4\text{-}37)$$

and

$$\int_{S_1} \mathbf{B}_2 \cdot d\mathbf{S}_1 = \int_{S_1} \mathbf{B}_1 \cdot d\mathbf{S}_1 + \Delta t \int_{S_1}\left[\frac{\partial \mathbf{B}}{\partial t}\right]_{t_1} \cdot d\mathbf{S}_1 + \cdots \qquad (4\text{-}38)$$

$$\oint_{C_1} \mathbf{B}_2 \cdot d\mathbf{l}_1 \times \mathbf{v}_1 \, \Delta t = \Delta t \oint_{C_1} \mathbf{B}_1 \cdot d\mathbf{l}_1 \times \mathbf{v}_1$$
$$+ (\Delta t)^2 \oint_{C_1}\left[\frac{\partial \mathbf{B}}{\partial t}\right]_{t_1} \cdot d\mathbf{l}_1 \times \mathbf{v}_1 + \cdots \qquad (4\text{-}39)$$

Substituting (4-38) and (4-39) into (4-36) and rearranging, we get

$$\int_{S_2} \mathbf{B}_2 \cdot d\mathbf{S}_2 - \int_{S_1} \mathbf{B}_1 \cdot d\mathbf{S}_1 = \Delta t \int_{S_1}\left[\frac{\partial \mathbf{B}}{\partial t}\right]_{t_1} \cdot d\mathbf{S}_1 - \Delta t \oint_{C_1} \mathbf{B}_1 \cdot d\mathbf{l}_1 \times \mathbf{v}_1$$
$$+ \text{ higher-order terms in } \Delta t \qquad (4\text{-}40)$$

Substituting (4-40) into (4-32), we obtain

$$\left[\frac{d}{dt}\int_S \mathbf{B} \cdot d\mathbf{S}\right]_{t_1} = \lim_{\Delta t \to 0}\frac{1}{\Delta t}\left\{\Delta t \int_{S_1}\left[\frac{\partial \mathbf{B}}{\partial t}\right]_{t_1} \cdot d\mathbf{S}_1 - \Delta t \oint_{C_1} \mathbf{B}_1 \cdot d\mathbf{l}_1 \times \mathbf{v}_1 \right.$$
$$\left. + \text{ higher-order terms in } \Delta t\right\}$$
$$= \int_{S_1}\left[\frac{\partial \mathbf{B}}{\partial t}\right]_{t_1} \cdot d\mathbf{S}_1 - \oint_{C_1} \mathbf{B}_1 \cdot d\mathbf{l}_1 \times \mathbf{v}_1$$
$$= \int_{S_1}\left[\frac{\partial \mathbf{B}}{\partial t}\right]_{t_1} \cdot d\mathbf{S}_1 - \oint_{C_1} [\mathbf{v} \times \mathbf{B}]_{t_1} \cdot d\mathbf{l}_1 \qquad (4\text{-}41)$$

Since Eq. (4-41) must be true for any time t_1, we have, in general,

$$\frac{d}{dt}\int_S \mathbf{B} \cdot d\mathbf{S} = \int_S \frac{\partial \mathbf{B}}{\partial t} \cdot d\mathbf{S} - \oint_C \mathbf{v} \times \mathbf{B} \cdot d\mathbf{l} \qquad (4\text{-}42)$$

where C is the contour and S is the surface bounded by C at any arbitrary time t.

To an observer moving with a point on the contour, the contour appears to be stationary and the observer will attribute the force experienced by a test charge at that point as due to an electric field alone. Denoting this electric field as $\mathbf{E}'$ and applying Faraday's law for the contour C and using (4-42), we have

$$\oint_C \mathbf{E}' \cdot d\mathbf{l} = -\frac{d}{dt}\int_S \mathbf{B} \cdot d\mathbf{S}$$

$$= -\int_S \frac{\partial \mathbf{B}}{\partial t} \cdot d\mathbf{S} + \oint_C \mathbf{v} \times \mathbf{B} \cdot d\mathbf{l} \qquad (4\text{-}43)$$

But, according to Stokes' theorem, we have

$$\oint_C \mathbf{E}' \cdot d\mathbf{l} = \int_S \mathbf{\nabla} \times \mathbf{E}' \cdot d\mathbf{S} \qquad (4\text{-}44\text{a})$$

and

$$\oint_C \mathbf{v} \times \mathbf{B} \cdot d\mathbf{l} = \int_C \mathbf{\nabla} \times (\mathbf{v} \times \mathbf{B}) \cdot d\mathbf{S} \qquad (4\text{-}44\text{b})$$

Substituting (4-44a) and (4-44b) into (4-43), we get

$$\int_S \mathbf{\nabla} \times \mathbf{E}' \cdot d\mathbf{S} = -\int_S \frac{\partial \mathbf{B}}{\partial t} \cdot d\mathbf{S} + \int_S \mathbf{\nabla} \times (\mathbf{v} \times \mathbf{B}) \cdot d\mathbf{S} \qquad (4\text{-}45)$$

or

$$\mathbf{\nabla} \times \mathbf{E}' = -\frac{\partial \mathbf{B}}{\partial t} + \mathbf{\nabla} \times (\mathbf{v} \times \mathbf{B}) \qquad (4\text{-}46)$$

Equation (4-46) is Faraday's law in differential form, where $\mathbf{E}'$ is the electric field as measured by an observer moving with a velocity $\mathbf{v}$, relative to the magnetic field $\mathbf{B}$.

On the other hand, a stationary observer views the force experienced by the test charge moving with the point on the contour as being composed of two parts, electric-type and magnetic-type, that is, one due to an electric field acting on the charge and the other due to a magnetic field acting on the charge. Since the magnetic force acting on the test charge is $q\mathbf{v} \times \mathbf{B}$, the observer will attribute a force of $\mathbf{F} - q\mathbf{v} \times \mathbf{B}$ only to the electric field where $\mathbf{F}$ is the total force acting on the charge. The total force acting on the charge must of course be the same whether viewed by an observer moving with the contour or by a stationary observer. Hence it is equal to $q\mathbf{E}'$. Thus the force attributed to the electric field by the stationary observer is $q\mathbf{E}' - q\mathbf{v} \times \mathbf{B} = q(\mathbf{E}' - \mathbf{v} \times \mathbf{B})$ or the electric field as viewed by the stationary observer is given by

$$\mathbf{E} = \mathbf{E}' - \mathbf{v} \times \mathbf{B} \tag{4-47}$$

Rearranging (4-46), we have

$$\nabla \times (\mathbf{E}' - \mathbf{v} \times \mathbf{B}) = -\frac{\partial \mathbf{B}}{\partial t} \tag{4-48}$$

which, with the aid of (4-47), becomes

$$\nabla \times \mathbf{E} = -\frac{\partial \mathbf{B}}{\partial t} \tag{4-31}$$

which is the same as the result obtained in the case of the stationary contour. Thus Eq. (4-31) holds, in general, where $\mathbf{E}$ is the induced electric field as viewed by an observer stationary relative to the time-varying magnetic field $\mathbf{B}$.

EXAMPLE 4-5.   For the test charge of Example 4-2, find the electric field as viewed by an observer moving with the test charge.

From Example 4-2, the electric and magnetic fields as viewed by a stationary observer are

$$\mathbf{E} = E_0 \mathbf{i}_y \quad \text{and} \quad \mathbf{B} = B_0 \mathbf{i}_z$$

The velocity of motion of the test charge is given by

$$\mathbf{v} = v_x \mathbf{i}_x + v_y \mathbf{i}_y$$
$$= \left(\frac{E_0}{B_0} - \frac{E_0}{B_0} \cos \omega_c t \right) \mathbf{i}_x + \left(\frac{E_0}{B_0} \sin \omega_c t \right) \mathbf{i}_y \tag{4-49}$$

where we have substituted for v_x and v_y from (4-8) and (4-9), respectively.

Rearranging (4-47), we note that the electric field $\mathbf{E}'$ as viewed by an observer moving with a velocity $\mathbf{v}$ relative to the magnetic field is given by

$$\mathbf{E}' = \mathbf{E} + \mathbf{v} \times \mathbf{B} \tag{4-50}$$

We can also obtain this result directly by noting that, for an observer moving with the test charge, the test charge appears to be stationary and hence the observer will attribute the force experienced by it to an electric field alone. Since the force experienced by the test charge is $\mathbf{F} = q(\mathbf{E} + \mathbf{v} \times \mathbf{B})$, the observer views an electric field of $\mathbf{F}/q = \mathbf{E} + \mathbf{v} \times \mathbf{B}$. Substituting for $\mathbf{E}$, $\mathbf{v}$, and $\mathbf{B}$ in (4-50), we obtain

$$\mathbf{E}' = E_0 \mathbf{i}_y + \left[\left(\frac{E_0}{B_0} - \frac{E_0}{B_0} \cos \omega_c t \right) \mathbf{i}_x + \left(\frac{E_0}{B_0} \sin \omega_c t \right) \mathbf{i}_y \right] \times B_0 \mathbf{i}_z \tag{4-51}$$
$$= E_0 \sin \omega_c t \, \mathbf{i}_x + E_0 \cos \omega_c t \, \mathbf{i}_y$$

Thus the electric field as viewed by an observer moving with the test charge is $(E_0 \sin \omega_c t \, \mathbf{i}_x + E_0 \cos \omega_c t \, \mathbf{i}_y)$.   ∎

EXAMPLE 4-6.   In Example 4-3, we obtained the circulation of the induced electric field around a rectangular loop moving in a time varying magnetic field by the direct application of Faraday's law in integral form given by (4-16). It is here desired to verify the result of Example 4-3 by using (4-43).

With reference to the notation of Fig. 4.4, the first integral on the right side of (4-43) is given by

$$-\int_{\substack{\text{plane surface } S \\ \text{bounded by } C}} \frac{\partial \mathbf{B}}{\partial t} \cdot d\mathbf{S} = \int_{S} B_0 \omega_1 \sin \omega_1 t \, \mathbf{i}_x \cdot (\cos \phi \, \mathbf{i}_x + \sin \phi \, \mathbf{i}_y) \, dS$$

$$= B_0 A \omega_1 \cos (\phi_0 + \omega_2 t) \sin \omega_1 t \qquad (4\text{-}52)$$

To evaluate the second integral on the right side of (4-43), we note that, along side *ef*,

$$\mathbf{v} \times \mathbf{B} = \frac{(fg)}{2} \omega_2 [\mathbf{i}_\phi]_{ef} \times B_0 \cos \omega_1 t \, \mathbf{i}_x$$

$$= -\frac{(fg)}{2} \omega_2 B_0 \cos \omega_1 t \sin \phi \, \mathbf{i}_z$$

so that

$$\int_e^f \mathbf{v} \times \mathbf{B} \cdot d\mathbf{l} = \frac{(ef)(fg)}{2} \omega_2 B_0 \cos \omega_1 t \sin \phi$$

$$= \frac{B_0 A \omega_2}{2} \cos \omega_1 t \sin \phi \qquad (4\text{-}53\text{a})$$

Along side *fg*, $\mathbf{v} \times \mathbf{B} \cdot d\mathbf{l} = 0$ so that

$$\int_f^g \mathbf{v} \times \mathbf{B} \cdot d\mathbf{l} = 0 \qquad (4\text{-}53\text{b})$$

Along side *gh*,

$$\mathbf{v} \times \mathbf{B} = \frac{(fg)}{2} \omega_2 [\mathbf{i}_\phi]_{gh} \times B_0 \cos \omega_1 t \, \mathbf{i}_x$$

$$= \frac{(fg)}{2} \omega_2 B_0 \cos \omega_1 t \sin \phi \, \mathbf{i}_z$$

so that

$$\int_g^h \mathbf{v} \times \mathbf{B} \cdot d\mathbf{l} = \frac{(gh)(fg)}{2} \omega_2 B_0 \cos \omega_1 t \sin \phi$$

$$= \frac{B_0 A \omega_2}{2} \cos \omega_1 t \sin \phi \qquad (4\text{-}53\text{c})$$

Along side *he*, $\mathbf{v} \times \mathbf{B} \cdot d\mathbf{l} = 0$ so that

$$\int_h^e \mathbf{v} \times \mathbf{B} \cdot d\mathbf{l} = 0 \qquad (4\text{-}53\text{d})$$

From (4-53a)–(4-53d), we have

$$\oint_C \mathbf{v} \times \mathbf{B} \cdot d\mathbf{l} = \oint_{efghe} \mathbf{v} \times \mathbf{B} \cdot d\mathbf{l}$$

$$= B_0 A \omega_2 \cos \omega_1 t \sin \phi$$

$$= B_0 A \omega_2 \cos \omega_1 t \sin (\phi_0 + \omega_2 t) \qquad (4\text{-}54)$$

Thus, from (4-52) and (4-54), we obtain

$$\oint_C \mathbf{E}' \cdot d\mathbf{l} = B_0 A \omega_1 \cos(\phi_0 + \omega_2 t) \sin \omega_1 t$$
$$+ B_0 A \omega_2 \cos \omega_1 t \sin(\phi_0 + \omega_2 t)$$

which agrees with (4-19). ∎

EXAMPLE 4-7.   In Example 4-4 we obtained the expression for the induced electric field due to a time-varying magnetic field possessing cylindrical symmetry about the z axis, by using Faraday's law in integral form. It is desired to verify the result by using Faraday's law in differential form given by (4-31).

From Example 4-4, we have the induced electric field given by

$$\mathbf{E} = \begin{cases} -\dfrac{B_0 r \omega}{2} \cos \omega t \, \mathbf{i}_\phi & \text{for } r < a \\[2ex] -\dfrac{B_0 a^2 \omega}{2r} \cos \omega t \, \mathbf{i}_\phi & \text{for } r > a \end{cases} \tag{4-25}$$

Hence

$$\mathbf{\nabla} \times \mathbf{E} = \begin{vmatrix} \dfrac{\mathbf{i}_r}{r} & \mathbf{i}_\phi & \dfrac{\mathbf{i}_z}{r} \\[2ex] \dfrac{\partial}{\partial r} & \dfrac{\partial}{\partial \phi} & \dfrac{\partial}{\partial z} \\[2ex] 0 & r E_\phi & 0 \end{vmatrix} = \dfrac{\mathbf{i}_z}{r}\left[\dfrac{\partial}{\partial r}(r E_\phi)\right] \tag{4-55}$$

$$= \begin{cases} -B_0 \omega \cos \omega t \, \mathbf{i}_z & \text{for } r < a \\ 0 & \text{for } r > a \end{cases}$$

From Faraday's law in differential form, we then have

$$\dfrac{\partial \mathbf{B}}{\partial t} = -\mathbf{\nabla} \times \mathbf{E} = \begin{cases} B_0 \omega \cos \omega t \, \mathbf{i}_z & \text{for } r < a \\ 0 & \text{for } r > a \end{cases} \tag{4-56}$$

Equation (4-56) is consistent with

$$\mathbf{B} = \begin{cases} B_0 \sin \omega t \, \mathbf{i}_z & \text{for } r < a \\ 0 & \text{for } r > a \end{cases}$$

which is the magnetic field specified in Example 4-4. ∎

Returning to Eq. (4-31) and taking the divergence of both sides, we have

$$\mathbf{\nabla} \cdot \mathbf{\nabla} \times \mathbf{E} = -\mathbf{\nabla} \cdot \dfrac{\partial \mathbf{B}}{\partial t} = -\dfrac{\partial}{\partial t}(\mathbf{\nabla} \cdot \mathbf{B}) \tag{4-57}$$

But, since $\mathbf{\nabla} \cdot \mathbf{\nabla} \times \mathbf{E} \equiv 0$, it follows from (4-57) that

$$\dfrac{\partial}{\partial t}(\mathbf{\nabla} \cdot \mathbf{B}) = 0 \tag{4-58}$$

or

$$\mathbf{\nabla} \cdot \mathbf{B} = \text{constant with time} \tag{4-59}$$

The constant on the right side of (4-59) must, however, be equal to zero since a nonzero value at any point in space requires the existence forever of isolated magnetic charge at that point. There is no experimental evidence of the existence of such magnetic charge. Thus, we note that Maxwell's equation for the divergence of the time-varying magnetic field given by

$$\mathbf{\nabla} \cdot \mathbf{B} = 0 \tag{4-60}$$

follows from the Maxwell's equation for the curl of $\mathbf{E}$ given by (4-31). As a consequence of (4-60), we have

$$\mathbf{B} = \mathbf{\nabla} \times \mathbf{A} \tag{4-61}$$

where $\mathbf{A}$ is a time-varying vector potential. Substituting (4-61) into (4-31), we get

$$\mathbf{\nabla} \times \mathbf{E} = -\frac{\partial}{\partial t}(\mathbf{\nabla} \times \mathbf{A}) = -\mathbf{\nabla} \times \frac{\partial \mathbf{A}}{\partial t}$$

or

$$\mathbf{\nabla} \times \left(\mathbf{E} + \frac{\partial \mathbf{A}}{\partial t}\right) = 0 \tag{4-62}$$

Thus $(\mathbf{E} + \partial \mathbf{A}/\partial t)$ can be expressed as the gradient of a time-varying scalar potential. In particular, we can write

$$\mathbf{E} + \frac{\partial \mathbf{A}}{\partial t} = -\mathbf{\nabla}V \tag{4-63}$$

where V is the time-varying scalar potential so that Eq. (4-63) reduces to $\mathbf{E} = -\mathbf{\nabla}V$ for the static case. Rearranging (4-63), we obtain

$$\mathbf{E} = -\mathbf{\nabla}V - \frac{\partial \mathbf{A}}{\partial t} \tag{4-64}$$

We will have an opportunity to study the time-varying scalar and vector potentials in Section 6.16.

4.4   The Dilemma of Ampere's Circuital Law and the Displacement Current Concept; Modified Ampere's Circuital Law in Integral Form

In Section 3.6 we introduced Ampere's circuital law in integral form, given by

$$\oint_C \mathbf{B} \cdot d\mathbf{l} = \mu_0(\text{current enclosed by } C) \tag{3-57}$$

In that connection we discussed the uniqueness of a closed path enclosing a current by considering the case of a straight filamentary wire of finite length along which charge flows from one end to the other end (Fig. 3.16) and the case of an infinitely long filamentary wire. We found that the current enclosed by a closed path C is not uniquely defined in the case of the finitely long wire, whereas it is uniquely defined for the case of the infinitely long wire. On the other hand, the magnetic field due to a current-carrying wire is uniquely given at every point through the Biot–Savart law and hence $\oint_C \mathbf{B} \cdot d\mathbf{l}$ for a given closed path C has a unique value. Thus it seems to be meaningless to apply Ampere's circuital law as given by (3-57) for the case of the finitely long wire. What then is the fallacy of the situation? Is there any modification required for (3-57) so that the dilemma is resolved?

To answer these questions, let us consider a semiinfinitely long, straight filamentary wire occupying the upper half of the z axis. Let there be a point source of charge Q C at the origin and let the current flowing along the wire to infinity be I amp as shown in Fig. 4.9 so that the charge Q is decreas-

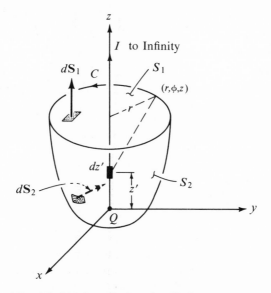

Fig. 4.9. For introducing the displacement current concept and deriving the modification to Ampere's circuital law.

ing at the rate of I C/sec. Let us consider a circular contour C of radius r in the plane normal to the wire and centered at a point on the wire a distance z from the origin, as shown in Fig. 4.9. The current enclosed by C is not uniquely defined since the current penetrating the plane surface S_1 bounded by the contour is I, whereas the current penetrating a bowl-shaped surface S_2 as shown in Fig. 4.9 is zero. On the other hand, $\oint_C \mathbf{B} \cdot d\mathbf{l}$ is unique since $\mathbf{B}$ along C is given by the application of the Biot–Savart law to the semiinfinitely long wire. According to the Biot–Savart law, the magnetic flux density at a point (r, ϕ, z) on the contour C due to an infinitesimal segment dz' of the wire at distance z' from the origin is given by

$$dB = \frac{\mu_0 I r \, dz'}{4\pi[(z - z')^2 + r^2]^{3/2}} \mathbf{i}_\phi \tag{4-65}$$

The magnetic flux density at (r, ϕ, z) due to the entire semiinfinitely long wire is given by

$$\mathbf{B} = \int_{z'=0}^{\infty} d\mathbf{B} = \frac{\mu_0 I r}{4\pi} \int_{z'=0}^{\infty} \frac{dz'}{[(z - z')^2 + r^2]^{3/2}} \mathbf{i}_\phi$$

$$= \frac{\mu_0 I}{4\pi r}\left(1 + \frac{z}{\sqrt{z^2 + r^2}}\right)\mathbf{i}_\phi \tag{4-66}$$

From (4-66), we have

$$\oint_C \mathbf{B} \cdot d\mathbf{l} = \int_{\phi=0}^{2\pi} \frac{\mu_0 I}{4\pi r}\left(1 + \frac{z}{\sqrt{z^2 + r^2}}\right) \mathbf{i}_\phi \cdot r \, d\phi \, \mathbf{i}_\phi$$

$$= \frac{\mu_0 I}{2}\left(1 + \frac{z}{\sqrt{z^2 + r^2}}\right) \tag{4-67}$$

If we apply Ampere's circuital law (3-57) to the contour C in conjunction with the surface S_1 without regard to the uniqueness of the current enclosed, we obtain

$$\oint_C \mathbf{B} \cdot d\mathbf{l} = \mu_0 I \tag{4-68}$$

Comparing (4-67) and (4-68), we note that the discrepancy between the right sides is by the amount

$$\frac{\mu_0 I}{2}\left(1 + \frac{z}{\sqrt{z^2 + r^2}}\right) - \mu_0 I = \frac{\mu_0 I}{2}\left(\frac{z}{\sqrt{z^2 + r^2}} - 1\right) \tag{4-69}$$

We have to resolve this discrepancy by some means. The only recourse seems to be the point charge at the origin whose value is decreasing at the rate of I C/sec. We have not as yet considered the electric field due to the point charge Q. As Q varies with time, the electric field flux due to it also varies with time. Let us consider the electric field flux through the surface S_1. Since the electric field intensity due to a point charge is spherically symmetric about the point charge, the electric field flux through any surface is equal to the solid angle subtended at the point charge by that surface times the point charge value divided by $4\pi\epsilon_0$.

To find the solid angle subtended by S_1 at Q, let us consider an infinitesimal area $dS_1 = r_1 \, dr_1 \, d\phi_1$ at the point (r_1, ϕ_1, z) on S_1. The projection of this area onto the plane normal to the line drawn from the origin to (r_1, ϕ_1, z) is $(r_1 z / \sqrt{r_1^2 + z^2}) \, dr_1 \, d\phi_1$. The projection of dS_1 onto the surface of a sphere of radius unity and centered at the origin or the infinitesimal solid angle subtended at the origin by dS_1 is given by

$$d\Omega_1 = \frac{r_1 z}{(r_1^2 + z^2)^{3/2}} \, dr_1 \, d\phi_1$$

The solid angle subtended at the origin by the entire area S_1 is then given by

$$\Omega_1 = \int_{S_1} d\Omega_1 = \int_{r_1=0}^{r} \int_{\phi_1=0}^{2\pi} \frac{r_1 z}{(r_1^2 + z^2)^{3/2}} \, dr_1 \, d\phi_1 = 2\pi\left(1 - \frac{z}{\sqrt{z^2 + r^2}}\right)$$

(4-70)

Since the normal to the surface S_1 drawn towards the direction of advance of a right-hand screw as it is turned in the sense of C is directed away from the point charge, the electric field flux passing through the surface S_1 towards the side of that normal is given by

$$\int_{S_1} \mathbf{E} \cdot d\mathbf{S}_1 = \frac{Q}{4\pi\epsilon_0}\Omega_1 = \frac{Q}{2\epsilon_0}\left(1 - \frac{z}{\sqrt{z^2 + r^2}}\right)$$

(4-71)

This electric field flux is changing with time. The rate at which it is changing with time is given by

$$\frac{d}{dt}\int_{S_1} \mathbf{E} \cdot d\mathbf{S}_1 = \frac{d}{dt}\left[\frac{Q}{2\epsilon_0}\left(1 - \frac{z}{\sqrt{z^2 + r^2}}\right)\right]$$
$$= \frac{1}{2\epsilon_0}\left(1 - \frac{z}{\sqrt{z^2 + r^2}}\right)\frac{dQ}{dt}$$

(4-72)

But, since the charge Q is decreasing at the rate of I C/sec, we have

$$\frac{dQ}{dt} = -I$$

(4-73)

Substituting (4-73) into (4-72), we obtain

$$\frac{d}{dt}\int_{S_1} \mathbf{E} \cdot d\mathbf{S}_1 = \frac{I}{2\epsilon_0}\left(\frac{z}{\sqrt{z^2 + r^2}} - 1\right)$$

(4-74)

The right side of (4-74) is exactly the same as the right side of (4-69) divided by $\mu_0\epsilon_0$. Suppose we now modify (3-57) to read

$$\oint_C \mathbf{B} \cdot d\mathbf{l} = \mu_0\Bigg(\text{current due to charges flowing through a}$$
$$\text{surface } S \text{ bounded by } C + \frac{d}{dt}\int_S \epsilon_0\mathbf{E} \cdot d\mathbf{S}\Bigg)$$

(4-75)

and apply it to the surface S_1, we obtain

$$\oint_C \mathbf{B} \cdot d\mathbf{l} = \mu_0\left[I + \frac{I}{2}\left(\frac{z}{\sqrt{z^2 + r^2}} - 1\right)\right] = \frac{\mu_0 I}{2}\left(1 + \frac{z}{\sqrt{z^2 + r^2}}\right)$$

which agrees with (4-67), deduced by using the Biot–Savart law. Thus our dilemma seems to be resolved!

Before we discuss the meaning of $\frac{d}{dt}\int_S \epsilon_0\mathbf{E} \cdot d\mathbf{S}$, let us apply (4-75) to the bowl-shaped surface S_2 bounded by C to see if it gives the correct result for $\oint_C \mathbf{B} \cdot d\mathbf{l}$. To do this, we note that the solid angle subtended by S_2 at Q is simply 4π minus the solid angle subtended by S_1 at Q. Thus the required solid angle Ω_2 is given by

$$\Omega_2 = (4\pi - \Omega_1) \tag{4-76}$$

where Ω_1 is given by (4-70). Substituting (4-70) into (4-76), we obtain

$$\Omega_2 = 2\pi\left(1 + \frac{z}{\sqrt{z^2 + r^2}}\right) \tag{4-77}$$

Now, noting that a right-hand screw advances into the bowl as it is turned in the sense of C from below the bowl whereas the electric field due to Q is directed away from Q, the electric field flux passing through the surface S_2 into the bowl is given by

$$\int_{S_2} \mathbf{E} \cdot d\mathbf{S}_2 = -\frac{Q}{4\pi\epsilon_0}\Omega_2 = -\frac{Q}{2\epsilon_0}\left(1 + \frac{z}{\sqrt{z^2 + r^2}}\right) \tag{4-78}$$

The rate at which this flux is changing with time is given by

$$\frac{d}{dt}\int_{S_2} \mathbf{E} \cdot d\mathbf{S}_2 = \frac{d}{dt}\left[-\frac{Q}{2\epsilon_0}\left(1 + \frac{z}{\sqrt{z^2 + r^2}}\right)\right]$$
$$= \frac{I}{2\epsilon_0}\left(1 + \frac{z}{\sqrt{z^2 + r^2}}\right) \tag{4-79}$$

Substituting this result into (4-75) applied to S_2, we obtain

$$\oint_C \mathbf{B} \cdot d\mathbf{l} = \mu_0\left(\text{current due to charges flowing through } S_2\right.$$
$$\left. + \frac{d}{dt}\int_{S_2} \epsilon_0\mathbf{E} \cdot d\mathbf{S}_2\right)$$
$$= \mu_0\left[0 + \frac{I}{2}\left(1 + \frac{z}{\sqrt{z^2 + r^2}}\right)\right]$$
$$= \frac{\mu_0 I}{2}\left(1 + \frac{z}{\sqrt{z^2 + r^2}}\right)$$

which agrees with (4-67), deduced by using the Biot–Savart law. Thus the modified law (4-75) gives the correct result for $\oint_C \mathbf{B} \cdot d\mathbf{l}$ irrespective of the surface bounded by C chosen to apply it.

We note that the quantity $\dfrac{d}{dt}\displaystyle\int_S \epsilon_0\mathbf{E} \cdot d\mathbf{S}$ has the units of current. This can be easily seen if we recognize from Gauss' law that $\int \mathbf{E} \cdot d\mathbf{S}$ has the units of Q/ϵ_0 and hence $\dfrac{d}{dt}\displaystyle\int_S \epsilon_0\mathbf{E} \cdot d\mathbf{S}$ has the units of dQ/dt or current. Equation (4-75) therefore suggests that there are two kinds of current penetrating a surface S bounded by C. The first kind is due to the actual flow of charges across the surface S. The second kind is due to the flux of $\epsilon_0\mathbf{E}$ penetrating S changing with time; Maxwell attributed to it the name "displacement current." Physically, the displacement current is not a current in the sense that there is no flow of a physical quantity, like charge, across the surface. Although the term "time rate of change of the flux of $\epsilon_0\mathbf{E}$" is more apt, we shall follow Maxwell's terminology and use the term "displacement current." The reason behind this terminology will become evident in Chapter 5.

To summarize the discussion thus far in this section, we have found that the dilemma of Ampere's circuital law given by (3-57) is resolved by modifying it to read

$$\oint_C \mathbf{B} \cdot d\mathbf{l} = \mu_0\{[I_c]_S + [I_d]_S\} \tag{4-80}$$

where $[I_c]_S$ is the current due to the actual flow of charges across the surface S bounded by C in the direction of advance of a right-hand screw as it is turned in the sense of C, and $[I_d]_S = \dfrac{d}{dt}\displaystyle\int_S \epsilon_0\mathbf{E} \cdot d\mathbf{S}$ is the displacement current penetrating the surface S in the same direction. We shall refer to Eq. (4-80) as the modified Ampere's circuital law in integral form. While Faraday's law was a consequence of experimental observations by Faraday, the modified Ampere's circuital law was a result of theoretical investigations by Maxwell.

Although we have here derived the modified Ampere's circuital law by considering a particular case, Maxwell provided a general proof based on Gauss' law and the law of conservation of charge. Since charge is conserved, the current due to flow of charge out of a closed surface S bounding a volume V must be equal to the time rate of decrease of the charge enclosed by the surface. This is the law of conservation of charge. If the current flowing out of the surface is $[I_c]_S$ and the charge enclosed by S is Q, we then have

$$[I_c]_S = -\frac{dQ}{dt} \tag{4-81}$$

But, from Gauss' law, we have

$$\oint_S \mathbf{E} \cdot d\mathbf{S} = \frac{Q}{\epsilon_0}$$

or

$$Q = \oint_S \epsilon_0\mathbf{E} \cdot d\mathbf{S} \tag{4-82}$$

Substituting (4-82) into (4-81) and rearranging, we obtain

$$[I_c]_S + \frac{d}{dt}\oint_S \epsilon_0\mathbf{E} \cdot d\mathbf{S} = 0 \tag{4-83}$$

or

$$[I_c]_S + [I_d]_S = 0 \tag{4-84}$$

Thus the law of conservation of charge states that the sum of the current due to the flow of charges and the displacement current across any closed surface must be equal to zero. We will now show that (4-80) is consistent but (3-57) is not consistent with (4-84). To do this, let us consider a closed path C in an electromagnetic field. Let S_1 and S_2 be two different surfaces bounded by C with their normals defined as shown in Fig. 4.7. The normal to S_1 is directed towards the side of advance of a right-hand screw as it is turned in the sense of C. Hence, from (4-80), we have

$$\oint_C \mathbf{B} \cdot d\mathbf{l} = \mu_0 \{ [I_c]_{S_1} + [I_d]_{S_1} \} \tag{4-85}$$

The normal to S_2 is directed opposite to the side of advance of a right-hand screw as it is turned in the sense of C. Hence, from (4-80), we have

$$\oint_C \mathbf{B} \cdot d\mathbf{l} = -\mu_0 \{ [I_c]_{S_2} + [I_d]_{S_2} \} \tag{4-86}$$

Now, since $\oint_C \mathbf{B} \cdot d\mathbf{l}$ is unique, the right sides of (4-85) and (4-86) are equal, giving us

$$[I_c]_{S_1+S_2} + [I_d]_{S_1+S_2} = 0 \tag{4-87}$$

which is consistent with (4-84), since $(S_1 + S_2)$ is a closed surface. On the other hand, if we use (3-57) we obtain, for the surface S_1,

$$\oint_C \mathbf{B} \cdot d\mathbf{l} = \mu_0 [I_c]_{S_1} \tag{4-88}$$

and for the surface S_2,

$$\oint_C \mathbf{B} \cdot d\mathbf{l} = -\mu_0 [I_c]_{S_2} \tag{4-89}$$

From (4-88) and (4-89), we have

$$[I_c]_{S_1+S_2} = 0 \tag{4-90}$$

which is inconsistent with (4-84) unless $[I_d]_{S_1+S_2}$ is equal to zero, which is true only in the static case. It is this inconsistency that prompted Maxwell to modify Ampere's circuital law by adding the displacement current term. A consequence of the displacement current term in the modified Ampere's circuital law is that the current enclosed by a closed path C in an electromagnetic field is generally not equal to $(1/\mu_0) \oint_C \mathbf{B} \cdot d\mathbf{l}$, unlike in the static magnetic field case.

EXAMPLE 4-8.   The arrangement shown in Fig. 4.10 is that of a V-shaped filamentary wire situated in the yz plane symmetrically about the z axis and with its vertex at the origin. Current flows along one leg from infinity to the origin at the rate of I_1 C/sec and leaves along another leg from the origin to infinity at the rate of I_2 C/sec. It is desired to find the values of $\oint \mathbf{B} \cdot d\mathbf{l}$ around two circular contours C_1 and C_2 of radii 1 m and centered at the origin, where (a) C_1 is in the xy plane and (b) C_2 is in the xz plane.

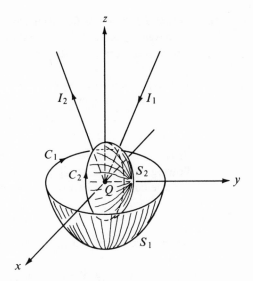

Fig. 4.10. For evaluating $\oint \mathbf{B} \cdot d\mathbf{l}$, around paths C_1 and C_2, due to a V-shaped filamentary wire with unequal currents in the two legs.

Since the current entering the origin is I_1 C/sec whereas the current leaving the origin is I_2 C/sec, there is a charge accumulation at the origin at the rate of $(I_1 - I_2)$ C/sec.

(a) To evaluate $\oint_{C_1} \mathbf{B} \cdot d\mathbf{l}$, let us choose the bowl-shaped surface S_1 bounded by C_1. $[I_c]_{S_1}$ is equal to zero since neither leg of the wire penetrates the surface. On the other hand, since half of the electric field flux emanating from the point charge penetrates the surfaces S_1 towards the side of advance of a right-hand screw as it is turned in the sense of C_1, $[I_d]_{S_1}$ is equal to $\frac{1}{2}(I_1 - I_2)$ C/sec. Thus, according to (4-80),

$$\oint_{C_1} \mathbf{B} \cdot d\mathbf{l} = \frac{\mu_0}{2}(I_1 - I_2)$$

(b) To evaluate $\oint_{C_2} \mathbf{B} \cdot d\mathbf{l}$, let us choose the bowl-shaped surface S_2 bounded by C_2. $[I_c]_{S_2}$ is equal to I_1 since that leg of the wire penetrates the surface with the current flowing towards the side of advance of a right-hand screw as it is turned in the sense of C_2. On the other hand, the electric field flux of the point charge penetrates S_2 in the opposite sense, and since half of the flux emanating from the point charge penetrates S_2, $[I_d]_{S_2}$ is equal to $-\frac{1}{2}(I_1 - I_2)$ C/sec. Thus, according to (4-80),

$$\oint_{C_2} \mathbf{B} \cdot d\mathbf{l} = \mu_0 \left[I_1 - \frac{1}{2}(I_1 - I_2) \right] = \frac{\mu_0}{2}(I_1 + I_2)$$

Note that if $I_1 = I_2 = I$, $\oint_{C_1} \mathbf{B} \cdot d\mathbf{l} = 0$ and $\oint_{C_2} \mathbf{B} \cdot d\mathbf{l} = \mu_0 I$. ∎

4.5   Modified Ampere's Circuital Law in Differential Form
(Maxwell's Second Curl Equation for the Electromagnetic Field) and the Continuity Equation

In the previous section we introduced the modified Ampere's circuital law in integral form, given by

$$\oint_C \mathbf{B} \cdot d\mathbf{l} = \mu_0\{[I_c]_S + [I_d]_S\} \tag{4-80}$$

where S is any surface bounded by C, $[I_c]_S$ is the current due to charges flowing across S, and $[I_d]_S$ is the displacement current through S. For a volume current of density $\mathbf{J}$, we have

$$[I_c]_S = \int_S \mathbf{J} \cdot d\mathbf{S} \tag{4-91}$$

Substituting for $[I_c]_S$ and $[I_d]_S$ in (4-80), we get

$$\oint_C \mathbf{B} \cdot d\mathbf{l} = \mu_0\left(\int_S \mathbf{J} \cdot d\mathbf{S} + \frac{d}{dt}\int_S \epsilon_0 \mathbf{E} \cdot d\mathbf{S}\right) \tag{4-92}$$

According to Stokes' theorem, we have

$$\oint_C \mathbf{B} \cdot d\mathbf{l} = \int_S (\nabla \times \mathbf{B}) \cdot d\mathbf{S}$$

where S is any surface bounded by the contour C. In particular, choosing the same surface as for the integrals on the right side of (4-92), we obtain

$$\int_S (\nabla \times \mathbf{B}) \cdot d\mathbf{S} = \mu_0\left(\int_S \mathbf{J} \cdot d\mathbf{S} + \frac{d}{dt}\int_S \epsilon_0 \mathbf{E} \cdot d\mathbf{S}\right) \tag{4-93}$$

If the surface S is stationary, that is, independent of time,

$$\frac{d}{dt}\int_S \epsilon_0 \mathbf{E} \cdot d\mathbf{S} = \int_S \frac{\partial}{\partial t}(\epsilon_0 \mathbf{E}) \cdot d\mathbf{S} \tag{4-94}$$

and (4-93) becomes

$$\int_S (\nabla \times \mathbf{B}) \cdot d\mathbf{S} = \int_S \mu_0\left[\mathbf{J} + \frac{\partial}{\partial t}(\epsilon_0 \mathbf{E})\right] \cdot d\mathbf{S} \tag{4-95}$$

Comparing the integrands on both sides of (4-95), we have

$$\nabla \times \mathbf{B} = \mu_0\left[\mathbf{J} + \frac{\partial}{\partial t}(\epsilon_0 \mathbf{E})\right] \tag{4-96}$$

Equation (4-96) is the differential form of the modified Ampere's circuital law and it is Maxwell's second curl equation for the electromagnetic field. While we have here derived (4-96) for a stationary S, it can be shown that it holds also for a time-varying surface S due to a moving C, where $\mathbf{E}$, $\mathbf{B}$, and $\mathbf{J}$ are the fields and the current density as viewed by a stationary observer. Following the terminology "displacement current" for the time rate of change of the flux of $\epsilon_0 \mathbf{E}$, the time rate of change of $\epsilon_0 \mathbf{E}$, that is $\frac{\partial}{\partial t}(\epsilon_0 \mathbf{E})$ is known as the "displacement current density."

EXAMPLE 4-9. In the previous section we deduced the magnetic field [Eq. (4-66)] due to a semiinfinitely long filamentary wire along which current flows to infinity from a source of point charge at the origin (Fig. 4.7). It is here desired to verify the result by using (4-96).

From the previous section, the magnetic field due to the wire is given at a point (r, ϕ, z) by

$$\mathbf{B} = \frac{\mu_0 I}{4\pi r}\left(1 + \frac{z}{\sqrt{z^2 + r^2}}\right)\mathbf{i}_\phi$$

Hence

$$\nabla \times \mathbf{B} = \frac{\mathbf{i}_r}{r}\left[-\frac{\partial}{\partial z}(rB_\phi)\right] + \frac{\mathbf{i}_z}{r}\left[\frac{\partial}{\partial r}(rB_\phi)\right]$$

$$= \frac{\mu_0 I}{4\pi r}\left[-\mathbf{i}_r\frac{\partial}{\partial z}\left(1 + \frac{z}{\sqrt{z^2 + r^2}}\right) + \mathbf{i}_z\frac{\partial}{\partial r}\left(1 + \frac{z}{\sqrt{z^2 + r^2}}\right)\right]$$

$$= -\frac{\mu_0 I}{4\pi(z^2 + r^2)^{3/2}}(r\mathbf{i}_r + z\mathbf{i}_z) \qquad (4\text{-}97)$$

Substituting $I = -dQ/dt$ in (4-97), we note that

$$\nabla \times \mathbf{B} = \mu_0\epsilon_0\frac{d}{dt}\left[\frac{Q}{4\pi\epsilon_0(z^2 + r^2)^{3/2}}(r\mathbf{i}_r + z\mathbf{i}_z)\right]$$

$$= \mu_0\epsilon_0\frac{\partial \mathbf{E}}{\partial t} \qquad (4\text{-}98)$$

thereby satisfying (4-96) since $\mathbf{J}$ is zero at (r, ϕ, z). ∎

Returning to Eq. (4-96) and taking the divergence of both sides, we have

$$\nabla \cdot \nabla \times \mathbf{B} = \nabla \cdot \mu_0\left[\mathbf{J} + \frac{\partial}{\partial t}(\epsilon_0 \mathbf{E})\right]$$

$$= \mu_0\left[\nabla \cdot \mathbf{J} + \frac{\partial}{\partial t}(\epsilon_0 \nabla \cdot \mathbf{E})\right] \qquad (4\text{-}99)$$

Since $\nabla \cdot \nabla \times \mathbf{B} \equiv 0$, (4-99) gives us

$$\nabla \cdot \mathbf{J} + \frac{\partial}{\partial t}(\epsilon_0 \nabla \cdot \mathbf{E}) = 0 \qquad (4\text{-}100)$$

But, according to the law of conservation of charge,

$$[I_c]_S = -\frac{dQ}{dt} \qquad (4\text{-}81)$$

where $[I_c]_S$ is the current due to the flow of charges out of a closed surface S and Q is the charge enclosed by S. In terms of current density $\mathbf{J}$ and charge density ρ, $[I_c]_S$ and Q are given by

$$[I_c]_S = \oint_S \mathbf{J} \cdot d\mathbf{S} \tag{4-101}$$

and

$$Q = \int_V \rho \, dv \tag{4-102}$$

where V is the volume bounded by S. Substituting (4-101) and (4-102) into (4-81), we obtain

$$\oint_S \mathbf{J} \cdot d\mathbf{S} = -\frac{d}{dt} \int_V \rho \, dv \tag{4-103}$$

Applying the divergence theorem to the left side of (4-103) and interchanging the differentiation and integration operations on the right side, we get

$$\int_V \mathbf{V} \cdot \mathbf{J} \, dv = -\int_V \frac{\partial \rho}{\partial t} \, dv \tag{4-104}$$

or

$$\int_V \left(\mathbf{V} \cdot \mathbf{J} + \frac{\partial \rho}{\partial t} \right) dv = 0 \tag{4-105}$$

Since (4-105) must be valid for any volume, it follows that

$$\mathbf{V} \cdot \mathbf{J} + \frac{\partial \rho}{\partial t} = 0 \tag{4-106}$$

Equation (4-106) is the law of conservation of charge in differential form. It is also known as the continuity equation. For static fields, $\partial \rho / \partial t = 0$ and (4-106) reduces to $\mathbf{V} \cdot \mathbf{J} = 0$, which agrees with (3-113). Comparing (4-100) with (4-106), we have

$$\frac{\partial}{\partial t}(\epsilon_0 \mathbf{V} \cdot \mathbf{E}) = \frac{\partial \rho}{\partial t}$$

or

$$\frac{\partial}{\partial t}(\epsilon_0 \mathbf{V} \cdot \mathbf{E} - \rho) = 0 \tag{4-107}$$

or

$$(\epsilon_0 \mathbf{V} \cdot \mathbf{E} - \rho) = \text{constant with time} \tag{4-108}$$

The constant on the right side of (4-108) must, however, be equal to zero since a nonzero value at any point in space requires the existence forever of a source of nonsolenoidal electric field flux other than electric charge at that point. Thus we note that Maxwell's equation for the divergence of the time-varying electric field given by

$$\mathbf{V} \cdot \mathbf{E} = \frac{\rho}{\epsilon_0} \tag{4-109}$$

follows from the Maxwell's equation for the curl of **B** given by (4-96) with the aid of the continuity equation (4-106).

4.6   Energy Storage in an Electric Field

In Section 2.8 we introduced the concept of potential difference between two points in an electric field as equal to the work done per unit charge in moving a test charge from one point to the other. In Section 2.9 we extended this to the concept of potential, which is simply the potential difference between two points, one of which is a reference point having zero potential. If we transfer a test charge from a point of higher potential to a point of lower potential, the field does the work and hence there is loss in potential energy of the system, which is supplied to the test charge. Where in the system does this energy come from? Alternatively, if we transfer the test charge from a point of lower potential to a point of higher potential, an external agent moving the charge has to do work, thus increasing the potential energy of the system. Where in the system does this energy expended by the external agent reside? Wherever in the system the energy may reside, a convenient way is to think of the energy as being stored in the electric field. In the first case, part of the stored energy in the field is expended in moving the test charge, whereas in the second case the energy expended by the external agent increases the stored energy.

Let us then consider a system of two point charges Q_1 and Q_2 situated an infinite distance apart so that no forces are exerted on either charge and hence the charges are in equilibrium. According to the definition of potential difference, an amount of work equal to Q_2 times the potential of Q_1 at Q_2 must be expended by an external agent to bring Q_2 close to Q_1 as shown in Fig. 4.11(a). Thus the potential energy of the system is increased by the amount

$$W_2 = Q_2 V_2^1 \tag{4-110}$$

where V_2^1 is the potential of Q_1 at the location of Q_2. If we start with a system of three charges Q_1, Q_2, Q_3 situated an infinite distance apart from each other, then the amount of work required to bring Q_2 and Q_3 close to Q_1 can be determined in two steps. First we bring Q_2 close to Q_1, for which the work required is given by (4-110). Then we bring Q_3 close to Q_1 as shown in Fig. 4.11(b). But, this time, we have to overcome not only the force exerted on Q_3 by Q_1 but also the force exerted by Q_2. Hence the required work is given by

$$W_3 = Q_3 V_3^1 + Q_3 V_3^2 \tag{4-111}$$

Thus the total work required to bring Q_2 and Q_3 close to Q_1 is

$$W_e = W_2 + W_3 = Q_2 V_2^1 + (Q_3 V_3^1 + Q_3 V_3^2) \tag{4-112}$$

The potential energy of the system is increased by the amount given by (4-112).

We can proceed in this manner and consider a system of n point charges $Q_1, Q_2, Q_3, \ldots, Q_n$ initially located infinitely far apart from each other.

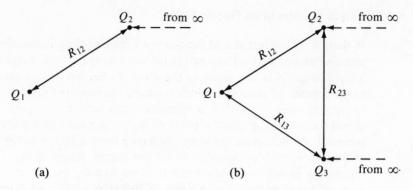

Fig. 4.11. Bringing point charges closer from infinity.

The total work required in bringing the charges close to each other is given by

$$W_e = W_2 + W_3 + \cdots + W_n$$
$$= Q_2 V_2^1 + (Q_3 V_3^1 + Q_3 V_3^2) + (Q_4 V_4^1 + Q_4 V_4^2 + Q_4 V_4^3) + \cdots$$
$$= \sum_{i=2}^{n} \sum_{j=1}^{i-1} Q_i V_i^j \tag{4-113}$$

where V_i^j is the potential of Q_j at the location of Q_i. However, we note that

$$Q_i V_i^j = Q_i \frac{Q_j}{4\pi\epsilon_0 R_{ji}} = Q_j \frac{Q_i}{4\pi\epsilon_0 R_{ij}} = Q_j V_j^i \tag{4-114}$$

Hence (4-113) may be written as

$$W_e = Q_1 V_1^2 + (Q_1 V_1^3 + Q_2 V_2^3) + (Q_1 V_1^4 + Q_2 V_2^4 + Q_3 V_3^4) + \cdots$$
$$= \sum_{i=2}^{n} \sum_{j=1}^{i-1} Q_j V_j^i \tag{4-115}$$

Adding (4-113) and (4-115), we have

$$2W_e = Q_1(V_1^2 + V_1^3 + V_1^4 + \cdots)$$
$$\quad + Q_2(V_2^1 + V_2^3 + V_2^4 + \cdots)$$
$$\quad + Q_3(V_3^1 + V_3^2 + V_3^4 + \cdots)$$
$$\quad + \cdots$$
$$= Q_1(\text{potential at } Q_1 \text{ due to all other charges})$$
$$\quad + Q_2(\text{potential at } Q_2 \text{ due to all other charges})$$
$$\quad + Q_3(\text{potential at } Q_3 \text{ due to all other charges})$$
$$\quad + \cdots$$
$$= Q_1 V_1 + Q_2 V_2 + Q_3 V_3 + \cdots$$
$$= \sum_{i=1}^{n} Q_i V_i \tag{4-116}$$

where V_i is the potential at Q_i due to all other charges. Dividing both sides of (4-116) by 2, we have

$$W_e = \frac{1}{2} \sum_{i=1}^{n} Q_i V_i \tag{4-117}$$

Thus the potential energy stored in the system of n point charges is given by (4-117).

EXAMPLE 4-10.   Three point charges of values 1, 2, and 3 C are situated at the corners of an equilateral triangle of sides 1 m. It is desired to find the work required to move these charges to the corners of an equilateral triangle of shorter sides $\frac{1}{2}$ m as shown in Fig. 4.12.

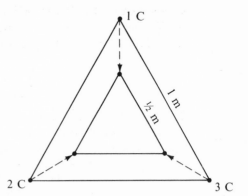

Fig. 4.12. Bringing three point charges from the corners of a larger equilateral triangle to the corners of a smaller equilateral triangle.

The potential energy stored in the system of three charges at the corners of the larger equilateral triangle is given by

$$\frac{1}{2}\sum_{i=1}^{3}Q_iV_i = \frac{1}{2}\left[1\left(\frac{2}{4\pi\epsilon_0}+\frac{3}{4\pi\epsilon_0}\right)+2\left(\frac{1}{4\pi\epsilon_0}+\frac{3}{4\pi\epsilon_0}\right)+3\left(\frac{1}{4\pi\epsilon_0}+\frac{2}{4\pi\epsilon_0}\right)\right]$$

$$= \frac{1}{2}\left[\frac{5+8+9}{4\pi\epsilon_0}\right]=\frac{11}{4\pi\epsilon_0}\ \text{N-m}$$

The potential energy stored in the system of three charges at the corners of the smaller equilateral triangle is equal to twice the above value since all distances are halved. The increase in potential energy of the system in going from the larger to the smaller equilateral triangle is equal to $11/4\pi\epsilon_0$ N-m. Obviously, this increase in energy must be supplied by an external agent and hence the work required to move the charges to the corners of the equilateral triangle of sides $\frac{1}{2}$ m from the corners of the equilateral triangle of sides 1 m is equal to $11/4\pi\epsilon_0$ N-m. ∎

If we have a continuous distribution of charge with density $\rho(r, \theta, \phi)$ instead of an assembly of discrete charges, we can treat it as a continuous collection of infinitesimal charges of value $\rho(r, \theta, \phi)\,\Delta v$, each of which can be considered as a point charge, and obtain the potential energy of the system as

$$W_e = \frac{1}{2}\lim_{\Delta v\to0}\sum[\rho(r,\theta,\phi)\,\Delta v]\,V(r,\theta,\phi)$$

$$= \frac{1}{2}\int_{\substack{\text{volume}\\\text{containing }\rho}}\rho V\,dv \tag{4-118a}$$

Similarly, for a surface charge distribution of density ρ_s on a surface S, we have

$$W_e = \frac{1}{2} \int_S \rho_s V \, dS \tag{4-118b}$$

Thus far, we have found the potential energy of the charge distribution by considering the work done in assembling the system. We stated at the beginning of this section that the potential energy can be thought of as being stored in the electric field set up by the system of charges. If so, we should be able to express the energy in terms of the electric field. To do this, we substitute for ρ in (4-118a) from (2-82) and obtain

$$W_e = \frac{1}{2} \int_{\substack{\text{volume} \\ \text{containing } \rho}} (\epsilon_0 \, \mathbf{\nabla} \cdot \mathbf{E}) V \, dv \tag{4-119}$$

Since $\mathbf{\nabla} \cdot \mathbf{E} = 0$ in the region not containing ρ, the value of the integral on the right side of (4-119) is not altered if we change the volume of integration from the volume containing ρ to the entire space. Thus

$$W_e = \frac{1}{2} \int_{\text{all space}} (\epsilon_0 \, \mathbf{\nabla} \cdot \mathbf{E}) V \, dv \tag{4-120}$$

We now use the vector identity

$$\mathbf{\nabla} \cdot V\mathbf{E} = V \mathbf{\nabla} \cdot \mathbf{E} + \mathbf{E} \cdot \mathbf{\nabla}V$$

to replace $V\mathbf{\nabla} \cdot \mathbf{E}$ on the right side of (4-120) by $\mathbf{\nabla} \cdot V\mathbf{E} - \mathbf{E} \cdot \mathbf{\nabla}V$ and obtain

$$\begin{aligned} W_e &= \frac{1}{2}\epsilon_0 \int_{\text{all space}} (\mathbf{\nabla} \cdot V\mathbf{E} - \mathbf{E} \cdot \mathbf{\nabla}V) \, dv \\ &= \frac{1}{2}\epsilon_0 \int_{\text{all space}} \mathbf{\nabla} \cdot V\mathbf{E} \, dv + \frac{1}{2}\epsilon_0 \int_{\text{all space}} \mathbf{E} \cdot \mathbf{E} \, dv \end{aligned} \tag{4-121}$$

where we have replaced $\mathbf{\nabla}V$ by $-\mathbf{E}$ in accordance with (2-138). Using the divergence theorem, we equate the first integral on the right side of (4-121) to a surface integral thus:

$$\int_{\text{all space}} \mathbf{\nabla} \cdot V\mathbf{E} \, dv = \int_{\substack{\text{surface} \\ \text{bounding} \\ \text{all space}}} V\mathbf{E} \cdot \mathbf{i}_n \, dS \tag{4-122}$$

However, as viewed from a surface bounding all space, a charge distribution of finite volume appears as a point charge, say Q. Hence, as $r \to \infty$, we can write

$$\mathbf{E} \longrightarrow \frac{Q}{4\pi\epsilon_0 r^2} \mathbf{i}_r$$

$$V \longrightarrow \frac{Q}{4\pi\epsilon_0 r}$$

$$\int_{\substack{\text{surface} \\ \text{bounding} \\ \text{all space}}} V\mathbf{E} \cdot \mathbf{i}_n \, dS = \lim_{r \to \infty} \int_{\theta=0}^{\pi} \int_{\phi=0}^{2\pi} \frac{Q}{4\pi\epsilon_0 r} \frac{Q}{4\pi\epsilon_0 r^2} \mathbf{i}_r \cdot r^2 \sin\theta \, d\theta \, d\phi \, \mathbf{i}_r$$

$$= \int_{\theta=0}^{\pi} \int_{\phi=0}^{2\pi} \lim_{r \to \infty} \frac{Q^2}{(4\pi\epsilon_0)^2 r} \sin\theta \, d\theta \, d\phi = 0 \qquad (4\text{-}123)$$

Equation (4-123) holds also for a charge distribution of infinite extent, provided the electric field due to the charge distribution falls off at least as $(1/r^2)\mathbf{i}_r$ and hence the potential falls off at least as $1/r$. Thus (4-121) reduces to

$$W_e = \frac{1}{2}\epsilon_0 \int_{\text{all space}} \mathbf{E} \cdot \mathbf{E} \, dv = \int_{\text{all space}} \left(\frac{1}{2}\epsilon_0 E^2\right) dv \qquad (4\text{-}124)$$

Equation (4-124) indicates clearly that the idea of energy residing in the electric field is a valid one provided we integrate $\frac{1}{2}\epsilon_0 E^2$ throughout the entire space. The quantity $\frac{1}{2}\epsilon_0 E^2$ is evidently the energy density in the electric field.

EXAMPLE 4-11.   A volume charge is distributed throughout a sphere of radius a meters, and centered at the origin, with uniform density ρ_0 C/m³. We wish to find the energy stored in the electric field of this charge distribution.

From Example 2-6, the electric field of the uniformly distributed spherical charge, having its center at the origin, is given by

$$\mathbf{E} = \begin{cases} \dfrac{\rho_0 a^3}{3\epsilon_0 r^2} \mathbf{i}_r & \text{for } r > a \\[2ex] \dfrac{\rho_0 r}{3\epsilon_0} \mathbf{i}_r & \text{for } r < a \end{cases}$$

Hence the energy density in the electric field is given by

$$\frac{1}{2}\epsilon_0 E^2 = \begin{cases} \dfrac{\rho_0^2 a^6}{18\epsilon_0 r^4} & \text{for } r > a \\[2ex] \dfrac{\rho_0^2 r^2}{18\epsilon_0} & \text{for } r < a \end{cases}$$

The energy stored in the electric field is

$$W_e = \int_{r=0}^{a} \int_{\theta=0}^{\pi} \int_{\phi=0}^{2\pi} \frac{\rho_0^2 r^2}{18\epsilon_0} r^2 \sin\theta \, dr \, d\theta \, d\phi$$

$$+ \int_{r=a}^{\infty} \int_{\theta=0}^{2\pi} \int_{\phi=0}^{2\pi} \frac{\rho_0^2 a^6}{18\epsilon_0 r^4} r^2 \sin\theta \, dr \, d\theta \, d\phi$$

$$= \frac{4\pi\rho_0^2 a^5}{15\epsilon_0} \quad \blacksquare$$

4.7   Energy Storage in a Magnetic Field

In the previous section we derived an expression for the energy density in an electric field by first finding the work required to be done by an external agent in assembling a system of point charges and then extending the result to a continuous distribution of charge. Just as work is required for gathering point charges from infinity, it requires work to gather a set of current loops from infinity. Just as we can interpret the energy expended by an external agent in assembling the charges as being stored in the electric field of the charges, we can think of the energy expended by an external agent in assembling the current loops as being stored in the magnetic field of the current loops. It is possible to derive an expression for the energy density in a magnetic field by starting with a set of current loops at infinity and proceeding in a similar manner as in the previous section. To simplify the derivation, we will, however, consider directly the building up of a solenoidal volume current distribution.

Let us then consider a solenoidal volume current distribution of density $\mathbf{J}$ in a volume V where $\mathbf{J}$ increases linearly with time from zero to a value $\mathbf{J}_0$ in a time t_0, that is, $\mathbf{J} = \mathbf{J}_0 t/t_0$. The magnetic field $\mathbf{B}$ associated with the current distribution also increases linearly with time, that is, $\mathbf{B} = \mathbf{B}_0 t/t_0$. The time varying magnetic field induces an electric field in accordance with Faraday's law. The induced electric field exerts forces on charges constituting the current flow. The work done by these forces must be balanced by an external agent to maintain the current density at $\mathbf{J}_0 t/t_0$ and hence is stored in the magnetic field as the potential energy associated with the current distribution.

To find this energy, let us divide the cross-sectional area S of the current distribution into a number of infinitesimal areas ΔS_i. Through each infinitesimal area, a current loop C_i can be defined by the direction line of the current density vector $\mathbf{J}_i = \mathbf{J}_{i0} t/t_0$ corresponding to that area as shown in Fig. 4-13. The current I_i flowing around the loop C_i is equal to $\mathbf{J}_i \cdot \Delta \mathbf{S}_i$. The amount of charge dQ_i crossing ΔS_i in time dt is equal to $I_i \, dt$. Denoting the induced electric field at the point occupied by ΔS_i to be $\mathbf{E}_i$, we obtain the force exerted by this field on the charge dQ_i to be $dQ_i \, \mathbf{E}_i = I_i \, dt \, \mathbf{E}_i$. The work done by this force as the charge dQ_i is displaced by the infinitesimal distance $d\mathbf{l}_i$ along $\mathbf{J}_i$ is $I_i \, dt \, \mathbf{E}_i \cdot d\mathbf{l}_i$. Hence, the work required to be done against the induced electric field around the loop C_i in time dt is

$$dW_m = - \oint_{C_i} I_i \, dt \, \mathbf{E}_i \cdot d\mathbf{l}_i = - I_i \, dt \oint_{C_i} \mathbf{E}_i \cdot d\mathbf{l}_i \qquad (4\text{-}125)$$

Using Faraday's law and substituting $\mathbf{B} = \nabla \times \mathbf{A}$ and then using Stoke's theorem, we have

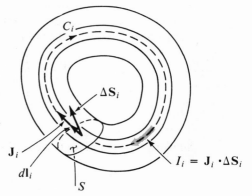

Fig. 4.13. Division of a solenoidal continuous distribution of current into a number of solenoidal current tubes having infinitesimal cross-sectional areas.

$$\oint_{C_i} \mathbf{E}_i \cdot d\mathbf{l}_i = -\frac{d}{dt} \int_{S_i} \mathbf{B} \cdot d\mathbf{S}$$

$$= -\frac{d}{dt} \int_{S_i} (\nabla \times \mathbf{A}) \cdot d\mathbf{S} = -\frac{d}{dt} \oint_{C_i} \mathbf{A}_i \cdot d\mathbf{l}_i \qquad (4\text{-}126)$$

where $\mathbf{A}$ is the magnetic vector potential associated with $\mathbf{B}$ and S_i is any surface bounded by C_i. In view of the linear increase of $\mathbf{B}$ with time, $\mathbf{A}$ also increases linearly with time. Thus, denoting $\mathbf{A}_i = \mathbf{A}_{i0} t/t_0$, we have

$$\oint_{C_i} \mathbf{E}_i \cdot d\mathbf{l}_i = -\frac{d}{dt} \oint_{C_i} \mathbf{A}_{i0} \frac{t}{t_0} \cdot d\mathbf{l}_i = -\oint_{C_i} \frac{\mathbf{A}_{i0}}{t_0} \cdot d\mathbf{l}_i \qquad (4\text{-}127)$$

Substituting (4-127) into (4-125), we obtain

$$dW_m = I_i \, dt \oint_{C_i} \frac{\mathbf{A}_{i0}}{t_0} \cdot d\mathbf{l}_i \qquad (4\text{-}128)$$

The total work required to be done by an external agent from $t = 0$ to t and for the entire current distribution is then given by

$$W_m = \sum_i \int_{t=0}^{t} I_i \, dt \oint_{C_i} \frac{\mathbf{A}_{i0}}{t_0} \cdot d\mathbf{l}_i$$

$$= \sum_i \oint_{C_i} \int_{t=0}^{t} \left(\frac{\mathbf{J}_{i0} t}{t_0} \cdot \Delta\mathbf{S}_i \right)\left(\frac{\mathbf{A}_{i0}}{t_0} \cdot d\mathbf{l}_i \right)$$

$$= \sum_i \frac{1}{2} \oint_{C_i} \left(\frac{\mathbf{J}_{i0} t}{t_0} \cdot \Delta\mathbf{S}_i \right)\left(\frac{\mathbf{A}_{i0} t}{t_0} \cdot d\mathbf{l}_i \right) \qquad (4\text{-}129)$$

$$= \sum_i \frac{1}{2} \oint_{C_i} (\mathbf{J}_i \cdot \Delta\mathbf{S}_i)(\mathbf{A}_i \cdot d\mathbf{l}_i)$$

$$= \sum_i \frac{1}{2} \oint_{C_i} (\mathbf{J}_i \cdot \mathbf{A}_i)(\Delta\mathbf{S}_i \cdot d\mathbf{l}_i)$$

since $\mathbf{J}_i$ and $d\mathbf{l}_i$ are parallel. Now, in the limit that all $\Delta \mathbf{S}_i \rightarrow 0$, the summation on the right side of (4-129) becomes an integral to give us the potential energy associated with the volume current distribution as

$$
\begin{aligned}
W_m &= \frac{1}{2} \int_S \oint_C (\mathbf{J} \cdot \mathbf{A})(d\mathbf{S} \cdot d\mathbf{l}) \\
&= \frac{1}{2} \int_{\substack{\text{volume} \\ \text{containing } \mathbf{J}}} \mathbf{J} \cdot \mathbf{A} \, dv
\end{aligned}
\tag{4-130a}
$$

Similarly, for a surface current distribution of density $\mathbf{J}_s$ on a surface S, we have

$$
W_m = \frac{1}{2} \int_S \mathbf{J}_s \cdot \mathbf{A} \, dS
\tag{4-130b}
$$

To express the energy in terms of the magnetic field, we substitute for $\mathbf{J}$ in (4-130a) from (3-76) [instead of from (4-96)], in view of the solenoidal nature of $\mathbf{J}$, and obtain

$$
W_m = \frac{1}{2} \int_{\substack{\text{volume} \\ \text{containing } \mathbf{J}}} \frac{1}{\mu_0} \boldsymbol{\nabla} \times \mathbf{B} \cdot \mathbf{A} \, dv
\tag{4-131}
$$

Since $\boldsymbol{\nabla} \times \mathbf{B} = 0$ in the region not containing $\mathbf{J}$ [using again (3-76) instead of (4-96) for the same reason], the value of the integral on the right side of (4-131) is not altered if we change the volume of integration from the volume containing $\mathbf{J}$ to the entire space. Thus

$$
W_m = \frac{1}{2} \int_{\text{all space}} \frac{1}{\mu_0} \boldsymbol{\nabla} \times \mathbf{B} \cdot \mathbf{A} \, dv
\tag{4-132}
$$

We now use the vector identity

$$
\boldsymbol{\nabla} \cdot (\mathbf{A} \times \mathbf{B}) = \mathbf{B} \cdot \boldsymbol{\nabla} \times \mathbf{A} - \mathbf{A} \cdot \boldsymbol{\nabla} \times \mathbf{B}
$$

to replace $\boldsymbol{\nabla} \times \mathbf{B} \cdot \mathbf{A}$ on the right side of (4-132) by $\mathbf{B} \cdot \boldsymbol{\nabla} \times \mathbf{A} - \boldsymbol{\nabla} \cdot (\mathbf{A} \times \mathbf{B})$ and obtain

$$
\begin{aligned}
W_m &= \frac{1}{2\mu_0} \int_{\text{all space}} [\mathbf{B} \cdot \boldsymbol{\nabla} \times \mathbf{A} - \boldsymbol{\nabla} \cdot (\mathbf{A} \times \mathbf{B})] \, dv \\
&= \frac{1}{2\mu_0} \int_{\text{all space}} \mathbf{B} \cdot \mathbf{B} \, dv - \frac{1}{2\mu_0} \int_{\text{all space}} \boldsymbol{\nabla} \cdot (\mathbf{A} \times \mathbf{B}) \, dv
\end{aligned}
\tag{4-133}
$$

where we have replaced $\boldsymbol{\nabla} \times \mathbf{A}$ by $\mathbf{B}$ in accordance with (3-82). Using the divergence theorem, we equate the second integral on the right side of (4-133) to a surface integral thus:

$$
\int_{\text{all space}} \boldsymbol{\nabla} \cdot (\mathbf{A} \times \mathbf{B}) \, dv = \int_{\substack{\text{surface} \\ \text{bounding} \\ \text{all space}}} (\mathbf{A} \times \mathbf{B}) \cdot \mathbf{i}_n \, dS
\tag{4-134}
$$

However, as viewed from a surface bounding all space, a solenoidal current distribution of finite volume appears as a dipole moment, say **m**. Hence, as $r \longrightarrow \infty$, we can write

$$\mathbf{B} \longrightarrow \frac{\mu_0 m}{4\pi r^3}(2 \cos \theta \, \mathbf{i}_r + \sin \theta \, \mathbf{i}_\theta)$$

$$\mathbf{A} \longrightarrow \frac{\mu_0 m}{4\pi r^2} \sin \theta \, \mathbf{i}_\phi$$

where the z axis is chosen to be along the direction of **m**. Thus

$$|\mathbf{A} \times \mathbf{B}| \sim \frac{1}{r^5}$$

whereas

$$dS \sim r^2$$

so that the integral on the right side of (4-134) is zero. This is true also for a current distribution of infinite extent, provided the magnetic flux density due to the current distribution falls off at least as $1/r^2$ and hence the magnetic vector potential falls off at least as $1/r$. Equation (4-133) then reduces to

$$W_m = \frac{1}{2\mu_0} \int_{\text{all space}} \mathbf{B} \cdot \mathbf{B} \, dv = \int_{\text{all space}} \left(\frac{1}{2} \frac{B^2}{\mu_0} \right) dv \qquad (4\text{-}135)$$

Equation (4-135) indicates clearly that the idea of energy residing in the magnetic field is a valid one provided we integrate $\frac{1}{2}B^2/\mu_0$ throughout the entire space. The quantity $\frac{1}{2}B^2/\mu_0$ is evidently the energy density in the magnetic field.

EXAMPLE 4-12. Current I flows in the $+z$ direction with uniform density on the cylindrical surface $r = a$ and returns in the $-z$ direction with uniform density on a second cylindrical surface $r = b$ so that the surface current distribution is given by

$$\mathbf{J}_s = \begin{cases} \dfrac{I}{2\pi a} \mathbf{i}_z & r = a \\[2ex] -\dfrac{I}{2\pi b} \mathbf{i}_z & r = b \end{cases}$$

We wish to find the energy stored in the magnetic field per unit length of the current distribution.

From application of Ampere's circuital law in integral form, we obtain the magnetic flux density due to the given current distribution as

$$\mathbf{B} = \begin{cases} 0 & r < a \\[1ex] \dfrac{\mu_0 I}{2\pi r} \mathbf{i}_\phi & a < r < b \\[1ex] 0 & r > b \end{cases} \qquad (4\text{-}136)$$

Since **B** is zero for $r > b$, the integral on the right side of (4-134) is zero so that we can use (4-135) for computing the energy. If we have a situation in which current flows on one surface in one direction and does not return on another surface, then the magnetic field will not be zero at $r = \infty$. In fact, it falls off as $1/r$ and the magnetic vector potential varies as $\ln r$ so that $|\mathbf{A} \times \mathbf{B}| \sim (1/r) \ln r$. But since $dS \sim r$, (4-134) does not reduce to zero. In such a case, we have to include the second term on the right side of (4-133) to compute W_m. However, in all physical situations, the current does return in the opposite direction on another surface and hence the magnetic field is zero at $r = \infty$. Now, returning to the solution of the example under consideration, we obtain, upon substitution of (4-136) into (4-135),

$$
W_m = \int_{z=-\infty}^{\infty} \int_{r=a}^{b} \int_{\phi=0}^{2\pi} \frac{1}{2\mu_0} \left(\frac{\mu_0 I}{2\pi r}\right)^2 r \, dr \, d\phi \, dz
$$
$$
= \int_{z=-\infty}^{\infty} \left(\frac{\mu_0 I^2}{4\pi} \ln \frac{b}{a}\right) dz
\tag{4-137}
$$

Thus the energy stored in the magnetic field per unit length of the current distribution is $(\mu_0 I^2/4\pi) \ln(b/a)$. ∎

4.8   Power Flow in an Electromagnetic Field; The Poynting Vector

In Section 4.6 we showed that the potential energy associated with a charge distribution can be thought of as residing in the electric field **E** set up by the charge distribution, with the energy density equal to $\frac{1}{2}\epsilon_0 E^2$. Similarly, in Section 4.7 we showed that the potential energy associated with a current distribution can be thought of as residing in the magnetic field **B** set up by the current distribution, with the energy density equal to $\frac{1}{2}B^2/\mu_0$. Let us now consider a point charge Q moving with a velocity **v** in a region of electromagnetic field characterized by electric and magnetic fields **E** and **B**. According to the Lorentz force equation, the force experienced by the point charge is given by

$$
\mathbf{F} = Q(\mathbf{E} + \mathbf{v} \times \mathbf{B})
$$

The work done by the force in displacing the charge by an infinitesimal distance $d\mathbf{l}$ is

$$
dW = \mathbf{F} \cdot d\mathbf{l} = Q(\mathbf{E} + \mathbf{v} \times \mathbf{B}) \cdot d\mathbf{l}
$$
$$
= Q\mathbf{E} \cdot d\mathbf{l} + Q\frac{d\mathbf{l}}{dt} \times \mathbf{B} \cdot d\mathbf{l}
\tag{4-138}
$$
$$
= Q\mathbf{E} \cdot d\mathbf{l} = Q\mathbf{E} \cdot \mathbf{v} \, dt
$$

This amount of work is done by the fields and the time rate at which it is done or the power supplied by the fields for the motion of the charge is

$$
\frac{dW}{dt} = Q\mathbf{E} \cdot \mathbf{v}
\tag{4-139}
$$

If we have a volume charge distribution of density ρ instead of a point charge Q, we can divide the volume into a number of infinitesimal volumes dv and consider the charge $\rho\,dv$ in each infinitesimal volume as a point charge. Substituting $Q = \rho\,dv$ in (4-139), we then have the power supplied by the field for the motion of the charge $\rho\,dv$ as

$$\frac{dW}{dt} = \rho\,dv\,\mathbf{E}\cdot\mathbf{v} \qquad (4\text{-}140)$$

The power supplied by the field to the entire volume charge distribution is given by the integral of (4-140) over the volume of the charge distribution. Thus, if a volume charge of density $\rho(r, \theta, \phi)$ is moving with a velocity $\mathbf{v}(r, \theta, \phi)$ in the region V of an electromagnetic field characterized by electric and magnetic fields $\mathbf{E}(r, \theta, \phi)$ and $\mathbf{B}(r, \theta, \phi)$, respectively, thereby constituting a current of density $\mathbf{J}(r, \theta, \phi)$, the power expended by the electromagnetic field is given by

$$P_d = \int_V \rho\,dv\,\mathbf{E}\cdot\mathbf{v} = \int_V \mathbf{E}\cdot\mathbf{J}\,dv \qquad (4\text{-}141)$$

where we have substituted $\mathbf{J}$ for $\rho\mathbf{v}$ in accordance with (3-10).

We now make use of the vector identity

$$\nabla\cdot(\mathbf{E}\times\mathbf{B}) = \mathbf{B}\cdot\nabla\times\mathbf{E} - \mathbf{E}\cdot\nabla\times\mathbf{B}$$

and Maxwell's curl equations

$$\nabla\times\mathbf{E} = -\frac{\partial\mathbf{B}}{\partial t}$$

$$\nabla\times\mathbf{B} = \mu_0\left(\mathbf{J} + \epsilon_0\frac{\partial\mathbf{E}}{\partial t}\right)$$

to obtain

$$\nabla\cdot(\mathbf{E}\times\mathbf{B}) = -\mathbf{B}\cdot\frac{\partial\mathbf{B}}{\partial t} - \mu_0\mathbf{E}\cdot\mathbf{J} - \mu_0\epsilon_0\mathbf{E}\cdot\frac{\partial\mathbf{E}}{\partial t} \qquad (4\text{-}142)$$

Noting that

$$\mathbf{B}\cdot\frac{\partial\mathbf{B}}{\partial t} = \frac{\partial}{\partial t}\left(\frac{1}{2}\mathbf{B}\cdot\mathbf{B}\right)$$

and

$$\mathbf{E}\cdot\frac{\partial\mathbf{E}}{\partial t} = \frac{\partial}{\partial t}\left(\frac{1}{2}\mathbf{E}\cdot\mathbf{E}\right)$$

(4-142) can be written as

$$\mathbf{E}\cdot\mathbf{J} + \frac{\partial}{\partial t}\left(\frac{1}{2\mu_0}\mathbf{B}\cdot\mathbf{B}\right) + \frac{\partial}{\partial t}\left(\frac{1}{2}\epsilon_0\mathbf{E}\cdot\mathbf{E}\right) = -\nabla\cdot\left(\mathbf{E}\times\frac{\mathbf{B}}{\mu_0}\right) \qquad (4\text{-}143)$$

Defining a vector $\mathbf{P}$ given by

$$\mathbf{P} = \mathbf{E}\times\frac{\mathbf{B}}{\mu_0} \qquad (4\text{-}144)$$

and taking the volume integral on both sides of (4-143) over the volume V, we obtain

$$\int_V \mathbf{E} \cdot \mathbf{J} \, dv + \int_V \frac{\partial}{\partial t}\left(\frac{1}{2\mu_0}\mathbf{B} \cdot \mathbf{B}\right) dv + \int_V \frac{\partial}{\partial t}\left(\frac{1}{2}\epsilon_0\mathbf{E} \cdot \mathbf{E}\right) dv$$
$$= -\int_V \mathbf{\nabla} \cdot \mathbf{P} \, dv \tag{4-145}$$

Interchanging the differentiation operation with time and integration over volume in the second and third terms on the left side of (4-145) and replacing the volume integral on the right side of (4-145) by a closed surface integral in accordance with the divergence theorem, we get

$$\int_V \mathbf{E} \cdot \mathbf{J} \, dv + \frac{\partial}{\partial t}\int_V \left(\frac{1}{2\mu_0}\mathbf{B} \cdot \mathbf{B}\right) dv + \frac{\partial}{\partial t}\int_V \left(\frac{1}{2}\epsilon_0\mathbf{E} \cdot \mathbf{E}\right) dv$$
$$= -\oint_S \mathbf{P} \cdot d\mathbf{S} \tag{4-146}$$

where S is the surface bounding the volume V.

On the left side of (4-146), the second and third terms represent the time rate of increase of energy stored in the magnetic and electric fields, respectively, in the volume V. Thus the left side is the sum of the power expended by the fields due to the motion of the charge and the time rate of increase of stored energy in the fields. Obviously then, the right side of (4-146) must represent the power flow into the volume V across the surface S, or

$$\text{the power flow out of volume } V \text{ across the surface } S = \oint_S \mathbf{P} \cdot d\mathbf{S} \tag{4-147}$$

It then follows that the vector $\mathbf{P}$ has the meaning of power density associated with the electromagnetic field at a point. The statement represented by (4-146) is known as Poynting's theorem after J. H. Poynting, who derived it in 1884, and the vector $\mathbf{P}$ is known as the Poynting vector. We note that the units of $\mathbf{P} = \mathbf{E} \times \mathbf{B}/\mu_0$ are

$$\frac{\text{newtons}}{\text{coulomb}} \times \frac{\text{newton-seconds}}{\text{coulomb-meter}} \div \frac{\text{newtons}}{(\text{ampere})^2}$$
$$= \frac{\text{newton-amperes}}{\text{coulomb-meter}} = \frac{\text{newtons}}{\text{second-meter}}$$
$$= \frac{\text{newton-meters}}{\text{second}} \times \frac{1}{(\text{meter})^2} = \frac{\text{watts}}{(\text{meter})^2}$$

and do indeed represent units of power density.

Caution must be exercised in the interpretation of the Poynting vector $\mathbf{P}$ as representing the power density at a point, since we can add to $\mathbf{P}$ on the right side of (4-146) any vector for which the surface integral over S vanishes, without affecting the equation. On the other hand, the interpretation of $\int_V (\mathbf{\nabla} \cdot \mathbf{P}) \, dv = \oint_S \mathbf{P} \cdot d\mathbf{S}$ as the power flow out of the volume V bounded by S should always give the correct answer. For example, let us consider a region free of charges and currents in which static electric and magnetic fields $\mathbf{E}$ and $\mathbf{B}$ exist. For such a situation, although $\mathbf{E} \times \mathbf{B}$ can be nonzero, $\nabla \cdot (\mathbf{E} \times \mathbf{B}) = \mathbf{B} \cdot (\nabla \times \mathbf{E}) - \mathbf{E} \cdot (\nabla \times \mathbf{B}) = 0$ since $\nabla \times \mathbf{E} = 0$ for a static electric field and $\nabla \times \mathbf{B} = 0$ for a static magnetic field in a current-free region. The fact that $\nabla \cdot (\mathbf{E} \times \mathbf{B}) = 0$ is consistent with the physical situation, since there is no change with time in the energy stored in the static electric and magnetic fields and hence there is no power flow associated with the fields. Thus the interpretation of the Poynting vector as the power density vector at a point in an electromagnetic field is strictly valid only in the sense that $\oint_S \mathbf{P} \cdot d\mathbf{S}$ gives the correct result for the power flow across the closed surface S.

EXAMPLE 4-13.   The electric field intensity $\mathbf{E}$ in the radiation field of an antenna located at the origin of a spherical coordinate system is given by

$$\mathbf{E} = \frac{E_0 \sin \theta}{r} \cos (\omega t - \beta r) \, \mathbf{i}_\theta$$

where E_0, ω, and $\beta (= \omega \sqrt{\mu_0 \epsilon_0})$ are constants. It is desired to find the magnetic field $\mathbf{B}$ associated with this electric field and then find the power radiated by the antenna by integrating the Poynting vector over a spherical surface of radius r centered at the antenna.

From Maxwell's equation for the curl of $\mathbf{E}$, we have

$$-\frac{\partial \mathbf{B}}{\partial t} = \nabla \times \mathbf{E}$$

$$= \begin{vmatrix} \dfrac{\mathbf{i}_r}{r^2 \sin \theta} & \dfrac{\mathbf{i}_\theta}{r \sin \theta} & \dfrac{\mathbf{i}_\phi}{r} \\[2mm] \dfrac{\partial}{\partial r} & \dfrac{\partial}{\partial \theta} & \dfrac{\partial}{\partial \phi} \\[2mm] 0 & E_0 \sin \theta \cos (\omega t - \beta r) & 0 \end{vmatrix}$$

$$= \frac{\beta}{r} E_0 \sin \theta \sin (\omega t - \beta r) \, \mathbf{i}_\phi$$

and

$$\mathbf{B} = \frac{\beta E_0}{\omega r} \sin \theta \cos (\omega t - \beta r) \, \mathbf{i}_\phi$$

The Poynting vector is then given by

$$\mathbf{P} = \mathbf{E} \times \frac{\mathbf{B}}{\mu_0}$$

$$= \frac{1}{\mu_0} \begin{vmatrix} \mathbf{i}_r & \mathbf{i}_\theta & \mathbf{i}_\phi \\[2mm] 0 & \dfrac{E_0 \sin \theta}{r} \cos (\omega t - \beta r) & 0 \\[2mm] 0 & 0 & \dfrac{\beta E_0}{\omega r} \sin \theta \cos (\omega t - \beta r) \end{vmatrix}$$

$$= \frac{\beta E_0^2 \sin^2 \theta}{\mu_0 \omega r^2} \cos^2 (\omega t - \beta r) \, \mathbf{i}_r$$

The power radiated by the antenna

$$= \oint_{\substack{\text{spherical surface} \\ \text{of radius } r}} \mathbf{P} \cdot d\mathbf{S}$$

$$= \int_{\theta=0}^{\pi} \int_{\phi=0}^{2\pi} \frac{\beta E_0^2 \sin^2 \theta}{\mu_0 \omega r^2} \cos^2 (\omega t - \beta r) \, \mathbf{i}_r \cdot r^2 \sin \theta \, d\theta \, d\phi \, \mathbf{i}_r$$

$$= \frac{2\pi \beta E_0^2 \cos^2 (\omega t - \beta r)}{\mu_0 \omega} \int_{\theta=0}^{\pi} \sin^3 \theta \, d\theta$$

$$= \frac{8\pi \beta E_0^2 \cos^2 (\omega t - \beta r)}{3\mu_0 \omega} \quad \blacksquare$$

4.9   The Phasor Concept and the Phasor Representation of Sinusoidally Time-Varying Fields and Maxwell's Equations for Sinusoidally Time-Varying Fields

In developing the electromagnetic field equations, we have thus far considered the time variation of the fields and the associated source quantities to be completely arbitrary. A very important special case of variation with time of the field and source quantities is the sinusoidal steady-state variation. Among the reasons for this importance are that, in practice, we do encounter such fields and that any function whose time variation is arbitrary can be expressed, in general, as an infinite sum of sinusoidal functions having a discrete or continuous spectrum of frequencies, depending upon whether the function is periodic or not. We therefore devote special attention to sinusoidally time-varying fields. In dealing with sinusoidally time-varying quantities, the phasor approach is convenient, as the student may have already learned in circuit analysis. However, we will here review the phasor concept and illustrate why it is convenient before applying it to electromagnetic fields.

A phasor is nothing but a complex number. It is represented graphically by the line drawn from the origin to the point, in the complex plane, corresponding to the complex number as shown in Fig. 4.14. The length of the line is equal to the magnitude of the complex number and the angle that the line makes with the positive real axis is the angle of the complex number. Sinusoidal functions of time are represented by phasors. In particular, when the sinusoidal function is expressed in cosinusoidal form, that is, in the form $A \cos (\omega t + \phi)$, the magnitude of the phasor is equal to the magnitude A of the cosinusoidal function and the angle of the phasor is equal to the phase angle ϕ of the cosinusoidal function for $t = 0$. The real part of the phasor is equal to $A \cos \phi$, which is the value of the function at $t = 0$. If we now imagine the phasor to be rotating about the origin in the counterclockwise direction at the rate of ω rad/sec as shown in Fig. 4.14, we can see that the instantaneous angle of the phasor is $(\omega t + \phi)$ and hence the

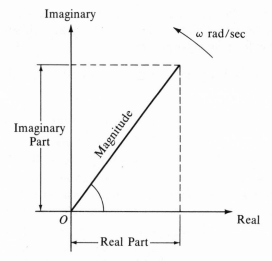

Fig. 4.14. Graphical representation of a phasor.

time variation of its projection on the real axis describes the time variation of the cosinusoidal function.

To illustrate why the phasor approach is convenient for solving sinusoidal steady-state problems, we consider the simple circuit shown in Fig. 4.15 in which a source of voltage $V(t) = V_m \cos(\omega t + \phi)$ drives a series combination of inductance L and resistance R. We will first find the solution for the current $I(t)$ in the steady state without using the phasor approach. Using Kirchhoff's voltage law, we have

$$L \frac{dI(t)}{dt} + RI(t) = V_m \cos(\omega t + \phi) \tag{4-148}$$

We know that the solution for the current in the steady state must also be a cosine function having the same frequency as that of the source voltage

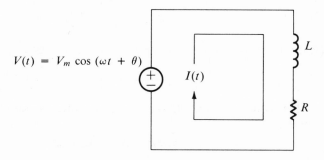

Fig. 4.15. A series RL circuit driven by a sinusoidally time-varying voltage source.

but having different magnitude and different phase angle in general. Thus let us assume the solution to be $I(t) = I_m \cos(\omega t + \theta)$. Substituting this solution in the differential equation, we have

$$L \frac{d}{dt}[I_m \cos(\omega t + \theta)] + RI_m \cos(\omega t + \theta) = V_m \cos(\omega t + \phi)$$

or

$$
\begin{aligned}
(-\omega L I_m \sin \omega t \cos \theta &- \omega L I_m \cos \omega t \sin \theta \\
+ RI_m \cos \omega t \cos \theta &- RI_m \sin \omega t \sin \theta) \\
= V_m \cos \omega t \cos \phi &- V_m \sin \omega t \sin \phi
\end{aligned}
\tag{4-149}
$$

Since (4-149) must be true for all values of time, the coefficients of $\sin \omega t$ on either side of it must be equal and, similarly, the coefficients of $\cos \omega t$ on either side of it must also be equal. Thus we have

$$-\omega L I_m \cos \theta - RI_m \sin \theta = -V_m \sin \phi \tag{4-150a}$$
$$-\omega L I_m \sin \theta + RI_m \cos \theta = V_m \cos \phi \tag{4-150b}$$

Squaring (4-150a) and (4-150b) and adding, we obtain

$$V_m^2 = \omega^2 L^2 I_m^2 + R^2 I_m^2$$

or

$$I_m = \frac{V_m}{\sqrt{R^2 + \omega^2 L^2}} \tag{4-151}$$

Multiplying (4-150a) by $\cos \theta$ and (4-150b) by $\sin \theta$ and adding, we get

$$\omega L I_m = V_m \sin(\phi - \theta) \tag{4-152a}$$

Similarly, multiplying (4-150a) by $-\sin \theta$ and (4-150b) by $\cos \theta$ and adding, we get

$$RI_m = V_m \cos(\phi - \theta) \tag{4-152b}$$

From (4-152a) and (4-152b), we have

$$\tan(\phi - \theta) = \frac{\omega L}{R}$$

or

$$\theta = \phi - \tan^{-1} \frac{\omega L}{R} \tag{4-153}$$

Hence the solution for $I(t)$ in the steady state is given by

$$I(t) = \frac{V_m}{\sqrt{R^2 + \omega^2 L^2}} \cos\left(\omega t + \phi - \tan^{-1}\frac{\omega L}{R}\right) \tag{4-154}$$

Let us now use the phasor concept to solve the same simple problem. Noting that

$$V_m \cos(\omega t + \phi) = \mathcal{R}e[V_m e^{j(\omega t + \phi)}] \tag{4-155a}$$

and

$$I_m \cos(\omega t + \theta) = \mathcal{R}e[I_m e^{j(\omega t + \theta)}] \tag{4-155b}$$

where $\Re e$ stands for "the real part of," we have from (4-148),

$$L\frac{d}{dt}\{\Re e[I_m e^{j(\omega t+\theta)}]\} + R\{\Re e[I_m e^{j(\omega t+\theta)}]\} = \Re e[V_m e^{j(\omega t+\phi)}] \quad (4\text{-}156)$$

However, since L and R are constants and also since d/dt and $\Re e$ can be interchanged, we can simplify (4-156) in accordance with the following steps:

$$\Re e\{\frac{d}{dt}[LI_m e^{j(\omega t+\theta)}]\} + \Re e[RI_m e^{j(\omega t+\theta)}] = \Re e[V_m e^{j(\omega t+\phi)}]$$

$$\Re e[j\omega LI_m e^{j(\omega t+\theta)}] + \Re e[RI_m e^{j(\omega t+\theta)}] = \Re e[V_m e^{j(\omega t+\phi)}] \quad (4\text{-}157)$$

$$\Re e[(R+j\omega L)I_m e^{j(\omega t+\theta)}] = \Re e[V_m e^{j(\omega t+\phi)}]$$

Equation (4-157) states that the real parts of two complex numbers are equal. Does this mean that the two complex numbers are equal? No, not in general! For example, consider $4+j2$ and $4+j5$. Their real parts are equal but the numbers themselves are not equal. However, (4-157) must hold for all values of time. Let us consider two times t_1 and t_2 corresponding to $(\omega t + \theta)$ equal to zero and $(\omega t + \theta)$ equal to $\pi/2$, respectively. Then, for time t_1, we have

$$\Re e[(R+j\omega L)I_m] = \Re e[V_m e^{j(\phi-\theta)}] \quad (4\text{-}158)$$

For time t_2, we have

$$\Re e\{[R+j\omega L]I_m e^{j(\pi/2)}\} = \Re e\{V_m e^{j[(\pi/2)-\theta+\phi]}\}$$

or

$$\Re e\{j[(R+j\omega L)I_m]\} = \Re e\{j[V_m e^{j(\phi-\theta)}]\}$$

or

$$\Im m[(R+j\omega L)I_m] = \Im m[V_m e^{j(\phi-\theta)}] \quad (4\text{-}159)$$

where $\Im m$ stands for "the imaginary part of." Equations (4-158) and (4-159) state that the real parts as well as the imaginary parts of two complex numbers are equal. Hence the two complex numbers must be equal. Thus we obtain

$$(R+j\omega L)I_m = V_m e^{j(\phi-\theta)}$$

or

$$(R+j\omega L)I_m e^{j\theta} = V_m e^{j\phi} \quad (4\text{-}160)$$

Multiplying both sides of (4-160) by $e^{j\omega t}$, we note that the two complex numbers in (4-157) are equal. Now, defining phasors $\bar{I}$ and $\bar{V}$ as

$$\bar{I} = I_m e^{j\theta} \quad \text{so that} \quad I(t) = \Re e(\bar{I}e^{j\omega t}) \quad (4\text{-}161a)$$

$$\bar{V} = V_m e^{j\phi} \quad \text{so that} \quad V(t) = \Re e(\bar{V}e^{j\omega t}) \quad (4\text{-}161b)$$

Eq. (4-160) can be written as

$$(R+j\omega L)\bar{I} = \bar{V} \quad (4\text{-}162)$$

Note that an overscore associated with a symbol represents the phasor (or complex) character of the quantity represented by the symbol. We can easily show that (4-162) leads to the same result as (4-154), since

$$I_m e^{j\theta} = \bar{I} = \frac{\bar{V}}{R + j\omega L}$$

$$= \frac{V_m e^{j\phi}}{\sqrt{R^2 + \omega^2 L^2} e^{j \tan^{-1} \omega L/R}} = \frac{V_m}{\sqrt{R^2 + \omega^2 L^2}} e^{j(\phi - \tan^{-1} \omega L/R)}$$

and

$$I(t) = I_m \cos(\omega t + \theta) = \Re e(I_m e^{j\theta} e^{j\omega t})$$

$$= \Re e\left[\frac{V_m}{\sqrt{R^2 + \omega^2 L^2}} e^{j(\phi - \tan^{-1} \omega L/R)} e^{j\omega t} \right]$$

$$= \frac{V_m}{\sqrt{R^2 + \omega^2 L^2}} \cos\left(\omega t + \phi - \tan^{-1} \frac{\omega L}{R} \right)$$

which is the same as (4-154).

In the foregoing illustration of the phasor technique, we have included several steps merely to understand the basis behind the phasor technique. It is clear that, hereafter, we can omit all steps up to (4-162) and write the phasor equation (4-162) directly from the differential equation (4-148) by simply replacing $I(t)$ and $V(t)$ by their phasors $\bar{I}$ and $\bar{V}$, respectively, and by replacing d/dt by $j\omega$. The phasor equation is then solved for the phasor $\bar{I}$ from which the time function $I(t)$ is obtained. Comparing with the trigonometric manipulations involved in the steps from (4-149) to (4-153) which have to be carried out for each different problem, we can now appreciate the simplicity of the phasor technique. As a numerical example, let us consider $V(t) = 10 \cos 1000t$, $L = 10^{-3}$ henry, and $R = 1$ ohm for the network of Fig. 4.15. The differential equation for $I(t)$ is given by

$$10^{-3} \frac{dI}{dt} + I = 10 \cos 1000t$$

Replacing the current and voltage by their phasors and d/dt by $j\omega$, we have

$$(j\omega 10^{-3} + 1)\bar{I} = 10 e^{j0}$$

or, since $\omega = 1000$ rad/sec,

$$(j1 + 1)\bar{I} = 10 e^{j0}$$

The phasor $\bar{I}$ is then given by

$$\bar{I} = \frac{10 e^{j0}}{1 + j1} = \frac{10 e^{j0}}{\sqrt{2} e^{j\pi/4}} = \frac{10}{\sqrt{2}} e^{-j\pi/4}$$

Finally,

$$I(t) = \Re e[\bar{I} e^{j(1000t)}]$$

$$= \Re e\left[\frac{10}{\sqrt{2}} e^{-j\pi/4} e^{j(1000t)} \right]$$

$$= 7.07 \cos(1000t - 45°)$$

The voltage and current phasors and the corresponding time functions are shown in Figs. 4.16(a) and 4.16(b), respectively. Note that in Fig. 4.16(a) we have turned the complex plane around by 90° in the counterclockwise direction to illustrate that the time variations of the projections of the phasors as they rotate in the counterclockwise direction describe the curves shown in Fig. 4.16(b).

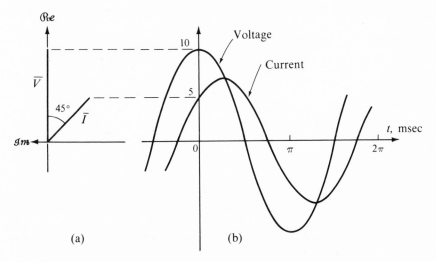

(a)   (b)

Fig. 4.16. (a) Voltage and current phasors for numerical values $V = 10$ volts, $\omega = 1000$ rad/sec, $L = 10^{-3}$ henry, and $R = 1$ ohm for the series *RL* circuit of Fig. 4.15. (b) Time functions corresponding to the voltage and current phasors of (a).

Extension of the phasor technique to vector quantities whose magnitudes vary sinusoidally with time follows from its application to the individual components of the vector along the coordinate axes. However, some confusion is bound to arise since both vectors and phasors are represented graphically in the same manner except that the vector has an arrowhead associated with it. A vector represents the magnitude and space direction of a quantity whereas a phasor represents the magnitude and phase angle of a sinusoidally varying function of time. Thus the angle which a phasor makes with the real axis of the complex plane has nothing to do with direction in space, and the angle which a vector makes with a reference axis in a spatial coordinate system has nothing to do with the phase angle which is associated with the time variation of the quantity. Nevertheless, there are certain similarities between vectors and phasors. These are pertinent to manipulations involving addition, subtraction, and multiplication by a constant. They both use the same graphical rules for carrying out these manipulations. Hence we must be careful, in performing these manipulations, not to get confused between the space angles associated with the vectors and the phase angles associated with the phasors. We will now consider an example to illustrate these differences and similarities between vectors and phasors.

EXAMPLE 4-14.   In the arrangement shown in Fig. 4.17(a), three line charges, infinitely long in the direction normal to the plane of the paper and having uniform densities varying sinusoidally with time, are situated at the corners of an equilateral triangle. The amplitudes of the sinusoidally time-varying charge

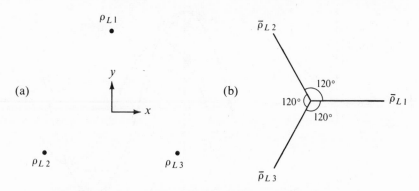

Fig. 4.17. (a) Geometrical arrangement of infinitely long uniform and sinusoidally time-varying line charges. (b) Phasor diagram of the sinusoidally time-varying line charge densities.

densities are such that, considered alone, each line charge produces unit peak electric field intensity at the geometric center of the triangle. The phasor diagram of the charge densities is shown in Fig. 4.17(b).

(a) Find the phasors representing the x and y components of the electric field intensity vector at the geometric center of the triangle.

(b) Determine how the magnitude and direction of the electric field intensity vector at the geometric center of the triangle vary with time.

The phasor diagram indicates that the line charge densities are given by

$$\rho_{L1} = \rho_{Lm} \cos \omega t$$
$$\rho_{L2} = \rho_{Lm} \cos (\omega t + 120°)$$
$$\rho_{L3} = \rho_{Lm} \cos (\omega t + 240°)$$

where ρ_{Lm} is the peak value of the charge densities.

(a) The electric field intensity vector due to an infinitely long line charge of uniform density is directed radially away from the line charge. Hence the field intensities due to the different line charges are directed as shown in Fig. 4. 18(a), with the complex numbers beside the vectors representing their phasors. For example, the phasor $1\underline{/0°}$ associated with the field intensity

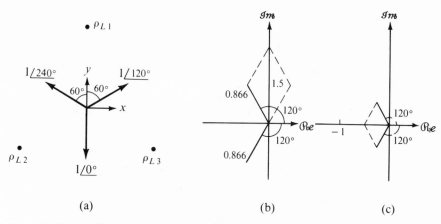

Fig. 4.18. For evaluating the phasors representing the x and y components of the electric field intensity vector at the geometric center of the line charge arrangement of Fig. 4.17.

vector due to the line charge of density ρ_{L1} indicates that the time variation of the magnitude of the vector is given by $1 \cos \omega t$. Thus, timewise, the vector oscillates back and forth along the y axis starting with a magnitude of 1 in the negative y direction, shrinking gradually to zero in a sinusoidal manner, then reversing its direction and growing in magnitude in the positive y direction until it reaches a maximum of unity, then shrinking back to zero, and so on.

Now, the x component of the phasor electric field intensity vector at the geometric center of the triangle is given by

$$\bar{E}_x = (1 \cos 30°)\underline{/120°} - (1 \cos 30°)\underline{/240°}$$
$$= 0.866\underline{/120°} - 0.866\underline{/240°} = 1.5\underline{/90°}$$

where we have used the construction shown in Fig. 4.18(b). We note that, in the above steps, certain manipulations are vector manipulations whereas certain other manipulations have to do with phasors. For example, in finding the x component of the phasor vector $1\underline{/120°}$ pointing away from the line charge of density ρ_{L2}, the phase angle 120° is preserved and the magnitude 1 is multiplied by the cosine of the angle which the vector makes with the x axis, giving us $(1 \cos 30°)\underline{/120°}$ or $0.866\underline{/120°}$. Similarly, the y component of the phasor electric field intensity vector at the geometric center of the triangle is given by

$$\bar{E}_y = -1\underline{/0°} + (1 \cos 60°)\underline{/120°} + (1 \cos 60°)\underline{/240°}$$
$$= -1\underline{/0°} + 0.5\underline{/120°} + 0.5\underline{/240°}$$
$$= 1\underline{/180°} + 0.5\underline{/180°} = 1.5\underline{/180°}$$

where we have used the construction shown in Fig. 4.18(c). The phasor diagram of the x and y components of the electric field intensity vector at the geometric center of the triangle relative to the phasor diagram of the line charge densities is shown in Fig. 4.19(a).

(b) From the results of part (a), we have

$$E_x(t) = \mathcal{R}e(\bar{E}_x e^{j\omega t}) = 1.5\cos(\omega t + 90°) = -1.5\sin\omega t$$
$$E_y(t) = \mathcal{R}e(\bar{E}_y e^{j\omega t}) = 1.5\cos(\omega t + 180°) = -1.5\cos\omega t$$

Now, since $\mathbf{E}(t) = E_x(t)\mathbf{i}_x + E_y(t)\mathbf{i}_y$, the magnitude of $\mathbf{E}(t)$ is given by

$$|\mathbf{E}(t)| = \sqrt{E_x^2(t) + E_y^2(t)}$$
$$= \sqrt{(-1.5\sin\omega t)^2 + (-1.5\cos\omega t)^2} = 1.5$$

The angle which the vector $\mathbf{E}(t)$ makes with the x axis is given by

$$\tan^{-1}\frac{E_y(t)}{E_x(t)} = \tan^{-1}\frac{-1.5\cos\omega t}{-1.5\sin\omega t} = \tan^{-1}\frac{-1.5\sin(\omega t + \pi/2)}{1.5\cos(\omega t + \pi/2)}$$
$$= \tan^{-1}[-\tan(\omega t + \pi/2)] = -(\omega t + \pi/2)$$

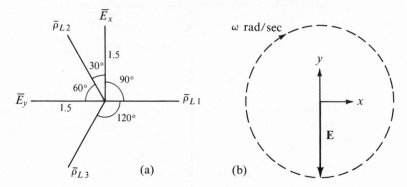

Fig. 4.19. (a) Phasor diagram of the x and y components of the electric field intensity vector at the geometric center of the line charge arrangement of Fig. 4.17, relative to the phasor diagram of the line charge densities. (b) For describing the time variation of the electric field intensity vector corresponding to the phasor diagram of (a).

Thus the magnitude of the electric field intensity vector at the geometric center of the triangle remains constant at 1.5 units and the angle which the vector makes with the x axis varies as $-(\omega t + \pi/2)$ with time; that is, the vector rotates with a constant magnitude and at a rate of ω rad/sec, with the direction at $t = 0$ along the negative y axis and in the sense shown in Fig. 4.19(b). The field is then said to be circularly polarized. ∎

We will now discuss briefly polarization of vector fields. Polarization is the characteristic by means of which we describe how the magnitude and the direction of the field vary with time. For an arbitrarily time-varying field characterized by random time-variations of its components along the coordinate axes at a point in space, the magnitude and direction of the field vary randomly with time. The field is then said to be unpolarized or randomly polarized. For a sinusoidally time-varying field at a particular frequency ω, the field vector is characterized by a well-defined polarization. In the most general case, the magnitude and direction of such a field vector at a point change with time in such a manner that the tip of the vector drawn at that point describes an ellipse as time progresses, as shown in Fig. 4.20(a). The field is then said to be elliptically polarized. There are two special cases of elliptical polarization. These are linear polarization and circular polarization.

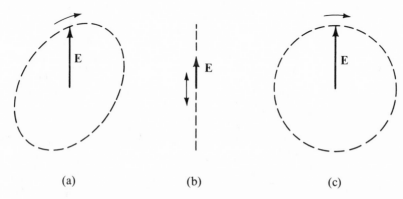

(a)   (b)   (c)

Fig. 4.20. For illustrating (a) elliptical polarization, (b) linear polarization, and (c) circular polarization of a field vector.

If the field vector at a point in space lies along the same straight line through that point as time progresses, as shown in Fig. 4.20(b), the field is said to be linearly polarized. Obviously, the components of a field vector along the coordinate axes are linearly polarized. If all the components of the field vector along the coordinate axes have the same phase, although possessing different magnitudes, then the field vector itself is linearly polarized. If the tip of a field vector drawn at a point in space describes a circle as time progresses, as shown in Fig. 4.20(c) the field is said to be circularly polarized. Circular polarization is realized by the superposition of two field components oriented perpendicular to each other and having the same magnitude but differing in phase by $\pi/2$ or 90° as in the case of the two components in Example 4-14. Elliptical polarization is realized by the superposition of two field components having in general different magnitudes as well as different phase angles. Since a circle and ellipse can be traversed in one of two senses, we have to distinguish between the opposite senses of rotation in the cases of circular and elliptical polarizations. The distinction is made as follows. Considering the vector to be the electric field intensity vector **E**, the field is said to be clockwise or right circularly (or elliptically) polarized if the vector rotates in the clockwise sense as seen looking along the direction of the Poynting vector $\mathbf{P} = \mathbf{E} \times \mathbf{B}/\mu_0$ where $\mathbf{B}$ is the magnetic field associated with $\mathbf{E}$. The field is said to be counterclockwise or left circularly (or elliptically) polarized if the vector rotates in the counterclockwise sense as seen looking along the direction of the Poynting vector.

Having illustrated the application of the phasor technique in dealing with sinusoidally time-varying vector fields, we now turn to the phasor representation of Maxwell's equations for sinusoidally time-varying fields. Maxwell's equations for time-varying fields are given by

$$\mathbf{V} \cdot \mathbf{E} = \frac{\rho}{\epsilon_0} \tag{4-163}$$

$$\mathbf{V} \cdot \mathbf{B} = 0 \tag{4-164}$$

$$\mathbf{V} \times \mathbf{E} = -\frac{\partial \mathbf{B}}{\partial t} \tag{4-165}$$

$$\mathbf{V} \times \mathbf{B} = \mu_0 \left[\mathbf{J} + \frac{\partial}{\partial t}(\epsilon_0 \mathbf{E}) \right] \tag{4-166}$$

whereas the continuity equation is given by

$$\mathbf{V} \cdot \mathbf{J} + \frac{\partial \rho}{\partial t} = 0 \tag{4-167}$$

In (4-163)–(4-167), the quantities $\mathbf{E}$, $\mathbf{B}$, ρ, and $\mathbf{J}$ are also functions of all three space coordinates in general. Thus we have

$$\mathbf{E} = \mathbf{E}(x, y, z, t) = E_x(x, y, z, t)\mathbf{i}_x + E_y(x, y, z, t)\mathbf{i}_y + E_z(x, y, z, t)\mathbf{i}_z$$

$$\mathbf{B} = \mathbf{B}(x, y, z, t) = B_x(x, y, z, t)\mathbf{i}_x + B_y(x, y, z, t)\mathbf{i}_y + B_z(x, y, z, t)\mathbf{i}_z$$

$$\rho = \rho(x, y, z, t)$$

$$\mathbf{J} = \mathbf{J}(x, y, z, t) = J_x(x, y, z, t)\mathbf{i}_x + J_y(x, y, z, t)\mathbf{i}_y + J_z(x, y, z, t)\mathbf{i}_z$$

For the particular case of sinusoidal variation with time, we have

$$\begin{aligned}
\mathbf{E} &= \mathbf{E}(x, y, z, t) \\
&= E_{x0}(x, y, z) \cos\left[\omega t + \phi_x(x, y, z)\right] \mathbf{i}_x \\
&\quad + E_{y0}(x, y, z) \cos\left[\omega t + \phi_y(x, y, z)\right] \mathbf{i}_y \\
&\quad + E_{z0}(x, y, z) \cos\left[\omega t + \phi_z(x, y, z)\right] \mathbf{i}_z \\
&= \mathcal{R}e[E_{x0}(x, y, z)e^{j\phi_x(x, y, z)}e^{j\omega t}\mathbf{i}_x \\
&\quad + E_{y0}(x, y, z)e^{j\phi_y(x, y, z)}e^{j\omega t}\mathbf{i}_y \\
&\quad + E_{z0}(x, y, z)e^{j\phi_z(x, y, z)}e^{j\omega t}\mathbf{i}_z] \\
&= \mathcal{R}e\{[\bar{E}_x(x, y, z)\mathbf{i}_x + \bar{E}_y(x, y, z)\mathbf{i}_y + \bar{E}_z(x, y, z)\mathbf{i}_z]e^{j\omega t}\} \\
&= \mathcal{R}e[\bar{\mathbf{E}}(x, y, z)e^{j\omega t}]
\end{aligned} \tag{4-168}$$

Similarly, we have

$$\mathbf{B} = \mathbf{B}(x, y, z, t) = \Re e[\bar{\mathbf{B}}(x, y, z)e^{j\omega t}] \qquad (4\text{-}169)$$

$$\rho = \rho(x, y, z, t) = \Re e[\bar{\rho}(x, y, z)e^{j\omega t}] \qquad (4\text{-}170)$$

$$\mathbf{J} = \mathbf{J}(x, y, z, t) = \Re e[\bar{\mathbf{J}}(x, y, z)e^{j\omega t}] \qquad (4\text{-}171)$$

In (4-168)–(4-171), the complex quantities $\bar{\mathbf{E}}(x, y, z)$, $\bar{\mathbf{B}}(x, y, z)$, $\bar{\rho}(x, y, z)$, and $\bar{\mathbf{J}}(x, y, z)$ are the phasor representations for the sinusoidally time-varying quantities $\mathbf{E}(x, y, z, t)$, $\mathbf{B}(x, y, z, t)$, $\rho(x, y, z, t)$, and $\mathbf{J}(x, y, z, t)$, respectively.

Substituting the respective phasors for the quantities $\mathbf{E}, \mathbf{B}, \rho$, and $\mathbf{J}$ and replacing $\partial/\partial t$ by $j\omega$ in (4-163)–(4-167), we obtain the phasor representations of Maxwell's equations as

$$\mathbf{\nabla} \cdot \bar{\mathbf{E}} = \frac{\bar{\rho}}{\epsilon_0} \qquad (4\text{-}172)$$

$$\mathbf{\nabla} \cdot \bar{\mathbf{B}} = 0 \qquad (4\text{-}173)$$

$$\mathbf{\nabla} \times \bar{\mathbf{E}} = -j\omega\bar{\mathbf{B}} \qquad (4\text{-}174)$$

$$\mathbf{\nabla} \times \bar{\mathbf{B}} = \mu_0(\bar{\mathbf{J}} + j\omega\epsilon_0\bar{\mathbf{E}}) \qquad (4\text{-}175)$$

whereas the corresponding continuity equation is given by

$$\mathbf{\nabla} \cdot \bar{\mathbf{J}} + j\omega\bar{\rho} = 0 \qquad (4\text{-}176)$$

In (4-172)–(4-176), we understand that $\bar{\mathbf{E}}, \bar{\mathbf{B}}, \bar{\rho}$, and $\bar{\mathbf{J}}$ are functions of x, y, z (but not t). Note that (4-173) follows from (4-174) whereas (4-172) follows from (4-175) with the aid of (4-176).

EXAMPLE 4-15. A sinusoidally time-varying electric field intensity vector is characterized by its phasor $\bar{\mathbf{E}}$, given by

$$\bar{\mathbf{E}} = (3e^{j\pi/2}\mathbf{i}_x + 5\mathbf{i}_y - 4e^{j\pi/2}\mathbf{i}_z)e^{-j0.02\pi(4x+3z)} \qquad (4\text{-}177)$$

(a) Show that the surfaces of constant phase of $\bar{\mathbf{E}}$ are planes. Find the equation of the planes.

(b) Show that the electric field is circularly polarized in the planes of constant phase.

(c) Obtain the magnetic flux density phasor $\bar{\mathbf{B}}$ associated with the given $\bar{\mathbf{E}}$ and determine if the field is right circularly polarized or left circularly polarized.

(a) The phase angle associated with $\bar{\mathbf{E}}$ is equal to $-0.02\pi(4x + 3z)$. Hence the surfaces of constant phase of $\bar{\mathbf{E}}$ are given by

$$-0.02\pi(4x + 3z) = \text{constant}$$

or

$$(4x + 3z) = \text{constant} \qquad (4\text{-}178)$$

Equation (4-178) represents planes and hence the surfaces of constant phase are planes.

(b) Combining the x and z components of $\bar{\mathbf{E}}$, we obtain

$$\bar{\mathbf{E}} = (5\mathbf{i}_y + 5\mathbf{i}_{xz}e^{j\pi/2})e^{-j0.02\pi(4x+3z)} \tag{4-179}$$

where $\mathbf{i}_{xz} = (3\mathbf{i}_x - 4\mathbf{i}_z)/5$ is the unit vector in the xz plane and making an angle of $-\tan^{-1}\frac{4}{3}$ or $-53.1°$ with the positive x axis. Thus the electric field is made up of two components perpendicular to each other and having equal magnitudes but differing in phase by $\pi/2$. Hence the field is circularly polarized. From (4-179), we observe that the field vector lies in planes defined by $\mathbf{i}_y$ and $\mathbf{i}_{xz}$. The equation of these planes is given by

$$\mathbf{i}_y \cdot \mathbf{i}_{xz} \times (\mathbf{r} - \mathbf{r}_0) = 0 \tag{4-180}$$

where $\mathbf{r}$ is the position vector of an arbitrary point (x, y, z) and $\mathbf{r}_0$ is the position vector of a reference point (x_0, y_0, z_0), both points lying in a particular plane. Simplifying (4-180), we obtain

$$4x + 3z = 4x_0 + 3z_0 = \text{constant}$$

which is the same as Eq. (4-178). Thus the field is circularly polarized in the planes of constant phase.

(c) The magnetic flux density phasor $\bar{\mathbf{B}}$ associated with the given $\bar{\mathbf{E}}$ can be obtained by using

$$\nabla \times \bar{\mathbf{E}} = -j\omega\bar{\mathbf{B}} \tag{4-174}$$

Substituting for $\bar{\mathbf{E}}$ in (4-174) from (4-177) and simplifying, we obtain

$$\bar{\mathbf{B}} = \frac{0.1\pi}{\omega}(-3\mathbf{i}_x + 5e^{j\pi/2}\mathbf{i}_y + 4\mathbf{i}_z)e^{-j0.02\pi(4x+3z)} \tag{4-181}$$

Let us now consider, for simplicity, the field vectors in the plane $4x + 3z = 0$. The phasor vectors $\bar{\mathbf{E}}_0$ and $\bar{\mathbf{B}}_0$ in this plane are given by

$$\bar{\mathbf{E}}_0 = 3e^{j\pi/2}\mathbf{i}_x + 5\mathbf{i}_y - 4e^{j\pi/2}\mathbf{i}_z \tag{4-182}$$

$$\bar{\mathbf{B}}_0 = \frac{0.1\pi}{\omega}(-3\mathbf{i}_x + 5e^{j\pi/2}\mathbf{i}_y + 4\mathbf{i}_z) \tag{4-183}$$

The corresponding real field vectors are given by

$$\begin{aligned}
\mathbf{E}_0 &= \mathfrak{Re}(\bar{\mathbf{E}}_0 e^{j\omega t}) \\
&= -3\sin\omega t\,\mathbf{i}_x + 5\cos\omega t\,\mathbf{i}_y + 4\sin\omega t\,\mathbf{i}_z
\end{aligned} \tag{4-184}$$

$$\begin{aligned}
\mathbf{B}_0 &= \mathfrak{Re}(\bar{\mathbf{B}}_0 e^{j\omega t}) \\
&= \frac{0.1\pi}{\omega}(-3\cos\omega t\,\mathbf{i}_x - 5\sin\omega t\,\mathbf{i}_y + 4\cos\omega t\,\mathbf{i}_z)
\end{aligned} \tag{4-185}$$

Substituting (4-184) and (4-185) into

$$\mathbf{P} = \mathbf{E}_0 \times \frac{\mathbf{B}_0}{\mu_0}$$

and simplifying, we obtain the Poynting vector $\mathbf{P}$ as

$$\mathbf{P} = \frac{0.1\pi}{\omega\mu_0}(20\mathbf{i}_x + 15\mathbf{i}_z) \tag{4-186}$$

Now, we note from (4-184) that the direction of $\mathbf{E}_0$ is along $5\mathbf{i}_y$ for $\omega t = 0$ and along $(-3\mathbf{i}_x + 4\mathbf{i}_z)$ for $\omega t = \pi/2$. These two directions and the direction of the Poynting vector are shown in Fig. 4.21. It can be seen that the electric field vector rotates in the clockwise sense as seen looking along the direction of the Poynting vector. Hence the field is right circularly polarized. ∎

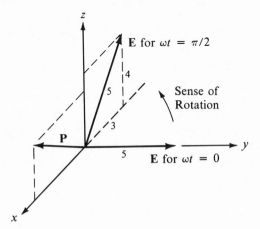

Fig. 4.21. For the determination of the sense of rotation of the circularly polarized vector of Example 4-15.

4.10 Power and Energy Considerations for Sinusoidally Time-Varying Electromagnetic Fields

In Section 4.8 we introduced the Poynting vector $\mathbf{P}$ given by

$$\mathbf{P} = \mathbf{E} \times \frac{\mathbf{B}}{\mu_0} \tag{4-144}$$

as the power density associated with the electromagnetic field at a point. The surface integral of the Poynting vector evaluated over a closed surface S always gives the correct result for the power flow across the surface out of the volume bounded by it. For a sinusoidally time-varying electromagnetic field characterized by complex field vectors,

$$\bar{\mathbf{E}} = \mathbf{E}_0 e^{j\phi}$$

$$\bar{\mathbf{B}} = \mathbf{B}_0 e^{j\theta}$$

the instantaneous Poynting vector $\mathbf{P}$ is given by

$$\mathbf{P} = \mathbf{E} \times \frac{\mathbf{B}}{\mu_0}$$

$$= (\mathfrak{Re}\, \bar{\mathbf{E}} e^{j\omega t}) \times \left(\frac{1}{\mu_0}\, \mathfrak{Re}\, \bar{\mathbf{B}} e^{j\omega t} \right)$$

$$= \mathbf{E}_0 \cos(\omega t + \phi) \times \frac{\mathbf{B}_0}{\mu_0} \cos(\omega t + \theta)$$

$$= \mathbf{E}_0 \times \frac{\mathbf{B}_0}{\mu_0} [\cos(\omega t + \phi) \cos(\omega t + \theta)] \qquad (4\text{-}187)$$

$$= \mathbf{E}_0 \times \frac{\mathbf{B}_0}{\mu_0} \left[\frac{1}{2} \cos(2\omega t + \phi + \theta) + \frac{1}{2} \cos(\phi - \theta) \right]$$

$$= \frac{1}{2} \mathbf{E}_0 \times \frac{\mathbf{B}_0}{\mu_0} \cos(\phi - \theta) + \frac{1}{2} \mathbf{E}_0 \times \frac{\mathbf{B}_0}{\mu_0} \cos(2\omega t + \phi - \theta)$$

The first term on the right side of (4-187) is independent of time whereas the second term varies sinusoidally with time. The time-average value of the second term obtained by integrating it through one period T and dividing by the period is equal to zero since the integral of a cosine or sine function over one period is equal to zero. Thus the time-average value of the Poynting vector $\mathbf{P}$, denoted as $\langle \mathbf{P} \rangle$ is given by

$$\langle \mathbf{P} \rangle = \frac{1}{T} \int_0^T \mathbf{P}\, dt$$

$$= \left\langle \frac{1}{2} \mathbf{E}_0 \times \frac{\mathbf{B}_0}{\mu_0} \cos(\phi - \theta) \right\rangle + \left\langle \frac{1}{2} \mathbf{E}_0 \times \frac{\mathbf{B}_0}{\mu_0} \cos(2\omega t + \phi - \theta) \right\rangle$$

$$= \frac{1}{2} \mathbf{E}_0 \times \frac{\mathbf{B}_0}{\mu_0} \cos(\phi - \theta)$$

$$= \mathfrak{Re}\left[\frac{1}{2} \mathbf{E}_0 \times \frac{\mathbf{B}_0}{\mu_0} e^{j(\phi - \theta)} \right] \qquad (4\text{-}188)$$

$$= \mathfrak{Re}\left(\frac{1}{2} \mathbf{E}_0 e^{j\phi} \times \frac{\mathbf{B}_0 e^{-j\theta}}{\mu_0} \right)$$

$$= \mathfrak{Re}\left(\frac{1}{2} \bar{\mathbf{E}} \times \frac{\bar{\mathbf{B}}^*}{\mu_0} \right)$$

where $\bar{\mathbf{B}}^*$ denotes the complex conjugate of $\bar{\mathbf{B}}$.

We now define the complex Poynting vector $\bar{\mathbf{P}}$ as

$$\bar{\mathbf{P}} = \frac{1}{2} \bar{\mathbf{E}} \times \frac{\bar{\mathbf{B}}^*}{\mu_0} \qquad (4\text{-}189)$$

so that the time-average Poynting vector $\langle \mathbf{P} \rangle$ can be written as

$$\langle \mathbf{P} \rangle = \mathfrak{Re}(\bar{\mathbf{P}}) \qquad (4\text{-}190)$$

We note that Eq. (4-189) is analogous to the expression for the complex power in sinusoidal steady-state circuit theory given by

$$\bar{P} = \frac{1}{2} \bar{V} \bar{I}^* \qquad (4\text{-}191)$$

where $\bar{V}$ and $\bar{I}$ are the complex voltage and complex current, respectively. By integrating the complex Poynting vector over a closed surface S, we obtain the complex power flowing across S out of the volume V bounded by it. Thus

$$\oint_S \bar{\mathbf{P}} \cdot d\mathbf{S} = \oint_S \frac{1}{2} \bar{\mathbf{E}} \times \frac{\bar{\mathbf{B}}^*}{\mu_0} \cdot d\mathbf{S}$$
$$= \frac{1}{2\mu_0} \int_V \mathbf{\nabla} \cdot (\bar{\mathbf{E}} \times \bar{\mathbf{B}}^*) \, dv \qquad (4\text{-}192)$$

where we have used the divergence theorem to replace the surface integral by a volume integral. Now, using the vector identity

$$\mathbf{\nabla} \cdot (\bar{\mathbf{E}} \times \bar{\mathbf{B}}^*) = \bar{\mathbf{B}}^* \cdot \mathbf{\nabla} \times \bar{\mathbf{E}} - \bar{\mathbf{E}} \cdot \mathbf{\nabla} \times \bar{\mathbf{B}}^* \qquad (4\text{-}193)$$

and Maxwell's curl equations for complex fields given by

$$\mathbf{\nabla} \times \bar{\mathbf{E}} = -j\omega\bar{\mathbf{B}} \qquad (4\text{-}174)$$

$$\mathbf{\nabla} \times \bar{\mathbf{B}} = \mu_0(\bar{\mathbf{J}} + j\omega\epsilon_0\bar{\mathbf{E}}) \qquad (4\text{-}175)$$

we have

$$\mathbf{\nabla} \cdot (\bar{\mathbf{E}} \times \bar{\mathbf{B}}^*) = \bar{\mathbf{B}}^* \cdot (-j\omega\bar{\mathbf{B}}) - \bar{\mathbf{E}} \cdot \mu_0(\bar{\mathbf{J}} + j\omega\epsilon_0\bar{\mathbf{E}})^*$$
$$= -j\omega\bar{\mathbf{B}}^* \cdot \bar{\mathbf{B}} - \mu_0(\bar{\mathbf{E}} \cdot \bar{\mathbf{J}}^* - j\omega\epsilon_0\bar{\mathbf{E}} \cdot \bar{\mathbf{E}}^*) \qquad (4\text{-}194)$$

However, the time-average stored energy density in the electric field is given by

$$\langle w_e \rangle = \left\langle \frac{1}{2}\epsilon_0 E^2 \right\rangle$$
$$= \left\langle \frac{1}{2}\epsilon_0 |\mathbf{E}_0|^2 \cos^2(\omega t + \phi) \right\rangle$$
$$= \left\langle \frac{1}{4}\epsilon_0 |\mathbf{E}_0|^2 + \frac{1}{4}\epsilon_0 |\mathbf{E}_0|^2 \cos 2(\omega t + \phi) \right\rangle \qquad (4\text{-}195)$$
$$= \frac{1}{4}\epsilon_0 |\mathbf{E}_0|^2 = \frac{1}{4}\epsilon_0 \mathbf{E}_0 e^{j\phi} \cdot \mathbf{E}_0 e^{-j\phi}$$
$$= \frac{1}{4}\epsilon_0 \bar{\mathbf{E}} \cdot \bar{\mathbf{E}}^*$$

Similarly, the time-average stored energy density in the magnetic field is given by

$$\langle w_m \rangle = \left\langle \frac{1}{2}\frac{B^2}{\mu_0} \right\rangle = \frac{1}{4}\bar{\mathbf{B}} \cdot \frac{\bar{\mathbf{B}}^*}{\mu_0} \qquad (4\text{-}196)$$

The time-average power density expended by the field due to the current flow is given by

$$\langle p_d \rangle = \langle \mathbf{E} \cdot \mathbf{J} \rangle = \mathfrak{Re}\left(\frac{1}{2}\bar{\mathbf{E}} \cdot \bar{\mathbf{J}}^* \right) \qquad (4\text{-}197)$$

so that the complex power density associated with the current flow is given by

$$\bar{p}_d = \frac{1}{2}\bar{\mathbf{E}} \cdot \bar{\mathbf{J}}^* \qquad (4\text{-}198)$$

Substituting (4-195), (4-196), and (4-198) into (4-194), we get

$$\mathbf{V} \cdot (\bar{\mathbf{E}} \times \bar{\mathbf{B}}^*) = -2\mu_0 \bar{p}_d - j4\omega\mu_0(\langle w_m \rangle - \langle w_e \rangle) \qquad (4\text{-}199)$$

Finally, substituting (4-199) into (4-192), we obtain

$$\oint_S \bar{\mathbf{P}} \cdot d\mathbf{S} = -\int_V \bar{p}_d \, dv - j2\omega \int_V (\langle w_m \rangle - \langle w_e \rangle) \, dv \qquad (4\text{-}200)$$

Equation (4-200) is known as the complex Poynting's theorem. Equating the real and imaginary parts on both sides of (4-200), we have

$$\mathcal{R}e\left(\int_V \bar{p}_d \, dv \right) = -\mathcal{R}e\left(\oint_S \bar{\mathbf{P}} \cdot d\mathbf{S} \right) \qquad (4\text{-}201)$$

or

$$\int_V \mathcal{R}e(\bar{p}_d) \, dv = -\oint_S [\mathcal{R}e(\bar{\mathbf{P}})] \cdot d\mathbf{S}$$

or

$$\int_V \langle p_d \rangle \, dv = -\oint_S \langle \mathbf{P} \rangle \cdot d\mathbf{S} \qquad (4\text{-}202)$$

and

$$\mathcal{I}m\left(\int_V \bar{p}_d \, dv \right) + 2\omega \int_V (\langle w_m \rangle - \langle w_e \rangle) \, dv = -\mathcal{I}m\left(\oint_S \bar{\mathbf{P}} \cdot d\mathbf{S} \right)$$

or

$$2\omega \int_V (\langle w_m \rangle - \langle w_e \rangle) \, dv = -\mathcal{I}m\left(\oint_S \bar{\mathbf{P}} \cdot d\mathbf{S} \right) - \mathcal{I}m\left(\int_V \bar{p}_d \, dv \right) \quad (4\text{-}203)$$

Equation (4-202) states that the time-average power expended by the field due to the current flow in the volume V is equal to the time-average power flowing into the volume V as given by the surface integral of the time-average Poynting vector over the surface S bounding V. If $\oint_S \langle \mathbf{P} \rangle \cdot d\mathbf{S}$ is zero, it means that there is no time-average power expended by the field in the volume V; whatever time-average power enters the volume V through part of the surface S leaves through the rest of that surface. Equation (4-203) provides a physical interpretation for the imaginary part of the complex Poynting vector. It relates the difference between the time-average magnetic and electric stored energies in the volume V to the reactive power flowing into the volume V as given by the imaginary part of the surface integral of the complex Poynting vector over the surface S and to the reactive power associated with the current flow in the volume V. We note that the complex Poynting theorem is analogous to a similar relationship in sinusoidal steady-state circuit theory given by

$$\frac{1}{2} \bar{V} \bar{I}^* = \langle P_d \rangle + j2\omega(\langle W_m \rangle - \langle W_e \rangle)$$

where $\langle P_d \rangle$ is the average power dissipated in the resistors, and $\langle W_m \rangle$ and $\langle W_e \rangle$ are the time-average stored energies in the inductors and capacitors, respectively.

TABLE 4.1. Summary of Electromagnetic Field Laws and Formulas

Description	Arbitrarily Time-Varying Fields	Sinusoidally Time-Varying Fields
Definition	$\mathbf{F} = q(\mathbf{E} + \mathbf{v} \times \mathbf{B})$	
Maxwell's equations and the continuity equation in differential form	$\nabla \cdot \mathbf{E} = \dfrac{\rho}{\epsilon_0}$	$\nabla \cdot \bar{\mathbf{E}} = \dfrac{\bar{\rho}}{\epsilon_0}$
	$\nabla \cdot \mathbf{B} = 0$	$\nabla \cdot \bar{\mathbf{B}} = 0$
	$\nabla \times \mathbf{E} = -\dfrac{\partial \mathbf{B}}{\partial t}$	$\nabla \times \bar{\mathbf{E}} = -j\omega\bar{\mathbf{B}}$
	$\nabla \times \mathbf{B} = \mu_0\left[\mathbf{J} + \dfrac{\partial}{\partial t}(\epsilon_0\mathbf{E})\right]$	$\nabla \times \bar{\mathbf{B}} = \mu_0(\bar{\mathbf{J}} + j\omega\epsilon_0\bar{\mathbf{E}})$
	$\nabla \cdot \mathbf{J} + \dfrac{\partial \rho}{\partial t} = 0$	$\nabla \cdot \bar{\mathbf{J}} + j\omega\bar{\rho} = 0$
Maxwell's equations and the continuity equation in integral form	$\oint_S \mathbf{E} \cdot d\mathbf{S} = \dfrac{1}{\epsilon_0}\int_V \rho\, dv$	$\oint_S \bar{\mathbf{E}} \cdot d\mathbf{S} = \dfrac{1}{\epsilon_0}\int \bar{\rho}\, dv$
	$\oint_S \mathbf{B} \cdot d\mathbf{S} = 0$	$\oint_S \bar{\mathbf{B}} \cdot d\mathbf{S} = 0$
	$\oint_C \mathbf{E} \cdot d\mathbf{l} = -\dfrac{d}{dt}\int_S \mathbf{B} \cdot d\mathbf{S}$	$\oint_C \bar{\mathbf{E}} \cdot d\mathbf{l} = -j\omega\int_S \bar{\mathbf{B}} \cdot d\mathbf{S}$
	$\oint_C \mathbf{B} \cdot d\mathbf{l} = \mu_0\left(\int_S \mathbf{J} \cdot d\mathbf{S} + \dfrac{d}{dt}\int_S \epsilon_0\mathbf{E} \cdot d\mathbf{S}\right)$	$\oint_C \bar{\mathbf{B}} \cdot d\mathbf{l} = \mu_0\left(\int_S \bar{\mathbf{J}} \cdot d\mathbf{S} + j\omega\int_S \epsilon_0\bar{\mathbf{E}} \cdot d\mathbf{S}\right)$
	$\oint_S \mathbf{J} \cdot d\mathbf{S} + \dfrac{d}{dt}\int_V \rho\, dv = 0$	$\oint_S \bar{\mathbf{J}} \cdot d\mathbf{S} + j\omega\int_V \bar{\rho}\, dv = 0$
Energy density in the electric field	$w_e = \dfrac{1}{2}\epsilon_0\mathbf{E} \cdot \mathbf{E}$	$\langle w_e \rangle = \dfrac{1}{4}\epsilon_0\bar{\mathbf{E}} \cdot \bar{\mathbf{E}}^*$
Energy density in the magnetic field	$w_m = \dfrac{1}{2}\mathbf{B} \cdot \dfrac{\mathbf{B}}{\mu_0}$	$\langle w_m \rangle = \dfrac{1}{4}\bar{\mathbf{B}} \cdot \dfrac{\bar{\mathbf{B}}^*}{\mu_0}$
Power density expended by the field due to current flow	$p_d = \mathbf{E} \cdot \mathbf{J}$	$\langle p_d \rangle = \dfrac{1}{2}\bar{\mathbf{E}} \cdot \bar{\mathbf{J}}^*$
Poynting vector	$\mathbf{P} = \mathbf{E} \times \dfrac{\mathbf{B}}{\mu_0}$	$\bar{\mathbf{P}} = \dfrac{1}{2}\bar{\mathbf{E}} \times \dfrac{\bar{\mathbf{B}}^*}{\mu_0}$

4.11 Summary of Electromagnetic Field Laws and Formulas

We now summarize in Table 4.1 the basic laws governing the electromagnetic field and the power and energy relations for the electromagnetic field. We recall that all four Maxwell's equations for time-varying fields are not independent. The divergence equation for the magnetic field follows from the curl equation for the electric field as shown in Section 4.3, whereas the divergence equation for the electric field follows from the curl equation for the magnetic field and the continuity equation as shown in Section 4.5.

Comparing Maxwell's equations for time-varying fields with those for the static fields discussed in Chapters 2 and 3, we observe a coupling between the time-varying electric field and the time-varying magnetic field. This is because the curl of the electric field is dependent on the time derivative of the magnetic field and the curl of the magnetic field is dependent on the time derivative of the electric field. Thus the solution for the electric field requires a knowledge of the magnetic field whereas the solution for the magnetic field requires a knowledge of the electric field. The two curl equations must therefore be solved simultaneously to obtain the solution for the electromagnetic field. It is precisely this two-way coupling between the time-varying electric and magnetic fields that gives rise to the phenomenon of electromagnetic wave propagation, as we will learn in Chapter 6.

PROBLEMS

4.1. The forces experienced by a test charge q C at a point in a region of electric and magnetic fields $\mathbf{E}$ and $\mathbf{B}$, respectively, are given as follows for three different velocities:

Velocity, m/sec	Force, N
$\mathbf{i}_x$	$q\mathbf{i}_x$
$\mathbf{i}_y$	$q(2\mathbf{i}_x + \mathbf{i}_y)$
$\mathbf{i}_z$	$q(\mathbf{i}_x + \mathbf{i}_y)$

Find $\mathbf{E}$ and $\mathbf{B}$ at that point.

4.2. The forces experienced by a test charge q C at a point in a region of electric and magnetic fields $\mathbf{E}$ and $\mathbf{B}$, respectively, are given as follows for three different velocities:

Velocity, m/sec	Force, N
$\mathbf{i}_x - \mathbf{i}_y$	0
$\mathbf{i}_x - \mathbf{i}_y + \mathbf{i}_z$	0
$\mathbf{i}_z$	$q(\mathbf{i}_x + \mathbf{i}_y)$

Find $\mathbf{E}$ and $\mathbf{B}$ at that point.

4.3. A region is characterized by crossed electric and magnetic fields $\mathbf{E} = E_0\mathbf{i}_y$ and $\mathbf{B} = B_0\mathbf{i}_z$, where E_0 and B_0 are constants. A test charge q having a mass m starts from the origin at $t = 0$ with an initial velocity v_0 in the y direction. Obtain the parametric equations of motion of the test charge. Sketch the path of the test charge.

4.4. A region is characterized by crossed electric and magnetic fields $\mathbf{E} = E_0\mathbf{i}_y$ and $\mathbf{B} = B_0\mathbf{i}_z$, where E_0 and B_0 are constants. A test charge q having a mass m starts from the origin at $t = 0$ with an initial velocity v_0 in the x direction. Obtain the parametric equations of motion of the test charge. Sketch the paths of the test charge for the following cases: (a) $v_0 = 0$, (b) $v_0 = E_0/2B_0$, (c) $v_0 = E_0/B_0$, (d) $v_0 = 2E_0/B_0$, and (e) $v_0 = 3E_0/B_0$.

4.5. A region is characterized by crossed electric and magnetic fields $\mathbf{E} = E_0 \cos \omega t \, \mathbf{i}_y$ and $\mathbf{B} = B_0\mathbf{i}_z$, where E_0 and B_0 are constants. A test charge q having a mass m starts from the origin at $t = 0$ with zero initial velocity. Obtain the parametric equations of motion of the test charge. Check your result with that of Example 4-2 by letting $\omega \longrightarrow 0$. Investigate the limiting case of $\omega \longrightarrow \omega_c$, where ω_c is equal to qB_0/m.

4.6. A region is characterized by crossed electric and magnetic fields given by

$$\mathbf{E} = E_0(-\sin \omega t \, \mathbf{i}_x + \cos \omega t \, \mathbf{i}_y) \qquad \mathbf{B} = B_0\mathbf{i}_z$$

where E_0 and B_0 are constants. A test charge q having a mass m starts from the origin at $t = 0$ with zero initial velocity. Obtain the parametric equations of motion of the test charge. Check your result with that of Example 4-2 by letting $\omega \longrightarrow 0$. Investigate the limiting case of $\omega \longrightarrow \omega_c$, where ω_c is equal to qB_0/m.

4.7. A magnetic field is given, in cylindrical coordinates, by

$$\mathbf{B} = \frac{B_0}{r}\mathbf{i}_\phi$$

where B_0 is a constant. A rectangular loop is situated in the yz plane and parallel to the z axis as shown in Fig. 4.22. If the loop is moving in that plane with a velocity $\mathbf{v} = v_0\mathbf{i}_y$, where v_0 is a constant, find the circulation of the induced electric field around the loop.

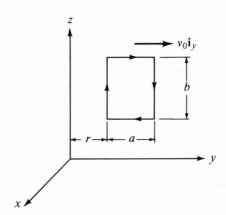

Fig. 4.22. For Problem 4.7.

4.8. For the rectangular loop arrangement of Fig. 4.22, find the circulation of the induced electric field around the loop if the loop is stationary but the magnetic field is varying with time in the manner

$$\mathbf{B} = \frac{B_0}{r} \cos \omega t \, \mathbf{i}_\phi$$

where B_0 is a constant.

4.9. For the rectangular loop arrangement of Fig. 4.22, find the circulation of the induced electric field around the loop if the loop is moving with a velocity $\mathbf{v} = v_0 \mathbf{i}_y$ and if the magnetic field is varying with time in the manner

$$\mathbf{B} = \frac{B_0}{r} \cos \omega t \, \mathbf{i}_\phi$$

where v_0 and B_0 are constants.

4.10. For each of the following magnetic fields, find the induced electric field everywhere, by using Faraday's law in integral form:

(a) $\mathbf{B} = \begin{cases} B_0 \sin \omega t \, \mathbf{i}_z & |x| < a \\ 0 & |x| > a \end{cases}$

(b) $\mathbf{B} = \begin{cases} 0 & r < a \\ B_0 \sin \omega t \, \mathbf{i}_z & a < r < b \\ 0 & r > b \end{cases}$

(c) $\mathbf{B} = \begin{cases} B_0 \left(1 - \dfrac{r^2}{a^2}\right) \sin \omega t \, \mathbf{i}_z & r < a \\ 0 & r > a \end{cases}$

where B_0 is a constant.

4.11. In a region characterized by a magnetic field $\mathbf{B} = B_0 \mathbf{i}_z$, where B_0 is a constant, a test charge q having a mass m is moving along a circular path of radius a and in the xy plane. Find the electric field as viewed by an observer moving with the test charge.

4.12. A region is characterized by crossed electric and magnetic fields $\mathbf{E} = E_0 \mathbf{i}_y$ and $\mathbf{B} = B_0 \mathbf{i}_z$, where E_0 and B_0 are constants. A test charge q having a mass m starts from the origin at $t = 0$ with an initial velocity $\mathbf{v} = (E_0/B_0)\mathbf{i}_x$. Find the electric field as viewed by an observer moving with the test charge.

4.13. Verify your answer to Problem 4.9 by using (4-43).

4.14. Verify your answers to Problem 4.10 by using Faraday's law in differential form.

4.15. A current I C/sec flows from a point charge Q_1 C situated at $(0, 0, -d)$ to a point charge Q_2 C situated at $(0, 0, d)$ along a straight filamentary wire as shown in Fig. 4.23. Find $\oint_C \mathbf{B} \cdot d\mathbf{l}$, where C is a circular path centered at $(0, 0, z)$ and lying in the plane normal to the z axis, in two ways: (a) by applying the Biot–Savart law to find the magnetic field due to the current-carrying wire and (b) by applying the modified Ampere's circuital law in integral form to the path C.

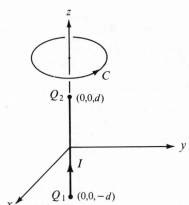

Fig. 4.23. For Problem 4.15.

4.16. Current flows away from a point charge Q C at the origin radially on the xy plane with density given by

$$\mathbf{J}_s = \frac{I}{2\pi r}\mathbf{i}_r \text{ amps/m}$$

Find $\oint_C \mathbf{B} \cdot d\mathbf{l}$ where C is a circular path centered at $(0, 0, z)$ and lying in the plane normal to the z axis in two ways: (a) by applying the Biot–Savart law to find the magnetic field due to the surface current and (b) by applying the modified Ampere's circuital law in integral form to the path C.

4.17. Current flows from a point charge Q_1 C at $(0, 0, a)$ to a point charge Q_2 coulombs at $(0, 0, -a)$ along a spherical surface of radius a and centered at the origin with density given by

$$\mathbf{J}_s = \frac{I}{2\pi a \sin \theta}\mathbf{i}_\theta \text{ amp/m}$$

Find $\oint_C \mathbf{B} \cdot d\mathbf{l}$, where C is a circular path centered at $(0, 0, z)$ and lying in the plane normal to the z axis. Consider both cases: path C outside the sphere and path C inside the sphere.

4.18. A point charge Q C moves along the z axis with a constant velocity v_0 m/sec. Assuming that the point charge crosses the origin at $t = 0$, find and sketch the variation with time of $\oint_C \mathbf{B} \cdot d\mathbf{l}$ where C is a circular path of radius a in the xy plane having its center at the origin, and traversed in the ϕ direction. From symmetry considerations, find $\mathbf{B}$ at points on C.

4.19. A point charge Q_1 C is situated at the origin. Current flows away from the point charge at the rate of I C/sec along a straight wire from the origin to infinity and passing through the point $(1, 1, 1)$. Find $\oint \mathbf{B} \cdot d\mathbf{l}$ around the closed path formed by the triangle having the vertices $(1, 0, 0)$, $(0, 1, 0)$, and $(0, 0, 1)$. Assume that the closed path is traversed in the clockwise direction as seen from the origin.

4.20. Repeat Prob. 4-19 if the straight wire, instead of extending to infinity, terminates on another point charge Q_2 C situated on the plane surface bounded by the triangular path and inside the closed path.

4.21. In the arrangement shown in Fig. 4.24, three point charges Q_1, Q_2, and Q_3 are situated along a straight line. A current of 2 amp flows from Q_1 to Q_2 whereas a current of 1 amp flows from Q_2 to Q_3. Find $\oint_C \mathbf{B} \cdot d\mathbf{l}$, where C is a circular path centered at Q_2 and in the plane normal to the line joining Q_1 to Q_3.

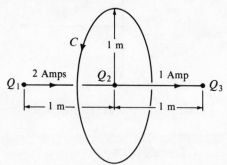

Fig. 4.24. For Problem 4.21.

4.22. Verify your result for the magnetic field due to the current-carrying wire of Problem 4.15, by using (4-96).

4.23. Verify your result for the magnetic field due to the moving charge of Problem 4.18, by using (4-96).

4.24. In a region containing no charges and currents, the magnetic field is given by

$$\mathbf{B} = B_0 \sin \beta z \sin \omega t \, \mathbf{i}_x$$

where B_0, β, and ω are constants. Using one of Maxwell's curl equations at a time, find two expressions for the associated electric field $\mathbf{E}$ and then find the relationship between β, ω, μ_0, and ϵ_0.

4.25. Four point charges having values 1, -2, 3, and 4 C are situated at the corners of a square of sides 1 m as shown in Fig. 4.25. Find the work required to move the point charges to the corners of a smaller square of sides $1/\sqrt{2}$ m.

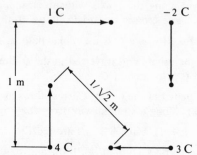

Fig. 4.25. For Problem 4.25.

4.26. Find the potential energy associated with the following volume charge distributions of density ρ in spherical coordinates using $W_e = \frac{1}{2} \int_{\text{vol}} \rho V \, dv$:

$$\text{(a) } \rho = \begin{cases} 0 & 0 < r < a \\ \rho_0 & a < r < b \\ 0 & b < r < \infty \end{cases}$$

$$\text{(b) } \rho = \begin{cases} \rho_0 \dfrac{r}{a} & 0 < r < a \\ 0 & a < r < \infty \end{cases}$$

where ρ_0 is a constant.

4.27. Verify your results for Problem 4.26 by performing volume integration of the electric energy densities associated with the charge distributions.

4.28. Two spherical charges, each of the same radius a m and the same uniform density ρ_0 C/m³ are situated infinitely apart.
  (a) The two spherical charges are now brought together and made into a single spherical charge having the same uniform density ρ_0 C/m³ as those of the original charges. Find the work required.
  (b) Instead of as in part (a), the two spherical charges are brought together and made into a single spherical charge of uniform density and of the same radius a as those of the original charges. Find the work required.

4.29. Show that the total energy stored in an electric field made up of two fields $\mathbf{E}_1$ and $\mathbf{E}_2$ is equal to the sum of the energies stored in the individual fields plus a coupling term, $\epsilon_0 \displaystyle\int_{\text{vol}} (\mathbf{E}_1 \cdot \mathbf{E}_2) \, dv$, that is,

$$W_e = \int_{\text{vol}} \left(\frac{1}{2}\epsilon_0 E_1^2 + \frac{1}{2}\epsilon_0 E_2^2 + \epsilon_0 \mathbf{E}_1 \cdot \mathbf{E}_2 \right) dv$$

4.30. Find the energy stored in the electric field set up by charges Q and $-Q$ uniformly distributed on concentric spherical surfaces of radii a and b, respectively, in three ways:
  (a) By using $W_e = \frac{1}{2} \displaystyle\int_{\text{vol}} \rho V \, dv$.
  (b) By performing volume integration of the energy density in the electric field set up by the charge distribution.
  (c) By considering the electric field as the superposition of the fields set up independently by the two spherical surface charges and using the result of Problem 4.29.

4.31. Find the energy associated with the following current distributions, in cylindrical coordinates, per unit length along the z axis, by using $W_m = \frac{1}{2} \displaystyle\int_{\text{vol}} \mathbf{J} \cdot \mathbf{A} \, dv$.

$$\text{(a) } \mathbf{J} = \begin{cases} \dfrac{I_0}{\pi a^2}\, \mathbf{i}_z & 0 < r < a \\ 0 & a < r < b \\ -\dfrac{I_0}{\pi(c^2 - b^2)}\, \mathbf{i}_z & b < r < c \\ 0 & c < r < \infty \end{cases}$$

$$\text{(b) } \mathbf{J} = \begin{cases} J_0 \dfrac{r}{a}\, \mathbf{i}_z & 0 < r < a \\ -\dfrac{J_0 a^2}{3b}\, \delta(r - b)\, \mathbf{i}_z & a < r < \infty \end{cases}$$

where I_0 and J_0 are constants.

4.32.   Find the energy associated with the following current distributions, per unit area in the $y = 0$ plane, by using $W_m = \frac{1}{2} \int_{vol} \mathbf{J} \cdot \mathbf{A} \, dv$.

(a) $\mathbf{J} = J_{s0}[\delta(y + a) - \delta(y - a)] \, \mathbf{i}_z$, where J_{s0} is a constant

(b) $\mathbf{J} = \begin{cases} y\mathbf{i}_z & |y| < a \\ 0 & |y| > a \end{cases}$

4.33.   Verify your results for Problems 4.31 and 4.32 by performing volume integration of the magnetic energy densities associated with the current distributions.

4.34.   Show that the total energy stored in a magnetic field made up of two fields $\mathbf{B}_1$ and $\mathbf{B}_2$ is equal to the sum of the energies stored in the individual fields plus a coupling term, $(1/\mu_0) \int_{vol} (\mathbf{B}_1 \cdot \mathbf{B}_2) \, dv$, that is,

$$W_m = \int_{vol} \left(\frac{B_1^2}{2\mu_0} + \frac{B_2^2}{2\mu_0} + \frac{\mathbf{B}_1 \cdot \mathbf{B}_2}{\mu_0} \right) dv$$

4.35.   A surface current distribution is given, in cylindrical coordinates, by

$$\mathbf{J}_s = \begin{cases} \dfrac{I_1}{a} \mathbf{i}_z & r = a \\[2mm] \dfrac{I_2}{b} \mathbf{i}_z & r = b \\[2mm] -\dfrac{I_1 + I_2}{c} \mathbf{i}_z & r = c \end{cases}$$

Find the energy stored in the magnetic field, set up by the current distribution, per unit length along the z axis in three ways:

(a) By using $W_m = \frac{1}{2} \int_S \mathbf{J}_s \cdot \mathbf{A} \, dS$.

(b) By performing volume integration of the energy density in the magnetic field set up by the current distribution.

(c) By considering the magnetic field as the superposition of the fields set up independently by two surface current distributions given by

$$\mathbf{J}_s = \begin{cases} -\dfrac{I_1}{a} \mathbf{i}_z & r = a \\[2mm] -\dfrac{I_1}{c} \mathbf{i}_z & r = c \end{cases}$$

$$\mathbf{J}_s = \begin{cases} \dfrac{I_2}{b} \mathbf{i}_z & r = b \\[2mm] -\dfrac{I_2}{c} \mathbf{i}_z & r = c \end{cases}$$

and using the result of Problem 4.34.

4.36.   An electric field intensity vector is given by

$$\mathbf{E} = 100 \cos(\omega t - \beta z) \, \mathbf{i}_x + 50 \sin(\omega t + \beta z) \, \mathbf{i}_y$$

where ω and β $(= \omega\sqrt{\mu_0 \epsilon_0})$ are constants. Find the associated magnetic flux density vector $\mathbf{B}$. Find the Poynting vector $\mathbf{E} \times \mathbf{B}/\mu_0$.

4.37. Electric and magnetic fields are given in cylindrical coordinates by

$$\mathbf{E} = \begin{cases} \dfrac{V_0}{r \ln b/a} \cos \beta z \cos \omega t \, \mathbf{i}_r, & a < r < b \\ 0 & \text{otherwise} \end{cases}$$

$$\mathbf{B} = \begin{cases} \dfrac{\mu_0 I_0}{2\pi r} \sin \beta z \sin \omega t \, \mathbf{i}_\phi & a < r < b \\ 0 & \text{otherwise} \end{cases}$$

where V_0, I_0, ω, and β $(= \omega\sqrt{\mu_0\epsilon_0})$ are constants. Find the expression for the power leaving the volume bounded by two constant z planes, one of which is the $z = 0$ plane. Draw a graph of the power versus z for $\omega t = \pi/4$.

4.38. In the region $r < a$ in cylindrical coordinates, charges are in motion under the combined influence of an electric field $\mathbf{E} = E_0 \mathbf{i}_z$ and a frictional mechanism, thereby constituting a current of density $\mathbf{J} = J_0 \mathbf{i}_z$, where E_0 and J_0 are constants. Obtain the magnetic field due to the current and show that $\mathbf{E} \times \mathbf{B}$ points everywhere towards the z axis, that is, in the $-\mathbf{i}_r$ direction. Show that $\oint_S (\mathbf{E} \times \mathbf{B}/\mu_0) \cdot d\mathbf{S}$, where S is the surface of a cylindrical volume of any radius r and length l, and with the z axis as its axis, gives the correct result for the power expended by the electric field in that volume.

4.39. The electric field intensity in the radiation field of an antenna located at the origin of a spherical coordinate system is given by

$$\mathbf{E} = E_0 \frac{\sin \theta \cos \theta}{r} \cos (\omega t - \beta r) \, \mathbf{i}_\theta$$

where E_0, ω, and β $(= \omega\sqrt{\mu_0\epsilon_0})$ are constants. Find the magnetic field associated with this electric field and then find the power radiated by the antenna by integrating the Poynting vector over a spherical surface of radius r centered at the origin.

4.40. Obtain the steady-state solution for the following differential equation in two ways: (a) without using the phasor technique, and (b) by using the phasor technique:

$$2 \times 10^{-3} \frac{dV}{dt} + V = 10 \sin \left(500t + \frac{\pi}{6} \right)$$

4.41. Repeat Problem 4.40 for the following integrodifferential equation:

$$\frac{dI}{dt} + 2I + \int I \, dt = 10 \cos \left(2t - \frac{\pi}{3} \right)$$

4.42. Two infinitely long, straight parallel wires carry currents $I_1 = I_0 \cos \omega t$ and $I_2 = I_0 \cos (\omega t + 90°)$ amp, respectively, as shown in Fig. 4.26. Determine the x and y components of the magnetic flux density vector at each of the three points A, B, and C. Describe how the magnitude and direction of the magnetic flux density vector varies with time at each of the three points A, B, and C.

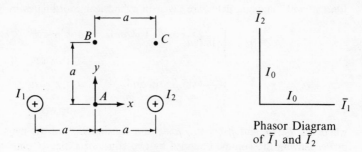

Fig. 4.26. For Problem 4.42.

4.43.   In the arrangement shown in Fig. 4.27(a), four line charges, infinitely long in the direction normal to the plane of the paper and having uniform charge densities varying sinusoidally with time are situated at the corners of a square. The amplitudes of the sinusoidally time-varying charge densities are such that, considered alone, each line charge produces unit peak electric field intensity at the center of the square. The phasor diagram of the charge densities is shown in Fig. 4.27(b).

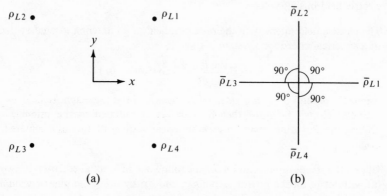

Fig. 4.27. For Problem 4.43.

(a) Find and sketch the phasor representing the x and y components of the electric field intensity vector at the center of the square.

(b) Determine how the magnitude and direction of the electric field intensity vector at the center of the square vary with time.

4.44.   Repeat Problem 4.43 for the rectangular arrangement of line charges shown in Fig. 4.28.

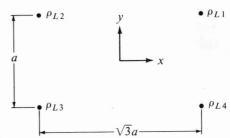

Fig. 4.28. For Problem 4.44.

4.45. A sinusoidally time-varying electric field intensity vector is characterized by its phasor $\bar{\mathbf{E}}$, given by

$$\bar{\mathbf{E}} = (-\mathbf{i}_x - 2\sqrt{3}\,\mathbf{i}_y + \sqrt{3}\,\mathbf{i}_z)e^{-j0.04\pi(\sqrt{3}x-2y-3z)}$$

(a) Show that the surfaces of constant phase of $\bar{\mathbf{E}}$ are planes. Find the equation of the planes.

(b) Show that the electric field is linearly polarized in the planes of constant phase.

(c) Find the direction of polarization.

4.46. A sinusoidally time-varying electric field intensity vector is characterized by its phasor $\bar{\mathbf{E}}$, given by

$$\bar{\mathbf{E}} = (-j1\mathbf{i}_x - 2\mathbf{i}_y + j\sqrt{3}\,\mathbf{i}_z)e^{-j0.05\pi(\sqrt{3}x+z)}$$

(a) Show that the surfaces of constant phase of $\bar{\mathbf{E}}$ are planes. Find the equation of the planes.

(b) Show that the electric field is circularly polarized in the planes of constant phase.

(c) Obtain the magnetic flux density phasor $\bar{\mathbf{B}}$ associated with the given $\bar{\mathbf{E}}$ and determine if the field is right circularly polarized or left circularly polarized.

4.47. Repeat Problem 4.46 for the following phasor electric field intensity vector:

$$\bar{\mathbf{E}} = \left[\left(-\sqrt{3} - j\frac{1}{2}\right)\mathbf{i}_x + \left(1 - j\frac{\sqrt{3}}{2}\right)\mathbf{i}_y + j\sqrt{3}\,\mathbf{i}_z\right]e^{-j0.02\pi(\sqrt{3}x+3y+2z)}$$

4.48. Show that a linearly polarized field vector can be expressed as the sum of left and right circularly polarized field vectors having equal magnitudes, and that an elliptically polarized field vector can be expressed as the sum of left and right circularly polarized field vectors having unequal magnitudes.

4.49. Find the time-average stored energy density in the electric field characterized by the phasor specified in Problem 4.47.

4.50. The electric field associated with a sinusoidally time-varying electromagnetic field is given by

$$\mathbf{E}(x, y, z, t) = 10 \sin \pi x \sin (6\pi \times 10^8 t - \sqrt{3}\,\pi z)\,\mathbf{i}_y \text{ volts/m}$$

Find (a) the time-average stored energy density in the electric field, (b) the time-average stored energy density in the magnetic field, (c) the time-average Poynting vector associated with the electromagnetic field, and (d) the imaginary part of the complex Poynting vector.

5

MATERIALS AND FIELDS

In this chapter we extend our study of fields in free space of the preceding three chapters to fields in the presence of materials. Materials contain charged particles which act as sources of electromagnetic fields. Under the application of external fields, these charged particles respond, giving rise to secondary fields comparable to the applied fields. While the properties of materials that produce these effects are determined on the atomic or "microscopic" scale, it is possible to develop a consistent theory based on "macroscopic" scale observations, that is, observations averaged over volumes large compared with atomic dimensions. We will learn that these macroscopic scale phenomena are equivalent to charge and current distributions acting as though they were situated in free space, so that the secondary fields can be found by using the knowledge gained in the preceding chapters. In fact, we have an interesting situation in which the equivalent charge and current distributions are related to the total fields in the material comprising the applied and the secondary fields, whereas the secondary fields are related to the equivalent charge and current distributions. We are thus faced with the simultaneous solution of two sets of equations governing these two relationships. Following this logic, we will introduce new vector fields and develop a new set of Maxwell's equations with associated constitutive relations which eliminate the necessity for the simultaneous solution by taking into account implicitly the equivalent charge and current distributions.

5.1 Conduction and Nonmagnetic Materials

Depending upon their response to an applied electric field, materials may be classified as conductors, semiconductors, or dielectrics. According to the classical model, an atom consists of a tightly bound, positively charged nucleus surrounded by a diffuse electron cloud having an equal and opposite charge to the nucleus, as shown in Fig. 5.1. While the electrons for the most

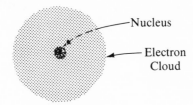

Nucleus

Electron
Cloud

Fig. 5.1. Classical model of an atom.

part are less tightly bound, the majority of them are associated with the nucleus and are known as "bound" electrons. These bound electrons can be displaced but not removed from the influence of the nucleus upon application of an electric field. Not taking part in this bonding mechanism are the "free" or "conduction" electrons. These electrons are constantly under thermal agitation, being released from the parent atom at one point and recaptured at another point. In the absence of an applied electric field, their motion is completely random; that is, the average thermal velocity on a macroscopic scale is zero so that there is no net current and the electron cloud maintains a fixed position. When an electric field is applied, an additional velocity due to the Coulomb force is superimposed on the random velocities, thereby causing a "drift" of the average position of the electrons along the direction opposite to that of the electric field. This process is known as "conduction." In certain materials, a large number of electrons may take part in this process. These materials are known as "conductors." In certain other materials, only very few or a negligible number of electrons may participate in conduction. These materials are known as "dielectrics" or insulators. We will later learn that a characteristic called polarization is more important than conduction in dielectrics. A class of materials for which conduction occurs not only by electrons but also by another type of carriers known as "holes"—vacancies created by detachment of electrons due to breaking of covalent bonds with other atoms—is intermediate to that of conductors and dielectrics. These materials are called "semiconductors."

The quantum theory describes the motion of the current carriers in terms of energy levels. According to this theory, the electrons in an atom can have associated with them only certain discrete values of energy. When a large number of atoms are packed together, as in a crystalline solid, each energy level in the individual atom splits into a number of levels with slightly different energies, with the degree of splitting governed by the interatomic spacing, thereby giving rise to alternate allowed and forbidden bands of energy levels as shown in Fig. 5.2. Each allowed band can be thought of as an

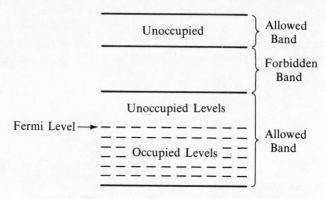

Fig. 5.2. Energy band structure for a crystalline solid.

almost continuous region of allowed energy levels. For example, for a typical solid having an atomic density of 10^{29} per m³, there will be almost 10^{29} levels in each band. A forbidden band consists of energy levels which no electron in any atom of the solid can occupy. According to Pauli's exclusion principle, each allowed energy level may not be occupied by more than one electron. Electrons naturally tend to occupy the lowest energy levels; at a temperature of absolute zero, all the levels below a certain level known as the Fermi level are occupied and all the levels above the Fermi level are unoccupied. Hence, depending upon the location of the Fermi level, we can have different cases as shown in Fig. 5.3.

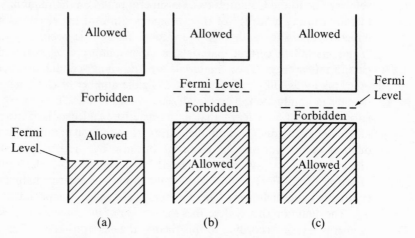

Fig. 5.3. Energy band diagrams for different cases: (a) Conductor. (b) Dielectric. (c) Semiconductor.

For case (a), the Fermi level lies within an allowed band. The band is therefore only partially filled at the temperature of absolute zero. At higher temperatures, the electron population in the band spreads out somewhat but only very few electrons reach above the Fermi level. Thus, since there are many unfilled levels in the same band, it is possible to increase the energy of the system by moving the electrons to these unoccupied levels very easily by the application of an electric field, thereby resulting in a drift velocity of the electrons in the direction opposite to that of the electric field. The material is then classified as a conductor. If the Fermi level is between two allowed bands as in (b) and (c) of Fig. 5.3, the lower band is completely filled whereas the next higher band is completely empty at the temperature of absolute zero. If the width of the forbidden band is very large as in (b), the situation at normal temperatures is essentially the same as at absolute zero and hence there are no neighboring empty energy levels for the electrons to move. The only way for conduction to take place is for the electrons in the filled band to get excited and move to the next higher band. But this is very difficult to achieve with reasonable electric fields and the material is then classified as a dielectric. Only by supplying a very large amount of energy can an electron be excited to move from the lower band to the higher band where it has available neighboring empty levels for causing conduction. The dielectric is said to break down under such conditions. If, on the other hand, the width of the forbidden band in which the Fermi level lies is not too large, as in (c), some of the electrons in the lower band move into the upper band at normal temperatures so that conduction can take place under the influence of an electric field, not only in the upper band but also in the lower band because of the vacancies (holes) left by the electrons which moved into the upper band. The material is then classified as a semiconductor. A semiconductor crystal in pure form is known as an intrinsic semiconductor. It is possible to alter the properties of an intrinsic crystal by introducing impurities into it. The crystal is then said to be an extrinsic semiconductor.

5.2 Conduction Current Density, Conductivity, and Ohm's Law

In Section 5.1 we classified materials on the basis of their ability to permit conduction of electrons under the application of an external electric field. For conductors, we are interested in knowing about the relationship between the "drift velocity" of the electrons and the applied electric field, since the predominant process is conduction. But for collisions with the atomic lattice, the electric field continuously accelerates the electrons in the direction opposite to it as they move about at random. Collisions with the atomic lattice, however, provide the frictional mechanism by means of which the electrons lose some of the momentum gained between collisions. The net effect is as though the electrons drift with an average drift velocity v_d, under the influence of the Coulomb force exerted by the applied electric field and an opposing force due to the frictional mechanism. This opposing force is proportional to the momentum of the electron and inversely proportional to the average time τ between collisions. Thus the equation of motion of an electron is given by

$$m\frac{d\mathbf{v}_d}{dt} = e\mathbf{E} - \frac{m\mathbf{v}_d}{\tau} \tag{5-1}$$

where e and m are the charge and mass of an electron.

Rearranging (5-1), we have

$$m\frac{d\mathbf{v}_d}{dt} + \frac{m}{\tau}\mathbf{v}_d = e\mathbf{E} \tag{5-2}$$

For the sudden application of a constant electric field $\mathbf{E}_0$ at $t = 0$, the solution for (5-2) is given by

$$\mathbf{v}_d = \frac{e\tau}{m}\mathbf{E}_0 - \frac{e\tau}{m}\mathbf{E}_0\exp{(-t/\tau)} \tag{5-3}$$

where we have evaluated the arbitrary constant of integration by using the initial condition that $\mathbf{v}_d = 0$ at $t = 0$. The values of τ for typical conductors such as copper are of the order of 10^{-14} sec so that the exponential term on the right side of (5-3) decays to negligible value in a time much shorter than that of practical interest. Thus, neglecting this term, we have

$$\mathbf{v}_d = \frac{e\tau}{m}\mathbf{E}_0 \tag{5-4}$$

and the drift velocity is proportional in magnitude and opposite in direction to the applied electric field since the value of e is negative.

In fact, since we can represent a time-varying field as a superposition of step functions starting at appropriate times, the exponential term in (5-3) may be neglected as long as the electric field varies slowly compared to τ. For fields varying sinusoidally with time, this means that as long as the period T of the sinusoidal variation is several times the value of τ, or the radian frequency $\omega \ll 2\pi/\tau$, the drift velocity follows the variations in the electric field. Since $1/\tau \approx 10^{14}$, this condition is satisfied even at frequencies up to several hundred gigahertz. Thus, for all practical purposes, we can assume that

$$\mathbf{v}_d = \frac{e\tau}{m}\mathbf{E} \tag{5-5}$$

Now, we define the "mobility," μ_e of the electron as the ratio of the magnitudes of the drift velocity and the applied electric field. Then we have

$$\mu_e = \frac{|\mathbf{v}_d|}{|\mathbf{E}|} = \frac{|e|\tau}{m} \tag{5-6}$$

and

$$\mathbf{v}_d = -\mu_e\mathbf{E} \qquad \text{for electrons} \tag{5-7a}$$

For values of τ typically of the order of 10^{-14} sec, we note by substituting for $|e|$ and m on the right side of (5-6) that the electron mobilities are of the order of 10^{-3} C-sec/kg. Alternative units for the mobility are square meters per volt-second. In semiconductors, conduction is due not only to the movement of electrons but also to the movement of holes. We can define the mobility μ_h of a hole similarly to μ_e as the ratio of the drift velocity of the hole to the applied electric field. Thus we have

$$\mathbf{v}_d = \mu_h \mathbf{E} \qquad \text{for holes} \tag{5-7b}$$

Note from (5-7b) that conduction of a hole takes place along the direction of the applied electric field since a hole is a vacancy created by the removal of an electron and hence a hole movement is equivalent to the movement of a positive charge of value equal to the magnitude of the charge of an electron. In general, the mobility of holes is lower than the mobility of electrons for a particular semiconductor. For example, for silicon, the values of μ_e and μ_h are

$$\mu_e = 0.125 \text{ m}^2/\text{volt-sec} \qquad \mu_h = 0.048 \text{ m}^2/\text{volt-sec}$$

The drift of electrons in a conductor and that of electrons and holes in a semiconductor is equivalent to a current flow. This current is known as the conduction current, in contrast to the convection current produced by the motion of charges in free space. The conduction current density may be obtained in the following manner. If there are N_e free electrons per cubic meter of the material, then the amount of charge ΔQ passing through an infinitesimal area ΔS at a point in the material in a time Δt is given by

$$\begin{aligned} \Delta Q &= N_e e (\Delta \mathbf{S} \cdot \mathbf{v}_d \, \Delta t) \\ &= N_e e (\Delta S \, \mathbf{i}_n \cdot \mathbf{v}_d) \, \Delta t \end{aligned} \tag{5-8}$$

where $\Delta \mathbf{S} = \Delta S \, \mathbf{i}_n$. The current ΔI flowing across ΔS is given by

$$\Delta I = \frac{\Delta Q}{\Delta t} = N_e e \, \Delta S \, \mathbf{i}_n \cdot \mathbf{v}_d \tag{5-9}$$

The magnitude of the current density at the point is the ratio of ΔI to ΔS for an orientation of ΔS which maximizes this ratio and as ΔS tends to zero. Obviously, the ratio is a maximum for an orientation of ΔS normal to $\mathbf{v}_d$ and is equal to $N_e |e| v_d$. Thus the conduction current density $\mathbf{J}_c$ resulting from the drift of electrons in the conductor is given by

$$\mathbf{J}_c = N_e e \mathbf{v}_d \tag{5-10}$$

Substituting for $\mathbf{v}_d$ from (5-7a), we have

$$\mathbf{J}_c = -\mu_e N_e e \mathbf{E} \tag{5-11}$$

Defining a quantity σ as

$$\sigma = -\mu_e N_e e = \mu_e N_e |e| \tag{5-12}$$

we obtain the simple and important relationship between $\mathbf{J}_c$ and $\mathbf{E}$

$$\mathbf{J}_c = \sigma \mathbf{E} \qquad (5\text{-}13)$$

The quantity σ is known as the electrical conductivity of the material and Eq. (5-13) is known as Ohm's law valid at a point. Equation (5-13) indicates that $\mathbf{J}_c$ is proportional to $\mathbf{E}$. Materials for which this relationship holds, that is, σ is independent of the magnitude as well as the direction of $\mathbf{E}$ are known as linear isotropic conductors. For certain conductors, each component of $\mathbf{J}_c$ can be dependent on all components of $\mathbf{E}$. In such cases, $\mathbf{J}_c$ is not parallel to $\mathbf{E}$ and the conductors are not isotropic. Such conductors are known as anisotropic conductors.

In a semiconductor we have two types of current carriers: electrons and holes. Accordingly, the current density in a semiconductor is the sum of the contributions due to the drifts of electrons and holes. If the densities of holes and electrons are N_h and N_e, respectively, the conduction current density is given by

$$\mathbf{J}_c = (\mu_h N_h |e| + \mu_e N_e |e|)\mathbf{E} \qquad (5\text{-}14)$$

Thus the conductivity of a semiconducting material is given by

$$\sigma = \mu_h N_h |e| + \mu_e N_e |e| \qquad (5\text{-}15a)$$

For an intrinsic semiconductor, $N_h = N_e$ so that (5-15a) reduces to

$$\sigma = (\mu_h + \mu_e)N_e |e| \qquad (5\text{-}15b)$$

The units of conductivity are (meter²/volt-second)(coulomb/meter³) or ampere/volt-meter, also commonly known as mhos per meter, where a mho ("ohm" spelled in reverse and having the symbol ℧) is an ampere per volt. The ranges of conductivities for conductors, semiconductors, and dielectrics are shown in Fig. 5.4. Values of conductivities for a few materials are listed in Table 5.1. The constant values of conductivities do not imply that the conduction current density is proportional to the applied electric

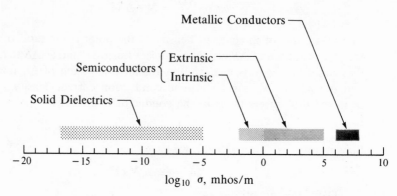

Fig. 5.4. Ranges of conductivities for conductors, semiconductors, and dielectrics.

TABLE 5.1. Conductivities of Some Materials

Material	Conductivity, mhos/m	Material	Conductivity, mhos/m
Silver	6.1×10^7	Sea water	4
Copper	5.8×10^7	Intrinsic germanium	2.2
Gold	4.1×10^7	Intrinsic silicon	1.6×10^{-3}
Aluminum	3.5×10^7	Fresh water	10^{-3}
Tungsten	1.8×10^7	Distilled water	2×10^{-4}
Brass	1.5×10^7	Dry earth	10^{-5}
Nickel	1.3×10^7	Wood	10^{-8}–10^{-11}
Solder	7.0×10^6	Bakelite	10^{-9}
Lead	4.8×10^6	Glass	10^{-10}–10^{-14}
Constantin	2.0×10^6	Porcelain	2×10^{-13}
Mercury	1.0×10^6	Mica	10^{-11}–10^{-15}
Nichrome	8.9×10^5	Fused quartz	0.4×10^{-17}

field intensity for all values of current density and field intensity. However, the range of current densities for which the material is linear, that is, for which the conductivity is a constant, is very large for conductors.

5.3 Conductors in Electric Fields

In Sections 5.1 and 5.2 we learned that the free electrons in a conductor drift under the influence of an electric field. Let us now consider an arbitrary-shaped conductor of uniform conductivity σ placed in a static electric field as shown in Fig. 5.5(a). The free electrons in the conductor move opposite to the direction lines of the electric field. If there is a way by means of which the flow of electrons can be continued to form a closed circuit, then a continuous flow of current takes place. In this section we will consider the conductor to be bounded by free space, in which case the electrons are held

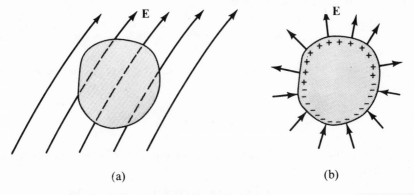

(a)   (b)

Fig. 5.5. For illustrating the surface charge formation at the boundary of a conductor placed in an electric field.

at the boundary from moving further by the atomic forces within the conductor and by the insulating property of free space. Thus a negative surface charge forms on that part of the boundary through which the electric field lines enter the conductor originally, as shown in Fig. 5.5(b). Now, since the conductor as a whole is neutral, an amount of positive charge equal in magnitude to the negative surface charge must exist somewhere in the conductor. Where in the conductor may this charge or, for that matter, any charge placed inside the conductor reside? We will answer this question in the following example.

EXAMPLE 5-1. Assume that, at $t = 0$, a charge distribution of density ρ_0 is created in a portion of a conductor of uniform conductivity σ. In the remaining portion of the conductor, the charge density is zero. It is desired to show that the charge density in the conductor decays exponentially to zero and appears as a surface charge at the boundary of the conductor.

Denoting the charge density and the electric field intensity at any time t in the interior of the conductor to be ρ and $\mathbf{E}$, respectively, we have, from Maxwell's divergence equation for the electric field,

$$\mathbf{V} \cdot \mathbf{E} = \frac{\rho}{\epsilon_0} \tag{2-82}$$

The time variation of charge density is governed by the continuity equation

$$\mathbf{V} \cdot \mathbf{J}_c + \frac{\partial \rho}{\partial t} = 0 \tag{5-16}$$

where $\mathbf{J}_c$ is the conduction current density due to the flow of charges in the conductor under the influence of $\mathbf{E}$. Equation (5-16) stated in integral form tells us that the total current leaving a volume of the conducting material is equal to the time rate of decrease of charge inside that volume. Substituting $\mathbf{J}_c = \sigma \mathbf{E}$ in (5-16), we have

$$\mathbf{V} \cdot \sigma \mathbf{E} + \frac{\partial \rho}{\partial t} = 0 \tag{5-17}$$

Since σ is uniform, we can take it outside the divergence operation in (5-17) to obtain

$$\sigma \mathbf{V} \cdot \mathbf{E} + \frac{\partial \rho}{\partial t} = 0 \tag{5-18}$$

Now, combining (5-18) and (2-82), we obtain a differential equation for ρ as given by

$$\frac{\partial \rho}{\partial t} + \frac{\sigma}{\epsilon_0} \rho = 0 \tag{5-19}$$

The solution to (5-19) is obtained by rearranging it and integrating as follows:

$$\int \frac{d\rho}{\rho} = -\int \frac{\sigma}{\epsilon_0} \, dt$$

$$\ln \rho = -\frac{\sigma}{\epsilon_0} t + \ln A \tag{5-20}$$

where $\ln A$ is the arbitrary constant of integration. Substituting the initial condition $\rho = \rho_0$ at $t = 0$ in (5-20) and rearranging, we obtain finally

$$\rho = \rho_0 \, e^{-(\sigma/\epsilon_0)t} = \rho_0 \, e^{-t/T} \tag{5-21}$$

where we define

$$T = \frac{\epsilon_0}{\sigma} \tag{5-22}$$

Thus the charge density inside the conductor decays exponentially with a time constant equal to ϵ_0/σ. In particular, if the charge density at any point is initially zero, it remains at zero. Hence no portion of the charge which decays in one region within the conductor can reappear in any other region within the conductor. On the other hand, the charge must be conserved. Thus the decaying charge can appear only as a surface charge at the boundary of the conductor. To see how fast the charge density at an interior point decays and appears simultaneously as a surface charge, let us consider the example of copper. For copper, $\sigma = 5.80 \times 10^7$ mhos/m so that

$$T = \frac{\epsilon_0}{\sigma} = \frac{10^{-9}}{36\pi \times 5.80 \times 10^7} = 1.5 \times 10^{-19} \text{ sec}$$

Thus, in a time equal to 1.5×10^{-19} sec, the charge density decays to e^{-1} times or about 37% of its initial value. We note that this time constant is extremely short so that we can assume that any charge density in the interior of a conductor disappears to the surface almost instantaneously. (Furthermore, we can assume that the surface charge formation follows any time variation in the electric field causing it so long as this time variation is slow compared to the time constant.) On the other hand, the time constant can be up to several days for dielectric materials. ▮

Returning now to the case of Fig. 5.5, we conclude that the positive charge equal in magnitude to the negative surface charge appears as a surface charge on that part of the boundary through which the electric field lines leave the conductor originally, as shown in Fig. 5.5(b). The surface charge distribution formed in this manner produces a secondary electric field which opposes the applied field inside the conductor. The secondary field should, in fact, cancel the applied field inside the conductor completely. If it does not, there will be further movement of charges to the surface until a distribution is achieved which produces a secondary field inside the conductor that cancels the applied field completely. All this adjustment should be governed by the time constant so that we can assume that a surface charge distribution which reduces the field inside the conductor to zero is formed almost instantaneously. The surface charge distribution will, in general, produce a secondary field outside the conductor which modifies the applied field.

Let us now investigate the properties of the electric field at the surface of a conductor. To do this, let us assume that the electric field intensity **E** on the free-space side of the boundary has a component E_t tangential to the boundary and a component E_n normal to the boundary. The electric field intensity inside the conductor is, of course, equal to zero. We now consider a rectangular path *abcda* of infinitesimal area in the plane normal to the boundary and with its sides *bc* and *ad* parallel to E_t and on either side of the boundary as shown in Fig. 5.6(a). Since the sides of the rectangle

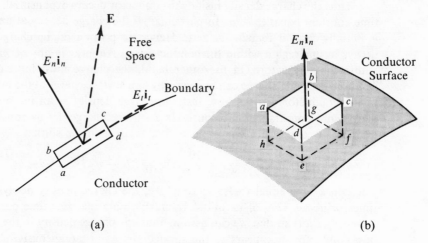

(a)   (b)

Fig. 5.6. For investigating the properties of the electric field intensity vector at the surface of a conductor.

are infinitesimally small, we can assume that E_t and E_n are constants along them. Applying $\oint \mathbf{E} \cdot d\mathbf{l} = 0$ to the path *abcda*, we have

$$\int_a^b \mathbf{E} \cdot d\mathbf{l} + \int_b^c \mathbf{E} \cdot d\mathbf{l} + \int_c^d \mathbf{E} \cdot d\mathbf{l} + \int_d^a \mathbf{E} \cdot d\mathbf{l} = 0 \tag{5-23}$$

The second integral in (5-23) is equal to $E_t(bc)$ and the fourth integral is zero. Now, if we let *ab* and *cd* tend to zero, shrinking the rectangle to the surface but still enclosing it, the first and third integrals in (5-23) go to zero, giving us

$$E_t(bc) = 0$$

or

$$E_t = 0 \tag{5-24}$$

Thus the tangential component of the electric field intensity at the boundary of a conductor placed in an electric field is zero. The electric field at the boundary is entirely normal to the surface. Note that we have not considered any time-varying magnetic flux enclosed by the rectangular path *abcda* since we are using static field laws. However, even if we do consider the time-varying magnetic flux, it will go to zero as *abcda* is shrunk to the surface, yielding the same result as (5-24).

We now suspect that the normal electric field at the boundary is related to the surface charge density. To investigate this, let us consider a rectangular box *abcdefgh* of infinitesimal volume enclosing an infinitesimal area of the boundary and parallel to it as shown in Fig. 5.6(b). Applying Gauss' law in integral form given by

$$\oint_S \mathbf{E} \cdot d\mathbf{S} = \frac{1}{\epsilon_0} (\text{charge enclosed by } S)$$

to the surface area of the box, we have

$$\underbrace{\int \mathbf{E} \cdot d\mathbf{S}}_{\substack{\text{top} \\ \text{surface} \\ abcd}} + \underbrace{\int \mathbf{E} \cdot d\mathbf{S}}_{\substack{\text{bottom} \\ \text{surface} \\ efgh}} + \underbrace{\int \mathbf{E} \cdot d\mathbf{S}}_{\substack{\text{side} \\ \text{surfaces}}} = \frac{1}{\epsilon_0} \begin{pmatrix} \text{charge enclosed in the} \\ \text{volume of the box} \end{pmatrix}$$

(5-25)

The second integral in (5-25) is zero since $\mathbf{E}$ is zero inside the conductor. Since the area *abcd* is infinitesimal, we assume $\mathbf{E}$ to be constant on it so that the first integral is equal to $E_n(abcd)$. Now, if we let the side surfaces tend to zero, shrinking the box to the surface but still enclosing it, the third integral goes to zero and the charge enclosed by the box tends to the surface charge density ρ_s times the area *abcd*, giving us

$$E_n(abcd) = \frac{1}{\epsilon_0} \rho_s(abcd)$$

or

$$E_n = \frac{\rho_s}{\epsilon_0} \tag{5-26}$$

Thus the electric field intensity at a point on the surface of a conductor placed in an electric field is entirely normal to the surface and equal to $1/\epsilon_0$ times the surface charge density at that point.

Finally, since the electric field on the conductor surface is entirely normal to it, we note that no work is required to move an imaginary test charge on the conductor surface or, for that matter, inside the conductor (since $\mathbf{E} = 0$). Thus the conductor surface as well as the interior of the conductor are equipotentials. We now summarize the properties associated with conductors in electric fields as follows:

(a) The charge density at any point in the interior of a conductor is zero. Any charge must reside on the surface only with an appropriate density to produce a secondary electric field inside the conductor which is exactly opposite to the applied electric field so that property (b) below is satisfied.

(b) The electric field intensity inside the conductor is zero.

(c) The electric field intensity at any point on the surface of the conductor is entirely normal to it and equal to $1/\epsilon_0$ times the surface charge density at that point.

(d) The conductor, including its surface, is an equipotential region.

EXAMPLE 5-2. An infinite plane conducting slab of thickness d occupies the region between $z = 0$ and $z = d$ as shown in Fig. 5.7(a). A uniform electric field $\mathbf{E} = E_0\mathbf{i}_z$, where E_0 is a constant is applied. It is desired to find the charge densities induced on the surfaces of the slab.

Since the applied electric field is uniform and is directed along the z direction, a negative charge of uniform density forms on the surface $z = 0$ due to the accumulation of free electrons at that surface. A positive charge

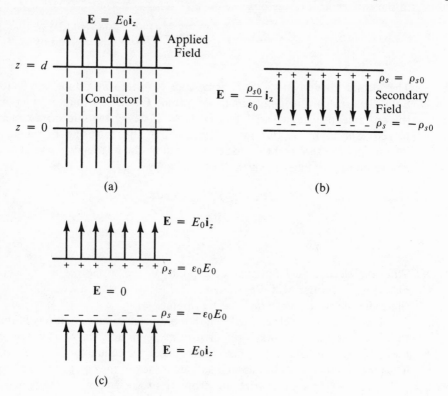

(a)

(b)

(c)

Fig. 5.7. (a) Infinite plane slab conductor in a uniform applied field. (b) Induced surface charge at the boundaries of the conductor and the secondary field. (c) Sum of the applied and the secondary fields.

of equal and opposite uniform density forms on the surface $z = d$ due to a deficiency of electrons at that surface. Let these surface charge densities be $-\rho_{s0}$ and ρ_{s0}, respectively. To satisfy the property that the field in the interior of the conductor is zero, the secondary field produced by the surface charges must be equal and opposite to the applied field; that is, it must be equal to $-E_0 i_z$. Now, each sheet of uniform charge density produces a field intensity directed normally away from it and having a magnitude $1/2\epsilon_0$ times the charge density so that the field due to the two surface charges together is equal to $-(\rho_{s0}/\epsilon_0)i_z$ inside the conductor and zero outside the conductor as shown in Fig. 5.7(b). Thus, for zero field inside the conductor,

$$-\frac{\rho_{s0}}{\epsilon_0} i_z = -E_0 i_z$$

or

$$\rho_{s0} = \epsilon_0 E_0 \tag{5-27}$$

The field outside the conductor remains the same as the applied field since the secondary field in that region due to the surface charges is zero. The induced surface charge distribution and the fields inside and outside the conductor are shown in Fig. 5.7(c). Note that the property that the field intensity at a point on the surface of the conductor is normal to it and equal to $1/\epsilon_0$ times the surface charge density at that point is satisfied on both surfaces $z = 0$ and $z = d$. ∎

5.4 Polarization in Dielectric Materials

We stated at the beginning of Section 5.1 that the bound electrons in an atom can be displaced but not removed from the influence of the parent nucleus upon application of an external electric field. When the centroids of the electron clouds surrounding the nucleii are displaced from the centroids of the nucleii, as shown in Fig. 5.8(a), to create a charge separation and hence form microscopic electric dipoles, the atoms are said to be "polarized." The schematic representation of an electric dipole formed in this manner is shown in Fig. 5.8(b). Such "polarization" may exist in the molecular structure of certain dielectric materials even under the application of no external electric field. The molecules are then said to be polar molecules. However, the polarization of individual atoms and molecules is randomly oriented and hence the material is not polarized on a macroscopic scale. In certain other dielectric materials, no polarization exists initially in the molecular structure. The molecules are then said to be nonpolar molecules.

Upon the application of an external electric field, the centroids of the electron clouds in the nonpolar molecules may become displaced from the centroids of the nucleii due to the Coulomb forces acting on the charges. This kind of polarization is known as electronic polarization. In the case of polar molecules, the electric field has the influence of exerting torques on

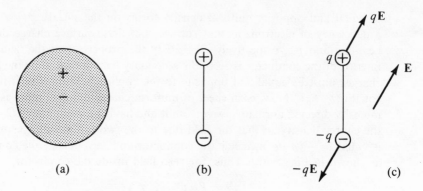

Fig. 5.8. (a) Polarization of bound charge in an atom under the influence of an electric field. (b) Schematic representation of electric dipole created due to polarization. (c) Torque acting on an electric dipole under the influence of an electric field.

the microscopic dipoles as shown in Fig. 5.8(c), to convert the initially random polarization into a partially coherent one along the field, on a macroscopic scale. This kind of polarization is known as orientational polarization. Certain materials, called "electrets," when allowed to solidify in the applied electric field, become permanently polarized in the direction of the field, that is, retain the polarization even after removal of the field. Certain other materials, known as "ferroelectric" materials, exhibit spontaneous, permanent polarization. A third kind of polarization, known as ionic polarization, results from the separation of positive and negative ions in molecules held together by ionic bonds formed by the transfer of electrons from one atom to another in the molecule. All three polarizations may occur simultaneously in a material.

The net dipole moment created due to polarization in a dielectric material will produce a field which opposes the applied electric field and changes its distribution both inside and outside the dielectric material, in general, from the one that existed in the absence of the material. This will be the topic of discussion in Section 5.5. In the remainder of this section, we will first derive the relationship between the dipole moments of the individual microscopic dipoles and the electric field responsible for the polarization by considering electronic polarization by means of an example. We will then define a new vector **P** which represents polarization on a macroscopic scale and relate it to the average macroscopic electric field.

EXAMPLE 5-3.   Assume that the nucleus of an atom is a point charge and that the electron cloud has originally a spherically symmetric, radially uniform charge distribution which is retained as it is displaced relative to the nucleus under the influence of a polarizing electric field. (This assumption is justified if the displacement between the centroids of the electron cloud and the nucleus is negligible compared to the radius of the electron cloud.) It is desired to find the dipole moment resulting from the polarizing field.

Let the electric field causing the displacement between the two centroids be $\mathbf{E}_p = E_0\mathbf{i}_z$, so that the displacement is along the z axis as shown in Fig. 5.9. Let this displacement be equal to d. The two forces which are acting on the nucleus are (a) the Coulomb force $\mathbf{F}_1$ due to the electric field $\mathbf{E}_p$ and (b) the restoring force $\mathbf{F}_2$ due to the electric field produced at the nucleus by the electron cloud.

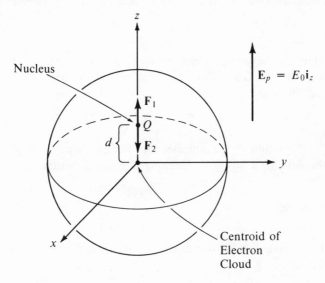

Fig. 5.9 For obtaining the dipole moment due to electronic polarization of an atom.

The force $\mathbf{F}_1$ is given by

$$\mathbf{F}_1 = Q\mathbf{E}_p = QE_0\mathbf{i}_z \qquad (5\text{-}28)$$

where Q is the charge of the nucleus. To find the restoring force $\mathbf{F}_2$, we take advantage of the spherical symmetry of the charge distribution in the electron cloud about its center and apply Gauss' law to a sphere of radius d centered at the origin to obtain the electric field $\mathbf{E}_2$ at the nucleus due to the electron cloud as

$$\mathbf{E}_2 = \frac{1}{\epsilon_0} \frac{\text{charge enclosed by spherical surface of radius } d}{\text{area of the spherical surface}} \mathbf{i}_z \qquad (5\text{-}29)$$

Now, since the total charge in the electron cloud is $-Q$ and since the charge density is uniform, the charge enclosed by the spherical surface of radius

d is $-Qd^3/a^3$, where a is the radius of the electron cloud. Thus we obtain

$$\mathbf{E}_2 = \frac{-Qd^3/a^3}{4\pi\epsilon_0 d^2}\mathbf{i}_z = -\frac{Qd}{4\pi\epsilon_0 a^3}\mathbf{i}_z \qquad (5\text{-}30)$$

Hence the restoring force on the nucleus is given by

$$\mathbf{F}_2 = Q\mathbf{E}_2 = -\frac{Q^2 d}{4\pi\epsilon_0 a^3}\mathbf{i}_z \qquad (5\text{-}31)$$

For equilibrium displacement d of the nucleus relative to the center of the electron cloud, the two forces $\mathbf{F}_1$ and $\mathbf{F}_2$ must add to zero, giving us

$$d = \frac{4\pi\epsilon_0 a^3}{Q} E_0 \qquad (5\text{-}32)$$

Thus the equilibrium displacement d is proportional to the electric field intensity E_0. The dipole moment $\mathbf{p}_e$ formed by the charge separation is then given by

$$\mathbf{p}_e = Qd\mathbf{i}_z = Q\frac{4\pi\epsilon_0 a^3}{Q} E_0\mathbf{i}_z = 4\pi\epsilon_0 a^3 \mathbf{E}_p \qquad (5\text{-}33)$$

Equation (5-33) indicates that the dipole moment $\mathbf{p}_e$ is proportional to the field $\mathbf{E}_p$ causing it. Defining a proportionality constant α_e as

$$\alpha_e = 4\pi\epsilon_0 a^3 \qquad (5\text{-}34)$$

we have

$$\mathbf{p}_e = \alpha_e \mathbf{E}_p \qquad (5\text{-}35)$$

The proportionality constant α_e is known as the "electronic polarizability" of the atom. ▐

It is found that the dipole moments due to orientational and ionic polarizations are also proportional to the polarizing field $\mathbf{E}_p$. The average dipole moment $\mathbf{p}$ per molecule is then given by

$$\mathbf{p} = \alpha\mathbf{E}_p \qquad (5\text{-}36)$$

where α is known as the molecular polarizability. Let us now consider a small volume Δv of the dielectric material. If N denotes the number of molecules per unit volume of the material, then there are $N\,\Delta v$ molecules in the volume Δv. We define a vector $\mathbf{P}$, called the "polarization vector," as

$$\mathbf{P} = \frac{1}{\Delta v}\sum_{j=1}^{N\,\Delta v}\mathbf{p}_j = N\mathbf{p} \qquad (5\text{-}37)$$

which has the meaning of "dipole moment per unit volume" or the "dipole moment density" in the material. Substituting (5-36) into (5-37), we have

$$\mathbf{P} = N\alpha\mathbf{E}_p \qquad (5\text{-}38)$$

The units of $\mathbf{P}$ are coulombs per square meter.

The field $\mathbf{E}_p$ in (5-36) and hence in (5-38) is the average electric field acting to polarize the individual molecule and is generally called the polarizing field or the local field. It is the average field that would exist in an imaginary cavity created by removing the molecule in question, keeping all the other molecules polarized in their locations. It is not the same as the average macroscopic field **E** at the molecule with all the molecules including the one in question remaining polarized in their locations. It is equal to the field **E** minus the average field produced by the dipole in the imaginary cavity. We have to find this average field to express $\mathbf{E}_p$ in terms of **E** so that **P** can be related to **E**. To determine this field rigorously, we need detailed information about the shape and charge distribution of the molecule. However, we will consider a simple special case of a spherical cavity and obtain the required field in the following example.

EXAMPLE 5-4. Two equal and opposite point charges Q and $-Q$ are situated at $(0, 0, d/2)$ and $(0, 0 - d/2)$, respectively, in cartesian coordinates as shown in Fig. 5.10, forming a dipole of moment $\mathbf{p} = Qd\mathbf{i}_z$. Obtain the average electric field intensity due to the dipole in a spherical volume of radius $a > d/2$ and centered at the origin.

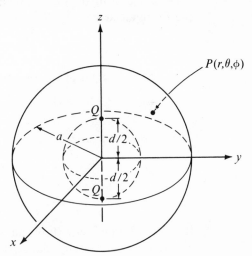

Fig. 5.10. For obtaining the average electric field intensity due to an electric dipole in a spherical volume.

Let us consider the fields due to the positive and negative point charges independently. Considering first the positive charge Q located at $(0, 0, d/2)$, we note that its electric field at an arbitrary point $P(r, \theta, \phi)$ is given by

$$\mathbf{E}_+ = \frac{Q}{4\pi\epsilon_0} \frac{1}{(r^2 + d^2/4 - rd\cos\theta)}\mathbf{i}_{QP} \qquad (5\text{-}39)$$

where $\mathbf{i}_{QP}$ is the unit vector along the line from the point charge Q to the point P. The volume integral of this field evaluated in the spherical volume V of radius *a* is given by

$$\int_V \mathbf{E}_+ \, dv = -Q \left[\int_V \frac{1}{4\pi\epsilon_0(r^2 + d^2/4 - rd\cos\theta)} \mathbf{i}_{PQ} \, dv \right] \quad (5\text{-}40)$$

where $\mathbf{i}_{PQ} = -\mathbf{i}_{QP}$. The quantity inside the brackets on the right side of (5-40) can be recognized as the electric field intensity produced at the location of the point charge by a volume charge distribution of uniform density 1 C/m³ in the spherical volume *V*. From Gauss' law, this electric field intensity is equal to

$$\frac{1}{\epsilon_0} \left(\frac{\text{charge enclosed within the sphere of radius } d/2}{\text{surface area of the sphere of radius } d/2} \right) \mathbf{i}_z$$

or $(d/6\epsilon_0)\mathbf{i}_z$.

Thus we obtain

$$\int_V \mathbf{E}_+ \, dv = -\frac{Qd}{6\epsilon_0} \mathbf{i}_z \quad (5\text{-}41a)$$

Similarly, the volume integral of the electric field due to the negative charge $-Q$ located at $(0, 0, -d/2)$ evaluated in the spherical volume *V* of radius *a* can be obtained as

$$\int_V \mathbf{E}_- \, dv = -\frac{Qd}{6\epsilon_0} \mathbf{i}_z \quad (5\text{-}41b)$$

The volume integral of the electric field due to the dipole is then given by

$$\int_V (\mathbf{E}_+ + \mathbf{E}_-) \, dv = -\frac{Qd}{3\epsilon_0} \mathbf{i}_z \quad (5\text{-}42)$$

Finally, the average field due to the dipole in the spherical volume is given by

$$\mathbf{E}_{av} = \frac{1}{V} \int_V (\mathbf{E}_+ + \mathbf{E}_-) \, dv$$

$$= \frac{1}{\frac{4}{3}\pi a^3} \left(-\frac{Qd}{3\epsilon_0} \mathbf{i}_z \right) = -\frac{\mathbf{p}}{4\pi\epsilon_0 a^3} \quad (5\text{-}43)$$

It is left as an exercise (Problem 5.16) for the student to show that (5-43) is true for any arbitrary charge distribution of dipole moment **p** situated in the spherical volume of radius *a*. ∎

From the result (5-43) of Example 5-4, we now relate the polarizing field $\mathbf{E}_p$ with the average macroscopic field **E** as

$$\mathbf{E}_p = \mathbf{E} - \mathbf{E}_{av} = \mathbf{E} - \left(\frac{-\mathbf{p}}{4\pi\epsilon_0 a^3} \right) = \mathbf{E} + \frac{\mathbf{P}}{3(\frac{4}{3}\pi a^3)N\epsilon_0} \quad (5\text{-}44)$$

where we have substituted $\mathbf{p} = \mathbf{P}/N$ from (5-37). Now, if we assume that the molecular volume is equal to the volume of the spherical cavity, then $(\frac{4}{3}\pi a^3)N$ is equal to 1 since *N* is the number of molecules per unit volume.

Equation (5-44) then reduces to

$$\mathbf{E}_p = \mathbf{E} + \frac{\mathbf{P}}{3\epsilon_0} \tag{5-45}$$

Although we have obtained (5-45) by making certain simplifying assumptions, it is found that the experimentally observed behavior of many dielectric materials agrees remarkably well with that following from (5-45). Substituting (5-45) into (5-38), we obtain

$$\mathbf{P} = N\alpha\left(\mathbf{E} + \frac{\mathbf{P}}{3\epsilon_0}\right) \tag{5-46}$$

Rearranging (5-46), we obtain the relationship between $\mathbf{P}$ and $\mathbf{E}$ as

$$\mathbf{P} = \frac{3\alpha N}{3\epsilon_0 - \alpha N}\,\epsilon_0\mathbf{E} \tag{5-47}$$

Defining a dimensionless parameter χ_e, known as the "electric susceptibility," as

$$\chi_e = \frac{3\alpha N}{3\epsilon_0 - \alpha N} \tag{5-48}$$

Eq. (5-47) can be written as

$$\mathbf{P} = \epsilon_0\chi_e\mathbf{E} \tag{5-49}$$

This simple relationship between the polarization vector $\mathbf{P}$ and the average macroscopic electric field $\mathbf{E}$ in the dielectric indicates that $\mathbf{P}$ is proportional to $\mathbf{E}$. Materials for which this relationship holds, that is, χ_e is independent of the magnitude as well as the direction of $\mathbf{E}$ are known as linear isotropic dielectric materials. For certain dielectric materials, each component of $\mathbf{P}$ can be dependent on all components of $\mathbf{E}$. In such cases, $\mathbf{P}$ is not parallel to $\mathbf{E}$ and the materials are not isotropic. Such materials are known as anisotropic dielectric materials.

5.5   Dielectrics in Electric Fields; Polarization Charge and Current

In Section 5.4 we learned that polarization occurs in dielectric materials under the influence of an applied electric field. We defined polarization by means of a polarization vector $\mathbf{P}$, which is the electric dipole moment per unit volume. The polarization vector is related to the electric field responsible for producing it, through Eq. (5-49). When a dielectric material is placed in an electric field, the induced polarization produces a secondary electric field, which reduces the applied field, which in turn causes a change in the polarization vector, and so on. When this adjustment process is complete, that is, when a steady state is reached, the sum of the originally applied field and the secondary field must be such that it produces a polarization which results in the secondary field. The situation is like a feedback loop as shown in Fig. 5.11. We will assume that the adjustment takes place instantaneously

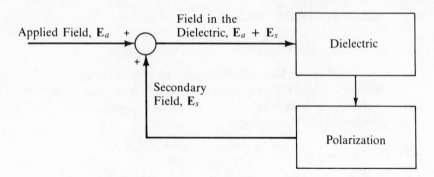

Fig. 5.11. Feedback loop illustrating the adjustment of polarization in a dielectric material to correspond to the sum of the applied field and the secondary field due to the polarization.

with the application of the field and investigate the different effects arising from the polarization. We do this by first considering some specific examples.

EXAMPLE 5-5.   An infinite plane dielectric slab of uniform electric susceptibility χ_{e0} and of thickness d occupies the region $0 < z < d$ as shown in Fig. 5.12(a). A uniform electric field $\mathbf{E}_a = E_0\mathbf{i}_z$ is applied. It is desired to investigate the effect of polarization induced in the dielectric.

The applied electric field induces dipole moments in the dielectric with the negative charges separated from the positive charges and pulled away

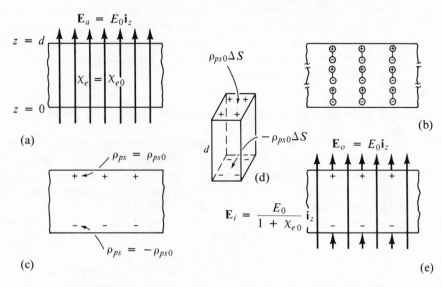

Fig. 5.12. For investigating the effects of polarization induced in a dielectric material of uniform susceptibility for a uniform applied electric field.

from the direction of the field. Since the electric field and the electric susceptibility are uniform, the density of the induced dipole moments, that is, the polarization vector **P**, is uniform as shown in Fig. 5.12(b). Such a distribution results in exact neutralization of all the charges except at the boundaries of the dielectric since, for each positive (or negative) charge not on the surface, there is the same amount of negative (or positive) charge associated with the dipole adjacent to it, thereby cancelling its effect. On the other hand, since the medium changes abruptly from dielectric to free space at the boundaries, no such neutralization of charges at the boundaries takes place. Thus the net result is the formation of a positive surface charge at the boundary $z = d$ and a negative surface charge at the boundary $z = 0$ as shown in Fig. 5.12(c). These surface charges are known as polarization surface charges since they are due to the polarization in the dielectric. In view of the uniform density of the dipole moments, the surface charge densities are uniform. Also, in the absence of a net charge in the interior of the dielectric, the surface charge densities must be equal in magnitude to preserve the charge neutrality of the dielectric.

Let us therefore denote the surface charge densities as

$$\rho_{ps} = \begin{cases} \rho_{ps0} & z = d \\ -\rho_{ps0} & z = 0 \end{cases} \tag{5-50}$$

where the subscript p in addition to the other subscripts stands for polarization. If we now consider a vertical column of infinitesimal rectangular cross-sectional area ΔS cut out from the dielectric as shown in Fig. 5.12(d), the equal and opposite surface charges make the column appear as a dipole of moment $(\rho_{ps0} \Delta S) d\mathbf{i}_z$. On the other hand, writing

$$\mathbf{P} = P_0 \mathbf{i}_z \tag{5-51}$$

where P_0 is a constant in view of the uniformity of the induced polarization, the dipole moment of the column is equal to **P** times the volume of the column, or $P_0(d \Delta S) \mathbf{i}_z$. Equating the dipole moments computed in the two different ways, we have

$$\rho_{ps0} = P_0 \tag{5-52}$$

Thus we have related the surface charge density to the magnitude of the polarization vector. Now, the surface charge distribution produces a secondary field $\mathbf{E}_s$ given by

$$\mathbf{E}_s = \begin{cases} -\dfrac{\rho_{ps0}}{\epsilon_0} \mathbf{i}_z = -\dfrac{P_0}{\epsilon_0} \mathbf{i}_z & \text{for } 0 < z < d \\ 0 & \text{otherwise} \end{cases} \tag{5-53}$$

When the secondary field $\mathbf{E}_s$ is superimposed on the applied field the net result is a reduction of the field inside the dielectric. Denoting the total field inside the dielectric as $\mathbf{E}_i$, we have

$$\mathbf{E}_i = \mathbf{E}_a + \mathbf{E}_s = E_0 \mathbf{i}_z - \frac{P_0}{\epsilon_0} \mathbf{i}_z = \left(E_0 - \frac{P_0}{\epsilon_0} \right) \mathbf{i}_z \qquad (5\text{-}54)$$

But, from (5-49),

$$\mathbf{P} = \epsilon_0 \chi_{e0} \mathbf{E}_i \qquad (5\text{-}55)$$

Substituting (5-51) and (5-54) into (5-55), we have

$$P_0 = \epsilon_0 \chi_{e0} \left(E_0 - \frac{P_0}{\epsilon_0} \right)$$

or

$$P_0 = \frac{\epsilon_0 \chi_{e0} E_0}{1 + \chi_{e0}} \qquad (5\text{-}56)$$

Thus the polarization surface charge densities are given by

$$\rho_{ps} = \begin{cases} \dfrac{\epsilon_0 \chi_{e0} E_0}{1 + \chi_{e0}} & z = d \\[2mm] -\dfrac{\epsilon_0 \chi_{e0} E_0}{1 + \chi_{e0}} & z = 0 \end{cases} \qquad (5\text{-}57)$$

and the electric field intensity inside the dielectric is

$$\mathbf{E}_i = \frac{E_0}{1 + \chi_{e0}} \mathbf{i}_z \qquad (5\text{-}58)$$

Since the secondary field produced outside the dielectric by the surface charge distribution is zero, the total field $\mathbf{E}_o$ outside the dielectric remains the same as the applied field. The field distribution both inside and outside the dielectric is shown in Fig. 5.12(e). Although we have demonstrated only the formation of a polarization surface charge in this example, it is easy to visualize that a nonuniform applied electric field or a nonuniform electric susceptibility of the material will result in the formation of a polarization volume charge in the dielectric due to imperfect cancellation of the charges associated with the dipoles. ∎

EXAMPLE 5-6.   An infinite plane dielectric slab of uniform electric susceptibility χ_{e0} and of thickness d occupies the region $0 < z < d$. A spatially uniform but time-varying electric field $\mathbf{E} = E_0 \cos \omega t \, \mathbf{i}_z$ is applied. It is desired to investigate the effect of polarization induced in the dielectric. Assume that the induced polarization follows exactly the time variations of the applied field.

Since the applied field and the electric susceptibility of the dielectric are spatially uniform, the induced polarization is such that only surface charges of equal and opposite density are formed at the boundaries of the dielectric, and no volume charge is formed inside the dielectric. At any particular time, the surface charge densities are given by (5-57), with the value of the applied field at that time substituted for E_0. Thus the time-varying surface charge densities are

$$\rho_{ps}(t) = \begin{cases} \dfrac{\epsilon_0 \chi_{e0} E_0}{1 + \chi_{e0}} \cos \omega t & z = d \\[2ex] -\dfrac{\epsilon_0 \chi_{e0} E_0}{1 + \chi_{e0}} \cos \omega t & z = 0 \end{cases} \tag{5-59}$$

But if the charge in a volume is varying with time, there must be a current flow out of or into that volume in accordance with the continuity equation, given in integral form by

$$\oint_S \mathbf{J} \cdot d\mathbf{S} + \frac{d}{dt} \int_V \rho \, dv = 0 \tag{4-103}$$

where S is the surface bounding the volume V. Obviously, in the present case the current flow must be inside the dielectric from one boundary to the other. This current is known as the polarization current since it is due to the polarization in the dielectric. For this example, the polarization current density must be entirely z-directed because of the uniformity of the polarization surface charge distributions and it must be uniform since the polarization volume charge density inside the dielectric is zero.

Let us therefore denote the polarization current density as

$$\mathbf{J}_p = J_{p0} \mathbf{i}_z \qquad 0 < z < d \tag{5-60}$$

where the subscript p stands for polarization. To find J_{p0} we apply (4-103) to a rectangular box enclosing an infinitesimal area ΔS of the surface $z = 0$ and parallel to it as shown in Fig. 5.13. Noting that the current outside the dielectric slab and the volume charge inside the slab are zero, we obtain

$$J_{p0} \, \Delta S + \frac{d}{dt} \{[\rho_{ps}]_{z=0} \Delta S\} = 0$$

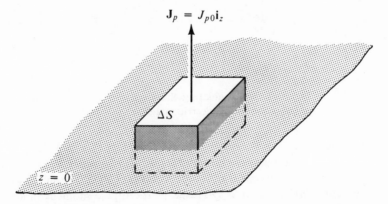

Fig. 5.13. For the determination of the polarization current density resulting from the time variation of the polarization charges induced in a dielectric material.

Thus

$$J_{p0} = -\frac{d}{dt}[\rho_{ps}]_{z=0} = -\frac{d}{dt}\left(-\frac{\epsilon_0 \chi_{e0} E_0}{1 + \chi_{e0}} \cos \omega t\right) = -\frac{\epsilon_0 \chi_{e0} E_0 \omega}{1 + \chi_{e0}} \sin \omega t$$

and

$$\mathbf{J}_p = -\frac{\epsilon_0 \chi_{e0} E_0 \omega}{1 + \chi_{e0}} \sin \omega t \, \mathbf{i}_z \qquad 0 < z < d \qquad (5\text{-}61)$$

It is left as an exercise for the student to verify that the same result is obtained for $\mathbf{J}_p$ by applying (4-103) to a rectangular box enclosing an infinitesimal area ΔS of the surface $z = d$ and parallel to it. Note that the polarization current density is out of phase by 90° with the applied electric field. ∎

We now derive general expressions for polarization surface and volume charge densities and polarization current density in terms of the polarization vector. To do this, let us consider a dielectric material of volume V' in which the polarization vector $\mathbf{P}$ is an arbitrary function of position as shown in Fig. 5.14. We divide the volume V' into a number of infinitesimal volumes dv'_i,

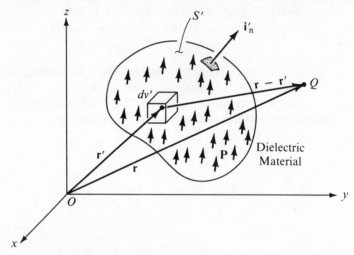

Fig. 5.14. For evaluating the electric potential due to induced polarization in a dielectric material.

$i = 1, 2, 3, \ldots, n$ defined by position vectors $\mathbf{r}'_i$, $i = 1, 2, 3, \ldots, n$, respectively. In each infinitesimal volume, we can consider $\mathbf{P}$ to be a constant so that the dipole moment in the ith volume is $\mathbf{P}_i \, dv'_i$. From (2-109), the scalar potential dV_i at a point $Q(\mathbf{r})$ due to the dipole moment in the ith volume is given by

$$dV_i = \frac{1}{4\pi\epsilon_0} \frac{\mathbf{P}_i \, dv'_i \cdot (\mathbf{r} - \mathbf{r}'_i)}{|\mathbf{r} - \mathbf{r}'_i|^3}$$

The total potential at $Q(\mathbf{r})$ due to the dipole moments in all the n infinitesimal volumes is then given by

$$V = \sum_{i=1}^{n} dV_i = \frac{1}{4\pi\epsilon_0} \sum_{i=1}^{n} \frac{\mathbf{P}_i \, dv'_i \cdot (\mathbf{r} - \mathbf{r}'_i)}{|\mathbf{r} - \mathbf{r}'_i|^3} \qquad (5\text{-}62)$$

Equation (5-62) is good only for $|\mathbf{r}| \gg |\mathbf{r}'_i|$, where $i = 1, 2, 3, \ldots, n$ since each dv'_i has a finite although infinitesimal volume. However, in the limit that $n \longrightarrow \infty$, all the infinitesimal volumes tend to zero; the right side of (5-62) becomes an integral and the expression is valid for any $\mathbf{r}$. Thus

$$V(\mathbf{r}) = \frac{1}{4\pi\epsilon_0} \int_{\text{volume } V'} \frac{\mathbf{P} \, dv' \cdot (\mathbf{r} - \mathbf{r}')}{|\mathbf{r} - \mathbf{r}'|^3}$$

$$= \frac{1}{4\pi\epsilon_0} \int_{\text{volume } V'} \mathbf{P} \cdot \mathbf{\nabla}' \frac{1}{|\mathbf{r} - \mathbf{r}'|} \, dv' \qquad (5\text{-}63)$$

Substituting the vector identity

$$\mathbf{\nabla}' \cdot \frac{\mathbf{P}}{|\mathbf{r} - \mathbf{r}'|} = \mathbf{P} \cdot \mathbf{\nabla}' \frac{1}{|\mathbf{r} - \mathbf{r}'|} + \frac{1}{|\mathbf{r} - \mathbf{r}'|} \mathbf{\nabla}' \cdot \mathbf{P}$$

in (5-63), we obtain

$$V(\mathbf{r}) = \frac{1}{4\pi\epsilon_0} \int_{\text{volume } V'} \mathbf{\nabla}' \cdot \frac{\mathbf{P}}{|\mathbf{r} - \mathbf{r}'|} \, dv' - \frac{1}{4\pi\epsilon_0} \int_{\text{volume } V'} \frac{1}{|\mathbf{r} - \mathbf{r}'|} \mathbf{\nabla}' \cdot \mathbf{P} \, dv' \qquad (5\text{-}64)$$

Applying the divergence theorem to the first integral on the right side of (5-64), we get

$$V(\mathbf{r}) = \frac{1}{4\pi\epsilon_0} \int_{\text{surface } S'} \frac{\mathbf{P} \cdot \mathbf{i}'_n}{|\mathbf{r} - \mathbf{r}'|} \, dS' + \frac{1}{4\pi\epsilon_0} \int_{\text{volume } V'} \frac{-\mathbf{\nabla}' \cdot \mathbf{P}}{|\mathbf{r} - \mathbf{r}'|} \, dv' \qquad (5\text{-}65)$$

where S' is the surface bounding the volume V' and $\mathbf{i}'_n$ is the unit normal vector to dS'.

The first integral on the right side of (5-65) represents the potential at $Q(\mathbf{r})$ due to a surface charge of density $\mathbf{P} \cdot \mathbf{i}'_n$ on the surface S' and the second integral is the potential at $Q(\mathbf{r})$ due to a volume charge of density $(-\mathbf{\nabla}' \cdot \mathbf{P})$ in the volume V'. Thus the potential at $Q(\mathbf{r})$ due to the polarization in the dielectric is the same as the sum of the potentials at $Q(\mathbf{r})$ due to a polarization surface charge of density

$$\rho_{ps}(\mathbf{r}') = \mathbf{P}(\mathbf{r}') \cdot \mathbf{i}'_n \qquad \text{on } S' \qquad (5\text{-}66a)$$

and due to a polarization volume charge of density

$$\rho_p(\mathbf{r}') = -\mathbf{\nabla}' \cdot \mathbf{P}(\mathbf{r}') \qquad \text{in } V' \qquad (5\text{-}66b)$$

We note that the total charge in V' is

$$\oint_{S'} \rho_{ps} \, dS' + \int_{V'} \rho_p \, dv' = \oint_{S'} (\mathbf{P} \cdot \mathbf{i}_n) \, dS' - \int_{V'} (\mathbf{\nabla}' \cdot \mathbf{P}) \, dv' = 0$$

according to the divergence theorem, so that the charge neutrality of the dielectric is satisfied. Thus the total polarization volume charge in V' is equal to the negative of the total polarization surface charge on S'. Omitting the primes in (5-66a) and (5-66b), we have

$$\rho_{ps} = \mathbf{P} \cdot \mathbf{i}_n \tag{5-67}$$

$$\rho_p = -\nabla \cdot \mathbf{P} \tag{5-68}$$

Now, the polarization current density $\mathbf{J}_p$ in the dielectric due to the time variation of the polarization charge density should satisfy the continuity equation

$$\nabla \cdot \mathbf{J}_p + \frac{\partial \rho_p}{\partial t} = 0 \tag{5-69}$$

Substituting for ρ_p in (5-69) from (5-68), we have

$$\nabla \cdot \mathbf{J}_p - \frac{\partial}{\partial t}(\nabla \cdot \mathbf{P}) = 0$$

or

$$\nabla \cdot \left(\mathbf{J}_p - \frac{\partial \mathbf{P}}{\partial t} \right) = 0$$

or

$$\mathbf{J}_p - \frac{\partial \mathbf{P}}{\partial t} = \text{constant with time} \tag{5-70}$$

The constant must, however, be zero since we know that $\mathbf{J}_p$ is zero when $\partial \mathbf{P}/\partial t$ is zero. Thus

$$\mathbf{J}_p = \frac{\partial \mathbf{P}}{\partial t} \tag{5-71}$$

Summarizing what we have learned in this section, the induced dipole moments due to polarization in a dielectric material placed in an electric field have the effect of creating in general the following:

(a) polarization surface charges, having densities given by (5-67), at the boundaries of the dielectric,

(b) polarization volume charge of density given by (5-68) in the dielectric and such that the total volume charge is exactly the negative of the total surface charge so as to preserve the charge neutrality of the material, and

(c) polarization current of density given by (5-71) resulting from the time variation of the polarization charges.

We have also shown that the polarization charges and currents alter the applied electric field in the material. Such a modification of the applied field occurs outside the material as well in the general case. The magnetic field associated with the applied electric field is also altered by the addition of the secondary magnetic field due to the polarization current and the time-variation of the secondary electric field.

5.6 Displacement Flux Density and Relative Permittivity

In Section 5.5 we learned that the electric field in a dielectric material is the superposition of an applied field $\mathbf{E}_a$ and a secondary field $\mathbf{E}_s$ which results from the polarization $\mathbf{P}$, which in turn is induced by the total field $(\mathbf{E}_a + \mathbf{E}_s)$, as shown in Fig. 5.11. Thus, from Fig. 5.11 and Eq. (5-49), we have

$$\mathbf{P} = \epsilon_0 \chi_e (\mathbf{E}_a + \mathbf{E}_s) \tag{5-72}$$

$$\mathbf{E}_s = f(\mathbf{P}) \tag{5-73}$$

where $f(\mathbf{P})$ denotes a function of $\mathbf{P}$. Determination of the secondary field $\mathbf{E}_s$ and hence the total field $(\mathbf{E}_a + \mathbf{E}_s)$ for a given applied field $\mathbf{E}_a$ requires a simultaneous solution of (5-72) and (5-73) which, in general, is very inconvenient. To circumvent this problem, we make use of the results of Section 5.5, in which we found that the induced polarization is equivalent to a polarization surface charge of density ρ_{ps}, a polarization volume charge of density ρ_p, and a polarization current of density $\mathbf{J}_p$, as given by (5-67), (5-68), and (5-71), respectively. The secondary electric and magnetic fields are the fields produced by these charges and current as if they were situated in free space, in the same way as the charges and currents responsible for the applied electric field and its associated magnetic field.

Thus the secondary electromagnetic field satisfies Maxwell's equations

$$\nabla \cdot \mathbf{E}_s = \frac{\rho_p}{\epsilon_0} \tag{5-74a}$$

$$\nabla \cdot \mathbf{B}_s = 0 \tag{5-74b}$$

$$\nabla \times \mathbf{E}_s = -\frac{\partial \mathbf{B}_s}{\partial t} \tag{5-74c}$$

$$\nabla \times \mathbf{B}_s = \mu_0 \left[\mathbf{J}_p + \frac{\partial}{\partial t}(\epsilon_0 \mathbf{E}_s) \right] \tag{5-74d}$$

where $\mathbf{B}_s$ is the secondary magnetic field. On the other hand, if the "true" charge and current densities responsible for the applied field $\mathbf{E}_a$ with its associated magnetic field $\mathbf{B}_a$ are ρ and $\mathbf{J}$, respectively, we have

$$\nabla \cdot \mathbf{E}_a = \frac{\rho}{\epsilon_0} \tag{5-75a}$$

$$\nabla \cdot \mathbf{B}_a = 0 \tag{5-75b}$$

$$\nabla \times \mathbf{E}_a = -\frac{\partial \mathbf{B}_a}{\partial t} \tag{5-75c}$$

$$\nabla \times \mathbf{B}_a = \mu_0 \left[\mathbf{J} + \frac{\partial}{\partial t}(\epsilon_0 \mathbf{E}_a) \right] \tag{5-75d}$$

Now, adding (5-74a)–(5-74d) to (5-75a)–(5-75d), respectively, we obtain

$$\mathbf{V} \cdot (\mathbf{E}_a + \mathbf{E}_s) = \frac{\rho + \rho_p}{\epsilon_0} \tag{5-76a}$$

$$\mathbf{V} \cdot (\mathbf{B}_a + \mathbf{B}_s) = 0 \tag{5-76b}$$

$$\mathbf{V} \times (\mathbf{E}_a + \mathbf{E}_s) = -\frac{\partial}{\partial t}(\mathbf{B}_a + \mathbf{B}_s) \tag{5-76c}$$

$$\mathbf{V} \times (\mathbf{B}_a + \mathbf{B}_s) = \mu_0 \left\{ \mathbf{J} + \mathbf{J}_p + \frac{\partial}{\partial t}[\epsilon_0(\mathbf{E}_a + \mathbf{E}_s)] \right\} \tag{5-76d}$$

Substituting

$$\mathbf{E} = \mathbf{E}_a + \mathbf{E}_s \tag{5-77a}$$

$$\mathbf{B} = \mathbf{B}_a + \mathbf{B}_s \tag{5-77b}$$

$$\rho_p = -\mathbf{V} \cdot \mathbf{P} \tag{5-68}$$

$$\mathbf{J}_p = \frac{\partial \mathbf{P}}{\partial t} \tag{5-71}$$

in (5-76a)–(5-76d), and rearranging, we obtain

$$\mathbf{V} \cdot (\epsilon_0 \mathbf{E} + \mathbf{P}) = \rho \tag{5-78a}$$

$$\mathbf{V} \cdot \mathbf{B} = 0 \tag{5-78b}$$

$$\mathbf{V} \times \mathbf{E} = -\frac{\partial \mathbf{B}}{\partial t} \tag{5-78c}$$

$$\mathbf{V} \times \mathbf{B} = \mu_0 \left[\mathbf{J} + \frac{\partial}{\partial t}(\epsilon_0 \mathbf{E} + \mathbf{P}) \right] \tag{5-78d}$$

where **E** and **B** are the total fields.

We now define a vector **D**, known as the displacement flux density vector, and given by

$$\mathbf{D} = \epsilon_0 \mathbf{E} + \mathbf{P} \tag{5-79}$$

Note that the units of **D** are the same as those of $\epsilon_0 \mathbf{E}$ and **P**, that is, coulombs per square meter, and hence it is a flux density vector. Substituting (5-79) into (5-78a)–(5-78d), we obtain

$$\mathbf{V} \cdot \mathbf{D} = \rho \tag{5-80}$$

$$\mathbf{V} \cdot \mathbf{B} = 0 \tag{5-81}$$

$$\mathbf{V} \times \mathbf{E} = -\frac{\partial \mathbf{B}}{\partial t} \tag{5-82}$$

$$\mathbf{V} \times \mathbf{B} = \mu_0 \left(\mathbf{J} + \frac{\partial \mathbf{D}}{\partial t} \right) \tag{5-83}$$

Thus the new field **D** results in a set of equations which does not explicitly contain the polarization charge and current densities, unlike the equations (5-76a)–(5-76d).

Substituting for **P** in (5-79) from (5-49), we have

$$\mathbf{D} = \epsilon_0 \mathbf{E} + \epsilon_0 \chi_e \mathbf{E} = \epsilon_0 (1 + \chi_e) \mathbf{E} = \epsilon_0 \epsilon_r \mathbf{E} = \epsilon \mathbf{E} \tag{5-84}$$

where we define

$$\epsilon_r = 1 + \chi_e \tag{5-85}$$

and

$$\epsilon = \epsilon_0 \epsilon_r \tag{5-86}$$

The quantity ϵ_r is known as the relative permittivity or dielectric constant of the dielectric and ϵ is the permittivity of the dielectric. Note that ϵ_r is dimensionless and that (5-84) is true only for linear dielectrics if ϵ is to be treated as a constant for a particular dielectric, whereas (5-79) holds in general. Substituting (5-84) into (5-80)–(5-83), we obtain

$$\nabla \cdot \mathbf{E} = \frac{\rho}{\epsilon} \tag{5-87a}$$

$$\nabla \cdot \mathbf{B} = 0 \tag{5-87b}$$

$$\nabla \times \mathbf{E} = -\frac{\partial \mathbf{B}}{\partial t} \tag{5-87c}$$

$$\nabla \times \mathbf{B} = \mu_0 \left[\mathbf{J} + \frac{\partial}{\partial t} (\epsilon \mathbf{E}) \right] \tag{5-87d}$$

Equations (5-87a)–(5-87d) are the same as Maxwell's equations for free space except that ϵ_0 is replaced by ϵ. Thus the electric and magnetic fields in the presence of a dielectric can be computed in exactly the same manner as for free space except that we have to use ϵ instead of ϵ_0 for permittivity. In fact, if $\chi_e = 0$, $\epsilon_r = 1$ and $\epsilon = \epsilon_0$ so that free space can be considered as a dielectric with $\epsilon = \epsilon_0$, and hence, Eqs. (5-87a)–(5-87d) can be used for free space as well. The permittivity ϵ takes into account the effects of polarization and there is no need to consider them when we use ϵ for ϵ_0, thereby eliminating the necessity for the simultaneous solution of (5-72) and (5-73). In the case of a boundary between two different dielectrics, the appropriate boundary conditions for **D** take into account implicitly the polarization surface charge. We will consider these boundary conditions in Section 5.12. The relative permittivity is an experimentally measurable parameter and its values for several dielectric materials are listed in Table 5.2.

EXAMPLE 5-7.   For the dielectric slab of Example 5-5, find and sketch the direction lines of the displacement flux density and the electric field intensity vectors both inside and outside the dielectric.

From Example 5-5, the electric field intensity inside the dielectric is given by

$$\mathbf{E}_i = \frac{E_0}{1 + \chi_{e0}} \mathbf{i}_z \tag{5-58}$$

TABLE 5.2. Relative Permittivities of Some Materials

Material	Relative Permittivity	Material	Relative Permittivity
Air	1.0006	Dry earth	5
Paper	2–3	Glass	5–10
Rubber	2–3.5	Mica	6
Teflon	2.1	Porcelain	6
Polyethylene	2.26	Neoprene	6.7
Polystyrene	2.56	Wet earth	10
Plexiglass	2.6–3.5	Ethyl alchohol	24.3
Nylon	3.5	Glycerol	42.5
Fused quartz	3.8	Distilled water	81
Bakelite	4.9	Titanium dioxide	100

The relative permittivity of the dielectric is $1 + \chi_{e0}$. Thus the displacement flux density inside the dielectric is

$$\mathbf{D}_i = \epsilon_0(1 + \chi_{e0})\mathbf{E}_i = \frac{\epsilon_0(1 + \chi_{e0})}{1 + \chi_{e0}} E_0\mathbf{i}_z = \epsilon_0 E_0\mathbf{i}_z$$

Outside the dielectric, the electric field intensity is the same as the applied value so that the displacement flux density is

$$\mathbf{D}_o = \epsilon_0\mathbf{E}_a = \epsilon_0 E_0\mathbf{i}_z$$

Thus, for this example, the displacement flux density vectors inside and outside the dielectric are the same and equal to the displacement flux density associated with the applied electric field intensity. Both $\mathbf{D}$ and $\mathbf{E}$ fields inside and outside the dielectric are shown in Fig. 5.15. We note that the direction lines of $\mathbf{D}$ do not begin or end on the polarization charges whereas the direction lines of $\mathbf{E}$ begin and end on them. The direction lines of $\mathbf{D}$ begin and

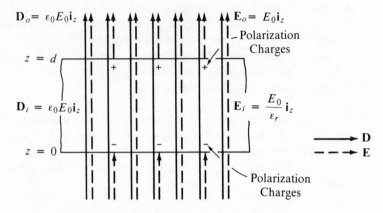

Fig. 5.15. Displacement flux density and electric field intensity vectors for the dielectric slab of Example 5-5.

end only on the charges other than the polarization charges whereas the direction lines of **E** begin and end on both kinds of charges. ∎

EXAMPLE 5-8. A point charge Q is situated at the center of a spherical dielectric shell of uniform permittivity ϵ and having inner and outer radii a and b, respectively, as shown in Fig. 5.16. The entire arrangement is enclosed by a grounded conducting shell of inner radius c and concentric with the dielectric shell. Find and sketch the **D** and **E** fields in three different regions: $0 < r < a$, $a < r < b$, and $b < r < c$. Also find and sketch the **P** field and the polarization charges in the dielectric and the charge induced on the conductor surface.

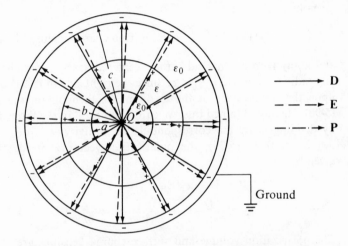

Fig. 5.16. Displacement flux density, electric field intensity, and polarization vectors for the arrangement of a point charge at the center of a spherical dielectric shell enclosed by a grounded spherical conductor concentric with the dielectric shell.

We make use of the spherical symmetry associated with the problem and apply the integral form of (5-87a) given by

$$\oint_S \mathbf{E} \cdot d\mathbf{S} = \frac{1}{\epsilon} \int_V \rho \, dv = \frac{1}{\epsilon} (\text{true charge enclosed by } S) \quad (5\text{-}88)$$

to three different spherical surfaces centered at the point charge and lying in the three different regions. Thus we obtain

$$\mathbf{E} = \begin{cases} \dfrac{Q}{4\pi\epsilon_0 r^2} \mathbf{i}_r & 0 < r < a \\[2ex] \dfrac{Q}{4\pi\epsilon r^2} \mathbf{i}_r & a < r < b \\[2ex] \dfrac{Q}{4\pi\epsilon_0 r^2} \mathbf{i}_r & b < r < c \end{cases} \quad (5\text{-}89)$$

The corresponding **D** field is given by

$$
\mathbf{D} = \begin{cases} \epsilon_0\mathbf{E} = \dfrac{Q}{4\pi r^2}\,\mathbf{i}_r, & 0 < r < a \\[2mm] \epsilon\mathbf{E} = \dfrac{Q}{4\pi r^2}\,\mathbf{i}_r, & a < r < b \\[2mm] \epsilon_0\mathbf{E} = \dfrac{Q}{4\pi r^2}\,\mathbf{i}_r, & b < r < c \end{cases}
$$

$$
= \frac{Q}{4\pi r^2}\,\mathbf{i}_r, \qquad 0 < r < c \tag{5-90}
$$

Alternatively and more elegantly, we can use the integral form of (5-80) given by

$$
\oint_S \mathbf{D} \cdot d\mathbf{S} = \int_V \rho\, dv = \text{(true charge enclosed by } S) \tag{5-91}
$$

and apply it to a spherical surface centered at the point charge and having any radius r, where $0 < r < c$. Since the right side of (5-91) does not depend upon the permittivity of the medium, we then obtain the result given by (5-90). Having obtained this, we can then find **E** in the three different regions by dividing **D** by the corresponding permittivity.

Now, from (5-79), the polarization vector **P** inside the dielectric is given by

$$
\mathbf{P} = \mathbf{D} - \epsilon_0[\mathbf{E}]_{a<r<b}
$$

$$
= \frac{Q}{4\pi r^2}\,\mathbf{i}_r - \epsilon_0\frac{Q}{4\pi\epsilon r^2}\,\mathbf{i}_r = \frac{Q}{4\pi r^2}\left(1 - \frac{\epsilon_0}{\epsilon}\right)\mathbf{i}_r \tag{5-92}
$$

The polarization volume and surface charge densities are

$$
\rho_p = -\nabla\cdot\mathbf{P} = -\frac{1}{r^2}\frac{\partial}{\partial r}(r^2 P_r)
$$

$$
= -\frac{1}{r^2}\frac{\partial}{\partial r}\left[\frac{Q}{4\pi}\left(1 - \frac{\epsilon_0}{\epsilon}\right)\right] = 0 \tag{5-93a}
$$

$$
[\rho_{ps}]_{r=a} = [\mathbf{P}]_{r=a}\cdot(-\mathbf{i}_r)
$$

$$
= \left[\frac{Q}{4\pi a^2}\left(1 - \frac{\epsilon_0}{\epsilon}\right)\mathbf{i}_r\right]\cdot(-\mathbf{i}_r) = -\frac{Q}{4\pi a^2}\left(1 - \frac{\epsilon_0}{\epsilon}\right) \tag{5-93b}
$$

$$
[\rho_{ps}]_{r=b} = [\mathbf{P}]_{r=b}\cdot(\mathbf{i}_r)
$$

$$
= \left[\frac{Q}{4\pi b^2}\left(1 - \frac{\epsilon_0}{\epsilon}\right)\mathbf{i}_r\right]\cdot(\mathbf{i}_r) = \frac{Q}{4\pi b^2}\left(1 - \frac{\epsilon_0}{\epsilon}\right) \tag{5-93c}
$$

The **D** and **E** fields in the three regions, the **P** field in the dielectric, and the polarization surface charge densities are shown in Fig. 5.16. From (5-26), the surface charge density induced on the conductor surface $r = c$ is given by

$$
[\rho_s]_{r=c} = \epsilon_0[\mathbf{E}]_{r=c}\cdot(-\mathbf{i}_r) = -\epsilon_0[E_r]_{r=c} = -\frac{Q}{4\pi c^2} \tag{5-94}
$$

so that the total charge induced on the conductor surface is $-Q$. These charges are shown in Fig. 5.16. We can obtain this result alternatively by recalling that **E** inside the conductor is zero. Then $\oint_S \mathbf{E} \cdot d\mathbf{S}$ for any surface S entirely within the conductor must be zero. For this to be true, an amount of charge equal and opposite to the sum of all kinds of charges (polarization or otherwise) enclosed by the conductor must be induced on the conductor surface. Since the sum of all kinds of charges enclosed by the conductor is

$$Q + [\rho_{ps}]_{r=a} 4\pi a^2 + [\rho_{ps}]_{r=b} 4\pi b^2 = Q$$

the induced charge on the conductor surface must be $-Q$. Alternatively and more elegantly, we note that $\mathbf{D} = \epsilon_0 \mathbf{E}$ is zero inside the conductor. Hence $\oint_S \mathbf{D} \cdot d\mathbf{S}$ for any surface S entirely within the conductor must be zero. For this to be true, an amount of charge equal and opposite to all charges other than polarization charges, enclosed by the conductor must be induced on the conductor surface. Since the charge, other than polarization charge, enclosed by the conductor is the point charge Q, the induced charge on the conductor surface must be $-Q$. This induced charge required to make the field inside the conductor equal to zero is acquired from the ground.

From Fig. 5.16, we once again note that the direction lines of **E** begin and end on all kinds of charges (polarization or otherwise) whereas the direction lines of **D** begin and end only on charges other than polarization charges. The gaps in the direction lines of **E** resulting from the polarization charges are filled by the direction lines of **P**. The flux of **E** through a spherical surface centered at the point charge varies from medium to medium, depending upon the permittivity of the medium in which the surface lies. On the other hand, the flux of **D** through that surface is always equal to only the true charges, that is, charges other than the polarization charges, enclosed by the surface, irrespective of the permittivities of the media bounded by the surface. Thus there is a displacement flux from the true charges which is independent of the medium as originally discovered by Faraday when he found experimentally that the induced charge on the conductor surface was independent of the medium. However, the vector **D** was introduced later by Maxwell, who called it the "displacement." This explains the name "displacement flux density" for **D**. In Section 4.4 we introduced the concept of displacement current as the time derivative of the flux of $\epsilon_0 \mathbf{E}$. We now recognize that $\epsilon_0 \mathbf{E}$ is simply the displacement flux density in free space and hence the name displacement current, again attributed to Maxwell, for the time derivative of the flux of $\epsilon_0 \mathbf{E}$. It follows that the generalization of the displacement current density of Section 4.5 to dielectric media is $\dfrac{\partial \mathbf{D}}{\partial t} = \dfrac{\partial}{\partial t}(\epsilon_0 \mathbf{E} + \mathbf{P})$, which reduces to $\dfrac{\partial}{\partial t}(\epsilon \mathbf{E})$ for linear dielectrics. ∎

5.7   Magnetization and Magnetic Materials

Thus far in this chapter, we have been concerned with the response of materials to electric fields. We now turn our attention to materials known as magnetic materials which, as the name implies, are classified according to their magnetic behavior. According to a simplified atomic model, the electrons associated with a particular nucleus orbit around the nucleus in circular paths while spinning about themselves. In addition, the nucleus itself has a spin motion associated with it. Since the movement of charge constitutes a current, these orbital and spin motions are equivalent to current loops of atomic dimensions. We learned in Chapter 3 that a circular current loop is the magnetic analog of the electric dipole. Thus each atom can be characterized by a superposition of magnetic dipole moments corresponding to the electron orbital motions, electron spin motions, and the nuclear spin. However, owing to the heavy mass of the nucleus, the angular velocity of the nuclear spin is much smaller than that of an electron spin and hence the equivalent current associated with the nuclear spin is much smaller than the equivalent current associated with an electron spin. The dipole moment due to the nuclear spin can therefore be neglected in comparison with the other two effects. The schematic representations of a magnetic dipole as seen from along its axis and from a point in its plane are shown in Figs. 5.17(a) and 5.17(b), respectively.

In many materials, the net magnetic moment of each atom is zero in the absence of an applied magnetic field. An applied magnetic field has the effect of inducing a net dipole moment or "magnetizing" the material by changing the angular velocities of the electron orbits. This induced "magnetization" is in opposition to the applied field so that there is a net reduction in the magnetic flux density in the material from the applied value. Such materials are said to be "diamagnetic." In fact, "diamagnetism," which is analogous to electronic polarization, is prevalent in all materials. We will illustrate the diamagnetic effect by means of the following example.

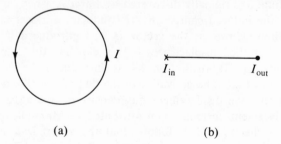

(a)   (b)

Fig. 5.17. Schematic representation of a magnetic dipole: (a) as seen from along its axis, and (b) as seen from a point in its plane.

EXAMPLE 5-9.   Assume that the nucleus of an atom is a point charge equal to $|e|$, where e is the charge of an electron. Consider an electron of mass m_e in a circular orbit of radius a around the nucleus with an angular velocity ω_0 rad/sec. It is desired to find the change in the dipole moment of the orbiting electron due to the application of a uniform external magnetic field perpendicular to the orbital plane of the electron, assuming that the radius of the orbit remains equal to a.

Let the nucleus be at the origin and the electronic orbit be in the xy plane as shown in Fig. 5.18, so that the angular velocity in the absence of the external field is $\pm\omega_0 \mathbf{i}_z$. Let the applied magnetic field be $\mathbf{B}_m = B_0 \mathbf{i}_z$ and the resulting angular velocity be $\pm\omega \mathbf{i}_z$. Under equilibrium conditions, the centripetal force $-m_e\omega^2 a\mathbf{i}_r$ acting on the electron is equal to the sum of two forces: (a) the Coulomb force $\mathbf{F}_1$ due to the attraction of the electron by the nucleus and (b) the magnetic force $\mathbf{F}_2$ due to the applied field acting on the orbiting electron. These forces are given by

$$\mathbf{F}_1 = -\frac{e^2}{4\pi\epsilon_0 a^2}\mathbf{i}_r$$

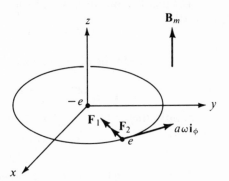

Fig. 5.18. For obtaining the change in the dipole moment of an electronic orbit around the nucleus due to an applied magnetic field.

and

$$\mathbf{F}_2 = \mp|e|\omega a\mathbf{i}_\phi \times B_0\mathbf{i}_z = \mp|e|\omega aB_0\mathbf{i}_r$$

Thus

$$-m_e\omega^2 a\mathbf{i}_r = -\frac{e^2}{4\pi\epsilon_0 a^2}\mathbf{i}_r \mp |e|\omega aB_0\mathbf{i}_r$$

or

$$\omega^2 = \frac{e^2}{4\pi m_e\epsilon_0 a^3} \pm \frac{|e|\omega B_0}{m_e} \tag{5-95}$$

In the absence of the external field, B_0 is zero, $\omega = \omega_0$ and we have

$$\omega_0^2 = \frac{e^2}{4\pi m_e\epsilon_0 a^3} \tag{5-96}$$

Substituting (5-96) into (5-95), we obtain

$$\omega^2 - \omega_0^2 = (\omega + \omega_0)(\omega - \omega_0) = \pm \frac{|e|\omega B_0}{m_e} \tag{5-97}$$

The perturbation in ω_0 by the external field is, however, so small that $\omega + \omega_0$ can be approximated by 2ω. Equation (5-97) then reduces to

$$\omega - \omega_0 \approx \pm \frac{|e|B_0}{2m_e} \tag{5-98}$$

Now, the equivalent current due to an orbiting electron is equal to the amount of charge passing through any point on the orbit in 1 sec, or e times the number of times that the electron passes through the point in 1 sec. For an angular velocity of $\omega\mathbf{i}_z$, the number of times is $\omega/2\pi$ so that the equivalent current is $|e|\omega/2\pi$. This current circulates in the sense opposite to that of the electron orbit since the electronic charge is negative. Thus the magnetic dipole moment due to the orbiting electron is given by

$$\mathbf{m} = \mp \frac{|e|\omega}{2\pi} \pi a^2 \mathbf{i}_z = \mp \frac{|e|\omega a^2}{2} \mathbf{i}_z \tag{5-99}$$

The dipole moment in the absence of the external field is

$$\mathbf{m}_0 = \mp \frac{|e|\omega_0 a^2}{2} \mathbf{i}_z \tag{5-100}$$

The change in the dipole moment due to application of $\mathbf{B}_m$ is

$$\Delta\mathbf{m} = \mathbf{m} - \mathbf{m}_0 = \mp \frac{|e|a^2}{2} (\omega - \omega_0)\mathbf{i}_z \tag{5-101}$$

Substituting (5-98) into (5-101), we obtain

$$\Delta\mathbf{m} = \mp \frac{|e|a^2}{2} \left(\pm \frac{|e|B_0}{2m_e} \right) \mathbf{i}_z = -\frac{e^2 a^2}{4m_e} \mathbf{B}_m \tag{5-102}$$

Thus the change in the dipole moment and hence the magnetic field resulting from the change is in opposition to the applied magnetic field and independent of the sense of the electron orbit. This is consistent with Lenz' law, discussed in Section 4.2, which states that the change in magnetic flux enclosed by a loop induces a current in the loop which opposes the change in the flux. In the present case, the application of the external magnetic field causes the change in flux enclosed by the electron orbit and the induced current is the current corresponding to the change in the angular velocity of the electron. ∎

The result of Example 5-9 illustrates the principle behind the diamagnetic property of materials without going into great detail. The change in the magnetic moment of each electronic orbit brought about by the applied magnetic field results in a net magnetization of the material which otherwise has a zero net moment.

In certain materials, diamagnetism is dominated by other effects known as paramagnetism, ferromagnetism, antiferromagnetism, and ferrimagnetism. Paramagnetism is similar to orientational polarization in dielectric materials. In "paramagnetic" materials, the individual atoms possess net nonzero magnetic moments even in the absence of an applied magnetic field. However, these "permanent" magnetic moments of the individual atoms are randomly oriented so that the net magnetization on a macroscopic scale is zero. An applied magnetic field has the influence of exerting torques on the permanent atomic magnetic dipoles as shown in Figure 5.19, to convert the initially random alignment into a partially coherent one along the field thereby inducing a net magnetization which results in an enhancement of the applied field.

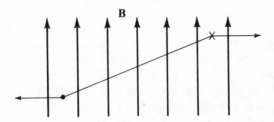

Fig. 5.19. Torque acting on a magnetic dipole under the influence of a magnetic field.

Ferromagnetism is the property by means of which a material can exhibit spontaneous magnetization, that is, magnetization even in the absence of an applied field, below a certain critical temperature known as the Curie temperature. Above the Curie temperature, the spontaneous magnetization vanishes and the ordinary paramagnetic behavior results. Ferromagnetic materials possess strong dipole moments owing to the predominance of the electron spin moments over the electron orbital moments. The theory of ferromagnetism is based on the concept of magnetic "domains," as formulated by Weiss in 1907. A magnetic domain is a small region in the material in which the atomic dipole moments are all aligned in one direction, due to strong interaction fields arising from the neighboring dipoles. In the absence of an external magnetic field, although each domain is magnetized to saturation, the magnetizations in various domains are randomly oriented as shown in Fig. 5.20(a) for a single crystal specimen. The random orientation results from minimization of the associated energy. The net magnetization is therefore zero on a macroscopic scale.

With the application of a weak external magnetic field, the volumes of the domains in which the original magnetizations are favorably oriented relative to the applied field grow at the expense of the volumes of the other domains, as shown in Fig. 5.20(b). This feature is known as domain wall

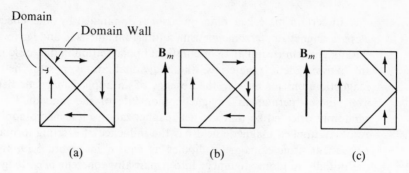

Fig. 5.20. For illustrating the different steps in the magnetization of a ferromagnetic specimen: (a) Unmagnetized state. (b) Domain wall motion. (c) Domain rotation.

motion. Upon removal of the applied field, the domain wall motion reverses, bringing the material close to its original state of magnetization. With the application of stronger external fields, the domain wall motion continues to such an extent that it becomes irreversible; that is, the material does not return to its original unmagnetized state on a macroscopic scale upon removal of the field. With the application of still stronger fields, the domain wall motion is accompanied by domain rotation, that is, alignment of the magnetizations in the individual domains with the applied field as shown in Fig. 5.20(c), thereby magnetizing the material to saturation. The material retains some magnetization along the direction of the applied field even after removal of the field. In fact, an external field opposite to the original direction has to be applied to bring the net magnetization back to zero. The phenomenon by means of which the present state of magnetization of the given material is dependent on its previous magnetic history is known as "hysteresis." We will discuss this topic further in Section 5.9. Unlike in the case of diamagnetic and paramagnetic materials, the magnetization in ferromagnetic materials is nonlinearly related to the applied field.

Antiferromagnetism and ferrimagnetism are modifications of ferromagnetism in materials which contain two interlocking sets of atoms. If the spin moments associated with these two sets of atoms are aligned parallel to each other, as shown in Fig. 5.21(a), the material behaves ferromagnetically. On the other hand, if the spin moments are aligned antiparallel to each other and are equal in magnitude as shown in Fig. 5.21(b), so that the net magnetic moment is zero even under the application of an external field, the material is said to be antiferromagnetic. If the antiparallel moments are unequal in magnitude as shown in Fig. 5.21(c), the net magnetic moment is not zero and the material is said to be ferrimagnetic. A subgroup of ferrimagnetic materials known as "ferrites" is of considerable importance technically because these materials have much lower conductivities than ferromagnetic

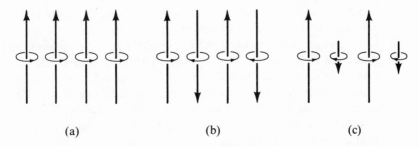

(a)   (b)   (c)

Fig. 5.21. Spin moments associated with interlocking sets of atoms for (a) ferromagnetic, (b) antiferromagnetic, and (c) ferrimagnetic materials.

materials while possessing comparable magnetization properties as ferromagnetic materials.

The net magnetic dipole moment created due to the magnetization of a material by an applied magnetic field produces a field which adds to the applied field (except in the case of materials for which the diamagnetic effect is the only one present) and changes its distribution both inside and outside the material in general from the one that exists in the absence of the material. This will be the topic of discussion in Section 5.8. In the remainder of this section, we will define a new vector $\mathbf{M}$, which represents the magnetization on a macroscopic scale, and relate it to the magnetic flux density. To do this let us consider a small volume Δv of a magnetic material. If N denotes the number of molecules per unit volume of the material, then there are $N \, \Delta v$ molecules in the volume Δv. We define a vector $\mathbf{M}$, called the "magnetization vector" as

$$\mathbf{M} = \frac{1}{\Delta v} \sum_{j=1}^{N\Delta v} \mathbf{m}_j = N\mathbf{m} \qquad (5\text{-}103)$$

where $\mathbf{m}$ is the average magnetic dipole moment per molecule. The magnetization vector $\mathbf{M}$ has the meaning of magnetic "dipole moment per unit volume" analogous to $\mathbf{P}$ in the case of dielectric materials. The units of $\mathbf{M}$ are ampere-meter^2/meter^3 or amperes per meter. We may relate the average dipole moment $\mathbf{m}$ to the magnetizing field $\mathbf{B}_m$ as given by

$$\mathbf{m} = \alpha_m \mathbf{B}_m \qquad (5\text{-}104)$$

where α_m, which may be called the magnetic polarizability, is a constant for linear magnetic materials but may be a function of $\mathbf{B}_m$ for nonlinear magnetic materials. Substituting (5-104) into (5-103), we have

$$\mathbf{M} = N\alpha_m \mathbf{B}_m \qquad (5\text{-}105)$$

The field $\mathbf{B}_m$ is the average magnetic field acting to magnetize the individual molecule and is generally called the local field, analogous to $\mathbf{E}_p$ in the case of dielectric polarization. It is the average field that would exist in an imaginary cavity created by removing the molecule under question while keeping all the other molecules magnetized in their locations. Thus it is not the same as the average macroscopic field **B** at the molecule with all the molecules including the one in question remaining magnetized in their locations. It is equal to the field **B** minus the average field produced by the dipole moment in the imaginary cavity. We have to find this average field to express $\mathbf{B}_m$ in terms of **B** so that **M** can be related to **B**. To do this, we once again consider a simple special case of a spherical cavity and obtain the required field in the following example.

EXAMPLE 5-10.   A circular loop of radius a and centered at the origin lies in the xy plane, as shown in Fig. 5.22. It carries a current I amp in the ϕ direction, thus forming a dipole of moment $\mathbf{m} = I\pi a^2 \mathbf{i}_z$. Obtain the average magnetic flux density due to the dipole in a spherical volume of radius $b > a$ and centered at the origin.

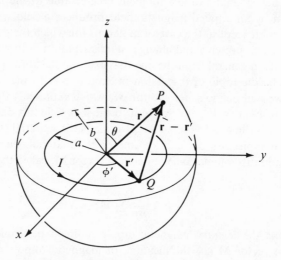

Fig. 5.22. For obtaining the average magnetic flux density due to a magnetic dipole in a spherical volume.

Let us consider an infinitesimal current element $Ia\,d\phi'\,\mathbf{i}_{\phi'}$ at the point $Q(a, \pi/2, \phi')$ on the current loop. The magnetic flux density $d\mathbf{B}$ at a point $P(r, \theta, \phi)$ due to this current element is given by

$$d\mathbf{B} = \frac{\mu_0}{4\pi} \frac{Ia\,d\phi'\,\mathbf{i}_{\phi'} \times (\mathbf{r} - \mathbf{r}')}{|\mathbf{r} - \mathbf{r}'|^3} \tag{5-106}$$

where **r** and **r**′ are position vectors corresponding to P and Q, respectively. The integral of $d\mathbf{B}$ evaluated in the spherical volume V of radius b can be written as

$$\int_V (d\mathbf{B}) \, dv = -\frac{\mu_0 I a \, d\phi' \, \mathbf{i}_{\phi'}}{4\pi} \times \left(-\int_V \frac{\mathbf{r} - \mathbf{r}'}{|\mathbf{r} - \mathbf{r}'|^3} \, dv \right) \qquad (5\text{-}107)$$

since the integration is with respect to the coordinates of the field point P. Now, the integral on the right side of (5-107) can be recognized as the electric field intensity at $(a, \pi/2, \phi')$ due to a volume charge distribution of uniform density $4\pi\epsilon_0$ C/m³ in the spherical volume V. From Gauss' law, this electric field intensity is $(4\pi a/3)(\mathbf{r}'/|\mathbf{r}'|)$. Substituting this result in (5-107), we have

$$\int_V (d\mathbf{B}) \, dv = -\frac{\mu_0 I a \, d\phi' \, \mathbf{i}_{\phi'}}{4\pi} \times \frac{4\pi a}{3} \frac{\mathbf{r}'}{|\mathbf{r}'|}$$

$$= \frac{\mu_0 I a^2}{3} \, d\phi' \, \mathbf{i}_z \qquad (5\text{-}108)$$

The volume integral of $\mathbf{B}$ in the volume V due to the entire current loop is then given by

$$\int_V \mathbf{B} \, dv = \int_{\phi'=0}^{2\pi} \int_V (d\mathbf{B}) \, dv$$

$$= \frac{\mu_0 I a^2}{3} \int_{\phi'=0}^{2\pi} d\phi' \, \mathbf{i}_z = \frac{2\mu_0 I \pi a^2}{3} \, \mathbf{i}_z \qquad (5\text{-}109)$$

Finally, the average field due to the dipole in the spherical volume is given by

$$\mathbf{B}_{av} = \frac{1}{V} \int_V \mathbf{B} \, dv$$

$$= \frac{1}{\frac{4}{3}\pi b^3} \left(\frac{2\mu_0 I \pi a^2}{3} \, \mathbf{i}_z \right) = \frac{\mu_0 \mathbf{m}}{2\pi b^3} \qquad (5\text{-}110)$$

It is left as an exercise (Problem 5.28) for the student to show that (5-110) is true for any arbitrary current distribution of dipole moment $\mathbf{m}$ situated in the spherical volume of radius b. ▐

From the result (5-110) of Example 5-10, we now relate the magnetizing field $\mathbf{B}_m$ with the average macroscopic field $\mathbf{B}$ as

$$\mathbf{B}_m = \mathbf{B} - \mathbf{B}_{av} = \mathbf{B} - \frac{\mu_0 \mathbf{m}}{2\pi b^3} = \mathbf{B} - \frac{\mu_0 \mathbf{M}}{\frac{3}{2}(\frac{4}{3}\pi b^3)N} \qquad (5\text{-}111)$$

where we have substituted $\mathbf{m} = \mathbf{M}/N$ from (5-103). Now, if we assume that the molecular volume is equal to the volume of the spherical cavity, then $(\frac{4}{3}\pi b^3)N$ is equal to 1 so that (5-11) reduces to

$$\mathbf{B}_m = \mathbf{B} - \tfrac{2}{3}\mu_0 \mathbf{M} \qquad (5\text{-}112)$$

Although we have obtained (5-112) by considering a spherical volume for the molecule, it is found that the general expression for $\mathbf{B}_m$ is of the form

$$\mathbf{B}_m = \mathbf{B} + (\gamma - 1)\mu_0 \mathbf{M} \qquad (5\text{-}113)$$

However, γ may be larger than the value $\frac{1}{3}$ in (5-112) by several orders of magnitude for some materials. Substituting (5-105) into (5-113), we obtain

$$\frac{\mathbf{M}}{N\alpha_m} = \mathbf{B} + (\gamma - 1)\mu_0\mathbf{M} \tag{5-114}$$

Rearranging (5-114), we obtain the relationship between $\mathbf{M}$ and $\mathbf{B}_m$ as

$$\mathbf{M} = \frac{N\alpha_m}{1 - (\gamma - 1)\mu_0 N\alpha_m}\mathbf{B} \tag{5-115}$$

Defining a dimensionless parameter χ_m, known as the "magnetic susceptibility," as

$$\chi_m = \frac{\mu_0 N\alpha_m}{1 - \gamma\mu_0 N\alpha_m} \tag{5-116}$$

Eq. (5-115) can be written as

$$\mathbf{M} = \frac{\chi_m}{1 + \chi_m}\frac{\mathbf{B}}{\mu_0} \tag{5-117}$$

We have thus established a simple relationship between the magnetization vector $\mathbf{M}$ and the average macroscopic magnetic field $\mathbf{B}$ in a magnetic material through the parameter χ_m. The parameter χ_m is, however, constant only for diamagnetic and paramagnetic materials and is dependent on $\mathbf{B}$ for ferromagnetic materials. Values of χ_m for some diamagnetic and paramagnetic materials are listed in Table 5.3. Also, comparing (5-117) with (5-49), we

TABLE 5.3. Magnetic Susceptibilities of Some Diamagnetic and Paramagnetic Materials

Diamagnetic Material	χ_m	Paramagnetic Material	χ_m
Nitrogen	-0.50×10^{-8}	Air	3.6×10^{-7}
Hydrogen	-0.21×10^{-8}	Oxygen	2.1×10^{-6}
Gold	-3.60×10^{-5}	Magnesium	1.2×10^{-5}
Mercury	-3.20×10^{-5}	Aluminum	2.3×10^{-5}
Silver	-2.60×10^{-5}	Tungsten	6.8×10^{-5}
Copper	-0.98×10^{-5}	Platinum	2.9×10^{-4}
Sodium	-0.24×10^{-5}	Palladium	8.2×10^{-4}
Bismuth	-1.66×10^{-4}	Liquid oxygen	3.5×10^{-3}

observe that whereas $\mathbf{M}$ and $\mathbf{B}$ are analogous to $\mathbf{P}$ and $\mathbf{E}$, respectively, χ_m is not analogous to χ_e owing to the manner in which χ_m is defined. We will discover the reason for this in Section 5.9. Equation (5-117) indicates that $\mathbf{M}$ is parallel to $\mathbf{B}$. Materials for which this relationship holds, that is, χ_m is independent of the direction of $\mathbf{B}$ are known as isotropic magnetic materials. For certain magnetic materials, each component of $\mathbf{M}$ can be dependent on all components of $\mathbf{B}$. In such cases, $\mathbf{M}$ is not parallel to $\mathbf{B}$ and the materials are not isotropic. Such materials are known as anisotropic magnetic materials.

5.8 Magnetic Materials in Magnetic Fields; Magnetization Current

In Section 5.7 we learned that magnetization occurs in magnetic materials under the influence of an applied magnetic field. We defined magnetization by means of a magnetization vector $\mathbf{M}$, which is the magnetic dipole moment per unit volume. The magnetization vector is related to the magnetic field responsible for producing it through Eq. (5-117). When a magnetic material is placed in a magnetic field, the resulting magnetization produces a secondary magnetic field, which increases the applied field, which in turn causes a change in the magnetization vector, and so on. When this adjustment process is complete, that is, when a steady state is reached, the sum of the originally applied field and the secondary field must be such that it produces a magnetization which results in the secondary field. The situation is like a feedback loop as shown in Fig. 5.23. We will assume that the adjustment takes place instantaneously with the application of the field and investigate the different effects arising from the magnetization. We do this by first considering an example.

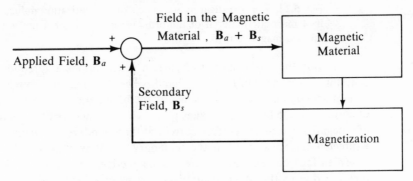

Fig. 5.23. Feedback loop illustrating the adjustment of magnetization in a magnetic material to correspond to the sum of the applied field and the secondary field due to the magnetization.

EXAMPLE 5-11. An infinite plane slab of magnetic material of uniform magnetic susceptibility χ_{m0} and of thickness d occupies the region $0 < z < d$, as shown in Fig. 5.24(a). A uniform magnetic field $\mathbf{B}_a = B_0 \mathbf{i}_x$ is applied. It is desired to investigate the effect of magnetization in the material.

The applied magnetic field results in magnetic dipole moments in the material which are oriented along the field. Since the magnetic field and the magnetic susceptibility are uniform, the density of the dipole moments, that is, the magnetization vector $\mathbf{M}$, is uniform as shown in Fig. 5.24(b). Such a distribution results in exact cancellation of currents everywhere except at the boundaries of the material since, for each current segment not on the surface, there is a current segment associated with the dipole adjacent to it

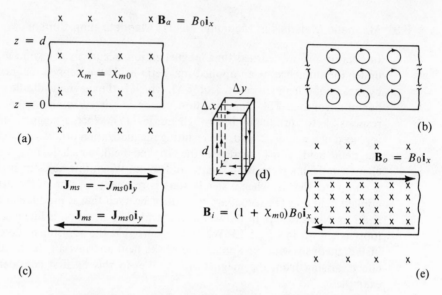

Fig. 5.24. For investigating the effects of magnetization induced in a magnetic material of uniform susceptibility for a uniform applied magnetic field.

and carrying the same amount of current in the opposite direction, thereby cancelling its effect. On the other hand, since the medium changes abruptly from magnetic material to free space at the boundaries, no such cancellation of currents at the boundaries takes place. Thus the net result is the formation of a negative y-directed surface current at the boundary $z = d$ and a positive y-directed surface current at the boundary $z = 0$ as shown in Fig. 5.24(c). These surface currents are known as magnetization surface currents since they are due to the magnetization in the material. In view of the uniform density of the dipole moments, the surface current densities are uniform. Also, in the absence of a net current in the interior of the magnetic material, the surface current densities must be equal in magnitude so that whatever current flows on one surface returns via the other surface.

Let us therefore denote the surface current densities as

$$\mathbf{J}_{ms} = \begin{cases} J_{ms0}\mathbf{i}_y & z = 0 \\ -J_{ms0}\mathbf{i}_y & z = d \end{cases} \tag{5-118}$$

where the subscript m in addition to the other subscripts stands for magnetization. If we now consider a vertical column of infinitesimal rectangular cross-sectional area $\Delta S = (\Delta x)(\Delta y)$ cut out from the magnetic material as shown in Fig. 5-24(d), the rectangular current loop of width Δx makes the column appear as a dipole of moment $(J_{ms0} \Delta x)(d \Delta y)\mathbf{i}_x$. On the other hand, writing

$$\mathbf{M} = M_0\mathbf{i}_x \tag{5-119}$$

where M_0 is a constant in view of the uniformity of the magnetization, the dipole moment of the column is equal to **M** times the volume of the column, or $M_0(d \, \Delta x \, \Delta y)\mathbf{i}_x$. Equating the dipole moments computed in the two different ways, we have

$$J_{ms0} = M_0 \tag{5-120}$$

Thus we have related the surface current density to the magnitude of the magnetization vector. Now, the surface current distribution produces a secondary field $\mathbf{B}_s$ given by

$$\mathbf{B}_s = \begin{cases} \mu_0 J_{ms0}\mathbf{i}_x = \mu_0 M_0\mathbf{i}_x & \text{for } 0 < z < d \\ 0 & \text{otherwise} \end{cases} \tag{5-121}$$

When the secondary field $\mathbf{B}_s$ is superimposed on the applied field, the net result is an increase in the field inside the material. Denoting the total field inside the material by $\mathbf{B}_i$, we have

$$\mathbf{B}_i = \mathbf{B}_a + \mathbf{B}_s = B_0\mathbf{i}_x + \mu_0 M_0\mathbf{i}_x = (B_0 + \mu_0 M_0)\mathbf{i}_x \tag{5-122}$$

But, from (5-117),

$$\mathbf{M} = \frac{\chi_{m0}}{1 + \chi_{m0}} \frac{\mathbf{B}_i}{\mu_0} \tag{5-123}$$

Substituting (5-119) and (5-122) into (5-123), we have

$$M_0 = \frac{\chi_{m0}}{1 + \chi_{m0}} \frac{B_0 + \mu_0 M_0}{\mu_0}$$

or

$$M_0 = \frac{\chi_{m0} B_0}{\mu_0} \tag{5-124}$$

Thus the magnetization surface current densities are given by

$$\mathbf{J}_{ms} = \begin{cases} \dfrac{\chi_{m0} B_0}{\mu_0} \mathbf{i}_y & z = 0 \\[2mm] -\dfrac{\chi_{m0} B_0}{\mu_0} \mathbf{i}_y & z = d \end{cases} \tag{5-125}$$

and the magnetic flux density inside the material is

$$\mathbf{B}_i = (1 + \chi_{m0})B_0\mathbf{i}_x \tag{5-126}$$

Since the secondary field produced outside the material by the surface current distribution is zero, the total field B_o outside the material remains the same as the applied field. The field distribution both inside and outside the magnetic material is shown in Fig. 5-24(e). Although we have demonstrated only the formation of a magnetization surface current in this example, it is easy to visualize that a nonuniform applied magnetic field or a nonuniform magnetic susceptibility of the material will result in the formation of a magnetization volume current in the magnetic material due to imperfect cancellation of the currents associated with the dipoles. ∎

We now derive general expressions for magnetization surface and volume current densities in terms of the magnetization vector. To do this, let us consider a magnetic material of volume V' in which the magnetization vector **M** is an arbitrary function of position, as shown in Fig. 5.25. We divide the

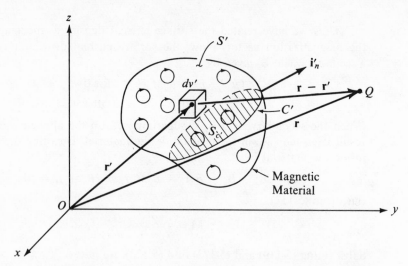

Fig. 5.25. For evaluating the magnetic vector potential due to induced magnetization in a magnetic material.

volume V' into a number of infinitesimal volumes dv'_i, $i = 1, 2, 3, \ldots, n$ defined by position vectors $\mathbf{r}'_i$, $i = 1, 2, 3, \ldots, n$, respectively. In each infinitesimal volume, we can consider **M** to be a constant so that the dipole moment in the ith volume is $\mathbf{M}_i \, dv'_i$. From (3-96), the magnetic vector potential $d\mathbf{A}_i$ at a point $Q(\mathbf{r})$ due to the dipole moment in the ith volume is given by

$$d\mathbf{A}_i = \frac{\mu_0}{4\pi} \frac{\mathbf{M}_i \, dv'_i \times (\mathbf{r} - \mathbf{r}'_i)}{|\mathbf{r} - \mathbf{r}'_i|^3}$$

The total vector potential at $Q(\mathbf{r})$ due to the dipole moments in all the n infinitesimal volumes is then given by

$$\mathbf{A}_i = \sum_{i=1}^{n} d\mathbf{A}_i = \frac{\mu_0}{4\pi} \sum_{i=1}^{n} \frac{\mathbf{M}_i \, dv'_i \times (\mathbf{r} - \mathbf{r}'_i)}{|\mathbf{r} - \mathbf{r}'_i|^3} \tag{5-127}$$

Equation (5-127) is good only for $|\mathbf{r}| \gg |\mathbf{r}'_i|$, where $i = 1, 2, 3, \ldots, n$ since each dv'_i has a finite although infinitesimal volume. However, in the limit that $n \to \infty$, all the infinitesimal volumes tend to zero; the right side of (5-127) becomes an integral and the expression is valid for any **r**. Thus

$$
\begin{aligned}
\mathbf{A}(\mathbf{r}) &= \frac{\mu_0}{4\pi} \int_{V'} \frac{\mathbf{M} \, dv' \times (\mathbf{r} - \mathbf{r}')}{|\mathbf{r} - \mathbf{r}'|^3} \\
&= \frac{\mu_0}{4\pi} \int_{V'} \mathbf{M} \times \nabla' \frac{1}{|\mathbf{r} - \mathbf{r}'|} \, dv'
\end{aligned}
\tag{5-128}
$$

Substituting the vector identity

$$\mathbf{V}' \times \frac{\mathbf{M}}{|\mathbf{r} - \mathbf{r}'|} = \mathbf{V}' \frac{1}{|\mathbf{r} - \mathbf{r}'|} \times \mathbf{M} + \frac{1}{|\mathbf{r} - \mathbf{r}'|} \mathbf{V}' \times \mathbf{M}$$

in (5-128), we obtain

$$\mathbf{A}(\mathbf{r}) = \frac{\mu_0}{4\pi} \int_{V'} \frac{\mathbf{V}' \times \mathbf{M}}{|\mathbf{r} - \mathbf{r}'|} \, dv' - \frac{\mu_0}{4\pi} \int_{V'} \mathbf{V}' \times \frac{\mathbf{M}}{|\mathbf{r} - \mathbf{r}'|} \, dv' \qquad (5\text{-}129)$$

Taking the dot product of the second integral on the right side of (5-129) with the unit vector $\mathbf{i}_x$ and using the divergence theorem, we have

$$\mathbf{i}_x \cdot \int_{V'} \mathbf{V}' \times \frac{\mathbf{M}}{|\mathbf{r} - \mathbf{r}'|} \, dv' = \int_{V'} \mathbf{i}_x \cdot \mathbf{V}' \times \frac{\mathbf{M}}{|\mathbf{r} - \mathbf{r}'|} \, dv'$$

$$= -\int_{V'} \mathbf{V}' \cdot \left(\mathbf{i}_x \times \frac{\mathbf{M}}{|\mathbf{r} - \mathbf{r}'|} \right) dv' \qquad (5\text{-}130)$$

$$= -\oint_{S'} \mathbf{i}_x \times \frac{\mathbf{M}}{|\mathbf{r} - \mathbf{r}'|} \cdot \mathbf{i}_n' \, dS'$$

where S' is the surface bounding the volume V' and $\mathbf{i}_n'$ is the unit normal vector to dS'. Proceeding further, we obtain

$$\mathbf{i}_x \cdot \int_{V'} \mathbf{V}' \times \frac{\mathbf{M}}{|\mathbf{r} - \mathbf{r}'|} \, dv' = -\oint_{S'} \mathbf{i}_x \times \frac{\mathbf{M}}{|\mathbf{r} - \mathbf{r}'|} \cdot \mathbf{i}_n' \, dS'$$

$$= -\oint_{S'} \mathbf{i}_x \cdot \frac{\mathbf{M} \times \mathbf{i}_n'}{|\mathbf{r} - \mathbf{r}'|} \, dS' \qquad (5\text{-}131\text{a})$$

$$= -\mathbf{i}_x \cdot \oint_{S'} \frac{\mathbf{M} \times \mathbf{i}_n'}{|\mathbf{r} - \mathbf{r}'|} \, dS'$$

Similarly, we can show that

$$\mathbf{i}_y \cdot \int_{V'} \mathbf{V}' \times \frac{\mathbf{M}}{|\mathbf{r} - \mathbf{r}'|} \, dv' = -\mathbf{i}_y \cdot \oint_{S'} \frac{\mathbf{M} \times \mathbf{i}_n'}{|\mathbf{r} - \mathbf{r}'|} \, dS' \qquad (5\text{-}131\text{b})$$

and

$$\mathbf{i}_z \cdot \int_{V'} \mathbf{V}' \times \frac{\mathbf{M}}{|\mathbf{r} - \mathbf{r}'|} \, dv' = -\mathbf{i}_z \cdot \oint_{S'} \frac{\mathbf{M} \times \mathbf{i}_n'}{|\mathbf{r} - \mathbf{r}'|} \, dS' \qquad (5\text{-}131\text{c})$$

It then follows from (5-131a)–(5-131c) that

$$\int_{V'} \mathbf{V}' \times \frac{\mathbf{M}}{|\mathbf{r} - \mathbf{r}'|} \, dv' = -\oint_{S'} \frac{\mathbf{M} \times \mathbf{i}_n'}{|\mathbf{r} - \mathbf{r}'|} \, dS' \qquad (5\text{-}132)$$

Substituting (5-132) into (5-129), we get

$$\mathbf{A}(\mathbf{r}) = \frac{\mu_0}{4\pi} \int_{V'} \frac{\mathbf{V}' \times \mathbf{M}}{|\mathbf{r} - \mathbf{r}'|} \, dv' + \frac{\mu_0}{4\pi} \oint_{S'} \frac{\mathbf{M} \times \mathbf{i}_n'}{|\mathbf{r} - \mathbf{r}'|} \, dS' \qquad (5\text{-}133)$$

The first integral on the right side of (5-133) represents the vector potential at $Q(\mathbf{r})$ due to a volume current of density $\mathbf{V}' \times \mathbf{M}$ in the volume V' and the second integral is the vector potential at $Q(\mathbf{r})$ due to a surface current of density $\mathbf{M} \times \mathbf{i}'_n$ on the surface S'. Thus the vector potential at $Q(\mathbf{r})$ due to the magnetization in the magnetic material is the same as the sum of the vector potentials at $Q(\mathbf{r})$ due to a magnetization volume current of density

$$\mathbf{J}_m(\mathbf{r}') = \mathbf{V}' \times \mathbf{M}(\mathbf{r}') \qquad \text{in } V' \tag{5-134}$$

and due to a magnetization surface current of density

$$\mathbf{J}_{ms}(\mathbf{r}') = \mathbf{M}(\mathbf{r}') \times \mathbf{i}'_n \qquad \text{on } S' \tag{5-135}$$

We note that the total volume current through any cross-sectional area $S_{c'}$ (of the volume V') bounded by the contour C' as shown in Fig. 5.25 is given by

$$\int_{S_{c'}} \mathbf{J}_m \cdot d\mathbf{S}_{c'} = \int_{S_{c'}} (\mathbf{V}' \times \mathbf{M}) \cdot d\mathbf{S}_{c'} = \oint_{C'} \mathbf{M} \cdot d\mathbf{l}'$$
$$= -\oint_{C'} (\mathbf{M} \times \mathbf{i}'_n) \cdot (\mathbf{i}'_n \times d\mathbf{l}') = -\oint_{C'} \mathbf{J}_{ms} \cdot (\mathbf{i}'_n \times d\mathbf{l}') \tag{5-136}$$

where we have used Stokes' theorem to transform the surface integration to line integration. The right side of (5-136) is exactly the surface current crossing the contour C' in the opposite direction to the volume current. Omitting the primes in (5-134) and (5-136), we have

$$\mathbf{J}_m = \mathbf{V} \times \mathbf{M} \tag{5-137}$$

$$\mathbf{J}_{ms} = \mathbf{M} \times \mathbf{i}_n \tag{5-138}$$

Summarizing what we have learned in this section, the magnetic dipole moments due to magnetization in a magnetic material placed in a magnetic field have the effect of creating in general the following:

(a) Magnetization surface currents, having densities given by (5-138), at the boundaries of the magnetic material.

(b) Magnetization volume current of density given by (5-137) in the magnetic material and such that the total volume current flowing through any cross-sectional area of the material is exactly opposite to the total surface current crossing the contour bounding the area.

We have also shown that the magnetization currents alter the applied magnetic field in the material. Such a modification of the applied field occurs outside the material as well in the general case. In the time-varying case, the electric field associated with the applied magnetic field is also altered by the addition of the secondary electric field due to the time variation of the secondary magnetic field.

5.9 Magnetic Field Intensity, Relative Permeability, and Hysteresis

In Section 5.8 we learned that the magnetic field in a magnetic material is the superposition of an applied field $\mathbf{B}_a$ and a secondary field $\mathbf{B}_s$ which results from the magnetization $\mathbf{M}$, which in turn is produced by the total field $(\mathbf{B}_a + \mathbf{B}_s)$, as shown in Fig. 5-23. Thus, from Fig. 5-23 and Eq. (5-117), we have

$$\mathbf{M} = \frac{\chi_m}{1 + \chi_m} \frac{\mathbf{B}_a + \mathbf{B}_s}{\mu_0} \tag{5-139}$$

$$\mathbf{B}_s = f(\mathbf{M}) \tag{5-140}$$

where $f(\mathbf{M})$ denotes a function of $\mathbf{M}$. Determination of the secondary field $\mathbf{B}_s$ and hence the total field $\mathbf{B}_a + \mathbf{B}_s$ for a given applied field $\mathbf{B}_a$ requires a simultaneous solution of (5-139) and (5-140) which, in general, is very inconvenient. To circumvent this problem, we make use of the results of Section 5.8, in which we found that the magnetization is equivalent to a magnetization surface current of density $\mathbf{J}_{ms}$ and a magnetization volume current of density $\mathbf{J}_m$ as given by (5-138) and (5-137), respectively. The secondary magnetic and electric fields are the fields produced by these currents as if they were situated in free space, in the same way as the currents responsible for the applied magnetic field and its associated electric field.

Thus the secondary electromagnetic field satisfies Maxwell's equations

$$\nabla \cdot \mathbf{D}_s = 0 \tag{5-141a}$$

$$\nabla \cdot \mathbf{B}_s = 0 \tag{5-141b}$$

$$\nabla \times \mathbf{E}_s = -\frac{\partial \mathbf{B}_s}{\partial t} \tag{5-141c}$$

$$\nabla \times \mathbf{B}_s = \mu_0 \left(\mathbf{J}_m + \frac{\partial \mathbf{D}_s}{\partial t} \right) \tag{5-141d}$$

where $\mathbf{E}_s$ is the secondary electric field intensity and $\mathbf{D}_s$ is its associated displacement flux density. On the other hand, if the "true" current and charge densities responsible for the applied field $\mathbf{B}_a$ with its associated electric field intensity $\mathbf{E}_a$ and displacement flux density $\mathbf{D}_a$ are $\mathbf{J}$ and ρ, respectively, we have

$$\nabla \cdot \mathbf{D}_a = \rho \tag{5-142a}$$

$$\nabla \cdot \mathbf{B}_a = 0 \tag{5-142b}$$

$$\nabla \times \mathbf{E}_a = -\frac{\partial \mathbf{B}_a}{\partial t} \tag{4-142c}$$

$$\nabla \times \mathbf{B}_a = \mu_0 \left(\mathbf{J} + \frac{\partial \mathbf{D}_a}{\partial t} \right) \tag{5-142d}$$

Now, adding (5-141a)–(5-141d) to (5-142a)–(5-142d), respectively, we obtain

$$\nabla \cdot (\mathbf{D}_a + \mathbf{D}_s) = \rho + 0 = \rho \tag{5-143a}$$

$$\nabla \cdot (\mathbf{B}_a + \mathbf{B}_s) = 0 \tag{5-143b}$$

$$\nabla \times (\mathbf{E}_a + \mathbf{E}_s) = -\frac{\partial}{\partial t} (\mathbf{B}_a + \mathbf{B}_s) \tag{5-143c}$$

$$\nabla \times (\mathbf{B}_a + \mathbf{B}_s) = \mu_0 \left[\mathbf{J} + \mathbf{J}_m + \frac{\partial}{\partial t} (\mathbf{D}_a + \mathbf{D}_s) \right] \tag{5-143d}$$

Substituting

$$\mathbf{B} = \mathbf{B}_a + \mathbf{B}_s \qquad (5\text{-}144a)$$

$$\mathbf{E} = \mathbf{E}_a + \mathbf{E}_s \qquad (5\text{-}144b)$$

$$\mathbf{D} = \mathbf{D}_a + \mathbf{D}_s \qquad (5\text{-}145)$$

$$\mathbf{J}_m = \mathbf{\nabla} \times \mathbf{M} \qquad (5\text{-}137)$$

in (5-143a)–(5-143d) and rearranging, we have

$$\mathbf{\nabla} \cdot \mathbf{D} = \rho \qquad (5\text{-}146a)$$

$$\mathbf{\nabla} \cdot \mathbf{B} = 0 \qquad (5\text{-}146b)$$

$$\mathbf{\nabla} \times \mathbf{E} = -\frac{\partial \mathbf{B}}{\partial t} \qquad (5\text{-}146c)$$

$$\mathbf{\nabla} \times \left(\frac{\mathbf{B}}{\mu_0} - \mathbf{M} \right) = \mathbf{J} + \frac{\partial \mathbf{D}}{\partial t} \qquad (5\text{-}146d)$$

where **E**, **B**, and **D** are the total fields.

We now define a vector **H**, known as the magnetic field intensity vector and given by

$$\mathbf{H} = \frac{\mathbf{B}}{\mu_0} - \mathbf{M} \qquad (5\text{-}147)$$

Note that the units of **H** are the same as those of $\mathbf{B}/\mu_0$ and **M**, that is, amperes per meter. Comparing with the units of volts per meter for the electric field intensity, we see the reason for referring to **H** as the magnetic field intensity. Substituting (5-147) into (5-146a)–(5-146d), we obtain

$$\mathbf{\nabla} \cdot \mathbf{D} = \rho \qquad (5\text{-}148)$$

$$\mathbf{\nabla} \cdot \mathbf{B} = 0 \qquad (5\text{-}149)$$

$$\mathbf{\nabla} \times \mathbf{E} = -\frac{\partial \mathbf{B}}{\partial t} \qquad (5\text{-}150)$$

$$\mathbf{\nabla} \times \mathbf{H} = \mathbf{J} + \frac{\partial \mathbf{D}}{\partial t} \qquad (5\text{-}151)$$

Thus the new field **H** results in a set of equations which do not explicitly contain the magnetization current density, unlike Eqs. (5-143a)–(5-143d). Substituting for **M** in (5-147) from (5-117), we have

$$\mathbf{H} = \frac{\mathbf{B}}{\mu_0} - \frac{\chi_m}{1 + \chi_m} \frac{\mathbf{B}}{\mu_0} = \frac{\mathbf{B}}{\mu_0(1 + \chi_m)} = \frac{\mathbf{B}}{\mu_0 \mu_r} = \frac{\mathbf{B}}{\mu} \qquad (5\text{-}152)$$

where we define

$$\mu_r = 1 + \chi_m \qquad (5\text{-}153)$$

and

$$\mu = \mu_0 \mu_r \qquad (5\text{-}154)$$

The quantity μ_r is known as the relative permeability of the magnetic material and μ is the permeability of the magnetic material. Note that μ_r is dimensionless and that (5-152) is true only for linear magnetic materials if μ is to be treated as a constant for a particular magnetic material, whereas (5-147) holds in general. Substituting (5-152) into (5-148)–(5-151), we obtain

$$\mathbf{V} \cdot \mathbf{D} = \rho \tag{5-155a}$$

$$\mathbf{V} \cdot \mathbf{B} = 0 \tag{5-155b}$$

$$\mathbf{V} \times \mathbf{E} = -\frac{\partial \mathbf{B}}{\partial t} \tag{5-155c}$$

$$\mathbf{V} \times \mathbf{B} = \mu \left(\mathbf{J} + \frac{\partial \mathbf{D}}{\partial t} \right) \tag{5-155d}$$

Equations (5-155a)–(5-155d) are the same as Maxwell's equations for nonmagnetic materials as given by (5-80)–(5-83) except that μ_0 is replaced by μ. Thus the electric and magnetic fields in the presence of a magnetic material can be computed in exactly the same manner as for nonmagnetic materials except that we have to use μ instead of μ_0 for permeability. In fact, if $\chi_m = 0$, $\mu_r = 1$ and $\mu = \mu_0$ so that a nonmagnetic material can be considered as a magnetic material with $\mu = \mu_0$ and hence Eqs. (5-155a)–(5-155d) can be used for nonmagnetic materials as well. The permeability μ takes into account the effects of magnetization and there is no need to consider them when we use μ for μ_0, thereby eliminating the necessity for the simultaneous solution of (5-139) and (5-140). In the case of a boundary between two different magnetic materials, the appropriate boundary conditions for **H** take into account implicitly the magnetization surface current. We will consider these boundary conditions in Section 5.12. Substituting for **B** in (5-117) in terms of **H** by using (5-152), we obtain

$$\mathbf{M} = \frac{\chi_m}{1 + \chi_m} \frac{\mathbf{B}}{\mu_0} = \frac{\chi_m}{1 + \chi_m} \frac{\mu_0(1 + \chi_m)}{\mu_0} \mathbf{H} = \chi_m \mathbf{H} \tag{5-156}$$

Equation (5-156) represents the traditional definition for χ_m, because of which we defined χ_m in Section 5.7 in a manner which is not analogous to the definition of χ_e in Section 5.4.

EXAMPLE 5-12.   For the slab of magnetic material in Example 5-11, find and sketch the magnetic field intensity and the magnetic flux density vectors both inside and outside the material.

From Example 5-11, the magnetic flux density inside the magnetic material is given by

$$\mathbf{B}_i = (1 + \chi_{m0})B_0 \mathbf{i}_x \tag{5-126}$$

The relative permeability of the material is $1 + \chi_{m0}$. Thus the magnetic field intensity inside the material is

$$\mathbf{H}_i = \frac{\mathbf{B}_i}{\mu_0(1 + \chi_{m0})} = \frac{(1 + \chi_{m0})B_0}{\mu_0(1 + \chi_{m0})} \mathbf{i}_x = \frac{B_0}{\mu_0} \mathbf{i}_x$$

Outside the magnetic material, the magnetic flux density is the same as the applied value so that the magnetic field intensity is

$$\mathbf{H}_o = \frac{\mathbf{B}_a}{\mu_0} = \frac{B_0}{\mu_0}\mathbf{i}_x$$

Thus, for this example, the magnetic field intensity vectors inside and outside the magnetic material are the same and equal to the magnetic field intensity associated with the applied magnetic flux density. Both **H** and **B** fields inside and outside the material are shown in Fig. 5.26(a). Now, considering a rectangle *abcda* in the *xz* plane and with the sides *ab* and *cd* parallel to the boundary $z = 0$ as shown in Fig. 5.26(b), we note that **H** is uniform along the contour *abcda* since it has the same value both inside and outside the material. Thus

$$\oint_{abcda} \mathbf{H} \cdot d\mathbf{l} = \mathbf{H} \cdot \oint_{abcda} d\mathbf{l} = 0 \tag{5-157}$$

On the other hand, noting that **B** is parallel to *ab* and *cd* but having unequal

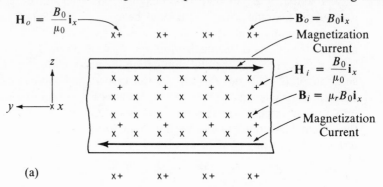

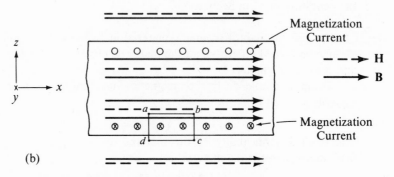

Fig. 5.26. Magnetic field intensity and magnetic flux density vectors for the magnetic material slab of Example 5-11.

magnitudes along them and perpendicular to bc and da, we obtain

$$
\oint_{abcda} \mathbf{B} \cdot d\mathbf{l} = \int_a^b \mathbf{B}_i \cdot d\mathbf{l} + \int_c^d \mathbf{B}_o \cdot d\mathbf{l}
$$

$$
= \int_a^b (1 + \chi_{m0}) B_0 \mathbf{i}_x \cdot dx\, \mathbf{i}_x + \int_c^d B_0 \mathbf{i}_x \cdot (-dx\, \mathbf{i}_x)
$$

$$
= (1 + \chi_{m0}) B_0(ab) - B_0(cd) \qquad (5\text{-}158)
$$

$$
= \chi_m B_0(ab) = \mu_0 \left(\frac{\chi_{m0} B_0}{\mu_0} \mathbf{i}_y \right) \cdot [(ab)\mathbf{i}_y]
$$

$$
= \mu_0 \text{ (magnetization surface current enclosed by } abcda)
$$

Comparing (5-157) and (5-158), we observe that the circulation of **H** is independent of magnetization currents whereas the circulation of **B** is not. The circulation of **H** depends only on those currents other than the magnetization currents, whereas the circulation of **B** depends on all kinds of currents. ∎

Returning now to Eq. (5-153), we note, from the values of χ_m for diamagnetic and paramagnetic materials listed in Table 5.3, that the relative permeabilities for these materials differ very little from unity. On the other hand, for ferromagnetic materials, the relative permeability can assume values of the order of several thousand. In fact, for these materials the relationship between **B** and **H** is nonlinear and is characterized by hysteresis so that there is no unique value of μ_r for a particular ferromagnetic material. The relationship between **B** and **H** is therefore presented in graphical form, as shown by a typical curve in Fig. 5.27. This curve is known as the hysteresis

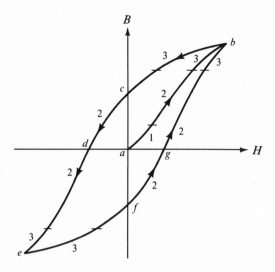

Fig. 5.27. Hysteresis curve for a ferromagnetic material.

curve or the **B-H** curve. To trace the development of the hysteresis effect, we start with an unmagnetized sample of ferromagnetic material in which both **B** and **H** are initially zero, corresponding to point *a* on the curve. As **H** is increased, the magnetization builds up, thereby increasing **B** gradually along the curve *ab* and finally to saturation at *b*, according to the following sequence of events as discussed in Section 5.7: (a) reversible motion of domain walls, (b) irreversible motion of domain walls, and (c) domain rotation. The regions corresponding to these events along the curve *ab* as well as other portions of the hysteresis curve are shown marked 1, 2, and 3, respectively, in Fig. 5-27. If the value of **H** is now decreased to zero, the value of **B** does not retrace the curve *ab* backwards but instead follows the curve *bc*, which indicates that a certain amount of magnetization remains in the material even after the magnetizing field is completly removed. In fact, it requires a magnetic field intensity in the opposite direction to bring **B** back to zero as shown by the portion *cd* of the curve. The value of **B** at the point *c* is known as the "remanence" or "retentivity," whereas the value of **H** at *d* is known as the "coercivity" of the material. Further increase in **H** in this direction results in the saturation of **B** in the direction opposite to that corresponding to *b* as shown by the portion *de* of the curve. If **H** is now decreased to zero, reversed in direction, and increased, the resulting variation of **B** occurs in accordance with the curve *efgb*, thereby completing the hysteresis loop. The characteristics of hysteresis curves for a few ferromagnetic materials are listed in Table 5.4. In view of the hysteresis effect, the incre-

TABLE 5.4. Characteristics of Hysteresis Curves for Some Ferromagnetic Materials

Material	Remanence, Wb/m^2	Coercivity, amp/m	Saturation Magnetization, Wb/m^2	Maximum μ_r
Cast iron	0.53	366	—	600
Permendur	1.4	160	2.4	5,000
Permalloy	—	24	1.6	25,000
Hypernik	0.73	3.2	1.65	70,000
Mumetal	—	4	0.65	100,000
Supermalloy	—	0.32	0.8	1,050,000

mental relative permeability defined by the slope of the hysteresis curve as given by

$$\mu_{ir} = \frac{1}{\mu_0} \frac{\Delta B}{\Delta H} \qquad (5\text{-}159)$$

is a useful parameter in addition to the relative permeability given by

$$\mu_r = \frac{1}{\mu_0} \frac{B}{H} \qquad (5\text{-}160)$$

for ferromagnetic materials.

5.10 Summary of Maxwell's Equations and Constitutive Relations

In the previous sections we introduced successively conductors, dielectrics, and magnetic materials. We discussed the various phenomena occurring in these materials in the presence of electric and magnetic fields. We learned several new concepts in this process. Important among these are the introduction of two new field vectors **D** and **H**. With the aid of these two new vectors, we developed a set of Maxwell's equations which permit us to solve field problems involving material media without explicitly taking into account the various phenomena occurring in them. These Maxwell's equations are given by

$$\mathbf{\nabla} \cdot \mathbf{D} = \rho \tag{5-161}$$

$$\mathbf{\nabla} \cdot \mathbf{B} = 0 \tag{5-162}$$

$$\mathbf{\nabla} \times \mathbf{E} = -\frac{\partial \mathbf{B}}{\partial t} \tag{5-163}$$

$$\mathbf{\nabla} \times \mathbf{H} = \mathbf{J} + \frac{\partial \mathbf{D}}{\partial t} \tag{5-164}$$

where ρ and **J** are the volume densities of the true charges and currents responsible for the fields characterized by the field intensity vectors **E** and **H** and the corresponding flux density vectors **D** and **B**. Equations (5-161)–(5-164) can as well be used for free space since they reduce to those of free space when the pertinent quantities are allowed to approach their free-space values.

The true charges are those which are free to move and not bound to their respective nucleii, as polarization charges are. Conduction charges in materials and space charges in vacuum tubes are examples of true charges. The true currents are those constituted by the movement of the free charges, as compared to polarization and magnetization currents which are due to the movement of charges bound to their respective nucleii. Conduction currents in materials and convection currents due to movement of space charge in vacuum tubes are examples of true currents. Thus **J** in (5-164) can represent conduction currents as well as convection currents. The charge and current densities are related via the continuity equation given by

$$\mathbf{\nabla} \cdot \mathbf{J} + \frac{\partial \rho}{\partial t} = 0 \tag{5-165}$$

The four Maxwell's equations given by (5-161)–(5-164) are not all independent; Eq. (5-162) follows from Eq. (5-163) whereas Eq. (5-161) follows from Eq. (5-164) with the aid of the continuity equation.

The vectors **E** and **B** are the fundamental field vectors which define the force **F** acting on a charge q moving with a velocity **v** in an electromagnetic field in accordance with the Lorentz force equation given by

$$\mathbf{F} = q(\mathbf{E} + \mathbf{v} \times \mathbf{B}) \tag{5-166}$$

The vectors **D** and **H** are mixed vectors which take into account the dielectric and magnetic properties of materials, respectively. They are related to **E** and **B**, respectively, via the equations

$$\mathbf{D} = \epsilon_0 \mathbf{E} + \mathbf{P} \tag{5-167}$$

$$\mathbf{H} = \frac{\mathbf{B}}{\mu_0} - \mathbf{M} \tag{5-168}$$

where **P** and **M** are the polarization and magnetization vectors, which define the state of polarization and magnetization, respectively, in the material. However, the relations which are more useful than (5-167) and (5-168) are

$$\mathbf{D} = \epsilon \mathbf{E} \tag{5-169}$$

$$\mathbf{H} = \frac{\mathbf{B}}{\mu} \tag{5-170}$$

where ϵ and μ are the permittivity and permeability, respectively, of the material which take into account implicitly the effects of **P** and **M**, respectively. Furthermore, for a material medium, the current density **J** is related to the electric field intensity **E** by

$$\mathbf{J} = \mathbf{J}_c = \sigma \mathbf{E} \tag{5-171}$$

where σ is the conductivity which takes into account the conductive property of the material. Equations (5-169), (5-170), and (5-171) are known as the constitutive relations. Together with the constitutive relations, Maxwell's equations form a sufficient set of equations to determine uniquely the electromagnetic field for a given ρ and **J** and in a medium for which ϵ, μ, and σ are specified.

For static fields, the time variations of all quantities are zero so that Maxwell's equations (5-161)–(5-164) reduce to

$$\nabla \cdot \mathbf{D} = \rho \tag{5-172}$$

$$\nabla \cdot \mathbf{B} = 0 \tag{5-173}$$

$$\nabla \times \mathbf{E} = 0 \tag{5-174}$$

$$\nabla \times \mathbf{H} = \mathbf{J} \tag{5-175}$$

whereas the continuity equation is given by

$$\nabla \cdot \mathbf{J} = 0 \tag{5-176}$$

In this case, we note that all four equations (5-172)–(5-175) are independent. For $\mathbf{J} = \mathbf{J}_c = \sigma \mathbf{E}$, Eq. (5-175) indicates coupling between electric and magnetic fields which is not present in the case of static fields in free space. We note, however, that this is a one-way coupling unlike the two-way coupling in the case of time-varying fields since the magnetic field depends upon the electric field through (5-175) but the electric field is independent of the magnetic field.

Returning to Maxwell's equations for arbitrarily time-varying fields given by (5-161)–(5-164), we obtain Maxwell's equations for sinusoidally time-varying fields by substituting the complex vectors $\bar{\mathbf{E}}$, $\bar{\mathbf{B}}$, $\bar{\mathbf{D}}$, and $\bar{\mathbf{H}}$ for the real vectors $\mathbf{E}$, $\mathbf{B}$, $\mathbf{D}$, and $\mathbf{H}$ and by replacing $\partial/\partial t$ by $j\omega$. Thus we have

$$\mathbf{\nabla} \cdot \bar{\mathbf{D}} = \rho \qquad (5\text{-}177)$$

$$\mathbf{\nabla} \cdot \bar{\mathbf{B}} = 0 \qquad (5\text{-}178)$$

$$\mathbf{\nabla} \times \bar{\mathbf{E}} = -j\omega\bar{\mathbf{B}} \qquad (5\text{-}179)$$

$$\mathbf{\nabla} \times \bar{\mathbf{H}} = \bar{\mathbf{J}} + j\omega\bar{\mathbf{D}} \qquad (5\text{-}180)$$

Writing (5-180) for a material medium as

$$\mathbf{\nabla} \times \bar{\mathbf{H}} = \sigma\bar{\mathbf{E}} + j\omega\epsilon\bar{\mathbf{E}} \qquad (5\text{-}181)$$

we observe that for $\sigma \gg \omega\epsilon$, the magnitude of the conduction current density term is greater than the magnitude of the displacement current density term so that $\mathbf{\nabla} \times \bar{\mathbf{H}} \approx \sigma\bar{\mathbf{E}}$. The material is then classified as a good conductor. On the other hand, for $\sigma \ll \omega\epsilon$, the magnitude of the displacement current density term is greater than the magnitude of the conduction current density term so that $\mathbf{\nabla} \times \bar{\mathbf{H}} \approx j\omega\epsilon\bar{\mathbf{E}}$. The material is then classified as a good dielectric. The critical frequency for which the two terms are equal is given by $\sigma = \omega\epsilon$ or $\omega = \sigma/\epsilon$. Thus, depending upon whether $\omega \ll \sigma/\epsilon$ or $\omega \gg \sigma/\epsilon$, the material can be regarded as a good conductor or a good dielectric. The situation, however, is not so simple because both σ and ϵ are in general functions of frequency.

With the understanding that σ and ϵ are frequency dependent, we now classify nonmagnetic materials as follows for the purpose of writing simplified sets of Maxwell's equations:

(a) Perfect dielectrics: These are idealizations of good dielectrics. These contain no true charges and currents. The corresponding Maxwell's equations are obtained by setting $\rho = 0$ and $\mathbf{J} = 0$.

(b) Good conductors: The magnitude of the conduction current density $\sigma\mathbf{E}$ is much greater than the magnitude of the displacement current density $\partial\mathbf{D}/\partial t$. To obtain the special set of Maxwell's equations, we set $\partial\mathbf{D}/\partial t = 0$. We also set $\rho = 0$ since, in accordance with the finding in Section 5.3, any charge density inside the conductor decays exponentially with a time constant equal to ϵ/σ, where we have replaced ϵ_0 in Section 5.3 by ϵ. For good conductors, $\sigma/\epsilon \gg \omega$ so that any initial charge density decays to a negligible fraction of its value in a fraction of a period.

(c) Perfect conductors: These are idealizations of good conductors obtained by letting $\sigma \rightarrow \infty$. The electric field inside a perfect conductor must be zero since otherwise, $\mathbf{J}_c = \sigma\mathbf{E}$ becomes infinite. Furthermore, for the time-varying case, the zero electric field results in $\partial\mathbf{B}/\partial t$ equal to zero or $\mathbf{B}$ equal to a constant with time. Thus a time-varying magnetic field cannot exist inside a perfect conductor. Hence we conclude that all fields inside a perfect conductor are zero for the time-varying case and the electric field is zero also for the static case. There remains only the possibility of a static magnetic field inside a perfect conductor.

We now list, in Table 5.5, Maxwell's equations and the continuity equation for the general case and for the special cases discussed above for both time-varying and static fields. Also listed are the corresponding integral forms of the equations. We note that, in certain cases, although certain terms on the right sides of the differential equations are set equal to zero, the corresponding terms on the right sides of the corresponding integral equations are not set equal to zero. This is because a differential equation is applicable at a point whereas the corresponding integral equation is appli-

TABLE 5.5. Summary of Maxwell's Equations and the Continuity Equation for Various Cases

Description	*Differential Form*	*Integral Form*
Time-varying fields; general case	$\nabla \cdot \mathbf{D} = \rho$	$\oint_S \mathbf{D} \cdot d\mathbf{S} = \int_V \rho \, dv$
	$\nabla \cdot \mathbf{B} = 0$	$\oint_S \mathbf{B} \cdot d\mathbf{S} = 0$
	$\nabla \times \mathbf{E} = -\dfrac{\partial \mathbf{B}}{\partial t}$	$\oint_C \mathbf{E} \cdot d\mathbf{l} = -\dfrac{d}{dt} \int_S \mathbf{B} \cdot d\mathbf{S}$
	$\nabla \times \mathbf{H} = \mathbf{J} + \dfrac{\partial \mathbf{D}}{\partial t}$	$\oint_C \mathbf{H} \cdot d\mathbf{l} = \int_S \mathbf{J} \cdot d\mathbf{S} + \dfrac{d}{dt} \int_S \mathbf{D} \cdot d\mathbf{S}$
	$\nabla \cdot \mathbf{J} + \dfrac{\partial \rho}{\partial t} = 0$	$\oint_S \mathbf{J} \cdot d\mathbf{S} + \dfrac{d}{dt} \int_V \rho \, dv = 0$
Static fields; general case	$\nabla \cdot \mathbf{D} = \rho$	$\oint_S \mathbf{D} \cdot d\mathbf{S} = \int_V \rho \, dv$
	$\nabla \cdot \mathbf{B} = 0$	$\oint_S \mathbf{B} \cdot d\mathbf{S} = 0$
	$\nabla \times \mathbf{E} = 0$	$\oint_C \mathbf{E} \cdot d\mathbf{l} = 0$
	$\nabla \times \mathbf{H} = \mathbf{J}$	$\oint_C \mathbf{H} \cdot d\mathbf{l} = \int_S \mathbf{J} \cdot d\mathbf{S}$
	$\nabla \cdot \mathbf{J} = 0$	$\oint_S \mathbf{J} \cdot d\mathbf{S} = 0$
Time-varying fields; perfect dielectrics $\rho = 0, \mathbf{J} = 0$	$\nabla \cdot \mathbf{D} = 0$	$\oint_S \mathbf{D} \cdot d\mathbf{S} = \int_V \rho \, dv$
	$\nabla \cdot \mathbf{B} = 0$	$\oint_S \mathbf{B} \cdot d\mathbf{S} = 0$
	$\nabla \times \mathbf{E} = -\dfrac{\partial \mathbf{B}}{\partial t}$	$\oint_C \mathbf{E} \cdot d\mathbf{l} = -\dfrac{d}{dt} \int_S \mathbf{B} \cdot d\mathbf{S}$
	$\nabla \times \mathbf{H} = \dfrac{\partial \mathbf{D}}{\partial t}$	$\oint_C \mathbf{H} \cdot d\mathbf{l} = \int_S \mathbf{J} \cdot d\mathbf{S} + \dfrac{d}{dt} \int_S \mathbf{D} \cdot d\mathbf{S}$

TABLE 5.5 (Cont'd.)

Description	*Differential Form*	*Integral Form*
Static fields; perfect dielectrics $\rho = 0, \mathbf{J} = 0$	$\nabla \cdot \mathbf{D} = 0$	$\oint_S \mathbf{D} \cdot d\mathbf{S} = \int_V \rho \, dv$
	$\nabla \cdot \mathbf{B} = 0$	$\oint_S \mathbf{B} \cdot d\mathbf{S} = 0$
	$\nabla \times \mathbf{E} = 0$	$\oint_C \mathbf{E} \cdot d\mathbf{l} = 0$
	$\nabla \times \mathbf{H} = 0$	$\oint_C \mathbf{H} \cdot d\mathbf{l} = \int_S \mathbf{J} \cdot d\mathbf{S}$
Time-varying fields; good conductors $\left\lvert \sigma \mathbf{E} \right\rvert \gg \left\lvert \dfrac{\partial \mathbf{D}}{\partial t} \right\rvert$ uniform σ	$\nabla \cdot \mathbf{D} = 0$	$\oint_S \mathbf{D} \cdot d\mathbf{S} = \int_V \rho \, dv$
	$\nabla \cdot \mathbf{B} = 0$	$\oint_S \mathbf{B} \cdot d\mathbf{S} = 0$
	$\nabla \times \mathbf{E} = -\dfrac{\partial \mathbf{B}}{\partial t}$	$\oint_C \mathbf{E} \cdot d\mathbf{l} = -\dfrac{d}{dt} \int_S \mathbf{B} \cdot d\mathbf{S}$
	$\nabla \times \mathbf{H} = \mathbf{J}_c = \sigma \mathbf{E}$	$\oint_C \mathbf{H} \cdot d\mathbf{l} = \int_S \mathbf{J} \cdot d\mathbf{S} + \dfrac{d}{dt} \int_S \mathbf{D} \cdot d\mathbf{S}$
	$\nabla \cdot \mathbf{J}_c = 0$	$\oint_S \mathbf{J}_c \cdot d\mathbf{S} + \dfrac{d}{dt} \int_V \rho \, dv = 0$
Static fields; conductors, uniform σ	$\nabla \cdot \mathbf{D} = 0$	$\oint_S \mathbf{D} \cdot d\mathbf{S} = \int_V \rho \, dv$
	$\nabla \cdot \mathbf{B} = 0$	$\oint_S \mathbf{B} \cdot d\mathbf{S} = 0$
	$\nabla \times \mathbf{E} = 0$	$\oint_C \mathbf{E} \cdot d\mathbf{l} = 0$
	$\nabla \times \mathbf{H} = \mathbf{J}_c = \sigma \mathbf{E}$	$\oint_C \mathbf{H} \cdot d\mathbf{l} = \int_S \mathbf{J} \cdot d\mathbf{S}$
	$\nabla \cdot \mathbf{J}_c = 0$	$\oint_S \mathbf{J}_c \cdot d\mathbf{S} = 0$
Perfect conductors	All fields are zero for the time-varying case; electric field is zero for the static case	

cable over a region. For example, although there is no true charge density associated with any point in a perfect dielectric medium, it is possible that a closed surface situated entirely within such a medium of finite extent can enclose a charge contained in that part of the volume bounded by the surface but lying outside the medium. Hence, although $\nabla \cdot \mathbf{D} = 0$ in the medium, we have to write the corresponding integral form as $\oint_S \mathbf{D} \cdot d\mathbf{S} = \int_V \rho \, dv$.

5.11 Power and Energy Considerations for Material Media

In Section 5.2 we learned that conductors are characterized by conduction current due to the movement of free charges under the influence of an applied electric field. In Section 4.8 we showed that the power expended by an electric field due to charges moving under its influence in a volume V is given by

$$P_d = \int_V \mathbf{E} \cdot \mathbf{J}\, dv \qquad (5\text{-}182)$$

where $\mathbf{J}$ is the current density resulting from the movement of the charges. If the motion of the charges is in free space, they are accelerated by the electric field and hence the power expended by the electric field is converted into kinetic energy. On the other hand, the free charges in a conductor drift with an average drift velocity because of the frictional mechanism provided by their collisions with the atomic lattice. Hence the power expended by the electric field is dissipated in the conductor in the form of heat. Replacing $\mathbf{J}$ in (5-182) by $\sigma\mathbf{E}$, we obtain the expression for the power dissipated in a volume V of a conductor as

$$P_d = \int_V \mathbf{E} \cdot \sigma\mathbf{E}\, dv \qquad (5\text{-}183)$$

It follows that the power dissipation density in a conductor is

$$p_d = \mathbf{E} \cdot \sigma\mathbf{E} = \sigma E^2 \qquad (5\text{-}184)$$

For sinusoidally time-varying fields, the time-average power dissipation density is

$$\langle p_d \rangle = \tfrac{1}{2}\sigma\bar{\mathbf{E}} \cdot \bar{\mathbf{E}}^* \qquad (5\text{-}185)$$

In Section 5.5 we learned that dielectrics in electric fields are characterized by induced polarization charges. From Section 4.6, the stored energy density associated with an electric field $\mathbf{E}$ in free space is given by

$$w_e = \tfrac{1}{2}\epsilon_0 E^2 = \tfrac{1}{2}\epsilon_0 \mathbf{E} \cdot \mathbf{E} \qquad (5\text{-}186)$$

This result was obtained by finding the work required to be done by an external agent to bring together a set of point charges from infinity and then extending the result to a volume charge distribution. We can do the same for a dielectric medium provided we take into account the polarization charges in finding the work required for assembling the charge distribution. The effect of the polarization charges is to neutralize partially the true charges. Hence, as we bring together charges from infinity, they are neutralized partially by the polarization charges. Thus, for the same electric field intensity in the dielectric as in free space, we have to actually assemble a true-charge distribution of greater density than in the free-space case. This requires more work to be done in the dielectric case so that more energy is stored in the dielectric case than in the free-space case for the same electric field intensity. From

$$\nabla \cdot \mathbf{D} = \nabla \cdot \epsilon\mathbf{E} = \rho \qquad (5\text{-}187)$$

the true-charge density which gives the same $\mathbf{E}$ in a dielectric medium of permittivity ϵ as in free space is ϵ/ϵ_0 times the charge density in the free-space case. From (4-118a), the work required to assemble a charge distribution is proportional to the charge density for a constant potential V and hence for a constant electric field intensity $\mathbf{E}$. The energy density associated with the electric field in the dielectric is therefore given by

$$w_e = \frac{\epsilon}{\epsilon_0}\left(\frac{1}{2}\epsilon_0 E^2\right) = \frac{1}{2}\epsilon E^2 = \frac{1}{2}\epsilon \mathbf{E} \cdot \mathbf{E} = \frac{1}{2}\mathbf{D} \cdot \mathbf{E} \qquad (5\text{-}188)$$

For sinusoidally time-varying fields, the time-average energy density is

$$\langle w_e \rangle = \tfrac{1}{4}\epsilon \bar{\mathbf{E}} \cdot \bar{\mathbf{E}}^* = \tfrac{1}{4}\bar{\mathbf{D}} \cdot \bar{\mathbf{E}}^* \qquad (5\text{-}189)$$

Substituting $\mathbf{D} = \epsilon_0 \mathbf{E} + \mathbf{P}$ in (5-188), we have

$$w_e = \tfrac{1}{2}(\epsilon_0 \mathbf{E} + \mathbf{P}) \cdot \mathbf{E} = \tfrac{1}{2}\epsilon_0 \mathbf{E} \cdot \mathbf{E} + \tfrac{1}{2}\mathbf{P} \cdot \mathbf{E} \qquad (5\text{-}190)$$

However,

$$\tfrac{1}{2}\mathbf{P} \cdot \mathbf{E} = \tfrac{1}{2}\int_0^{\mathbf{P},\,\mathbf{E}} d(\mathbf{P} \cdot \mathbf{E}) = \tfrac{1}{2}\int_0^{\mathbf{P},\,\mathbf{E}} (\mathbf{P} \cdot d\mathbf{E} + \mathbf{E} \cdot d\mathbf{P})$$

$$= \int_0^{\mathbf{P}} \mathbf{E} \cdot d\mathbf{P} \qquad (5\text{-}191)$$

where we have used the substitution $\mathbf{P} \cdot d\mathbf{E} = \mathbf{E} \cdot d\mathbf{P}$ in view of the linear relationship between $\mathbf{P}$ and $\mathbf{E}$. Substituting (5-191) into (5-190), we get

$$w_e = \tfrac{1}{2}\epsilon_0 \mathbf{E} \cdot \mathbf{E} + \int_0^{\mathbf{P}} \mathbf{E} \cdot d\mathbf{P} \qquad (5\text{-}192)$$

We note that the first term on the right side of (5-192) is the energy density in the electric field if the medium is free space. The second term on the right side of (5-192) is the work done per unit volume by the $\mathbf{E}$ field in the dielectric as the polarization is built up gradually from zero to the final value $\mathbf{P}$. This is known as the polarization energy density.

In Section 5.8 we learned that magnetic materials in magnetic fields are characterized by magnetization currents. From Section 4.7, the stored energy density associated with a magnetic field $\mathbf{B}$ in free space is given by

$$w_m = \frac{1}{2}\frac{B^2}{\mu_0} = \frac{1}{2}\frac{\mathbf{B}}{\mu_0} \cdot \mathbf{B} = \frac{1}{2}\mu_0 H^2 = \frac{1}{2}\mu_0 \mathbf{H} \cdot \mathbf{H} \qquad (5\text{-}193)$$

This result was obtained by finding the work required to be done for building up a volume current distribution. We can do the same for a magnetic material medium, provided we take into account the magnetization currents in finding the work required for building up the current distribution. The effect of the magnetization currents is to aid the true currents (for $\mu > \mu_0$). Hence, as the current is built up, it is aided by the magnetization current. Thus, for the same magnetic flux density in the magnetic material as in free space, it is sufficient if we actually build up a true current distribution of lesser density than in the free-space case. This requires less work to be done in the magnetic material case so that less energy is stored in the magnetic material case than in the free-space case for the same magnetic flux density. From

$$\nabla \times \mathbf{H} = \nabla \times \frac{\mathbf{B}}{\mu} = \mathbf{J} \qquad (5\text{-}194)$$

the true current density which gives the same **B** in a magnetic material medium of permeability μ as in free space is μ_0/μ times the current density in the free-space case. From (4-130a), the work required to build up a current distribution is proportional to the current density for a constant vector potential **A** and hence for a constant magnetic flux density **B**. The energy density associated with the magnetic field in the magnetic material is therefore given by

$$w_m = \frac{\mu_0}{\mu}\left(\frac{1}{2}\frac{B^2}{\mu_0}\right) = \frac{1}{2}\frac{B^2}{\mu} = \frac{1}{2}\frac{\mathbf{B}}{\mu}\cdot\mathbf{B} = \frac{1}{2}\mathbf{H}\cdot\mathbf{B} \qquad (5\text{-}195)$$

For sinusoidally time-varying fields, the time-average energy density is

$$\langle w_m \rangle = \frac{1}{4}\frac{\bar{\mathbf{B}}}{\mu}\cdot\bar{\mathbf{B}}^* = \frac{1}{4}\bar{\mathbf{H}}\cdot\bar{\mathbf{B}}^* \qquad (5\text{-}196)$$

Substituting $\mathbf{B} = \mu_0\mathbf{H} + \mu_0\mathbf{M}$ in (5-195), we have

$$w_m = \tfrac{1}{2}\mathbf{H}\cdot(\mu_0\mathbf{H} + \mu_0\mathbf{M}) = \tfrac{1}{2}\mu_0\mathbf{H}\cdot\mathbf{H} + \tfrac{1}{2}\mu_0\mathbf{H}\cdot\mathbf{M} \qquad (5\text{-}197)$$

However,

$$\tfrac{1}{2}\mu_0\mathbf{H}\cdot\mathbf{M} = \tfrac{1}{2}\int_0^{M,H} d(\mu_0\mathbf{H}\cdot\mathbf{M}) = \tfrac{1}{2}\int_0^{M,H}(\mu_0\mathbf{H}\cdot d\mathbf{M} + \mu_0\mathbf{M}\cdot d\mathbf{H})$$

$$= \int_0^M \mu_0\mathbf{H}\cdot d\mathbf{M} \qquad (5\text{-}198)$$

where we have used the substitution $\mathbf{H}\cdot d\mathbf{M} = \mathbf{M}\cdot d\mathbf{H}$ in view of the linear relationship between **M** and **H**. Substituting (5-198) into (5-197), we get

$$w_m = \tfrac{1}{2}\mu_0\mathbf{H}\cdot\mathbf{H} + \int_0^M \mu_0\mathbf{H}\cdot d\mathbf{M} \qquad (5\text{-}199)$$

We note that the first term on the right side of (5-199) is the energy density in the **H** field if the medium is free space. The second term on the right side of (5-199) is the work done by the **H** field in the magnetic material as the magnetization is built up from zero to the final value **M**. This is known as the magnetization energy density. Note that for the same magnetic field intensity in the magnetic material as in free space, we have to actually build up a true current distribution of greater density than in the free-space case.

For nonlinear magnetic materials, we cannot use the result $\tfrac{1}{2}\mathbf{H}\cdot\mathbf{B}$ given by (5-195) for finding the magnetic energy stored in the material since μ is not constant for a particular material but is dependent on **H**. To obtain the correct expression, we write (5-199) as

$$w_m = \int_0^H d(\tfrac{1}{2}\mu_0\mathbf{H}\cdot\mathbf{H}) + \int_0^M \mu_0\mathbf{H}\cdot d\mathbf{M}$$

$$= \int_0^{\mu_0 H} \mathbf{H}\cdot d(\mu_0\mathbf{H}) + \int_0^{\mu_0 M} \mathbf{H}\cdot d(\mu_0\mathbf{M}) \qquad (5\text{-}200)$$

$$= \int_0^{\mu_0 H + \mu_0 M} \mathbf{H}\cdot d(\mu_0\mathbf{H} + \mu_0\mathbf{M}) = \int_0^B \mathbf{H}\cdot d\mathbf{B}$$

It follows from (5-200) that the increase in stored energy density corresponding to an infinitesimal increment $d\mathbf{B}$ is $\mathbf{H} \cdot d\mathbf{B}$, where $\mathbf{H}$ is the magnetic field intensity at which $d\mathbf{B}$ is achieved.

Let us now consider the vector identity given by

$$\nabla \cdot (\mathbf{E} \times \mathbf{H}) = \mathbf{H} \cdot (\nabla \times \mathbf{E}) - \mathbf{E} \cdot (\nabla \times \mathbf{H}) \qquad (5\text{-}201)$$

Substituting for $\nabla \times \mathbf{E}$ and $\nabla \times \mathbf{H}$ on the right side of (5-201) from Maxwell's equations (5-163) and (5-164) respectively, we obtain

$$\nabla \cdot (\mathbf{E} \times \mathbf{H}) = \mathbf{H} \cdot \left(-\frac{\partial \mathbf{B}}{\partial t}\right) - \mathbf{E} \cdot \left(\mathbf{J} + \frac{\partial \mathbf{D}}{\partial t}\right)$$

$$= -\mathbf{E} \cdot \sigma\mathbf{E} - \mathbf{E} \cdot \frac{\partial \mathbf{D}}{\partial t} - \mathbf{H} \cdot \frac{\partial \mathbf{B}}{\partial t} \qquad (5\text{-}202)$$

$$= -\sigma\mathbf{E} \cdot \mathbf{E} - \frac{\partial}{\partial t}\left(\frac{1}{2}\mathbf{D} \cdot \mathbf{E}\right) - \frac{\partial}{\partial t}\left(\frac{1}{2}\mathbf{H} \cdot \mathbf{B}\right)$$

or

$$-\nabla \cdot (\mathbf{E} \times \mathbf{H}) = p_d + \frac{\partial}{\partial t}(w_e + w_m) \qquad (5\text{-}203)$$

where p_d, w_e, and w_m are, respectively, the power dissipation density due to the conductivity of the medium, the electric stored energy density, and the magnetic stored energy density. Integrating both sides of (5-203) in a volume V of the material and applying the divergence theorem to the left-side integral, we get

$$-\oint_S (\mathbf{E} \times \mathbf{H}) \cdot d\mathbf{S} = \int_V p_d \, dv + \frac{\partial}{\partial t}\int_V (w_e + w_m) \, dv \qquad (5\text{-}204)$$

where S is the surface bounding the volume V and $d\mathbf{S}$ is directed out of the volume V. The right side of (5-204) represents the time rate of increase of energy stored in the electric and magnetic fields in the volume V plus the time rate of energy dissipated in V due to conduction current flow. Thus $\oint_S (\mathbf{E} \times \mathbf{H}) \cdot d\mathbf{S}$ represents the power flow across S out of the volume V. It follows that the density of power flow or the Poynting vector associated with the electromagnetic field in a material medium is given by

$$\mathbf{P} = \mathbf{E} \times \mathbf{H} \qquad (5\text{-}205)$$

For sinusoidally time-varying fields, the complex Poynting's theorem is

$$\oint_S \bar{\mathbf{P}} \cdot d\mathbf{S} = -\int_V \langle p_d \rangle \, dv - j2\omega \int_V (\langle w_m \rangle - \langle w_e \rangle) \, dv \qquad (5\text{-}206)$$

where $\bar{\mathbf{P}}$ is the complex Poynting vector given by

$$\bar{\mathbf{P}} = \tfrac{1}{2}\bar{\mathbf{E}} \times \bar{\mathbf{H}}^* \qquad (5\text{-}207)$$

and $\langle p_d \rangle$, $\langle w_m \rangle$, and $\langle w_e \rangle$ are given by (5-185), (5-196), and (5-189), respectively.

EXAMPLE 5-13. Current flows axially with uniform density $\mathbf{J}_c = J_0 \mathbf{i}_z$ amp/m² along a cylindrical conductor of radius a and length l and having a uniform conductivity σ_0 mhos/m as shown in Fig. 5.28(a), by the application of a potential difference between the ends of the conductor. It is desired to verify that the total power dissipated in the conductor is correctly given by the surface integral of the Poynting vector over the surface of the conductor.

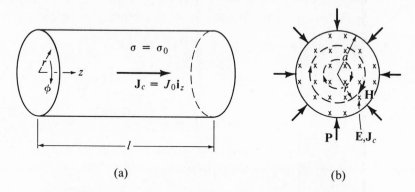

(a)   (b)

Fig. 5.28. For showing that the power dissipated in a conductor due to conduction current flow is correctly given by the surface integral of the Poynting vector over the surface of the conductor.

The power dissipation density inside the conductor is given by

$$p_d = \sigma E^2 = \sigma_0 \left(\frac{J_0}{\sigma_0}\right)^2 = \frac{J_0^2}{\sigma_0}$$

Since p_d is uniform over the volume of the conductor, the total power dissipated in the conductor is

$$P_d = p_d \text{ (volume of the conductor)} = \frac{J_0^2}{\sigma_0} (\pi a^2 l) = \frac{J_0^2 \pi a^2 l}{\sigma_0}$$

Applying $\oint_C \mathbf{H} \cdot d\mathbf{l} = \int_S \mathbf{J} \cdot d\mathbf{S}$ to a circular path of radius r in the cross-sectional plane of the conductor and centered at the axis of the conductor as shown in Fig. 5-28(b), we obtain, from symmetry considerations,

$$\mathbf{H} = H_\phi \mathbf{i}_\phi = \frac{J_0 \pi r^2}{2 \pi r} \mathbf{i}_\phi = \frac{J_0 r}{2} \mathbf{i}_\phi \qquad \text{for } r \leq a$$

The Poynting vector is then given by

$$\mathbf{P} = \mathbf{E} \times \mathbf{H} = \frac{J_0}{\sigma_0} \mathbf{i}_z \times \frac{J_0 r}{2} \mathbf{i}_\phi = -\frac{J_0^2 r}{2\sigma_0} \mathbf{i}_r \qquad \text{for } r \leq a$$

We note that the Poynting vector is directed radially towards the axis of the conductor. Hence, to find the total power flow into the conductor, it is sufficient if we perform the surface integration of the Poynting vector over the cylindrical surface $r = a$. Over this surface, we have

$$[\mathbf{P}]_{r=a} = -\frac{J_0^2 a}{2\sigma_0}\mathbf{i}_r$$

Since the magnitude of $[\mathbf{P}]_{r=a}$ is uniform, the surface integral of the Poynting vector over $r = a$ is

$$P = \frac{J_0^2 a}{2\sigma_0}(\text{surface area}) = \frac{J_0^2 a}{2\sigma_0}(2\pi a l) = \frac{J_0^2 \pi a^2 l}{\sigma_0}$$

which is the same as the result obtained by volume integration of the power dissipation density. This merely shows that the surface integral of the Poynting vector gives the correct result for the power dissipated and does not mean that the power is entering through the cylindrical surface. The actual power must be supplied by the source which maintains the potential difference between the ends of the conductor. ∎

5.12 Boundary Conditions

In Section 5.10 we summarized Maxwell's equations for the general case of a medium characterized by arbitrary values of σ, ϵ, and μ and for different special cases. For electromagnetic field problems involving several different media, the fields in each medium must satisfy separately the Maxwell's equations in differential form for that medium. On the other hand, the integral forms of Maxwell's equations must be satisfied collectively by the fields in all the media associated with the contours, surfaces, and volumes over which the integrals are evaluated. Thus the sets of solutions for the fields in different media obtained by solving the corresponding Maxwell's equations in differential form are tied together by a set of conditions determined by the integral forms of Maxwell's equations. These conditions are known as the "boundary conditions" since they relate the fields on one side of a boundary between two media to the fields on the other side of that boundary. The boundary conditions take into account any surface charges and currents existing on the boundaries, which the differential equations ignore.

In fact, we already introduced certain boundary conditions in previous sections without mentioning the name. An example of this is in Section 5.3, where we found that the tangential component of the electric field intensity at the boundary between a conductor placed in an electric field is zero, whereas the normal component of the electric field intensity at a point on the boundary is equal to $1/\epsilon_0$ times the surface charge density at that point in order to satisfy the criterion of zero electric field inside the conductor in the absence of a mechanism to permit the flow of a current. In this section we will derive the boundary conditions for the most general case of time-varying fields in two media characterized by different sets of values of σ, ϵ, and μ by using the integral forms of Maxwell's equations and the continuity equation. From the experience gained in this process, we will then write simplified sets of boundary conditions for the different special cases. From Table 5.5, the integral forms of Maxwell's equations are

$$\oint_S \mathbf{D} \cdot d\mathbf{S} = \int_V \rho \, dv \tag{5-208}$$

$$\oint_S \mathbf{B} \cdot d\mathbf{S} = 0 \tag{5-209}$$

$$\oint_C \mathbf{E} \cdot d\mathbf{l} = -\frac{d}{dt} \int_S \mathbf{B} \cdot d\mathbf{S} \tag{5-210}$$

$$\oint_C \mathbf{H} \cdot d\mathbf{l} = \int_S \mathbf{J} \cdot d\mathbf{S} + \frac{d}{dt} \int_S \mathbf{D} \cdot d\mathbf{S} \tag{5-211}$$

and the integral form of the continuity equation is

$$\oint_S \mathbf{J} \cdot d\mathbf{S} + \frac{d}{dt} \int_V \rho \, dv = 0 \tag{5-212}$$

Let us consider two semiinfinite media separated by the plane boundary $z = 0$ as shown in Fig. 5.29. Let us denote all quantities pertinent to medium 1 by subscript 1 and all quantities pertinent to medium 2 by subscript 2. Let $\mathbf{i}_n$ be the unit normal vector to the surface $z = 0$ directed into medium

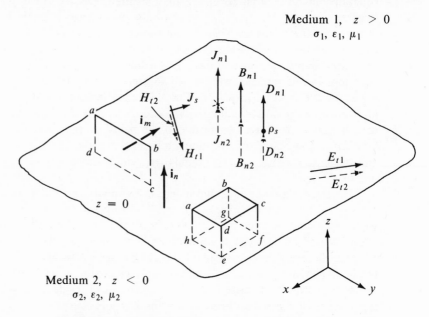

Fig. 5.29. For deriving the boundary conditions at the interface between two arbitrary media.

1 as shown in Fig. 5.29 and let all normal components of fields at the boundary in both media denoted by an additional subscript n be directed along $\mathbf{i}_n$. Let the surface charge density and the surface current density on $z = 0$ be ρ_s and $\mathbf{J}_s$, respectively. Note that the fields at the boundary in both media and the surface charge and current densities are, in general, functions of x, y, and t whereas the fields away from the boundary are, in general, functions of x, y, z, and t.

First we consider a rectangular box *abcdefgh* of infinitesimal volume enclosing an infinitesimal area of the boundary and parallel to it as shown in Fig. 5-29. Applying (5-208) to this box in the limit that the side surfaces (abbreviated ss) tend to zero, thereby shrinking the box to the surface, we have

$$\lim_{ss \to 0} \oint_{\Delta S} \mathbf{D} \cdot d\mathbf{S} = \lim_{ss \to 0} \int_{\Delta v} \rho \, dv \qquad (5\text{-}213)$$

where ΔS and Δv are the surface area and the volume, respectively, of the box. In the limit that the box shrinks to the surface, the contribution from the side surfaces to the integral on the left side of (5-213) approaches zero. The sum of the contributions from the top and bottom surfaces becomes $[D_{n1}(abcd) - D_{n2}(efgh)]$ since *abcd* and *efgh* are infinitesimal areas. The quantity on the right side of (5-213) would be zero but for the surface charge on the boundary, since shrinking the box to the surface reduces only the volume charge enclosed by it to zero, keeping the surface charge still enclosed by it. This surface charge is equal to $\rho_s(abcd)$. Thus Eq. (5-213) gives

$$D_{n1}(abcd) - D_{n2}(efgh) = \rho_s(abcd)$$

or

$$D_{n1} - D_{n2} = \rho_s \qquad (5\text{-}214)$$

since the two areas *abcd* and *efgh* are equal. In vector notation, (5-214) is written as

$$\mathbf{i}_n \cdot (\mathbf{D}_1 - \mathbf{D}_2) = \rho_s \qquad (5\text{-}215)$$

In words, Eqs. (5-214) and (5-215) state that, at any point on the boundary, the components of $\mathbf{D}_1$ and $\mathbf{D}_2$ normal to the boundary are discontinuous by the amount of the surface charge density at that point.

Similarly, applying (5-209) to the box *abcdefgh* in the limit that the box shrinks to the surface, we obtain

$$\lim_{ss \to 0} \oint_{\Delta S} \mathbf{B} \cdot d\mathbf{S} = 0 \qquad (5\text{-}216)$$

Using the same argument as for the left side of (5-213), the quantity on the left side of (5-216) is equal to $[B_{n1}(abcd) - B_{n2}(efgh)]$. Thus Eq. (5-216) gives

$$B_{n1}(abcd) - B_{n2}(efgh) = 0$$

or

$$B_{n1} - B_{n2} = 0 \tag{5-217}$$

In vector notation, Eq. (5-217) is written as

$$\mathbf{i}_n \cdot (\mathbf{B}_1 - \mathbf{B}_2) = 0 \tag{5-218}$$

In words, Eqs. (5-217) and (5-218) state that, at any point on the boundary, the components of $\mathbf{B}_1$ and $\mathbf{B}_2$ normal to the boundary are equal.

Now, we consider a rectangular contour *abcda* of infinitesimal area in the plane normal to the boundary and with its sides *ab* and *cd* parallel to the boundary as shown in Fig. 5-29. Applying (5-210) to this contour in the limit that *ad* and *bc* $\longrightarrow$ 0, thereby shrinking the rectangle to the surface, we have

$$\lim_{\substack{ad \to 0 \\ bc \to 0}} \oint_{abcda} \mathbf{E} \cdot d\mathbf{l} = -\lim_{\substack{ad \to 0 \\ bc \to 0}} \frac{d}{dt} \int_{\substack{area \\ abcd}} \mathbf{B} \cdot d\mathbf{S} \tag{5-219}$$

In the limit that the rectangle shrinks to the surface, the contribution from *ad* and *bc* to the integral on the left side of (5-219) approaches zero. Since *ab* and *cd* are infinitesimal, the sum of the contributions from *ab* and *cd* becomes $[E_{ab}(ab) + E_{cd}(cd)]$, where E_{ab} and E_{cd} are the components of $\mathbf{E}_1$ and $\mathbf{E}_2$ along *ab* and *cd*, respectively. The right side of (5-219) is equal to zero since the magnetic flux crossing the area *abcd* approaches zero as the area *abcd* tends to zero. Thus Eq. (5-219) gives

$$E_{ab}(ab) + E_{cd}(cd) = 0$$

or, since *ab* and *cd* are equal,

$$\mathbf{i}_{ab} \cdot (\mathbf{E}_1 - \mathbf{E}_2) = 0 \tag{5-220}$$

where $\mathbf{i}_{ab}$ is the unit vector along *ab*. Let us now define $\mathbf{i}_m$ to be the unit vector normal to the area *abcd* and in the direction of advance of a right-hand screw as it is turned in the sense of the path *abcda*. Note that $\mathbf{i}_m$ is tangential to the boundary. We then have

$$\mathbf{i}_{ab} = \mathbf{i}_m \times \mathbf{i}_n \tag{5-221}$$

Substituting (5-221) into (5-220) and rearranging the order of the scalar triple product, we obtain

$$\mathbf{i}_m \cdot \mathbf{i}_n \times (\mathbf{E}_1 - \mathbf{E}_2) = 0 \tag{5-222}$$

Since we can choose the rectangle *abcd* to be in any plane normal to the boundary, (5-222) must be true for all orientations of $\mathbf{i}_m$. It then follows that

$$\mathbf{i}_n \times (\mathbf{E}_1 - \mathbf{E}_2) = 0 \tag{5-223}$$

or, in scalar form,

$$E_{t1} - E_{t2} = 0 \tag{5-224}$$

where E_{t1} and E_{t2} are the tangential components of $\mathbf{E}_1$ and $\mathbf{E}_2$, respectively, at the boundary. In words, Eqs. (5-223) and (5-224) state that, at any point on the boundary, the components of $\mathbf{E}_1$ and $\mathbf{E}_2$ tangential to the boundary are equal.

Similarly, applying (5-211) to the contour *abcda* in the limit that *ad* and *bc* $\longrightarrow$ 0, we have

$$\lim_{\substack{ad\to0\\bc\to0}} \oint_{abcda} \mathbf{H} \cdot d\mathbf{l} = \lim_{\substack{ad\to0\\bc\to0}} \int_{\substack{area\\abcd}} \mathbf{J} \cdot d\mathbf{S} + \lim_{\substack{ad\to0\\bc\to0}} \frac{d}{dt} \int_{\substack{area\\abcd}} \mathbf{D} \cdot d\mathbf{S} \qquad (5\text{-}225)$$

Using the same argument as for the left side of (5-219), the quantity on the left side of (5-225) is equal to $[H_{ab}(ab) + H_{cd}(cd)]$, where H_{ab} and H_{cd} are the components of $\mathbf{H}_1$ and $\mathbf{H}_2$ along *ab* and *cd*, respectively. The second integral on the right side of (5-225) is zero since the displacement flux crossing the area *abcd* approaches zero as the area *abcd* tends to zero. The first integral on the right side of (5-225) would also be equal to zero but for a contribution from the surface current on the boundary because shrinking the rectangle to the surface reduces only the volume current enclosed by it to zero, keeping the surface current still enclosed by it. This contribution is the surface current flowing normal to the line which *abcd* approaches when it shrinks to the surface, that is, the current crossing this line along the direction of $\mathbf{i}_m$. This quantity is equal to $[\mathbf{J}_s \cdot \mathbf{i}_m](ab)$. Thus Eq. (5-225) gives

$$H_{ab}(ab) + H_{cd}(cd) = (\mathbf{J}_s \cdot \mathbf{i}_m)(ab)$$

or, since *ab* and *cd* are equal,

$$\mathbf{i}_{ab} \cdot (\mathbf{H}_1 - \mathbf{H}_2) = \mathbf{J}_s \cdot \mathbf{i}_m \qquad (5\text{-}226)$$

Substituting (5-221) into (5-226) and rearranging the order of the scalar triple product, we obtain

$$\mathbf{i}_n \times (\mathbf{H}_1 - \mathbf{H}_2) \cdot \mathbf{i}_m = \mathbf{J}_s \cdot \mathbf{i}_m \qquad (5\text{-}227)$$

Since (5-227) must be true for all orientations of $\mathbf{i}_m$, that is, for a rectangle *abcd* in any plane normal to the boundary, it follows that

$$\mathbf{i}_n \times (\mathbf{H}_1 - \mathbf{H}_2) = \mathbf{J}_s \qquad (5\text{-}228)$$

or, in scalar form,

$$H_{t1} - H_{t2} = J_s \qquad (5\text{-}229)$$

where H_{t1} and H_{t2} are the tangential components of $\mathbf{H}_1$ and $\mathbf{H}_2$, respectively, at the boundary. In words, Eqs. (5-228) and (5-229) state that, at any point on the boundary, the components of $\mathbf{H}_1$ and $\mathbf{H}_2$ tangential to the boundary are discontinuous by the amount equal to the surface current density at that point.

Finally, applying (5-212) to the box *abcdefgh* in the limit that the box shrinks to the surface, we have

$$\lim_{ss\to0} \oint_{\Delta S} \mathbf{J} \cdot d\mathbf{S} + \lim_{ss\to0} \frac{d}{dt} \int_{\Delta v} \rho \, dv = 0 \qquad (5\text{-}230)$$

In the limit that the box shrinks to the surface, the contribution to the first integral on the left side of (5-230) from the side surfaces of the box would be zero but for the surface current on the boundary, since although the volume current emanating from the side surfaces reduces to zero, there still remains the surface current emanating from them. This current is

$$\oint_{abcda} \mathbf{J}_s \cdot (d\mathbf{l} \times \mathbf{i}_n) = \int_{abcd} (\mathbf{\nabla}_s \cdot \mathbf{J}_s)\, dS = (\mathbf{\nabla}_s \cdot \mathbf{J}_s)(abcd)$$

where the subscript s in $\mathbf{\nabla}_s$ denotes that the divergence is computed in the two dimensions tangential to the surface. The sum of the contributions from the top and bottom surfaces is equal to $[J_{n1}(abcd) - J_{n2}(efgh)]$ or $[\mathbf{i}_n \cdot (\mathbf{J}_1 - \mathbf{J}_2)](abcd)$. The second integral on the left side of (5-230) is equal to $(\partial\rho_s/\partial t)(abcd)$. Thus Eq. (5-230) gives

$$[\mathbf{\nabla}_s \cdot \mathbf{J}_s + \mathbf{i}_n \cdot (\mathbf{J}_1 - \mathbf{J}_2)](abcd) + \frac{\partial\rho_s}{\partial t}(abcd) = 0$$

or

$$\mathbf{i}_n \cdot (\mathbf{J}_1 - \mathbf{J}_2) = -\mathbf{\nabla}_s \cdot \mathbf{J}_s - \frac{\partial\rho_s}{\partial t} \qquad (5\text{-}231)$$

In words, Eq. (5-231) states that, at any point on the boundary, the components of $\mathbf{J}_1$ and $\mathbf{J}_2$ normal to the boundary are discontinuous by the amount equal to the negative of the sum of the two-dimensional divergence of the surface current density and the time derivative of the surface charge density at that point.

Equations (5-215), (5-218), (5-223), (5-228), and (5-231) form the set of boundary conditions for the most general case of time-varying fields in two arbitrary media. Although we have derived these boundary conditions by considering a plane surface, it should be obvious that we can consider any arbitrary-shaped boundary and obtain the same results by letting the box and the rectangle shrink to points on the boundary. We can now write the boundary conditions for various special cases by inspection of the corresponding sets of Maxwell's equations and continuity equation listed in Table 5.5. These boundary conditions are listed in Table 5.6, together with the general boundary conditions. Depending upon the problem, only some of the boundary conditions need to be used in the determination of the fields, whereas some or all of the remaining boundary conditions are automatically satisfied and the rest determine the surface charge and current densities on the boundary. Before we consider some examples, let us investigate the boundary condition for the power flow normal to the boundary. Letting $\mathbf{P}_1$ and $\mathbf{P}_2$ be the Poynting vectors corresponding to the fields in media 1 and 2, respectively, we have

$$\begin{aligned}
\mathbf{i}_n \cdot (\mathbf{P}_1 - \mathbf{P}_2) &= \mathbf{i}_n \cdot (\mathbf{E}_1 \times \mathbf{H}_1 - \mathbf{E}_2 \times \mathbf{H}_2) \\
&= (\mathbf{i}_n \times \mathbf{E}_1) \times (\mathbf{i}_n \times \mathbf{H}_1) \cdot \mathbf{i}_n - (\mathbf{i}_n \times \mathbf{E}_2) \times (\mathbf{i}_n \times \mathbf{H}_2) \cdot \mathbf{i}_n
\end{aligned} \qquad (5\text{-}232)$$

TABLE 5.6. Summary of Boundary Conditions for Various Cases (i_n is the unit vector normal to the boundary and drawn towards medium 1)

Description	Boundary Conditions
Time-varying fields Medium 1: arbitrary, $\sigma_1 \neq \infty$ Medium 2: arbitrary, $\sigma_2 \neq \infty$	$i_n \cdot (\mathbf{D}_1 - \mathbf{D}_2) = \rho_s$ $i_n \cdot (\mathbf{B}_1 - \mathbf{B}_2) = 0$ $i_n \times (\mathbf{E}_1 - \mathbf{E}_2) = 0$ $i_n \times (\mathbf{H}_1 - \mathbf{H}_2) = \mathbf{J}_s$ $i_n \cdot (\mathbf{J}_1 - \mathbf{J}_2) = -\nabla_s \cdot \mathbf{J}_s - \dfrac{\partial \rho_s}{\partial t}$
Static fields Medium 1: arbitrary, $\sigma_1 \neq \infty$ Medium 2: arbitrary, $\sigma_2 \neq \infty$	$i_n \cdot (\mathbf{D}_1 - \mathbf{D}_2) = \rho_s$ $i_n \cdot (\mathbf{B}_1 - \mathbf{B}_2) = 0$ $i_n \times (\mathbf{E}_1 - \mathbf{E}_2) = 0$ $i_n \times (\mathbf{H}_1 - \mathbf{H}_2) = \mathbf{J}_s$ $i_n \cdot (\mathbf{J}_1 - \mathbf{J}_2) = -\nabla_s \cdot \mathbf{J}_s$
Time-varying fields Medium 1: perfect dielectric, $\sigma_1 = 0$ Medium 2: perfect dielectric, $\sigma_2 = 0$	$i_n \cdot (\mathbf{D}_1 - \mathbf{D}_2) = 0$ $i_n \cdot (\mathbf{B}_1 - \mathbf{B}_2) = 0$ $i_n \times (\mathbf{E}_1 - \mathbf{E}_2) = 0$ $i_n \times (\mathbf{H}_1 - \mathbf{H}_2) = 0$
Static fields Medium 1: perfect dielectric, $\sigma_1 = 0$ Medium 2: perfect dielectric, $\sigma_2 = 0$	$i_n \cdot (\mathbf{D}_1 - \mathbf{D}_2) = 0$ $i_n \cdot (\mathbf{B}_1 - \mathbf{B}_2) = 0$ $i_n \times (\mathbf{E}_1 - \mathbf{E}_2) = 0$ $i_n \times (\mathbf{H}_1 - \mathbf{H}_2) = 0$
Time-varying fields Medium 1: perfect dielectric, $\sigma_1 = 0$ Medium 2: perfect conductor, $\sigma_2 = \infty$	$i_n \cdot \mathbf{D}_1 = \rho_s$ $i_n \cdot \mathbf{B}_1 = 0$ $i_n \times \mathbf{E}_1 = 0$ $i_n \times \mathbf{H}_1 = \mathbf{J}_s$
Static electric field Medium 1: perfect dielectric, $\sigma_1 = 0$ Medium 2: perfect conductor, $\sigma_2 = \infty$	$i_n \cdot \mathbf{D}_1 = \rho_s$ $i_n \times \mathbf{E}_1 = 0$

Substituting (5-223) and (5-228) into (5-232), we get

$$
\begin{aligned}
i_n \cdot (\mathbf{P}_1 - \mathbf{P}_2) &= (i_n \times \mathbf{E}_2) \times [(i_n \times \mathbf{H}_2) + \mathbf{J}_s] \cdot i_n \\
&\quad - (i_n \times \mathbf{E}_2) \times (i_n \times \mathbf{H}_2) \cdot i_n \\
&= [(i_n \times \mathbf{E}_2) \times \mathbf{J}_s] \cdot i_n \\
&= [(i_n \times \mathbf{E}_1) \times \mathbf{J}_s] \cdot i_n \\
&= [(i_n \cdot \mathbf{J}_s)\mathbf{E}_1 - (\mathbf{J}_s \cdot \mathbf{E}_1)i_n] \cdot i_n \\
&= -\mathbf{J}_s \cdot \mathbf{E}_1
\end{aligned}
\tag{5-233}
$$

since $(i_n \cdot \mathbf{J}_s)$ is equal to zero. Thus, at any point on the boundary, the components of the Poynting vector normal to the boundary are discontinuous by the amount equal to the power density associated with the surface current density at that point. In the absence of a surface current, the normal components of the Poynting vector are continuous.

EXAMPLE 5-14. In Fig. 5.30, medium 1 comprises the region $z > 0$ and medium 2 comprises the region $z < 0$. All fields are spatially uniform in both media

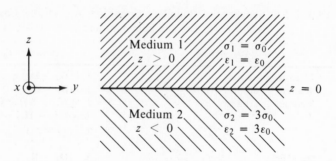

Fig. 5.30. For Example 5-14.

and independent of time. The quantities σ_0 and ϵ_0 are constants. If the current density in medium 1 is given by

$$\mathbf{J}_1 = J_0(\mathbf{i}_x + 2\mathbf{i}_y + 6\mathbf{i}_z)$$

where J_0 is a constant, find (a) the electric field intensity vector $\mathbf{E}_2$ in medium 2, and (b) the surface charge density ρ_s on the interface $z = 0$.

The electric field intensity $\mathbf{E}_1$ in medium 1 is given by

$$\mathbf{E}_1 = \frac{\mathbf{J}_1}{\sigma_1} = \frac{J_0}{\sigma_0}(\mathbf{i}_x + 2\mathbf{i}_y + 6\mathbf{i}_z)$$

From (5-223), the tangential component of $\mathbf{E}_2$ is equal to the tangential component of $\mathbf{E}_1$. Thus $E_{2x} = J_0/\sigma_0$ and $E_{2y} = 2J_0/\sigma_0$. Since all fields are spatially uniform and independent of time, $\mathbf{V}_s \cdot \mathbf{J}_s = 0$ and $\partial\rho_s/\partial t = 0$. Then, from (5-231) the normal component of $\mathbf{J}_2$ is equal to the normal component of $\mathbf{J}_1$. Thus $J_{2z} = 6J_0$ and $E_{2z} = J_{2z}/\sigma_2 = 6J_0/3\sigma_0 = 2J_0/\sigma_0$. The electric field intensity $\mathbf{E}_2$ in medium 2 is therefore given by

$$\mathbf{E}_2 = \frac{J_0}{\sigma_0}(\mathbf{i}_x + 2\mathbf{i}_y + 2\mathbf{i}_z)$$

From (5-215), the surface charge density ρ_s on the interface $z = 0$ is given by

$$\rho_s = \mathbf{i}_z \cdot (\mathbf{D}_1 - \mathbf{D}_2)$$
$$= D_{1z} - D_{2z} = \epsilon_1 E_{1z} - \epsilon_2 E_{2z}$$
$$= \epsilon_0 \frac{6J_0}{\sigma_0} - 2\epsilon_0 \frac{2J_0}{\sigma_0} = 2\frac{\epsilon_0 J_0}{\sigma_0} \quad \blacksquare$$

EXAMPLE 5-15.   In Fig. 5.31, a perfect dielectric medium $x < 0$ is bounded by a perfect conductor ($x = 0$). The electric field intensity for $x < 0$ is given by

$$\mathbf{E}(x, y, z, t) = [E_1 \cos(\omega t - \beta x \cos\theta - \beta z \sin\theta) \\ + E_2 \cos(\omega t + \beta x \cos\theta - \beta z \sin\theta)]\mathbf{i}_y$$

where E_1, E_2, ω, β, and θ are constants. Find the relationship between E_1 and E_2. Then find the surface current density on the surface $z = 0$.

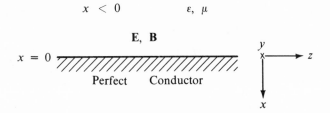

Fig. 5.31. For Example 5-15.

From the boundary conditions listed in Table 5.6, the tangential component of the electric field intensity at the surface of a perfect conductor must be zero. Thus

$$[E_y]_{x=0} = (E_1 + E_2) \cos(\omega t - \beta z \sin\theta) = 0$$

For this to be true for all values of z and t, $E_1 + E_2$ must be zero. Hence

$$E_2 = -E_1$$

The electric field intensity for $x < 0$ is then given by

$$\begin{aligned}\mathbf{E} &= [E_1 \cos(\omega t - \beta x \cos\theta - \beta z \sin\theta) \\ &\quad - E_1 \cos(\omega t + \beta x \cos\theta - \beta z \sin\theta)]\mathbf{i}_y \\ &= 2E_1 \sin(\beta x \cos\theta) \sin(\omega t - \beta z \sin\theta)\,\mathbf{i}_y\end{aligned} \qquad (5\text{-}234)$$

The corresponding magnetic flux density **B** can be obtained by using Maxwell's curl equation for **E**, given by

$$-\frac{\partial \mathbf{B}}{\partial t} = \nabla \times \mathbf{E} = -\frac{\partial E_y}{\partial z}\mathbf{i}_x + \frac{\partial E_y}{\partial x}\mathbf{i}_z \qquad (5\text{-}235)$$

Substituting for E_y in (5-235) from (5-234) and integrating with respect to time, we obtain

$$\begin{aligned}\mathbf{B} &= -\frac{2E_1\beta}{\omega}[\sin\theta \sin(\beta x \cos\theta) \sin(\omega t - \beta z \sin\theta)\,\mathbf{i}_x \\ &\quad -\cos\theta \cos(\beta x \cos\theta) \cos(\omega t - \beta z \sin\theta)\,\mathbf{i}_z]\end{aligned}$$

The magnetic flux density at the surface of the perfect conductor is given by

$$[\mathbf{B}]_{x=0} = \frac{2E_1\beta}{\omega} \cos\theta \cos(\omega t - \beta z \sin\theta)\,\mathbf{i}_z$$

Note that the condition of zero normal component of **B** at the surface of the perfect conductor is automatically satisfied by the zero tangential component of **E**. This is because the boundary condition for the tangential component of **E** is obtained from the integral form of $\nabla \times \mathbf{E} = -\partial \mathbf{B}/\partial t$ whereas the boundary condition for the normal component of **B** is obtained from the integral form of $\nabla \cdot \mathbf{B} = 0$. However, $\nabla \cdot \mathbf{B} = 0$ follows from $\nabla \times \mathbf{E} = -\partial \mathbf{B}/\partial t$. Hence the two boundary conditions are not independent. Finally, the surface current density at the surface of the perfect conductor is given by

$$\mathbf{J}_s = -\mathbf{i}_x \times [\mathbf{H}]_{x=0}$$

$$= \frac{2E_1\beta}{\omega\mu} \cos\theta \cos(\omega t - \beta z \sin\theta)\, \mathbf{i}_y \quad \blacksquare$$

PROBLEMS

5.1. Consider two electrons moving under thermal agitation with equal and opposite velocities. A uniform electric field is applied along the direction of motion of one of the electrons. Show that the gain in kinetic energy by the accelerating electron is greater than the loss in kinetic energy by the decelerating electron.

5.2. (a) For a sinuosidally time-varying electric field $\mathbf{E} = \mathbf{E}_0 \cos\omega t$, where $\mathbf{E}_0$ is a constant, show that the steady-state solution for Eq. (5-2) is given by

$$\mathbf{v}_d = \frac{\tau e}{m\sqrt{1 + \omega^2\tau^2}} \mathbf{E}_0 \cos(\omega t - \tan^{-1}\omega\tau)$$

(b) Based on the assumption of one free electron per atom, the free electron density N_e in silver is 5.86×10^{28} m^{-3}. Using the conductivity for silver given in Table 5.1, find the frequency at which the drift velocity lags the applied field by $\pi/4$ rad. What is the ratio of the mobility at this frequency to the mobility at zero frequency?

5.3. The plane surfaces $x = 0$, $y > 0$ and $y = 0$, $x > 0$, and the curved surface $xy = 2$ form the boundaries of conductors extending away from the region between them. If the electrostatic potential in the region between the surfaces is given by $V = 50\,xy$ volts, find the surface charge densities on all three surfaces.

5.4. The region $z < -d$ is occupied by a conductor. An infinitely long line charge of uniform density ρ_{L0} C/m is situated along the x axis. From the secondary field required to make the total field inside the conductor equal to zero and from symmetry considerations as illustrated in Fig. 5.32, show that the induced charge

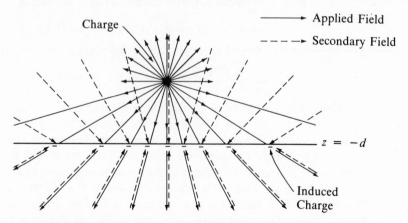

Fig. 5.32. For Problems 5.4 and 5.5. Charge is line charge for Problem 5.4 and point charge for Problem 5.5.

density on the surface of the conductor is given in cartesian coordinates by $-\rho_{L0}d/\pi(y^2 + d^2)$ C/m². Show that the induced surface charge per unit length along the x direction is equal to $-\rho_{L0}$. Show that the field outside the conductor is the same as the field due to the line charge along the x axis and an image line charge of uniform density $-\rho_{L0}$ C/m situated parallel to the actual line charge and passing through $(0, 0, -2d)$.

5.5. The region $z < -d$ is occupied by a conductor. A point charge Q C is situated at the origin. From the secondary field required to make the total field inside the conductor equal to zero and from symmetry considerations as illustrated in Fig. 5.32, show that the induced charge density on the surface of the conductor is given in cylindrical coordinates by $-Qd/2\pi(r_c^2 + d^2)^{3/2}$ C/m². Show that the total induced surface charge is $-Q$ C. Show that the field outside the conductor is the same as the field due to the point charge Q at the origin and an image point charge $-Q$ situated at $(0, 0, -2d)$.

5.6. (a) An infinite plane conducting slab carries uniformly distributed surface charges on both of its surfaces. If the net surface charge density, that is, the sum of the surface charge densities on the two surfaces, is ρ_{s0} C/m², find the surface charge densities on the two surfaces.

(b) Two infinite plane parallel conducting slabs carry uniformly distributed surface charges on all four of their surfaces. If the net surface charge densities are ρ_{s1} and ρ_{s2} C/m², respectively, for the two slabs, find the surface charge densities on all four surfaces.

5.7. Two infinitely long, coaxial, hollow cylindrical conductors of inner radii a and c, respectively, and outer radii $b (< c)$ and d, respectively, carry uniformly distributed surface charges on all four of their surfaces. If the net surface charges per unit length are ρ_{L1} and ρ_{L2} C for the inner and outer conductors, respectively, find the surface charge densities on all four surfaces.

5.8. Two concentric, spherical conducting shells of inner radii a and c, respectively, and outer radii $b (< c)$ and d, respectively, carry uniformly distributed surface charges on all four of their surfaces. If the net surface charges are Q_1 and Q_2 C for the inner and outer conductors, respectively, find the surface charge densities on all four surfaces.

5.9. Figure 5.33 shows the electric field intensities on either side of a point on a surface charge layer in free space.

(a) Using the integral form of Maxwell's curl equation for $\mathbf{E}$, show that the tangential components E_{t1} and E_{t2} are equal.

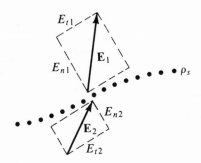

Fig. 5.33. For Problem 5.9.

(b) Using the integral form of Maxwell's divergence equation for **E**, show that the normal components E_{n1} and E_{n2} are related in the manner

$$E_{n1} - E_{n2} = \frac{\rho_s}{\epsilon_0}$$

where ρ_s is the surface charge density at the point.

5.10. Figure 5.34 shows the magnetic flux densities on either side of a point on a surface current layer in free space.

(a) Using the integral form of Maxwell's divergence equation for **B**, show that the normal components B_{n1} and B_{n2} are equal.

(b) Using the integral form of Maxwell's curl equation for **B**, show that the tangential components B_{t1} and B_{t2} are related in the manner

$$B_{t1} - B_{t2} = \mu_0 J_s$$

where $\mathbf{J}_s$ is the surface current density at the point. Note that $\mathbf{J}_s$ is directed into the paper whereas B_{t1} and B_{t2} are in the plane of the paper.

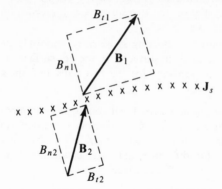

Fig. 5.34. For Problem 5.10.

5.11. The electric field intensity outside a conducting sphere of radius a and centered at the origin is given by

$$\mathbf{E} = E_0\left(1 + \frac{2a^3}{r^3}\right)\cos\theta\, \mathbf{i}_r - E_0\left(1 - \frac{a^3}{r^3}\right)\sin\theta\, \mathbf{i}_\theta$$

(a) Show that **E** satisfies Maxwell's equations.

(b) Show that the tangential component of **E** is zero at the conductor surface.

(c) Find the charge density on the conductor surface.

(d) Find the applied field by letting $a \longrightarrow 0$ and then find the secondary field both inside and outside $r = a$.

(e) Show that the secondary field on either side of the boundary satisfies the conditions (a) and (b) stated in Problem 5.9.

5.12. The radius of the electron cloud in a helium atom is approximately equal to 10^{-10} m. Compute the relative displacement between the centroids of the nucleus and the electron cloud under the influence of an electric field $E_0 = 5 \times 10^6$ volts/m. Compare your result with the radius of the electron cloud.

5.13. In Example 5-3, assume that the charge distribution in the electron cloud is a function of the radial distance from the centroid. If the relative displacement d between the centroids of the nucleus and the electron cloud is very small compared to the effective atomic radius, show that the electronic polarizability is approximately given by

$$\alpha_e \approx \frac{3\epsilon_0 Q}{|\rho(0)|}$$

where $\rho(0)$ is the charge density at the center of the electron cloud and Q is the magnitude of the total charge in the electron cloud. Verify the result for the uniformly charged cloud.

5.14. Show that the torque acting on an electric dipole of moment $\mathbf{p}$ in a uniform electric field $\mathbf{E}_p$ is equal to $\mathbf{p} \cdot \mathbf{E}_p$. Show that the torque tends to align the dipole moment with the field.

5.15. Two infinitely long line charges of uniform densities ρ_{L0} and $-\rho_{L0}$ are situated parallel to the z axis and pass through the points $(d/2, 0, 0)$ and $(-d/2, 0, 0)$, respectively. Show that the average electric field intensity in a cylindrical volume of radius $a > d/2$ and having the z axis as its axis is equal to $-(\rho_{L0}d/2\pi\epsilon_0 a^2)\mathbf{i}_x$.

5.16. Show that the average electric field intensity due to an arbitrary volume charge distribution of dipole moment $\mathbf{p}$ in a spherical volume of radius a is given by

$$\mathbf{E}_{av} = -\frac{\mathbf{p}}{4\pi\epsilon_0 a^3}$$

5.17. The region $a < r < b$ in spherical coordinates is filled with a dielectric material of uniform susceptibility χ_{e0}. A point charge Q is situated at the origin.

(a) Show that the polarization volume charge density is zero and that the polarization surface charge densities are given by

$$\rho_{ps} = \begin{cases} \dfrac{\chi_{e0}}{1 + \chi_{e0}} \dfrac{Q}{4\pi a^2} & r = a \\[3mm] \dfrac{\chi_{e0}}{1 + \chi_{e0}} \dfrac{Q}{4\pi b^2} & r = b \end{cases}$$

(b) Find the electric field intensities in the three different regions $r < a$, $a < r < b$, and $r > b$.

(c) Discuss your results for the limiting case $a \rightarrow 0$ and $b \rightarrow \infty$.

5.18. The region $z < -d$ is occupied by a dielectric of uniform electric susceptibility χ_{e0}. A point charge Q is situated at the origin. Show that the polarization surface charge density is equal to

$$-\frac{Q\chi_{e0}d}{2\pi(2 + \chi_{e0})(r_c^2 + d^2)^{3/2}}$$

and that the polarization volume charge density is zero. Show that the electric field intensity inside the dielectric is the same as that due to a point charge equal to $2Q/(2 + \chi_{e0})$ at the origin. Show that the electric field intensity outside the dielectric is the same as that due to the point charge Q at the origin and an image point charge $-\chi_{e0}Q/(2 + \chi_{e0})$ at $(0, 0, -2d)$.

5.19. A dielectric sphere of radius a and having uniform electric susceptibility χ_{e0} is centered at the origin. The electric field intensities outside and inside the sphere are given in spherical coordinates by

$$\mathbf{E}_o = \left(1 + \frac{2\chi_{e0}}{3 + \chi_{e0}} \frac{a^3}{r^3}\right) E_0 \cos\theta \, \mathbf{i}_r - \left(1 - \frac{\chi_{e0}}{3 + \chi_{e0}} \frac{a^3}{r^3}\right) E_0 \sin\theta \, \mathbf{i}_\theta$$

$$\mathbf{E}_i = \frac{3}{3 + \chi_{e0}} (E_0 \cos\theta \, \mathbf{i}_r - E_0 \sin\theta \, \mathbf{i}_\theta)$$

where E_0 is a constant.

(a) Show that $\mathbf{E}_o$ and $\mathbf{E}_i$ satisfy Maxwell's equations.

(b) Find the applied field by letting $a \longrightarrow 0$ and then find the secondary field both inside and outside $r = a$.

(c) Show that the tangential components of the secondary field on either side of $r = a$ are equal.

(d) From the normal components of the secondary field on either side of $r = a$, obtain the polarization surface charge density at $r = a$, using condition (b) stated in Problem 5.9.

(e) Show that the surface charge density found in part (d) is consistent with the polarization vector corresponding to $\mathbf{E}_i$.

5.20. An infinite plane dielectric slab of thickness d and having a nonuniform electric susceptibility given by

$$\chi_e(z) = \frac{z}{4 - z}$$

occupies the region $1 < z < 2$. A uniform electric field $\mathbf{E}_a = E_0 \mathbf{i}_z$ is applied. Show that the induced polarization volume and surface charge densities are given by

$$\rho_p = -\tfrac{1}{4}\epsilon_0 E_0 \qquad 1 < z < 2$$

$$\rho_{ps} = \begin{cases} \tfrac{1}{2}\epsilon_0 E_0 & z = 2 \\ -\tfrac{1}{4}\epsilon_0 E_0 & z = 1 \end{cases}$$

Find the secondary and total electric fields both inside and outside the dielectric. Obtain the polarization current density in the dielectric if the applied electric field is time-varying in the manner $\mathbf{E}_a = E_0 \cos \omega t \, \mathbf{i}_z$.

5.21. Two perfectly conducting, infinite plane parallel sheets separated by a distance d carry uniformly distributed surface charges of equal and opposite densities ρ_{s0} and $-\rho_{s0}$, respectively. For each of the following cases, find the potential difference between the two plates:

(a) The medium between the two plates is free space.

(b) The medium between the two plates is a dielectric of uniform permittivity $\epsilon = 4\epsilon_0$.

(c) The medium between the two plates consists of two dielectric slabs of thicknesses t and $d - t$ and having permittivities $\epsilon_1 = 2\epsilon_0$ and $\epsilon_2 = 4\epsilon_0$, respectively.

(d) The medium between the two plates is a dielectric of nonuniform permittivity which varies linearly from a value of ϵ_1 near one plate to a value of ϵ_2 near the second plate.

5.22. Two perfectly conducting, infinite plane parallel sheets separated by a distance d carry uniformly distributed surface charges of equal and opposite densities. For each of the cases listed in Problem 5.21, find the required surface charge densities if the potential difference between the two plates is to be V_0.

5.23. An infinite plane dielectric slab of thickness d and having a nonuniform permittivity given by

$$\epsilon = \frac{4\epsilon_0}{(1 + z/d)^2}$$

occupies the region $0 < z < d$. A uniform electric field $\mathbf{E}_a = E_0\mathbf{i}_z$ is applied. Find the following quantities:

(a) $\mathbf{D}$ outside the dielectric.

(b) $\mathbf{D}$ inside the dielectric.

(c) $\mathbf{E}$ inside the dielectric.

(d) $\mathbf{P}$ inside the dielectric.

(e) ρ_{ps} on the surfaces $z = 0$ and $z = d$.

(f) ρ_p inside the dielectric.

5.24. The region $a < r < b$ in spherical coordinates is occupied by a dielectric material. A point charge Q is situated at the origin. It is found that the electric field intensity inside the dielectric is given by

$$\mathbf{E} = \frac{Q}{4\pi\epsilon_0 b^2}\mathbf{i}_r, \qquad a < r < b$$

Find the following quantities:

(a) The permittivity of the dielectric.

(b) ρ_{ps} on the surfaces $r = a$ and $r = b$.

(c) ρ_p inside the dielectric.

5.25. Show that the result given by (5-98) for the change in the angular velocity of an electron in a circular orbit of radius a under the influence of an applied magnetic field follows from the application of Faraday's law in integral form to the electronic orbit.

5.26. Show that the torque acting on an arbitrary current loop of dipole moment $\mathbf{m}$ in a uniform magnetic field $\mathbf{B}_m$ is equal to $\mathbf{m} \times \mathbf{B}_m$. Show that the torque tends to align the dipole moment with the field.

5.27. Two infinitely long filamentary wires situated parallel to the z axis and passing through the points $(d/2, 0, 0)$ and $(-d/2, 0, 0)$ carry currents I amp in the positive and negative z directions, respectively. Show that the average magnetic flux density in a cylindrical volume of radius $a > d/2$ and having the z axis as its axis is equal to $-(\mu_0 I d/2\pi a^2)\mathbf{i}_y$.

5.28. Show that the average magnetic flux density due to an arbitrary volume current distribution of dipole moment **m** in a spherical volume of radius b is given by

$$\mathbf{B}_{av} = \frac{\mu_0 \, \mathbf{m}}{2\pi b^3}$$

5.29. The region $a < r < b$ in cylindrical coordinates is filled with a magnetic material of uniform susceptibility χ_{m0}. A filamentary wire situated along the z axis carries current I amp in the z direction.

(a) Show that the magnetization volume current density is zero and that the magnetization surface current densities are given by

$$\mathbf{J}_{ms} = \begin{cases} \chi_{m0} \dfrac{I}{2\pi a} \mathbf{i}_z & r = a \\[2mm] -\chi_{m0} \dfrac{I}{2\pi b} \mathbf{i}_z & r = b \end{cases}$$

(b) Find the magnetic flux densities in the three different regions $r < a$, $a < r < b$, and $r > b$.

(c) Discuss your results for the limiting case $a \longrightarrow 0$ and $b \longrightarrow \infty$.

5.30. The region $z < -d$ is occupied by a magnetic material of uniform susceptibility χ_{m0}. An infinitely long filamentary wire carrying current I amp in the x direction is situated along the x axis. Show that the magnetization surface current density is equal to

$$\frac{\chi_{m0} dI}{\pi(2 + \chi_{m0})(y^2 + d^2)} \mathbf{i}_x \text{ amp/m}$$

and that the magnetization volume current density is equal to zero. Show that the magnetic flux density inside the magnetic material is the same as that due to a filamentary wire along the x axis carrying $[(2 + 2\chi_{m0})/(2 + \chi_{m0})]I$ amp in the x direction. Show that the magnetic flux density outside the magnetic material is the same as that due to the filamentary wire along the x axis carrying I amp in the x direction and an image filamentary wire parallel to the x axis and passing through $(0, 0, -2d)$ and carrying a current $\chi_{m0} I/(2 + \chi_{m0})$ amp in the x direction.

5.31. A sphere of magnetic material of radius a and having uniform susceptibility χ_{m0} is centered at the origin. The magnetic flux densities outside and inside the sphere are given in spherical coordinates by

$$\mathbf{B}_o = \left(1 + \frac{2\chi_{m0}}{3 + \chi_{m0}} \frac{a^3}{r^3}\right) B_0 \cos\theta \, \mathbf{i}_r - \left(1 - \frac{\chi_{m0}}{3 + \chi_{m0}} \frac{a^3}{r^3}\right) B_0 \sin\theta \, \mathbf{i}_\theta$$

$$\mathbf{B}_i = \frac{3(1 + \chi_{m0})}{3 + \chi_{m0}} (B_0 \cos\theta \, \mathbf{i}_r - B_0 \sin\theta \, \mathbf{i}_\theta)$$

where B_0 is a constant.

(a) Show that $\mathbf{B}_o$ and $\mathbf{B}_i$ satisfy Maxwell's equations.

(b) Find the applied field by letting $a \longrightarrow 0$ and then find the secondary field both inside and outside $r = a$.

(c) Show that the normal components of the secondary field on either side of $r = a$ are equal.

(d) From the tangential components of the secondary field on either side of $r = a$, obtain the magnetization surface current density at $r = a$, using condition (b) stated in Problem 5.10.

(e) Show that the surface current density found in part (d) is consistent with the magnetization vector corresponding to $\mathbf{B}_i$.

5.32. An infinite plane slab of magnetic material of thickness d and having a nonuniform magnetic susceptibility given by

$$\chi_m(z) = \frac{z}{4}$$

occupies the region $1 < z < 2$. A uniform magnetic field $\mathbf{B}_a = B_0 \mathbf{i}_x$ is applied. Show that the induced magnetization volume and surface current densities are given by

$$\mathbf{J}_m = \frac{B_0}{4\mu_0} \mathbf{i}_y \qquad 1 < z < 2$$

$$\mathbf{J}_{ms} = \begin{cases} \dfrac{B_0}{4\mu_0} \mathbf{i}_y & z = 1 \\[2mm] -\dfrac{B_0}{2\mu_0} \mathbf{i}_y & z = 2 \end{cases}$$

Find the secondary and total magnetic fields both inside and outside the magnetic material.

5.33. Two perfectly conducting, infinite plane parallel sheets separated by a distance d carry uniformly distributed surface currents having equal and opposite densities $\mathbf{J}_{s0}$ and $-\mathbf{J}_{s0}$, respectively. For each of the following cases, find the magnetic flux between the current sheets per unit length along the direction of flow of the current.

(a) The medium between the two plates is free space.

(b) The medium between the two plates is a magnetic material of uniform permeability $\mu = 4\mu_0$.

(c) The medium between the two plates consists of two magnetic material slabs of thicknesses t and $d - t$ and having permeabilities $\mu_1 = 2\mu_0$ and $\mu_2 = 4\mu_0$, respectively.

(d) The medium between the two plates is a magnetic material of nonuniform permeability which varies linearly from a value of μ_1 near one plate to a value of μ_2 near the second plate.

5.34. Two perfectly conducting, infinite plane parallel sheets separated by a distance d carry uniformly distributed surface currents having equal and opposite densities. For each of the cases listed in Problem 5.33, find the required surface current densities if the magnetic flux between the current sheets per unit length along the direction of flow of the current is to be ψ_0.

5.35. An infinite plane magnetic material slab of thickness d and having a nonuniform permeability given by

$$\mu = \mu_0 \left(1 + \frac{z}{d} \right)^2$$

occupies the region $0 < z < d$. A uniform magnetic field $\mathbf{B}_a = B_0 \mathbf{i}_y$ is applied. Find the following quantities:

(a) $\mathbf{H}$ outside the magnetic material.

(b) $\mathbf{H}$ inside the magnetic material.

(c) $\mathbf{B}$ inside the magnetic material.

(d) $\mathbf{M}$ inside the magnetic material.

(e) $\mathbf{J}_{ms}$ on the surfaces $z = 0$ and $z = d$.

(f) $\mathbf{J}_m$ inside the magnetic material.

5.36. The region $a < r < b$ in cylindrical coordinates is occupied by a magnetic material. A filamentary wire situated along the z axis carries current I amp in the z direction. It is found that the magnetic flux density inside the magnetic material is given by

$$\mathbf{B} = \frac{\mu_0 I}{2\pi a} \mathbf{i}_\phi \qquad a < r < b$$

Find the following quantities:

(a) The permeability of the magnetic material.

(b) $\mathbf{J}_{ms}$ on the surfaces $r = a$ and $r = b$.

(c) $\mathbf{J}_m$ inside the magnetic material.

5.37. A portion of the **B-H** curve for a ferromagnetic material can be approximated by the analytical expression

$$\mathbf{B} = \mu_0 k H \mathbf{H}$$

where k is a constant having the units of meters per ampere. Find μ_r, μ_{ir}, χ_m, and $\mathbf{M}$.

5.38. Show that Eq. (5-162) follows from Eq. (5-163) whereas Eq. (5-161) follows from Eqs. (5-164) and (5-165).

5.39. Two infinite plane conducting sheets separated by a distance d carry uniformly distributed surface charges of densities ρ_{s0} and $-\rho_{s0}$, respectively. Find the electric stored energy per unit area of the plates if the medium between the plates is (a) free space, and (b) a dielectric of uniform permittivity $\epsilon = 4\epsilon_0$.

5.40. The region between two infinite plane conducting sheets separated by a distance d is characterized by a uniform electric field intensity E_0 directed normal to the plates. Find the electric stored energy per unit area of the plates if the medium between the plates is (a) free space, and (b) a dielectric of uniform permittivity $\epsilon = 4\epsilon_0$.

5.41. Two infinite plane conducting sheets separated by a distance d carry uniformly distributed surface currents of densities $\mathbf{J}_{s0}$ and $-\mathbf{J}_{s0}$, respectiveiy. Find the magnetic stored energy per unit area of the plates if the medium between the plates is (a) free space, and (b) a magnetic material of uniform permeability $\mu = 4\mu_0$.

5.42. The region between two infinite plane conducting sheets separated by a distance d is characterized by a uniform magnetic flux density B_0 directed tangential to the plates. Find the magnetic stored energy per unit area of the plates if the medium between the plates is (a) free space, and (b) a magnetic material of uniform permeability $\mu = 4\mu_0$.

5.43. For the **B-H** curve of Problem 5.37, find the work done per unit volume in magnetizing the material from zero to a certain value $B_0 = \mu_0 k H_0^2$.

5.44. The region $r \leq a$ in cylindrical coordinates is occupied by a magnetic material of uniform permeability μ. The magnetic field intensity is given by

$$\mathbf{H} = \begin{cases} H_0 \cos \omega t \, \mathbf{i}_z & r \leq a \\ 0 & \text{otherwise} \end{cases}$$

where H_0 is a constant. Show that the time rate of change of energy stored in the magnetic field per length l of the magnetic material is correctly given by the power flow into the material obtained by evaluating the surface integral of the Poynting vector over the surface of the cylindrical volume of length l and bounded by $r = a$.

5.45. The region $0 < z < d$ is occupied by a dielectric material of uniform permittivity ϵ. The electric field intensity is given by

$$\mathbf{E} = \begin{cases} E_0 \cos \omega t \, \mathbf{i}_z & 0 < z < d \\ 0 & \text{otherwise} \end{cases}$$

where E_0 is a constant. Assume cylindrical symmetry and show that the time rate of change of energy stored in the cylindrical volume $r < a$ of the dielectric material is correctly given by the power flow into the material obtained by evaluating the surface integral of the Poynting vector over the surface of that volume.

5.46. Medium 1, comprising the region $r < a$ in spherical coordinates, is a perfect dielectric of permittivity $\epsilon_1 = 2\epsilon_0$ whereas medium 2, comprising the region $r > a$, is a perfect dielectric of permittivity $\epsilon_2 = 4\epsilon_0$. The electric field intensity in medium 1 is given by $\mathbf{E}_1 = E_0 \, \mathbf{i}_z$, where E_0 is a constant. Find the electric field intensity at $r = a$ in medium 2.

5.47. Medium 1, comprising the region $z > 0$, is characterized by $\sigma_1 = 0$, $\epsilon_1 = \epsilon_0$, and $\mu_1 = 4\mu_0$ whereas medium 2, comprising the region $z < 0$, is characterized by $\sigma_2 = 0$, $\epsilon_2 = \epsilon_0$, and $\mu_2 = 2\mu_0$. All fields are spatially uniform in both media and independent of time. The magnetic flux density vector $\mathbf{B}_1$ in medium 1 is given by

$$\mathbf{B}_1 = B_0(2\mathbf{i}_x + 4\mathbf{i}_y + 5\mathbf{i}_z) \, \text{Wb/m}^2$$

where B_0 is a constant. The boundary $z = 0$ between the two media carries a surface current of density $\mathbf{J}_s$ given by

$$\mathbf{J}_s = \frac{B_0}{\mu_0}(\mathbf{i}_x - 2\mathbf{i}_y) \, \text{amp/m}$$

Determine the magnetic flux density vector $\mathbf{B}_2$ in medium 2.

5.48. Two infinite, perfectly conducting plates occupy the planes $x = 0$ and $x = a$. An electric field given by

$$\mathbf{E} = E_0 \sin \frac{\pi x}{a} \cos \frac{\pi t}{a\sqrt{\mu_0 \epsilon_0}} \mathbf{i}_z$$

where E_0 is a constant, exists in the medium between the plates, which is free space.

(a) Using one of Maxwell's curl equations, obtain the magnetic field associated with the given $\mathbf{E}$.

(b) Determine the surface current densities on the two plates.

5.49. The region $z < 0$ is free space and the region $z > 0$ is a perfect dielectric of permittivity $\epsilon = 4\epsilon_0$. The electric field intensities $\mathbf{E}_1$ and $\mathbf{E}_2$ in the two media are given by

$$\mathbf{E}_1 = [E_i \cos \omega(t - \sqrt{\mu_0\epsilon_0}\, z) + E_r \cos \omega(t + \sqrt{\mu_0\epsilon_0}\, z)]\mathbf{i}_x \qquad \text{for } z < 0$$

$$\mathbf{E}_2 = E_t \cos \omega(t - 2\sqrt{\mu_0\epsilon_0}\, z)\, \mathbf{i}_x \qquad\qquad\qquad \text{for } z > 0$$

where E_i, E_r, and E_t are constants.

(a) Find $\mathbf{H}_1$ and $\mathbf{H}_2$ associated with $\mathbf{E}_1$ and $\mathbf{E}_2$, respectively.

(b) Find the relationships between E_r and E_i and between E_t and E_i.

5.50. Show that, for time-varying fields, the boundary condition for the normal component of $\mathbf{B}$ follows from the boundary condition for the tangential component of $\mathbf{E}$, whereas the boundary condition for the normal component of $\mathbf{D}$ follows from the boundary conditions for the tangential component of $\mathbf{H}$ and the normal component of $\mathbf{J}$.

6

APPLIED ELECTROMAGNETICS

In Chapter 2 we set our goal to learn how to interpret Maxwell's equations and the associated constitutive relations and to use them to discuss various applications. In the preceding chapters we achieved the first task, namely that of introducing and understanding Maxwell's equations,

$$\mathbf{V} \cdot \mathbf{D} = \rho$$

$$\mathbf{V} \cdot \mathbf{B} = 0$$

$$\mathbf{V} \times \mathbf{E} = -\frac{\partial \mathbf{B}}{\partial t}$$

$$\mathbf{V} \times \mathbf{H} = \mathbf{J} + \frac{\partial \mathbf{D}}{\partial t}$$

and the various related concepts. We now have the basic electromagnetic theory necessary to venture into the realm of applied electromagnetic theory to which this chapter serves as an introduction. The topics of applied electromagnetic theory are varied, but perhaps the most important among them are concerned with the field basis of circuit theory and with electromagnetic waves. This is reflected in the topics covered in this chapter.

PART I.   Statics, Quasistatics, and Distributed Circuits

6.1   Poisson's Equation

Maxwell's equations for the static electric field are given by

$$\mathbf{V} \cdot \mathbf{D} = \rho \tag{6-1}$$

$$\mathbf{V} \times \mathbf{E} = 0 \tag{6-2}$$

In view of (6-2), $\mathbf{E}$ can be expressed as the gradient of a scalar potential V as we learned in Section 2.12. Thus we have

$$\mathbf{E} = -\mathbf{V}V \tag{6-3}$$

and

$$\mathbf{D} = \epsilon\mathbf{E} = -\epsilon\,\mathbf{V}V \tag{6-4}$$

Substituting (6-4) into (6-1), we obtain

$$\mathbf{V} \cdot \epsilon\,\mathbf{V}V = -\rho$$

or

$$\mathbf{V}V \cdot \mathbf{V}\epsilon + \epsilon\mathbf{V}^2 V = -\rho \tag{6-5}$$

where $\mathbf{V}^2 V$ is the Laplacian of V. Equation (6-5) is the differential equation for the electrostatic potential V in a region of volume charge density ρ. If we assume that ϵ is a constant in the region, $\mathbf{V}\epsilon$ is equal to zero so that (6-5) reduces to

$$\mathbf{V}^2 V = -\frac{\rho}{\epsilon} \tag{6-6}$$

Equation (6-6) is known as Poisson's equation. If the medium is charge free, then $\rho = 0$ and (6-6) reduces to

$$\mathbf{V}^2 V = 0 \tag{6-7}$$

which is known as Laplace's equation. In this section we discuss the applications of Poisson's equation by considering two examples.

EXAMPLE 6-1.   Charge is distributed with uniform density ρ_0 C/m³ throughout a sphere of radius a centered at the origin. It is desired to find the electrostatic potential and hence the electric field intensity both inside and outside the sphere by using Poisson's equation for $r < a$ and Laplace's equation for $r > a$.

From Poisson's and Laplace's equations, we have

$$\mathbf{V}^2 V = \begin{cases} -\dfrac{\rho_0}{\epsilon} & \text{for } r < a \\[2mm] 0 & \text{for } r > a \end{cases} \tag{6-8}$$

Because of the spherical symmetry of the charge distribution, the potential is a function of r only. Thus all derivatives of V with respect to θ and ϕ are

348

zero, so that (6-8) becomes

$$\frac{\partial}{\partial r}\left(r^2 \frac{\partial V}{\partial r}\right) = \begin{cases} -\dfrac{\rho_0 r^2}{\epsilon} & \text{for } r < a \\ 0 & \text{for } r > a \end{cases} \tag{6-9}$$

Integrating both sides of (6-9) with respect to r and then dividing by r^2 and integrating again with respect to r, we obtain

$$V = \begin{cases} -\dfrac{\rho_0 r^2}{6\epsilon} - \dfrac{A}{r} + B & \text{for } r < a \\ -\dfrac{C}{r} + D & \text{for } r > a \end{cases} \tag{6-10}$$

where A, B, C, and D are arbitrary constants of integration.

We now have to evaluate the arbitrary constants A, B, C, and D by using the boundary conditions and other considerations. First, the potential can be arbitrarily set equal to zero at $r = \infty$, so that $D = 0$. Second, from $\mathbf{E} = -\nabla V$, we have

$$\mathbf{E} = \begin{cases} \left(\dfrac{\rho_0 r}{3\epsilon} - \dfrac{A}{r^2}\right)\mathbf{i}_r, & \text{for } r < a \\ -\dfrac{C}{r^2}\mathbf{i}_r, & \text{for } r > a \end{cases}$$

so that, from Gauss' law, the charge contained within a sphere of radius $r\,(< a)$ centered at the origin is $4\pi r^2 \epsilon E_r = 4\pi r^2 \epsilon[(\rho_0 r/3\epsilon) - (A/r^2)]$. For the charge distribution under consideration, this quantity must approach zero as $r \rightarrow 0$. This is possible only if A is equal to zero. The solution for V is thus reduced to

$$V = \begin{cases} -\dfrac{\rho_0 r^2}{6\epsilon} + B & \text{for } r < a \\ -\dfrac{C}{r} & \text{for } r > a \end{cases} \tag{6-11}$$

Next, we note that, at the boundary $r = a$, the potential must be continuous in the absence of an impulse type of discontinuity in the normal component of electric field intensity (such discontinuities can exist in the presence of an electric dipole layer at the surface). Also, at $r = a$, D_r must be continuous in the absence of a surface charge. Using these two boundary conditions, we have

$$-\frac{\rho_0 a^2}{6\epsilon} + B = -\frac{C}{a} \tag{6-12a}$$

$$\epsilon \frac{\rho_0 a}{3\epsilon} = -\epsilon \frac{C}{a^2} \tag{6-12b}$$

Solving (6-12a) and (6-12b) for B and C and substituting in (6-11), we obtain the required solution for V as

$$V = \begin{cases} -\dfrac{\rho_0 r^2}{6\epsilon} + \dfrac{\rho_0 a^2}{2\epsilon} & \text{for } r < a \\[2ex] \dfrac{\rho_0 a^3}{3\epsilon r} & \text{for } r > a \end{cases} \tag{6-13}$$

The corresponding solution for $\mathbf{E}$ is

$$\mathbf{E} = \begin{cases} \dfrac{\rho_0 r}{3\epsilon}\mathbf{i}_r & \text{for } r < a \\[2ex] \dfrac{\rho_0 a^3}{3\epsilon r^2}\mathbf{i}_r & \text{for } r > a \end{cases} \tag{6-14}$$

We note that this solution is in agreement with the result of Example 2-6. ∎

In the preceding example we illustrated the method of solving for the potential for a given charge distribution, using Poisson's and Laplace's equations. However, Poisson's equation is more useful for another class of problems, in which the charge distribution is the quantity to be determined. We now consider an example of this type.

EXAMPLE 6-2.   A simplified model of a vacuum diode consists of two parallel conducting plates occupying the planes $x = 0$ and $x = d$, between which an electric field is established by maintaining a potential difference of V_0 volts as shown in Fig. 6.1. The plate at the lower potential is called the cathode and the plate at the higher potential is called the anode. The cathode is heated so that it emits electrons into the space between the plates, to be collected by the anode and thereby establish a current flow. Let us assume for simplicity that (a) the electrons are emitted from the cathode with zero

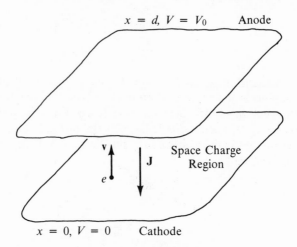

Fig. 6.1. Simplified model of a vacuum diode.

initial velocity and (b) the number of electrons emitted from the cathode is limited not by the cathode temperature but by the space charge between the cathode and the anode. For steady current flow under these conditions, the electric field at the cathode is zero. If it were some nonzero value and directed towards the cathode, the electrons would be emitted from the cathode with some acceleration and the current would then be temperature limited but not space-charge limited. (In the actual case, the field intensity is slightly nonzero and directed towards the cathode, since no electrons would leave the cathode otherwise.) If the field intensity were some nonzero value and directed towards the anode, there would be no space charge, since the electrons could not leave the cathode. It is desired to find the potential distribution and hence the space charge distribution between the cathode and the anode.

Let V be the potential at a distance x from the cathode, which is considered to be at zero potential. Then the work done by the electric field in moving an electron through a distance x from the cathode is equal to $|e|V$, where e is the charge of the electron. This work must be equal to the kinetic energy acquired by the electron. Thus, denoting $\mathbf{v} = v(x)\mathbf{i}_x$ as the velocity of the electron, we have

$$|e|V = \tfrac{1}{2}mv^2 \tag{6-15}$$

where m is the electronic mass. From (6-15), we get $v = \sqrt{2|e|V/m}$ so that

$$\mathbf{v} = \sqrt{\frac{2|e|V}{m}}\,\mathbf{i}_x \tag{6-16}$$

If $\rho(x)$ is the density of the space charge constituted by the electrons, the current density $\mathbf{J}$ is given by

$$\mathbf{J} = \rho\mathbf{v} = \rho\sqrt{\frac{2|e|}{m}}V^{1/2}\mathbf{i}_x \tag{6-17}$$

For steady current flow,

$$\mathbf{J} = J_0\mathbf{i}_x \tag{6-18}$$

where J_0 is a constant. Comparing the right sides of (6-17) and (6-18) we obtain

$$\rho = J_0\sqrt{\frac{m}{2|e|}}V^{-1/2}$$

From Poisson's equation, we now have

$$\frac{d^2V}{dx^2} = -\frac{\rho}{\epsilon_0} = -\left[\frac{J_0}{\epsilon_0}\sqrt{\frac{m}{2|e|}}\right]V^{-1/2} = kV^{-1/2} \tag{6-19}$$

where $k = -(J_0/\epsilon_0)\sqrt{m/|2e|}$ is a constant. Equation (6-19) is the differential equation for V in the region between the cathode and the anode. To solve for V, we multiply the left and right sides of (6-19) by $2(dV/dx)\,dx$ and $2\,dV$, respectively, to obtain

$$2\frac{dV}{dx}\,d\!\left(\frac{dV}{dx}\right) = 2kV^{-1/2}\,dV \tag{6-20}$$

Integrating both sides of (6-20), we get

$$\left(\frac{dV}{dx}\right)^2 = 4kV^{1/2} + A$$

where A is the constant of integration to be evaluated from the boundary condition for dV/dx at the cathode. But dV/dx is the negative of the electric field intensity. Since the electric field intensity as well as the potential are zero at the cathode, A is equal to zero. Thus

$$\frac{dV}{dx} = 2\sqrt{k}\,V^{1/4}$$

or

$$V^{-1/4}\,dV = 2\sqrt{k}\,dx \tag{6-21}$$

Integrating both sides of (6-21), we obtain

$$\tfrac{4}{3}V^{3/4} = 2\sqrt{k}\,x + B$$

where B is the constant of integration. To evaluate B, we note that $V = 0$ for $x = 0$. Hence $B = 0$, giving us

$$V = (\tfrac{3}{2}\sqrt{k}\,x)^{4/3}$$

Finally, from the condition that $V = V_0$ for $x = d$, we have

$$V_0 = (\tfrac{3}{2}\sqrt{k}\,d)^{4/3}$$

so that

$$V = V_0\!\left(\frac{x}{d}\right)^{4/3} \tag{6-22}$$

Equation (6-22) is the required solution for the potential between the two plates. The electric field intensity is given by

$$\mathbf{E} = -\nabla V = -\frac{\partial V}{\partial x}\mathbf{i}_x = -\frac{4}{3}\frac{V_0}{d}\!\left(\frac{x}{d}\right)^{1/3}\mathbf{i}_x$$

The space charge density is given by

$$\rho = \epsilon_0\,\nabla\cdot\mathbf{E} = \epsilon_0\frac{\partial E_x}{\partial x} = -\frac{4}{9}\frac{\epsilon_0 V_0}{d^2}\!\left(\frac{x}{d}\right)^{-2/3}$$

The current density is given by

$$\mathbf{J} = \rho\sqrt{\frac{2\,|e|}{m}}\,V^{1/2}\mathbf{i}_x = -\frac{4}{9}\epsilon_0\sqrt{\frac{2\,|e|}{m}}\,\frac{V_0^{3/2}}{d^2}\mathbf{i}_x$$

This equation is known as the Child–Langmuir law. The negative sign for $\mathbf{J}$ arises from the fact that the current flow is opposite to the direction of motion of the electrons.

6.2 Laplace's Equation

A very important class of problems encountered in practice are those for which the charges are confined to the surfaces of conductors. For such problems, either the charge distribution on the surfaces of the conductors, or the potentials of the conductors, or a combination of the two are specified and the problem consists of finding the potential and hence the electric field in the charge-free region bounded by the conductors. Obviously, the potential in the charge-free region satisfies Laplace's equation

$$\nabla^2 V = 0 \qquad (6\text{-}7)$$

assuming ϵ to be constant. Hence the solution consists of finding a potential that satisfies Laplace's equation and the specified boundary conditions. Since the right side of Laplace's equation is zero irrespective of the boundary conditions, we can obtain a general solution for the potential that satisfies a particular simplified form of Laplace's equation once and for all. The general solution consists of arbitrary constants of integration, which are evaluated by using the boundary conditions for the specific problem.

Let us consider the cartesian coordinate system. In the general case for which the potential is a function of all three coordinates x, y, and z, Laplace's equation is given by

$$\frac{\partial^2 V}{\partial x^2} + \frac{\partial^2 V}{\partial y^2} + \frac{\partial^2 V}{\partial z^2} = 0 \qquad (6\text{-}23)$$

However, if the potential is a function of only one of the coordinates, say x, and independent of the other two, we obtain a simplified version of Laplace's equation as

$$\frac{\partial^2 V}{\partial x^2} = \frac{d^2 V}{dx^2} = 0 \qquad (6\text{-}24)$$

Integrating (6-24) with respect to x twice, we obtain

$$V = Ax + B \qquad (6\text{-}25)$$

where A and B are the arbitrary constants of integration. Equation (6-25) is the general solution for the electrostatic potential in a charge-free region for the case in which the potential is a function of x only. In other words, all problems for which the potential varies with x only but having different boundary conditions must have solutions of the form given by (6-25). Only the values of the arbitrary constants A and B differ from one problem to the other. Thus, having found the general solution once and for all, it is a matter of fitting the given boundary conditions to evaluate the arbitrary constants for obtaining the particular solution to the problem. Let us consider a simple example.

EXAMPLE 6-3. Two parallel conducting plates occupying the planes $x = 0$ and $x = d$ are kept at potentials $V = 0$ and $V = V_0$, respectively, as shown in

Fig. 6.2. We wish to find the solution for the potential and hence for the electric field intensity between the plates and evaluate the charge densities on the plates.

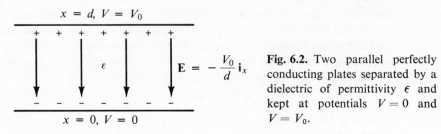

$x = d$, $V = V_0$

$$E = -\frac{V_0}{d}\mathbf{i}_x$$

$x = 0$, $V = 0$

Fig. 6.2. Two parallel perfectly conducting plates separated by a dielectric of permittivity ϵ and kept at potentials $V = 0$ and $V = V_0$.

The general solution for the potential between the two plates is given by (6-25). The boundary conditions are

$$V = 0 \qquad \text{for } x = 0$$
$$V = V_0 \qquad \text{for } x = d$$

Substituting these boundary conditions in (6-25), we have

$$0 = A(0) + B \qquad \text{or} \qquad B = 0$$
$$V_0 = A(d) + B = A(d) + 0 \qquad \text{or} \qquad A = \frac{V_0}{d}$$

Thus the required solution for the potential is

$$V = \frac{V_0}{d}x \qquad 0 < x < d$$

The electric field intensity is given by

$$\mathbf{E} = -\nabla V = -\frac{\partial V}{\partial x}\mathbf{i}_x = -\frac{V_0}{d}\mathbf{i}_x \qquad 0 < x < d$$

The field is shown sketched in Fig. 6.2. The surface charge densities on the two plates are given by

$$[\rho_s]_{x=0} = [\mathbf{D}]_{x=0} \cdot \mathbf{i}_x = -\frac{\epsilon V_0}{d}\mathbf{i}_x \cdot \mathbf{i}_x = -\frac{\epsilon V_0}{d}$$

$$[\rho_s]_{x=d} = [\mathbf{D}]_{x=d} \cdot (-\mathbf{i}_x) = \left(-\frac{\epsilon V_0}{d}\mathbf{i}_x\right) \cdot (-\mathbf{i}_x) = \frac{\epsilon V_0}{d} \quad \blacksquare$$

EXAMPLE 6-4. Let the region between the two plates in Example 6-3 consist of two dielectric media having permittivities ϵ_1 for $0 < x < t$ and ϵ_2 for $t < x < d$ as shown in Fig. 6.3. It is desired to find the solutions for the potentials in the two regions $0 < x < t$ and $t < x < d$.

Since the permittivities of the two regions are different, the solutions for the potentials in the two regions must be different although having the same general form as given by (6-25). We therefore choose different arbitrary constants for the two different regions. Thus the general solutions for the

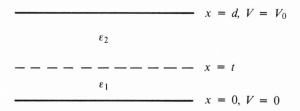

Fig. 6.3. Two parallel perfectly conducting plates separated by two dielectric media of permittivities ϵ_1 and ϵ_2 and kept at potentials $V = 0$ and $V = V_0$.

potentials in the two regions are

$$V_1 = A_1 x + B_1 \qquad 0 < x < t \tag{6-26a}$$

$$V_2 = A_2 x + B_2 \qquad t < x < d \tag{6-26b}$$

The boundary conditions specified in the problem are

$$V_1 = 0 \qquad \text{for } x = 0 \tag{6-27a}$$

$$V_2 = V_0 \qquad \text{for } x = d \tag{6-27b}$$

However, we have four arbitrary constants A_1, B_1, A_2, and B_2 to be determined. Hence we need two more boundary conditions. Obviously, we turn our attention to the boundary $x = t$ between the two dielectrics for these two conditions, which are

$$V_1 = V_2 \qquad \text{for } x = t \tag{6-27c}$$

and

$$D_{x_1} = D_{x_2}$$

or

$$\epsilon_1 \frac{\partial V_1}{\partial x} = \epsilon_2 \frac{\partial V_2}{\partial x} \qquad \text{for } x = t \tag{6-27d}$$

Substituting the four boundary conditions (6-27a)–(6-27d) into (6-26a) and (6-26b), we obtain

$$0 = A_1(0) + B_1$$

$$V_0 = A_2(d) + B_2$$

$$A_1(t) + B_1 = A_2(t) + B_2$$

$$\epsilon_1 A_1 = \epsilon_2 A_2$$

Solving these four equations for the four arbitrary constants and substituting the resulting values in (6-26a) and (6-26b), we find the required solutions for V_1 and V_2 as

$$V_1 = \frac{\epsilon_2 x}{\epsilon_2 t + \epsilon_1 (d - t)} V_0 \qquad 0 < x < t$$

$$V_2 = \frac{\epsilon_2 t + \epsilon_1 (x - t)}{\epsilon_2 t + \epsilon_1 (d - t)} V_0 \qquad t < x < d$$

The potential at the interface $x = t$ is

$$\frac{\epsilon_2 t}{\epsilon_2 t + \epsilon_1 (d - t)} V_0 \quad \blacksquare$$

Thus far we have considered the one-dimensional case for which the potential is a function of x only. The one-dimensional problems for which the potentials are a function of y only and z only are not any different from the case considered, since the differential equations for V are the same as (6-24) except that x is replaced by y or z. Thus there is only one one-dimensional problem in the cartesian coordinate system although there are three coordinates. Considering the three commonly used coordinate systems, that is, cartesian, cylindrical, and spherical coordinate systems and arguing in this manner, we note that there are only five different one-dimensional problems in all although there are nine coordinates. There is not much to be gained by considering in detail the remaining four one-dimensional problems. Hence we simply list in Table 6.1 the general solutions for each case, a particular set of boundary conditions and the corresponding particular solution. It is left as an exercise (Problem 6.3) for the student to verify these.

TABLE 6.1. General Solutions for One-Dimensional Laplace's Equations and Particular Solutions for Particular Sets of Boundary Conditions

Coordinate with Which V Varies	General Solution	Boundary Conditions	Particular Solution
x	$Ax + B$	$V = 0, \quad x = 0$ $V = V_0, x = d$	$\dfrac{V_0}{d} x$
r (cylindrical)	$A \ln r + B$	$V = 0, \quad r = a$ $V = V_0, r = b$	$\dfrac{V_0}{\ln b/a} \ln \dfrac{r}{a}$
ϕ	$A\phi + B$	$V = 0, \quad \phi = 0$ $V = V_0, \phi = \alpha$	$\dfrac{V_0}{\alpha} \phi$
r (spherical)	$\dfrac{A}{r} + B$	$V = 0, \quad r = a$ $V = V_0, r = b$	$\dfrac{V_0}{(1/b) - (1/a)} \left(\dfrac{1}{r} - \dfrac{1}{a} \right)$
θ	$A \ln \left(\tan \dfrac{\theta}{2} \right) + B$	$V = 0, \quad \theta = \alpha$ $V = V_0, \theta = \beta$	$V_0 \dfrac{\ln [(\tan \theta/2)/(\tan \alpha/2)]}{\ln [(\tan \beta/2)/(\tan \alpha/2)]}$

Before we take up the discussion of Laplace's equation in two dimensions, we consider briefly the use of analogy in solving magnetic field problems involving permanent magnetization. From Maxwell's curl equation for the static magnetic field, we have, for a region free of true currents, that is, for $\mathbf{J} = 0$,

$$\nabla \times \mathbf{H} = 0$$

We can then express $\mathbf{H}$ as the gradient of a scalar magnetic potential V_m, that is,

$$\mathbf{H} = -\nabla V_m \tag{6-28}$$

Substituting $\mathbf{B} = \mu_0(\mathbf{H} + \mathbf{M})$ in Maxwell's divergence equation for $\mathbf{B}$, we have

$$\nabla \cdot \mathbf{B} = \nabla \cdot \mu_0(\mathbf{H} + \mathbf{M}) = 0$$

or

$$\nabla \cdot \mathbf{H} = -\nabla \cdot \mathbf{M} \tag{6-29}$$

Substituting (6-28) into (6-29), we obtain

$$\nabla^2 V_m = \nabla \cdot \mathbf{M} \tag{6-30}$$

Comparing (6-28) and (6-30) with (6-3) and (6-6), respectively, we observe the following analogy:

$$\mathbf{H} \longleftrightarrow \mathbf{E} \tag{6-31a}$$

$$V_m \longleftrightarrow V \tag{6-31b}$$

$$\nabla \cdot \mathbf{M} \longleftrightarrow -\frac{\rho}{\epsilon} \tag{6-31c}$$

If $\mathbf{M}$ is discontinuous at a boundary, then $\nabla \cdot \mathbf{M}$ results in an impulse function. To find the appropriate analogy, we consider a rectangular box of infinitesimal volume Δv and enclosing a portion of the boundary at which $\mathbf{M}$ is discontinuous as shown in Fig. 6.4. Then we have

$$\int_{\Delta v} \nabla \cdot \mathbf{M} \, dv \longleftrightarrow -\int_{\Delta v} \frac{\rho}{\epsilon} \, dv$$

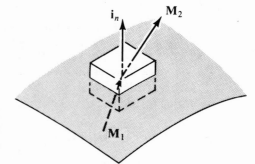

Fig. 6.4. For showing that a discontinuity in $\mathbf{M}$ at a boundary is analogous to a surface charge density.

From the divergence theorem, $\int_{\Delta v} \nabla \cdot \mathbf{M} \, dv$ is equal to $\oint_S \mathbf{M} \cdot \mathbf{i}_n \, dS$, where S is the surface area of the box. Now, if we let the box shrink to the boundary, this integral becomes $(\mathbf{M}_2 - \mathbf{M}_1) \cdot \mathbf{i}_n \Delta S$ whereas $\int_{\Delta v} (\rho/\epsilon) \, dv$ becomes $(\rho_s/\epsilon) \Delta S$, where ΔS is the surface area on the boundary to which the box shrinks and ρ_s is the surface charge density. Thus we have

$$(\mathbf{M}_2 - \mathbf{M}_1) \cdot \mathbf{i}_n \Delta S \longleftrightarrow -\frac{\rho_s}{\epsilon} \Delta S$$

or

$$(\mathbf{M}_2 - \mathbf{M}_1) \cdot \mathbf{i}_n \longleftrightarrow -\frac{\rho_s}{\epsilon} \qquad (6\text{-}31\text{d})$$

Making use of the analogy indicated by (6-31a)–(6-31d), we can solve magnetostatic problems involving permanent magnetization. Let us consider an example.

EXAMPLE 6-5.   The region $0 < x < d$ is occupied by a medium characterized by the magnetization vector $\mathbf{M} = M_0\mathbf{i}_x$, where M_0 is a constant, as shown in Fig. 6.5(a). It is desired to find $\mathbf{H}$ and $\mathbf{B}$ both inside and outside the region $0 < x < d$.

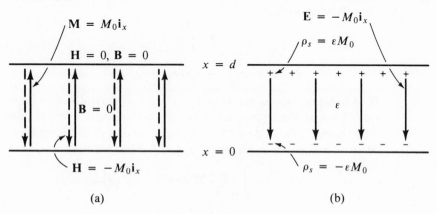

(a)   (b)

Fig. 6.5. (a) A medium characterized by magnetization vector $\mathbf{M} = M_0\mathbf{i}_x$. (b) Electrostatic analog of (a).

Since there are no true currents associated with the medium, we can use the analogy developed above. For the given $\mathbf{M}$, $\nabla \cdot \mathbf{M} = 0$ so that the analogous volume charge density is zero. However,

$$(\mathbf{M}_2 - \mathbf{M}_1) \cdot \mathbf{i}_n = \begin{cases} (0 - M_0\mathbf{i}_x) \cdot (-\mathbf{i}_x) = M_0 & \text{for } x = 0 \\ (0 - M_0\mathbf{i}_x) \cdot \mathbf{i}_x = -M_0 & \text{for } x = d \end{cases}$$

The analogous surface charge density is therefore given by

$$\rho_s = \begin{cases} -\epsilon M_0 & \text{for } x = 0 \\ \epsilon M_0 & \text{for } x = d \end{cases}$$

From the solution to Example 6-3, the electrostatic potential and the electric field intensity for this surface charge distribution are

$$V = M_0 x$$

$$\mathbf{E} = \begin{cases} -M_0\mathbf{i}_x & 0 < x < d \\ 0 & \text{otherwise} \end{cases}$$

The surface charge distribution and the electric field lines are shown in Fig. 6.5(b). Now, from (6-31a), the required magnetic field intensity is given by

$$\mathbf{H} = \begin{cases} -M_0\mathbf{i}_x & 0 < x < d \\ 0 & \text{otherwise} \end{cases}$$

The corresponding magnetic flux density is

$$\mathbf{B} = \mu_0(\mathbf{H} + \mathbf{M}) = \begin{cases} \mu_0(-M_0\mathbf{i}_x + M_0\mathbf{i}_x) & 0 < x < d \\ \mu_0(0 + 0) & \text{otherwise} \end{cases}$$
$$= 0 \qquad\qquad\qquad\qquad\qquad \text{everywhere}$$

These are shown in Fig. 6.5(a). ∎

We now consider the solution of Laplace's equation in two dimensions. If the potential is a function of the two coordinates x and y and independent of z, then it satisfies the equation

$$\frac{\partial^2 V}{\partial x^2} + \frac{\partial^2 V}{\partial y^2} = 0 \tag{6-32}$$

Equation (6-32) is a partial differential equation in two dimensions x and y. The technique by means of which it is solved is known as the "separation of variables" technique. It consists of assuming that the solution for the potential is the product of two functions, one of which is a function of x only and the second, a function of y only. Denoting these functions to be X and Y, respectively, we have

$$V(x, y) = X(x)\,Y(y) \tag{6-33}$$

Substituting this assumed solution into the differential equation, we obtain

$$Y\frac{d^2 X}{dx^2} + X\frac{d^2 Y}{dy^2} = 0 \tag{6-34}$$

Dividing both sides of (6-34) by XY and rearranging, we get

$$\frac{1}{X}\frac{d^2 X}{dx^2} = -\frac{1}{Y}\frac{d^2 Y}{dy^2} \tag{6-35}$$

The left side of (6-35) involves x only whereas the right side involves y only. Thus Eq. (6-35) states that a function of x only is equal to a function of y only. A function of x only other than a constant cannot be equal to a function of y only other than the same constant for all values of x and y. For example, $2x$ is equal to $4y$ for only those pairs of values of x and y for which $x = 2y$. But we are seeking a solution which is good for all pairs of x and y. Thus the only solution which satisfies (6-35) is that each side of (6-35) must be equal to a constant. Denoting this constant as α^2, we have

$$\frac{d^2 X}{dx^2} = \alpha^2 X \tag{6-36a}$$

and

$$\frac{d^2Y}{dy^2} = -\alpha^2 Y \qquad (6\text{-}36\text{b})$$

Note that we have obtained two ordinary differential equations involving the separate independent variables x and y, respectively, starting with the partial differential equation involving both of the variables x and y. It is for this reason that the method is known as the separation of variables technique. The constant α^2 is known as the separation constant.

For a nonzero α^2, the solutions for Eq. (6-36a) must be functions of x which when differentiated twice result in the same functions multiplied by α^2. The functions that satisfy this property are the exponential functions $e^{\alpha x}$ and $e^{-\alpha x}$. Since (6-36a) is a linear differential equation, the general solution consists of a superposition of the two solutions multiplied by arbitrary constants. For $\alpha^2 = 0$, the solution for (6-36a) can be obtained by integrating it twice. Thus

$$X(x) = \begin{cases} Ae^{\alpha x} + Be^{-\alpha x} & \text{for } \alpha \neq 0 \\ A_0 x + B_0 & \text{for } \alpha = 0 \end{cases} \qquad (6\text{-}37\text{a})$$

where A, B, A_0, and B_0 are the arbitrary constants. Similarly, for $\alpha^2 \neq 0$, the solutions for Eq. (6-36b) must be functions of y which when differentiated twice result in the same functions multiplied by $-\alpha^2$. The functions that satisfy this property are $\cos \alpha y$ and $\sin \alpha y$. Again, since (6-36b) is a linear differential equation, the general solution consists of a superposition of the two solutions multiplied by arbitrary constants. For $\alpha^2 = 0$, the solution for (6-36b) can be obtained by integrating it twice. Thus

$$Y(y) = \begin{cases} C \cos \alpha y + D \sin \alpha y & \text{for } \alpha \neq 0 \\ C_0 y + D_0 & \text{for } \alpha = 0 \end{cases} \qquad (6\text{-}37\text{b})$$

where C, D, C_0, and D_0 are the arbitrary constants. Substituting (6-37a) and (6-37b) into (6-33), we obtain the required solution for (6-32) as

$$V(x, y) = \begin{cases} (Ae^{\alpha x} + Be^{-\alpha x})(C \cos \alpha y + D \sin \alpha y) & \text{for } \alpha \neq 0 \\ (A_0 x + B_0)(C_0 y + D_0) & \text{for } \alpha = 0 \end{cases} \qquad (6\text{-}38)$$

We now consider an example of the application of (6-38).

EXAMPLE 6-6. Let us consider the idealized problem of an infinitely long rectangular slot cut in a semiinfinite plane conducting slab held at zero potential as shown in Fig. 6.6. With reference to the coordinate system shown in the figure, assume that a potential distribution given by $V = V_0 \sin(\pi y/b)$, where V_0 is a constant, is created at the mouth $x = a$ of the slot by the application of a potential to an appropriately shaped conductor away from the mouth of the slot not shown in the figure. It is desired to find the potential distribution in the slot.

The problem is two dimensional in x and y and hence the general solution

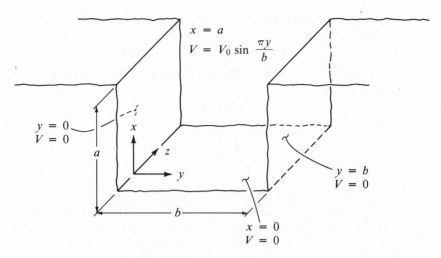

Fig. 6.6. An infinitely long rectangular slot cut in a semiinfinite plane conducting slab at zero potential. The potential at the mouth of the slot is $V_0 \sin(\pi y/b)$ volts.

for V is given by (6-38). The boundary conditions are

$V = 0$	$y = 0, 0 < x < a$	(6-39a)
$V = 0$	$y = b, 0 < x < a$	(6-39b)
$V = 0$	$x = 0, 0 < y < b$	(6-39c)
$V = V_0 \sin \dfrac{\pi y}{b}$	$x = a, 0 < y < b$	(6-39d)

The solution corresponding to $\alpha = 0$ does not fit the boundary conditions since V is required to be zero for two values of y and in the range $0 < x < a$. Hence we can ignore that solution and consider only the solution for $\alpha \neq 0$. Applying the boundary condition (6-39a), we have

$$0 = (Ae^{\alpha x} + Be^{-\alpha x})(C) \qquad \text{for } 0 < x < a$$

The only way of satisfying this equation for a range of values of x is by setting $C = 0$. Next, applying the boundary condition (6-39c), we have

$$0 = (A + B)D \sin \alpha y \qquad \text{for } 0 < y < b$$

This requires that $(A + B)D = 0$, which can be satisfied by either $D = 0$ or $A + B = 0$. However, $D = 0$ results in a trivial solution of zero for the potential. Hence we set

$$A + B = 0 \qquad \text{or} \qquad B = -A$$

Thus the solution for V reduces to

$$\begin{aligned} V(x, y) &= (Ae^{\alpha x} - Ae^{-\alpha x})D \sin \alpha y \\ &= A' \sinh \alpha x \sin \alpha y \end{aligned} \qquad (6\text{-}40)$$

where $A' = 2AD$. Next, applying boundary condition (6-39b) to (6-40), we have

$$0 = A' \sinh \alpha x \sin \alpha b \qquad \text{for } 0 < x < a$$

To satisfy this equation without obtaining a trivial solution of zero for the potential, we set

$$\sin \alpha b = 0$$

or

$$\alpha b = n\pi \qquad n = 1, 2, 3, \ldots$$

$$\alpha = \frac{n\pi}{b} \qquad n = 1, 2, 3, \ldots \tag{6-41}$$

Since several values of α given by (6-41) satisfy the boundary condition, several solutions are possible for the potential. To take this fact into account, we write the solution as the superposition of all these solutions multiplied by different arbitrary constants. In this manner we obtain

$$V(x, y) = \sum_{n=1, 2, 3, \ldots}^{\infty} A'_n \sinh \frac{n\pi x}{b} \sin \frac{n\pi y}{b} \qquad \text{for } 0 < y < b \tag{6-42}$$

Finally, applying the boundary condition (6-39d) to (6-42), we get

$$V_0 \sin \frac{\pi y}{b} = \sum_{n=1, 2, 3, \ldots}^{\infty} A'_n \sinh \frac{n\pi a}{b} \sin \frac{n\pi y}{b} \qquad \text{for } 0 < y < b \tag{6-43}$$

On the right side of (6-43), we have an infinite series of sine terms in y whereas on its left side, we have only one sine term in y. Equating the coefficients of the sine terms having the same arguments, we obtain

$$A'_n \sinh \frac{n\pi a}{b} = \begin{cases} V_0 & \text{for } n = 1 \\ 0 & \text{for } n \neq 1 \end{cases}$$

or

$$A'_1 = \frac{V_0}{\sinh (\pi a/b)}$$

$$A'_n = 0 \qquad \text{for } n \neq 1$$

Substituting this result in (6-42), we obtain the required solution for V as

$$V(x, y) = V_0 \frac{\sinh (\pi x/b)}{\sinh (\pi a/b)} \sin \frac{\pi y}{b} \tag{6-44}$$

Having found the solution, it is always worthwhile to check if it satisfies Laplace's equation and the given boundary conditions to make sure that no error was made in obtaining the solution. The above solution does satisfy these two criteria. ∎

If the solution, irrespective of how it is obtained, satisfies Laplace's equation and the specified boundary conditions, it is *the* solution according to the uniqueness theorem. To prove this theorem, let us assume to the contrary that two solutions V_1 and V_2 are possible for the same problem.

Then each of these must satisfy Laplace's equation so that

$$\nabla^2 V_1 = 0 \tag{6-45a}$$

$$\nabla^2 V_2 = 0 \tag{6-45b}$$

The difference $V_d = V_1 - V_2$ must also satisfy Laplace's equation. Thus

$$\nabla^2 V_d = \nabla^2(V_1 - V_2) = \nabla^2 V_1 - \nabla^2 V_2 = 0 \tag{6-46}$$

Also, both V_1 and V_2 must satisfy the same boundary conditions, so that

$$[V_d]_S = [V_1 - V_2]_S = [V_1]_S - [V_2]_S = 0 \tag{6-47}$$

where S represents the boundary surface. Now, using the vector identity

$$\nabla \cdot (V\mathbf{A}) = V\nabla \cdot \mathbf{A} + \nabla V \cdot \mathbf{A}$$

we have

$$\nabla \cdot (V_d \nabla V_d) = V_d \nabla^2 V_d + |\nabla V_d|^2 \tag{6-48}$$

Integrating both sides of (6-48) throughout the volume enclosed by the boundary S, we have

$$\int_{\text{vol}} (\nabla \cdot V_d \nabla V_d) \, dv = \int_{\text{vol}} (V_d \nabla^2 V_d) \, dv + \int_{\text{vol}} |\nabla V_d|^2 \, dv \tag{6-49}$$

However, from the divergence theorem and from (6-47),

$$\int_{\text{vol}} (\nabla \cdot V_d \nabla V_d) \, dv = \oint_S (V_d \nabla V_d) \cdot d\mathbf{S} = 0$$

Also, noting that $\nabla^2 V_d = 0$ in accordance with (6-46), Eq. (6-49) reduces to

$$\int_{\text{vol}} |\nabla V_d|^2 \, dv = 0 \tag{6-50}$$

Since $|\nabla V_d|^2$ is positive everywhere, the only way that (6-50) can be satisfied is if $|\nabla V_d|^2$ is equal to zero throughout the volume of interest. Thus

$$\nabla V_d = 0$$

or

$$V_d = V_1 - V_2 = \text{constant} \tag{6-51}$$

However, V_d is equal to zero on the boundaries and hence the constant on the right side of (6-51) must be zero, giving us $V_1 = V_2$ throughout the volume of interest and thereby proving the uniqueness theorem.

EXAMPLE 6-7. The rectangular slot of Fig. 6.6 is covered at the mouth $x = a$ by a conducting plate which is kept at a potential $V = V_0$, a constant, making sure that the edges touching the corners of the slot are insulated as shown by the cross-sectional view in Fig. 6.7(a). We wish to find the potential in the slot for this new boundary condition.

Since the boundary conditions (6-39a)–(6-39c) remain the same, all we have to do to find the required solution for the potential is to substitute

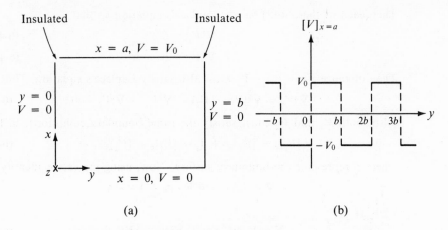

(a)   (b)

Fig. 6.7. (a) Cross section of an infinitely long rectangular slot cut in a semiinfinite plane conducting slab held at zero potential and covered at the mouth by a conducting plate held at a potential of V_0 volts. (b) Choice of potential to create an odd periodic function of period $2b$ in y for $[V]_{x=a}$.

the new boundary condition

$$V = V_0 \qquad x = a, 0 < y < b$$

in (6-42) and evaluate the coefficients A'_n. Thus we have

$$V_0 = \sum_{n=1, 2, 3, \ldots}^{\infty} A'_n \sinh \frac{n\pi a}{b} \sin \frac{n\pi y}{b} \qquad \text{for } 0 < y < b \qquad (6\text{-}52)$$

We have an infinite series of sine terms in y having periods $2b/n$ on the right side of (6-52) whereas the left side of (6-52) is a constant. Thus we cannot hope to obtain A'_n simply by comparing the coefficients of the sine terms having like arguments. If we do so, we get the ridiculous answer of $V_0 = 0$ and all $A'_n = 0$ since there is no constant term on the right side and there are no sine terms on the left side. The correct way of evaluating A'_n is to make use of the so-called orthogonality property of sine functions, which reads

$$\int_{y=0}^{p} \sin \frac{m\pi y}{p} \sin \frac{n\pi y}{p} \, dy = \begin{cases} 0 & m \neq n \\ \dfrac{p}{2} & m = n \end{cases}$$

where m and n are integers. Multiplying both sides of (6-52) by $\sin (m\pi y/b) \, dy$ and integrating between the limits 0 and b, we have

$$\int_{y=0}^{b} V_0 \sin \frac{m\pi y}{b} \, dy = \int_{y=0}^{b} \sum_{n=1, 2, 3, \ldots}^{\infty} A'_n \sinh \frac{n\pi a}{b} \sin \frac{n\pi y}{b} \sin \frac{m\pi y}{b} \, dy \qquad (6\text{-}53)$$

The integration and summation on the right side of (6-53) can be interchanged, giving us

$$\int_{y=0}^{b} V_0 \sin \frac{m\pi y}{b} \, dy = \sum_{n=1,2,3,\ldots}^{\infty} A'_n \sinh \frac{n\pi a}{b} \int_{y=0}^{b} \sin \frac{m\pi y}{b} \sin \frac{n\pi y}{b} \, dy$$

or

$$\frac{V_0 b}{m\pi}(1 - \cos m\pi) = \left(A'_m \sinh \frac{m\pi a}{b} \right) \frac{b}{2}$$

or

$$A'_m = \begin{cases} \dfrac{4V_0}{m\pi} \dfrac{1}{\sinh (m\pi a/b)} & \text{for } m \text{ odd} \\ 0 & \text{for } m \text{ even} \end{cases} \tag{6-54}$$

Substituting this result in (6-42), we obtain the required solution for the potential inside the slot as

$$V = \sum_{n=1,3,5,\ldots}^{\infty} \frac{4V_0}{n\pi} \frac{\sinh (n\pi x/b)}{\sinh (n\pi a/b)} \sin \frac{n\pi y}{b} \tag{6-55}$$

The above procedure for evaluating the constants A'_n can also be appreciated by recognizing that the right side of (6-52) is the Fourier series for an odd periodic function in y having the period $2b$. We must then have an odd periodic function of period $2b$ on the left side of (6-52). To achieve this, we note that, since the solution is for inside the slot only, it is sufficient if we satisfy the boundary condition for $[V]_{x=a}$ for the range $0 < y < b$. We are therefore at liberty to choose $[V]_{x=a}$ for the remainder of y so that an odd periodic function of period $2b$ is obtained. Obviously, the choice must be as shown in Fig. 6.7(b). The evaluation of A'_n then consists of finding the coefficients of the Fourier series for this function and comparing these with the coefficients of the series on the right side of (6-52). The steps leading from (6-53) to (6-54) are essentially equivalent to this procedure.

Another class of problems for which Laplace's equation is applicable is those involving the determination of steady current in a conducting slab under the application of potential difference between different surfaces of the slab. For the steady-current condition we have

$$\mathbf{\nabla} \cdot \mathbf{J}_c = 0$$

where $\mathbf{J}_c$ is the current density. Replacing $\mathbf{J}_c$ by $\sigma\mathbf{E}$, where σ is the conductivity of the slab, we have

$$\mathbf{\nabla} \cdot \sigma\mathbf{E} = 0$$

Substituting for $\mathbf{E}$ in terms of V, we get

$$-\mathbf{\nabla} \cdot \sigma \, \mathbf{\nabla} V = 0 \tag{6-56}$$

If σ is constant, Eq. (6-56) reduces to

$$\nabla^2 V = 0$$

Thus the potential associated with the steady current flow satisfies Laplace's equation. Hence the solution for this potential can be obtained in exactly the same manner as for the charged conductor problems. In fact, the solution for the potential for a particular steady-current problem can be written down by inspection if the solution for the potential for an analogous charged conductor problem is already known and vice versa. Having found the solution for the potential, the current density can be found by using

$$\mathbf{J}_c = \sigma\mathbf{E} = -\sigma\,\nabla V \tag{6-57}$$

EXAMPLE 6-8.   A thin rectangular slab of uniform conductivity σ_0 mhos/m has its edges coated with perfectly conducting material. One of the edges is kept at a potential V_0 relative to the other three by appropriate placement of insulators as shown in Fig. 6.8(a). It is desired to find the steady-current distribution in the conductor.

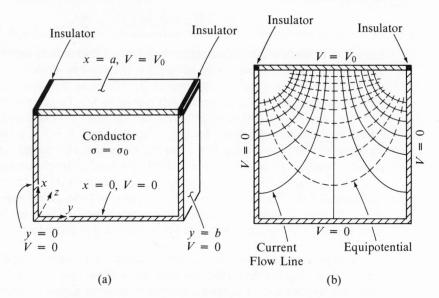

(a)   (b)

Fig. 6.8. (a) A rectangular slab of conductivity σ_0 with one of its edges kept at a potential V_0 relative to the other three. (b) Equipotentials and direction lines of current density for the conducting slab for the case $b/a = 1$.

The problem is exactly analogous to the rectangular slot problem of Example 6-7. Hence, from the solution for the potential found in that problem and given by (6-55), we obtain the required current density as

$$\mathbf{J}_c = -\sigma_0\,\nabla\left(\sum_{n=1,3,5,\ldots}^{\infty} \frac{4V_0}{n\pi}\frac{\sinh\,(n\pi x/b)}{\sinh\,(n\pi a/b)}\sin\frac{n\pi y}{b}\right)$$

$$= -\frac{4V_0\sigma_0}{b}\sum_{n=1,3,5,\ldots}^{\infty}\frac{1}{\sinh\,(n\pi a/b)}\left(\cosh\frac{n\pi x}{b}\sin\frac{n\pi y}{b}\,\mathbf{i}_x\right.$$

$$\left. + \sinh\frac{n\pi x}{b}\cos\frac{n\pi y}{b}\,\mathbf{i}_y\right)$$

The approximate shapes of the equipotentials and the direction lines of $\mathbf{J}_c$ are sketched in Fig. 6.8(b) for $b = a$, that is, for a square conducting slab. ∎

We have illustrated the solution of the two-dimensional Laplace's equation in the cartesian coordinates x and y and its applications. The technique of solution in the other coordinate systems or even in three dimensions is the same, that is, the separation of variables technique except that we get some complicated functions in certain cases. Hence, instead of pursuing this topic further, we will discuss a numerical method of solving Laplace's equation which is well suited for adaptation to a digital computer. To illustrate the principle behind the method, let us pose the following problem: Supposing we know the potentials $V_1, V_2, \ldots, V_6$ at six points which are equidistant from a point P $(0, 0, 0)$ and lying on mutually perpendicular axes (which we call x, y, and z) passing through P as shown in Fig. 6.9, how do we

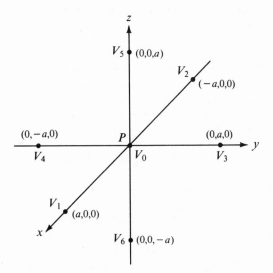

Fig. 6.9. For showing that the potential at a point P is approximately equal to the average of the potentials at six points equidistant from P and lying along mutually perpendicular axes through P.

evaluate approximately the potential at the point P consistent with Laplace's equation? To answer this question, we recognize that

$$[\nabla^2 V]_P = [\nabla^2 V]_{(0,0,0)} = \left[\frac{\partial^2 V}{\partial x^2} + \frac{\partial^2 V}{\partial y^2} + \frac{\partial^2 V}{\partial z^2}\right]_{(0,0,0)} = 0 \qquad (6\text{-}58)$$

However,

$$
\begin{aligned}
\left[\frac{\partial^2 V}{\partial x^2}\right]_{(0,0,0)} &\approx \frac{1}{a}\left\{\left[\frac{\partial V}{\partial x}\right]_{(a/2,0,0)} - \left[\frac{\partial V}{\partial x}\right]_{(-a/2,0,0)}\right\} \\
&\approx \frac{1}{a}\left\{\frac{[V]_{(a,0,0)} - [V]_{(0,0,0)}}{a} - \frac{[V]_{(0,0,0)} - [V]_{(-a,0,0)}}{a}\right\} \\
&= \frac{1}{a^2}(V_1 - V_0 - V_0 + V_2) \\
&= \frac{1}{a^2}(V_1 + V_2 - 2V_0)
\end{aligned}
\tag{6-59a}
$$

Similarly,

$$
\left[\frac{\partial^2 V}{\partial y^2}\right]_{(0,0,0)} \approx \frac{1}{a^2}(V_3 + V_4 - 2V_0)
\tag{6-59b}
$$

and

$$
\left[\frac{\partial^2 V}{\partial z^2}\right]_{(0,0,0)} \approx \frac{1}{a^2}(V_5 + V_6 - 2V_0)
\tag{6-59c}
$$

Substituting (6-59a)–(6-59c) into (6-58) and rearranging, we have

$$
V_0 \approx \frac{1}{6}(V_1 + V_2 + V_3 + V_4 + V_5 + V_6)
\tag{6-60}
$$

Thus the potential at P is approximately equal to the average of the potentials at the six equidistant points lying along mutually perpendicular axes through P. The result becomes more and more accurate as the spacing a becomes less and less. If the potential is a function of two dimensions x and y only, we then have $V_5 = V_6 = V_0$ and (6-60) reduces to

$$
V_0 \approx \frac{1}{4}(V_1 + V_2 + V_3 + V_4)
\tag{6-61}
$$

To illustrate the application of (6-61), we now consider an example.

EXAMPLE 6-9.   Two sides of an infinitely long box having a right-angled equilateral triangular cross section are kept at zero potential whereas the third side is kept at a potential of 100 volts as shown in Fig. 6.10. The region inside the box is charge free. It is divided into squares and right-angled equilateral triangles as shown in the figure. It is desired to find the potentials at the points a, b, and c using (6-61).

The solution consists of finding a set of values for the potentials at a, b, and c which, together with the potentials at points on the boundaries, are consistent with (6-61). By averaging the potentials at d, f, h, and j which are equidistant from a and lie on mutually perpendicular lines passing through it, we find an initial value of $\frac{1}{4}(0 + 0 + 0 + 100)$ or 25 volts for the potential at a. Using this value and the potentials at d, i, and j, we then find the potential at b to be $\frac{1}{4}(25 + 0 + 100 + 100)$ or 56.25. However, we round off all numbers to the nearest tenth of a volt. In rounding off a decimal

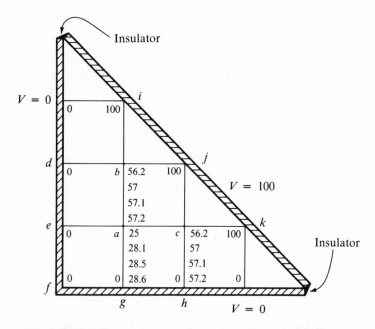

Fig. 6.10. For illustrating a numerical method of solving Laplace's equation.

ending exactly with 5, we increase the previous number by 1 if it is odd and keep it unchanged if it is even. Thus the potential at b is rounded off to 56.2 volts. Similarly, using the potentials at a, j, k, and h, we obtain a value of 56.2 volts for the potential at c. Since we now have potentials at points b and c which together with points e and g are closer to point a than the set of points d, f, h, and j, we recompute the potential at a by averaging the potentials at b, c, e, and g to obtain a value of $\frac{1}{4}(56.2 + 56.2 + 0 + 0) = 28.1$ volts. We now note that the potentials at b and c have to be recomputed because they are inconsistent with the newly computed potential at point a and the potentials at the boundary points. We thus obtain a value of $\frac{1}{4}(28.1 + 0 + 100 + 100) \approx 57.0$ volts for the potentials at b and c. This requires a revision of the potential at a to $\frac{1}{4}(57 + 57 + 0 + 0) = 28.5$ volts. This process of iteration is continued until a set of values for the potentials at a, b, and c are obtained which, together with the potentials at the boundary points, are consistent with (6-61). The final values obtained in this manner are 28.6, 57.2, and 57.2 volts for a, b, and c, respectively. Obviously, these values are approximate because of the finite spacing between the grid points. By dividing the region inside the box into smaller squares and triangles, more accurate values can be obtained. In cases where the potentials at the insulated corners are required for the computation of initial values of potentials at grid points inside, average values of potentials on either side of the corners are used. ∎

6.3   The Method of Images

We learned in Chapter 5 that a conductor surface is an equipotential. We also learned that the electric field at the conductor surface is entirely normal to it. In fact, these two properties are equivalent. In this section we will make use of this property to develop a method for computing the electric field due to charges in the presence of conductors. This method is called the "method of images." We will illustrate the method of images by means of two examples.

EXAMPLE 6-10.   A point charge Q is situated at a distance d from a grounded infinite plane conductor. We wish to find the electric field due to the point charge and the induced surface charge density on the conductor.

First, let us consider two point charges Q and $-Q$ situated at a distance $2d$ apart as shown in Fig. 6.11. The potential at any point P located at a distance r_1 from Q and r_2 from $-Q$ is given by

$$V = \frac{Q}{4\pi\epsilon r_1} - \frac{Q}{4\pi\epsilon r_2} \tag{6-62}$$

If the point P lies in the plane normal to and bisecting the line joining the point charges, r_1 is equal to r_2 and the potential is zero. Thus this plane is an equipotential. In particular, it is at zero potential. If we insert an infinite

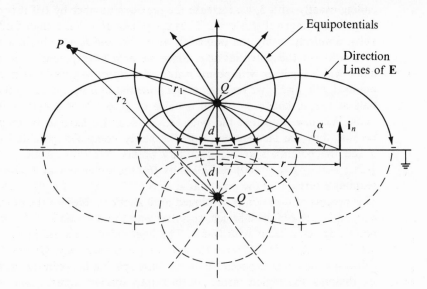

Fig. 6.11. For illustrating that the field due to a point charge Q near a grounded infinite plane conductor is the same as that due to the point charge and an "image" charge $(-Q)$ situated at the mirror image of Q in the plane conductor.

plane conductor into this plane, the field distribution due to the point charges will remain unaltered since the conductor satisfies the boundary condition. Conversely, the field due to a point charge Q situated at a distance d from a grounded infinite plane conductor is exactly the same as the field due to the charge Q plus an "image" charge $-Q$ situated at the mirror image of Q in the plane. The direction lines of the electric field intensity and the equipotential surfaces due to the dipole formed by Q and $-Q$ can be found by using the methods of Chapter 2. These are sketched in Fig. 6.11. The image charge is only a virtual charge. The field due to the real charge Q exists only on the side of that charge, with the field lines terminating on the induced charge formed on the surface of the grounded conductor. The virtual nature of the image charge is shown by the broken field lines and equipotentials on the side of the image charge.

The induced surface charge density is equal to the normal component (which is the only component present) of the displacement flux density at the conductor surface. With reference to the geometry shown in Fig. 6.11, the displacement flux density at a point on the conductor surface situated at a distance r from the projection of Q onto the surface is given by

$$\mathbf{D} = -2\frac{Q}{4\pi(d^2 + r^2)}\sin\alpha \,\mathbf{i}_n$$
$$= -\frac{Qd}{2\pi(d^2 + r^2)^{3/2}}\mathbf{i}_n \tag{6-63}$$

where $\mathbf{i}_n$ is the unit vector normal to the conductor surface. The induced surface charge density is thus given by

$$\rho_s = \mathbf{D} \cdot \mathbf{i}_n = -\frac{Qd}{2\pi(d^2 + r^2)^{3/2}} \tag{6-64}$$

The total induced surface charge Q_i is given by

$$Q_i = \int_{\substack{\text{conductor} \\ \text{surface}}} \rho_s \, dS = \int_{r=0}^{\infty} \int_{\phi=0}^{2\pi} \frac{-Qd}{2\pi(d^2 + r^2)^{3/2}} r \, dr \, d\phi$$
$$= Q \int_{\alpha=\pi/2}^{0} \cos\alpha \, d\alpha = -Q \tag{6-65}$$

Thus the total induced surface charge is equal to the image charge. This is to be expected since all field lines ending on the conductor would end on the image charge if the conductor were not present, but an actual charge of $-Q$ is present at the image point. ∎

EXAMPLE 6-11.   An infinitely long line charge of uniform density ρ_{L0} C/m is situated parallel to and at a distance d from the axis of an infinitely long grounded conducting cylinder of radius a ($<d$) as shown by the cross-sectional view in Fig. 6.12. We wish to find the image charge required for computing the field outside the conducting cylinder.

Let us postulate an infinitely long image line charge of uniform density ρ'_{L0} at a distance b from the axis of the conducting cylinder and in the plane containing the axis of the cylinder and the real line charge, as shown in Fig. 6.12. Choosing the line through point P_1 and parallel to the axis of the cylinder

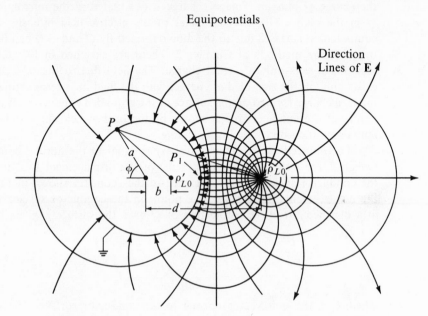

Fig. 6.12. For finding the image charge required for computing the field due to a line charge of uniform density parallel to an infinitely long grounded conducting cylinder.

as the reference for zero potential, the potential at any arbitrary point P on the conductor surface can be written as

$$V = -\frac{\rho_{L0}}{2\pi\epsilon_0} \ln \frac{\sqrt{d^2 + a^2 + 2ad \cos \phi}}{(d-a)} - \frac{\rho'_{L0}}{2\pi\epsilon_0} \ln \frac{\sqrt{b^2 + a^2 + 2ab \cos \phi}}{(a-b)}$$

$$(6\text{-}66)$$

But this quantity must be equal to zero since the conductor is an equipotential and the potential at P_1 is zero. This requires that

$$\rho'_{L0} = -\rho_{L0} \tag{6-67}$$

and

$$\ln \left[\frac{\sqrt{d^2 + a^2 + 2ad \cos \phi}}{(d-a)} \frac{(a-b)}{\sqrt{b^2 + a^2 + 2ab \cos \phi}} \right] = 0$$

or

$$\frac{\sqrt{d^2 + a^2 + 2ad \cos \phi}}{(d-a)} \frac{(a-b)}{\sqrt{b^2 + a^2 + 2ab \cos \phi}} = 1 \tag{6-68}$$

To find the solution for (6-68), let us consider $\phi = 0$. We then have

$$\left(\frac{d+a}{d-a}\right)\left(\frac{a-b}{a+b}\right) = 1$$

or

$$a^2 = bd \qquad\qquad (6\text{-}69)$$

which satisfies (6-68) for all ϕ. Thus, an image line charge of uniform density $-\rho_{L0}$ and located at a distance $b = a^2/d$ from the axis of the cylinder satisfies the equipotential requirement for the grounded conducting cylinder. The field outside the cylinder is therefore exactly the same as the field set up by the actual line charge of density ρ_{L0} at distance d from the axis and the image line charge of density $-\rho_{L0}$ at distance a^2/d from the axis. The direction lines of the electric field intensity and the associated equipotential surfaces can be obtained by the methods learned in Chapter 2. These are shown sketched in Fig. 6.12. It is left as an exercise (Problem 6.15) for the student to show that the total induced surface charge per unit length of the cylinder is equal to the image charge density $-\rho_{L0}$. The field inside the cylinder is, of course, equal to zero since the image charge is only a virtual charge. ∎

Proceeding in the same manner as in the preceding example, we can obtain the image charge for a point charge near a grounded spherical conductor. If the point charge Q is situated at a distance d from the center of the spherical conductor of radius a, the image charge is a point charge of value $-Qa/d$. It lies at a distance a^2/d from the center of the sphere, along the line joining the center to the charge Q and on the side of Q. We leave the derivation as an exercise (Problem 6.16) for the student. The method of images can also be applied for finding fields due to charges in the presence of dielectrics. We will, however, not pursue that topic here.

6.4 Conductance, Capacitance, and Inductance

In Chapter 5 we introduced conductors, dielectrics, and magnetic materials. Let us now consider three different arrangements, each consisting of two parallel perfectly conducting plates as shown in Figs. 6.13(a), (b), and (c). For the structure of Fig. 6.13(a), the medium between the parallel plates is filled with a conducting material of uniform conductivity σ. For the structure of Fig. 6.13(b), the medium between the parallel plates is filled with a perfect dielectric of uniform permittivity ϵ. For the structure of Fig. 6.13(c), the two parallel plates are joined at one end of the structure by a perfectly conducting plate and the medium between the plates is filled with a magnetic material of uniform permeability μ. Note that free space may be considered as a perfect dielectric of permittivity ϵ_0 and a magnetic material of permeability μ_0. We apply a potential difference of V_0 volts between the parallel plates

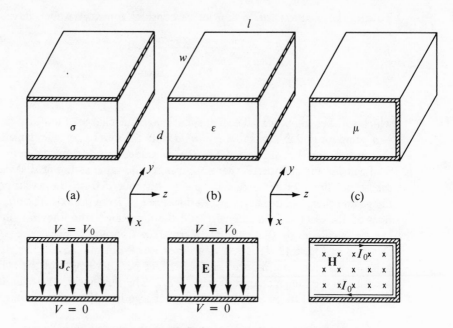

Fig. 6.13. Three different structures each consisting of two parallel perfectly conducting plates. The medium between the plates is a conductor for structure (a), a dielectric for structure (b), and a magnetic material for structure (c). The two plates are joined at one end by another perfectly conducting plate for structure (c).

of structures (a) and (b) by connecting appropriate constant voltage sources which are not shown in the figure. We pass a z-directed surface current I_0 uniformly distributed in the y direction along the upper plate of structure (c) and return it in the opposite direction along the lower plate by connecting an appropriate constant current source which is not shown in the figure.

The medium between the plates of structure (a) is then characterized by an electric field from the upper to the lower plate and hence by a conduction current in the same direction. The medium between the plates of structure (b) is characterized by an electric field only from the upper to the lower plate and no current. The medium between the plates of structure (c) is characterized by a magnetic field parallel to the plates and towards the direction of advance of a right-hand screw as it is turned in the sense of the current flow. Since the conduction current cannot leave the conductor, it has to be tangential to the conductor surface. This forces the electric field for structure (a) to be in the x direction. On the other hand, the electric field at the surface of a dielectric need not be tangential to it. This results in fringing of the electric field in the case of structure (b). However, by assuming that *d* is very small compared to *w* and *l*, or by assuming that the structure is actually part of a much larger structure, we can neglect fringing and consider the electric field to be entirely in the *x* direction. For the same assumption in the case of structure (c), the magnetic field can be considered to be entirely in the *y* direction.

From the result of Example 6-3, the electric field in the case of structures (a) and (b) is then given by

$$\mathbf{E} = \frac{V_0}{d}\mathbf{i}_x \tag{6-70}$$

The current density $\mathbf{J}_c$ for structure (a) is given by

$$\mathbf{J}_c = \sigma\mathbf{E} = \frac{\sigma V_0}{d}\mathbf{i}_x \tag{6-71}$$

The total current I_c flowing from the upper plate to the lower plate is given by the surface integral of the current density over the cross section of the conductor. However, since the current density is uniform and directed normal to the plates, we can obtain this current by simply multiplying the magnitude of the current density by the area of the plates. Thus

$$I_c = J_c(wl) = \frac{\sigma V_0}{d}wl \tag{6-72}$$

We now define a quantity known as the "conductance" (○──ⱳⱳ──○), denoted by the symbol *G*, as the ratio of the current flowing from one plate to the other to the potential difference between the plates. From (6-72), the the conductance of the conducting slab arrangement of Fig. 6.13(a) is given by

$$G = \frac{I_c}{V_0} = \frac{\sigma wl}{d} \tag{6-73}$$

We note from (6-73) that the conductance is a function purely of the dimensions of the conductor and its conductivity. The units of conductance are (mhos/meter)(meter²/meter) or mhos. The reciprocal of the "conductance" is the "resistance" (○──ⱳⱳ──○), which is denoted by the symbol *R* and has the units of ohms. Thus

$$R = \frac{V_0}{I_c}$$

or

$$V_0 = I_c R$$

which is the familiar form of Ohm's law applicable to a finite region of conducting material. The resistance of the slab conductor is given by

$$R = \frac{d}{\sigma wl} = \frac{d}{\sigma A}$$

where *A* is the area of the plates.

The phenomenon associated with conduction current is power dissipation. From Chapter 5, the power dissipation density is given by

$$p_d = \mathbf{J}_c \cdot \mathbf{E} = \sigma \mathbf{E} \cdot \mathbf{E} = \sigma E^2 \qquad (6\text{-}74)$$

Performing volume integration of the power dissipation density over the volume of the conductor of Fig. 6.13(a), we obtain the total power dissipated in the conductor as

$$
\begin{aligned}
P_d &= \int_{\text{vol}} p_d \, dv = \int_{\text{vol}} \sigma E^2 \, dv \\
&= \int_{\text{vol}} \frac{\sigma V_0^2}{d^2} \, dv \\
&= \frac{\sigma V_0^2}{d^2} (\text{volume of the conductor}) \\
&= \frac{\sigma V_0^2}{d^2} (dwl) = \frac{\sigma wl}{d} V_0^2 = G V_0^2
\end{aligned}
\qquad (6\text{-}75)
$$

Equation (6-75) gives the physical interpretation that conductance is the parameter associated with power dissipation in a conductor.

Turning our attention to the structure of Fig. 6.13(b), the displacement flux density is given by

$$\mathbf{D} = \epsilon \mathbf{E} = \frac{\epsilon V_0}{d} \mathbf{i}_x \qquad (6\text{-}76)$$

The surface charge density on the upper plate is given by

$$[\rho_s]_{x=0} = [\mathbf{D}]_{x=0} \cdot (\mathbf{i}_x) = \frac{\epsilon V_0}{d} \qquad (6\text{-}77a)$$

The surface charge density on the lower plate is given by

$$[\rho_s]_{x=d} = [\mathbf{D}]_{x=d} \cdot (-\mathbf{i}_x) = -\frac{\epsilon V_0}{d} \qquad (6\text{-}77b)$$

The total charge on either plate is given by the surface integral of the charge density on that plate over the area of the plate. However, since the charge densities here are uniform, we can obtain the total charge simply by multiplying the charge density by the area of the plate. Thus the magnitude Q of the charge on either plate is given by

$$Q = \rho_s(wl) = \frac{\epsilon V_0}{d} wl \qquad (6\text{-}78)$$

We now define a quantity known as the "capacitance" (o——|⊢—o), denoted by the symbol C, as the ratio of the magnitude of the charge on either plate to the potential difference between the plates. From (6-78), the capacitance of the dielectric slab arrangement of Fig. 6.13(b) is given by

$$C = \frac{Q}{V_0} = \frac{\epsilon wl}{d} \qquad (6\text{-}79)$$

We note from (6-79) that the capacitance is a function purely of the dimensions of the dielectric slab and its permittivity. The units of capacitance are (farads/meter)(meter²/meter) or farads.

The phenomenon associated with the electric field in a dielectric medium is energy storage. From Chapter 5, the electric stored energy density is given by

$$w_e = \frac{1}{2}\mathbf{D} \cdot \mathbf{E} = \frac{1}{2}\epsilon E^2 \tag{6-80}$$

Performing volume integration of the electric stored energy density over the volume of the dielectric of Fig. 6.13(b), we obtain the total electric stored energy in the dielectric as

$$
\begin{aligned}
W_e &= \int_{\text{vol}} w_e \, dv = \int_{\text{vol}} \frac{1}{2}\epsilon E^2 \, dv \\
&= \int_{\text{vol}} \frac{1}{2}\frac{\epsilon V_0^2}{d^2} \, dv \\
&= \frac{1}{2}\frac{\epsilon V_0^2}{d^2}(\text{volume of the dielectric}) \\
&= \frac{1}{2}\frac{\epsilon V_0^2}{d^2}(dwl) = \frac{1}{2}\frac{\epsilon wl}{d}V_0^2 = \frac{1}{2}CV_0^2
\end{aligned}
\tag{6-81}
$$

Equation (6-81) gives the physical interpretation that capacitance is the parameter associated with storage of electric energy in a dielectric.

Turning our attention to the structure of Fig. 6.13(c) and neglecting fringing, the magnetic field intensity between the plates is the same as that due to infinite plane current sheets of densities given by

$$
\mathbf{J} = \begin{cases} \dfrac{I_0}{w}\mathbf{i}_z & \text{for } x = 0 \\[2ex] -\dfrac{I_0}{w}\mathbf{i}_z & \text{for } x = d \end{cases}
$$

Hence the magnetic field intensity is uniform between the plates and zero outside the plates, that is,

$$
\mathbf{H} = \begin{cases} H_0\mathbf{i}_y & 0 < x < d \\ 0 & \text{otherwise} \end{cases}
$$

From the boundary condition for the tangential magnetic field intensity, the value of H_0 is equal to the surface current density I_0/w since the field is zero outside the plates. Thus

$$\mathbf{H} = \frac{I_0}{w}\mathbf{i}_y \qquad \text{for } 0 < x < d$$

and

$$\mathbf{B} = \mu\mathbf{H} = \frac{\mu I_0}{w}\mathbf{i}_y \qquad \text{for } 0 < x < d \tag{6-82}$$

The magnetic flux ψ linking the current I_0 is given by the surface integral of the magnetic flux density over the area bounded by any contour along which the current flows. This area is simply the cross-sectional area of the magnetic material normal to the magnetic field lines. Since the magnetic field lines are straight, it may seem like they do not link the current. However, straight lines are circles of infinite radii and hence the magnetic field does link the current. For the structure of Fig. 6.13(c), since the magnetic flux density is uniform, we can obtain the required magnetic flux ψ by simply multiplying the magnetic flux density by the cross-sectional area normal to it. The quantity ψ is known as the magnetic flux linkage associated with the current I_0. Thus

$$\psi = B_y(dl) = \frac{\mu I_0}{w} dl \tag{6-83}$$

We now define a quantity known as the "inductance" ($\circ\!\!-\!\!\curvearrowright\!\!\curvearrowright\!\!-\!\!\circ$), denoted by the symbol L, as the ratio of the magnetic flux linkage associated with the current I_0 to the current I_0. From (6-83), the inductance of the magnetic material slab arrangement of Fig. 6.13(c) is given by

$$L = \frac{\psi}{I_0} = \frac{\mu dl}{w} \tag{6-84}$$

We note from (6-84) that the inductance is a function purely of the dimensions of the magnetic material and its permeability. The units of inductance are (henrys/meter)(meter^2/meter) or henrys.

The phenomenon associated with magnetic field in a magnetic material medium is energy storage. From Chapter 5, the magnetic stored energy density is given by

$$w_m = \frac{1}{2}\mathbf{H} \cdot \mathbf{B} = \frac{1}{2}\mu H^2 \tag{6-85}$$

Performing volume integration of the magnetic stored energy density over the volume of the magnetic material of Fig. 6.13(c), we obtain the total magnetic stored energy in the magnetic material as

$$
\begin{aligned}
W_m &= \int_{\text{vol}} w_m \, dv = \int_{\text{vol}} \frac{1}{2}\mu H^2 \, dv \\
&= \int_{\text{vol}} \frac{1}{2}\frac{\mu I_0^2}{w^2} \, dv \\
&= \frac{1}{2}\frac{\mu I_0^2}{w^2}(\text{volume of the magnetic material}) \\
&= \frac{1}{2}\frac{\mu I_0^2}{w^2}(dwl) = \frac{1}{2}\frac{\mu dl}{w}I_0^2 = \frac{1}{2}LI_0^2
\end{aligned}
\tag{6-86}
$$

Equation (6-86) gives the physical interpretation that inductance is the parameter associated with storage of magnetic energy in a magnetic material.

To write general expressions for the conductance, capacitance and inductance in terms of the fields, let us consider the three structures shown by the cross-sectional views in Figs. 6.14(a), (b), and (c), which consist of identical pairs of parallel, infinitely long, perfect conductors having arbitrary but uniform cross sections. Let the medium between the two conductors of structures (a), (b), and (c) be characterized by uniform conductivity σ,

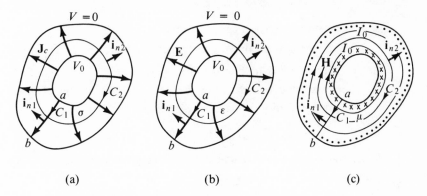

Fig. 6.14. For writing general expressions for (a) conductance, (b) capacitance, and (c) inductance.

uniform permittivity ϵ, and uniform permeability μ, respectively. As in the case of the structures of Fig. 6.13, we apply a potential difference between the two conductors of structures (a) and (b) and pass a current into the plane of the paper along one conductor of structure (c), returning it out of the plane of the paper along its second conductor. Structures (a) and (b) are then characterized by an electric field whose direction lines originate normal to the inner conductor and terminate normal to the outer conductor. The electric field results in a conduction current in structure (a). Structure (c) is characterized by a magnetic field, whose direction lines lie in the plane of the paper and surround the inner conductor.

Let us now consider unit lengths of the three structures normal to the plane of the paper. We can then write the following quantities:

For structure (a),

V_0, potential difference between the conductors $= \int_a^b \mathbf{E} \cdot d\mathbf{l}$ (6-87a)

I_c, current flowing from the inner to the outer conductor
$=$ current crossing the area formed by the contour C_2 and length unity in the axial direction

$$= \oint_{C_2} \mathbf{J}_c \cdot \mathbf{i}_{n2} \, dl = \sigma \oint_{C_2} \mathbf{E} \cdot \mathbf{i}_{n2} \, dl \qquad (6\text{-}87b)$$

For structure (b),

V_0, potential difference between the conductors $= \int_a^b \mathbf{E} \cdot d\mathbf{l}$ (6-88a)

Q, magnitude of surface charge on either conductor

= displacement flux crossing the area formed by the contour C_2 and length unity in the axial direction

$$= \oint_{C_2} \mathbf{D} \cdot \mathbf{i}_{n2} \, dl = \epsilon \oint_{C_2} \mathbf{E} \cdot \mathbf{i}_{n2} \, dl \tag{6-88b}$$

For structure (c),

I_0, surface current flowing on either conductor $= \oint_{C_2} \mathbf{H} \cdot d\mathbf{l}$ (6-89a)

ψ, magnetic flux linking the current I_0

= magnetic flux crossing the area formed by the contour C_1 and length unity in the axial direction

$$= \int_a^b \mathbf{B} \cdot \mathbf{i}_{n1} \, dl = \mu \int_a^b \mathbf{H} \cdot \mathbf{i}_{n1} \, dl \tag{6-89b}$$

From (6-87a)–(6-89b), we can write the general expressions for G, C, and L per unit length, denoted by $\mathcal{G}, \mathcal{C}$, and $\mathcal{L}$, as

$$\mathcal{G} = \frac{I_c}{V_0} = \frac{\sigma \oint_{C_2} \mathbf{E} \cdot \mathbf{i}_{n2} \, dl}{\int_a^b \mathbf{E} \cdot d\mathbf{l}} \tag{6-90}$$

$$\mathcal{C} = \frac{Q}{V_0} = \frac{\epsilon \oint_{C_2} \mathbf{E} \cdot \mathbf{i}_{n2} \, dl}{\int_a^b \mathbf{E} \cdot d\mathbf{l}} \tag{6-91}$$

$$\mathcal{L} = \frac{\psi}{I_0} = \frac{\mu \int_a^b \mathbf{H} \cdot \mathbf{i}_{n1} \, dl}{\oint_{C_2} \mathbf{H} \cdot d\mathbf{l}} \tag{6-92}$$

From (6-90) and (6-91), we note that

$$\frac{\mathcal{G}}{\mathcal{C}} = \frac{\sigma}{\epsilon} \text{mhos/farads} \tag{6-93}$$

From the discussion of Section 3.10, the electric field lines of structure (b) and the magnetic field lines of structure (c) are everywhere orthogonal to each other and their magnitudes are proportional, since the conductor cross sections for the two structures are the same. Thus we can write

$$\mathbf{E} = k\mathbf{H} \times \mathbf{i}_z \tag{6-94}$$

where k is the constant of proportionality and $\mathbf{i}_z$ is directed into the plane of the paper. Substituting (6-94) into (6-91), we have

$$\mathcal{C} = \frac{\epsilon \oint_{C_2} k\mathbf{H} \times \mathbf{i}_z \cdot \mathbf{i}_{n2} \, dl}{\int_a^b k\mathbf{H} \times \mathbf{i}_z \cdot d\mathbf{l}}$$

$$= \frac{\epsilon k \oint_{C_2} \mathbf{H} \cdot \mathbf{i}_z \times \mathbf{i}_{n2} \, dl}{k \int_a^b \mathbf{H} \cdot \mathbf{i}_z \times d\mathbf{l}} = \frac{\epsilon \oint_{C_2} \mathbf{H} \cdot d\mathbf{l}}{\int_a^b \mathbf{H} \cdot \mathbf{i}_{n1} \, dl} \tag{6-95}$$

From (6-92) and (6-95), we note that

$$\mathcal{LC} = \mu\epsilon \text{ henry-farad/m}^2 \tag{6-96}$$

Equations (6-93) and (6-96) provide simple relationships between the conductance per unit length, capacitance per unit length, and inductance per unit length of a structure consisting of two infinitely long, parallel perfect conductors having arbitrary but uniform cross sections. Expressions for these three quantities are listed in Table 6.2 for some common configurations of conductors having the cross sections shown in Fig. 6.15.

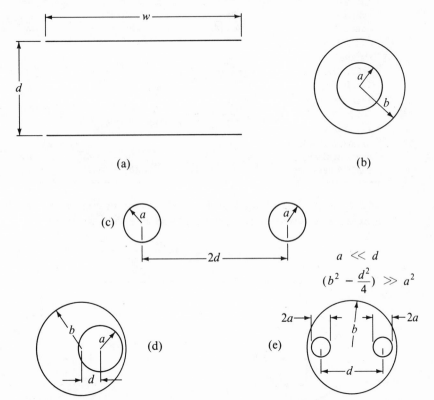

(a)

(b)

(c)

(d)

(e)

$a \ll d$

$(b^2 - \dfrac{d^2}{4}) \gg a^2$

Fig. 6.15. Cross sections of some common configurations of parallel infinitely long conductors.

EXAMPLE 6-12.   It is desired to obtain the conductance, capacitance, and inductance per unit length of the parallel cylindrical wire arrangement of Fig. 6.15(c).

In view of (6-93) and (6-96), it is sufficient if we find one of the three quantities. Hence we choose to find the capacitance per unit length. To do this, we refer to Example 6-11 and Fig. 6.12 and note that placing a cylindrical conductor coinciding with the equipotential cylindrical surface having its axis at a distance b from the line charge ρ_{L0} and on the side opposite to the grounded conductor will not alter the field. Hence the field of the parallel wire arrangement of Fig. 6.15(c) is exactly the same as the field due to equal

TABLE 6.2. Conductance, Capacitance, and Inductance per Unit Length for Some Structures Consisting of Infinitely Long Conductors Having the Cross Sections Shown in Fig. 6.15

Description	$\mathcal{G}$, Conductance per Unit Length	$\mathcal{C}$, Capacitance per Unit Length	$\mathcal{L}$, Inductance per Unit Length
Parallel plane conductors, Fig. 6.15(a)	$\sigma \dfrac{w}{d}$	$\epsilon \dfrac{w}{d}$	$\mu \dfrac{d}{w}$
Coaxial cylindrical conductors, Fig. 6.15(b)	$\dfrac{2\pi\sigma}{\ln(b/a)}$	$\dfrac{2\pi\epsilon}{\ln(b/a)}$	$\dfrac{\mu}{2\pi} \ln \dfrac{b}{a}$
Parallel cylindrical wires, Fig. 6.15(c)	$\dfrac{\pi\sigma}{\cosh^{-1}(d/a)}$	$\dfrac{\pi\epsilon}{\cosh^{-1}(d/a)}$	$\dfrac{\mu}{\pi} \cosh^{-1} \dfrac{d}{a}$
Eccentric inner conductor, Fig. 6.15(d)	$\dfrac{2\pi\sigma}{\cosh^{-1}\left(\dfrac{a^2+b^2-d^2}{2ab}\right)}$	$\dfrac{2\pi\epsilon}{\cosh^{-1}\left(\dfrac{a^2+b^2-d^2}{2ab}\right)}$	$\dfrac{\mu}{2\pi}\cosh^{-1}\dfrac{a^2+b^2-d^2}{2ab}$
Shielded parallel cylindrical wires, Fig. 6.15(e)	$\dfrac{\pi\sigma}{\ln\left[\dfrac{d(b^2-d^2/4)}{a(b^2+d^2/4)}\right]}$	$\dfrac{\pi\epsilon}{\ln\left[\dfrac{d(b^2-d^2/4)}{a(b^2+d^2/4)}\right]}$	$\dfrac{\mu}{\pi} \ln \dfrac{d(b^2-d^2/4)}{a(b^2+d^2/4)}$

and opposite line charges situated as shown in Fig. 6.16. The potential difference between the two points A and B is then given by

$$V_0 = 2\frac{\rho_{L0}}{2\pi\epsilon} \ln \frac{2d-a-b}{a-b} \qquad (6\text{-}97)$$

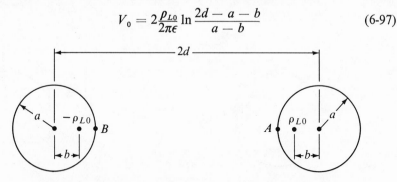

Fig. 6.16. For the determination of $\mathcal{G}$, $\mathcal{L}$, and $\mathcal{C}$ for the parallel cylindrical wire arrangement of Fig. 6.15(c).

However, from Example 6-11,

$$b = \frac{a^2}{2d-b}$$

or

$$b = d \pm \sqrt{d^2-a^2} \qquad (6\text{-}98)$$

Ignoring the plus sign on the right side of (6-98), since b has to be less than

d, and substituting for b in (6-97) we have

$$V_0 = \frac{\rho_{L0}}{\pi\epsilon} \ln \frac{2d - a - d + \sqrt{d^2 - a^2}}{a - d + \sqrt{d^2 - a^2}}$$

$$= \frac{\rho_{L0}}{\pi\epsilon} \ln \frac{\sqrt{d^2 - a^2} + (d - a)}{\sqrt{d^2 - a^2} - (d - a)} \tag{6-99}$$

$$= \frac{\rho_{L0}}{\pi\epsilon} \ln \frac{d + \sqrt{d^2 - a^2}}{a} = \frac{\rho_{L0}}{\pi\epsilon} \cosh^{-1} \frac{d}{a}$$

Finally, the capacitance per unit length is given by

$$\mathcal{C} = \frac{\rho_{L0}}{V_0} = \frac{\pi\epsilon}{\cosh^{-1}(d/a)} \tag{6-100}$$

which agrees with the expression given in Table 6.2. The corresponding expressions for $\mathcal{G}$ and $\mathcal{L}$ obtained by using (6-93) and (6-96), respectively, are given in Table 6.2. ∎

For volume current distributions, we have to consider the magnetic field internal to the current distribution in addition to the magnetic field external to it. The inductance associated with the internal field is known as the "internal inductance" as compared to the "external inductance" associated with the external field. The inductance we defined by (6-84) and (6-92) is the external inductance. To obtain the internal inductance, we have to take into account the fact that different flux lines in the volume occupied by the current distribution link different partial amounts of the total current. We will illustrate this by means of an example.

EXAMPLE 6-13.   A current I amp flows with uniform volume density $\mathbf{J} = J_0 \mathbf{i}_z$ amp/m² along an infinitely long, solid cylindrical conductor of radius a and returns with uniform surface density in the opposite direction along the surface of an infinitely long, perfectly conducting cylinder of radius b ($> a$) and coaxial with the inner conductor. It is desired to find the internal inductance per unit length of the inner conductor.

The cross-sectional view of the conductor arrangement is shown in Fig. 6.17(a). From symmetry considerations, the magnetic field is entirely in the ϕ direction and independent of ϕ. Applying Ampere's circuital law to a circular contour of radius r ($< a$) as shown in Fig. 6.17(a), we have

$$2\pi r H_\phi = \pi r^2 J_0$$

or

$$\mathbf{H} = H_\phi \mathbf{i}_\phi = \frac{J_0 r}{2} \mathbf{i}_\phi \qquad r < a$$

The corresponding magnetic flux density is given by

$$\mathbf{B} = \mu\mathbf{H} = \frac{\mu J_0 r}{2} \mathbf{i}_\phi \qquad r < a$$

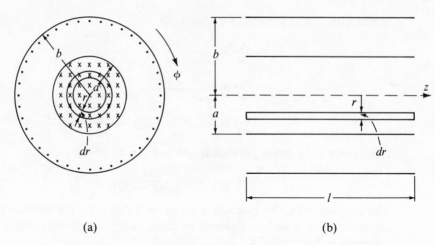

(a)   (b)

Fig. 6.17. For evaluating the internal inductance per unit length associated with a volume current of uniform density along an infinitely long cylindrical conductor.

where μ is the permeability of the conductor. Let us now consider a rectangle of infinitesimal width dr in the r direction and length l in the z direction at a distance r from the axis as shown in Fig. 6.17(b). The magnetic flux $d\psi_i$ crossing this rectangular surface is given by

$$d\psi_i = B_\phi(\text{area of the rectangle})$$
$$= \frac{\mu J_0 r l \, dr}{2}$$

where the subscript i denotes flux internal to the conductor. This flux surrounds only the current flowing within the radius r, as can be seen from Fig. 6.17(a). Let N be the fraction of the total current I linked by this flux. Then

$$N = \frac{\text{current flowing within radius } r \ (< a)}{\text{total current } I}$$
$$= \frac{J_0 \pi r^2}{J_0 \pi a^2} = \left(\frac{r}{a}\right)^2$$

The contribution from the flux $d\psi_i$ to the internal flux linkage associated with the current I is the product of N and the flux itself, that is, $N \, d\psi_i$. To obtain the internal flux linkage associated with I, we integrate $N \, d\psi_i$ between the limits $r = 0$ and $r = a$, taking into account the dependence of N upon $d\psi_i$. Thus

$$\psi_i = \int_{r=0}^{a} N \, d\psi_i = \int_{r=0}^{a} \left(\frac{r}{a}\right)^2 \frac{\mu J_0 l r}{2} \, dr = \frac{\mu J_0 l a^2}{8}$$

Finally, the required internal inductance per unit length is

$$\mathcal{L}_i = \frac{\psi_i}{lI} = \frac{(\mu J_0 a^2/8)}{(J_0 \pi a^2)} = \frac{\mu}{8\pi} \tag{6-101}$$

Alternatively, we can obtain $\mathcal{L}_i$ from energy considerations by making use of the result (6-86) that the magnetic stored energy is equal to $\frac{1}{2}LI^2$. For $\mathcal{L}_i$, we have to consider the energy stored in the volume internal to the current distribution. For unit length of the conductor, this is given by

$$W_{mi} = \int_{vol} \frac{1}{2} \mu H^2 \, dv$$

$$= \int_{r=0}^{a} \int_{\phi=0}^{2\pi} \int_{z=0}^{1} \frac{1}{2} \mu \left(\frac{J_0 r}{2}\right)^2 r \, dr \, d\phi \, dz = \frac{\pi \mu J_0^2 a^4}{16}$$

The internal inductance is then given by

$$\mathcal{L}_i = \frac{2W_{mi}}{I^2} = \frac{(\pi \mu J_0^2 a^4/8)}{(J_0^2 \pi^2 a^4)} = \frac{\mu}{8\pi}$$

which is the same as (6-101). Finally, to find the total inductance per unit length of the arrangement of Fig. 6.17(a), we have to add the external inductance due to the flux in the region $a < r < b$ to the internal inductance given by (6-101). This external inductance is given in Table 6.2. ∎

From the steps involved in the solution of Example 6-13, we observe that the general expression for the internal inductance is

$$L_{int} = \frac{1}{I} \int_{S} N \, d\psi \qquad (6\text{-}102a)$$

where S is any surface through which the internal magnetic flux associated with I passes. We note that (6-102a) is also good for computing the external inductance since for external inductance N is independent of $d\psi$. Hence

$$L_{ext} = \frac{N}{I} \int_{S} d\psi = N \frac{\psi}{I} \qquad (6\text{-}102b)$$

In Eq. (6-102b), the value of N is unity if I is a surface current as for the structures of Figs. 6.13(c) and 6.14(c). On the other hand, for a filamentary wire wound on a core, N is equal to the number of turns of the winding in which case ψ represents the flux through the core, that is, the flux crossing the surface formed by one turn. To explain this, let us consider a two-turn winding *abcdefghi* carrying current I as shown in Fig. 6.18(a) and imagine the flux lines penetrating the surface formed by the two-turn winding. According to definition, the magnetic flux linking I is the flux crossing the surface formed by the two-turn winding. Let us twist the portion *cdef* of the winding and stretch the winding to the shape shown in Fig. 6.18(b). We can now see that the flux lines come from underneath the surface of the first turn (*abcd*), go below the surface of the second turn (*efgh*), and come out of it again as shown in Fig. 6.18(b) so that the flux linking I is equal to twice the flux passing through one of the surfaces *abcd* and *efgh*.

The discussion pertaining to inductance thus far has been concerned with "self inductance," that is, inductance associated with a current distribu-

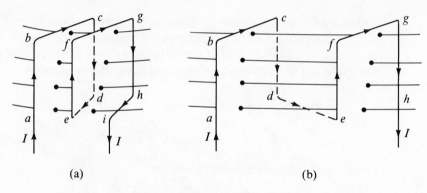

(a)   (b)

Fig. 6.18. For illustrating that the flux linking a filamentary wire of N turns is equal to N times the flux crossing the surface formed by one turn.

tion by virtue of its own flux linking it. On the other hand, if we have two independent currents I_1 and I_2, we can talk of the flux due to one current linking the second current. This leads to the concept of "mutual inductance." The mutual inductance denoted as L_{12} is defined as

$$L_{12} = N_1 \frac{\psi_{12}}{I_2} \qquad (6\text{-}103)$$

where ψ_{12} is the magnetic flux produced by I_2 but linking one turn of the N_1-turn winding carrying current I_1. Similarly,

$$L_{21} = N_2 \frac{\psi_{21}}{I_1} \qquad (6\text{-}104)$$

where ψ_{21} is the magnetic flux produced by I_1 but linking one turn of the N_2-turn winding carrying current I_2. It is left as an exercise (Problem 6.24) for the student to show that $L_{21} = L_{12}$. We will now consider a simple example illustrating the computation of mutual inductance.

EXAMPLE 6-14. A single straight wire, infinitely long and carrying current I_1, lies below to the left and parallel to a two-wire telephone line carrying current I_2, as shown by the cross-sectional and plan views in Figs. 6.19(a) and 6-19(b), respectively. It is desired to obtain the mutual inductance between the single wire and the telephone line per unit length of the wires. The thickness of the telephone wire is assumed to be negligible.

Choosing a coordinate system with the axis of the single wire as the z axis and applying Ampere's circuital law to a circular path around the single wire, we obtain the magnetic flux density due to the single wire as

$$\mathbf{B} = \frac{\mu_0 I_1}{2\pi r} \mathbf{i}_\phi$$

The flux $d\psi_{21}$ crossing a rectangular surface of length unity and width dy

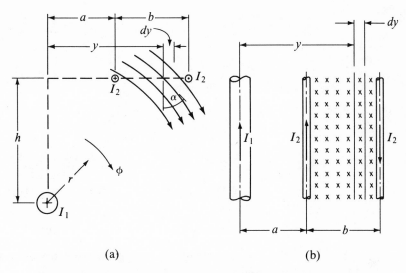

(a)   (b)

Fig. 6.19. For the computation of mutual inductance per unit length between a two-wire telephone line and a single wire parallel to it.

lying between the telephone wires as shown in Fig. 6.19(b) is then given by

$$d\psi_{21} = B \, dy \cos \alpha = \frac{\mu_0 I_1 y}{2\pi(h^2 + y^2)} \, dy$$

where α is the angle between the flux lines and the normal to the rectangular surface as shown in Fig. 6.19(a). The total flux ψ_{21} crossing the rectangular surface of length unity and extending from one telephone wire to the other is

$$\psi_{21} = \int_{y=a}^{a+b} d\psi_{21} = \int_{y=a}^{a+b} \frac{\mu_0 I_1 y}{2\pi(h^2 + y^2)} \, dy$$

$$= \frac{\mu_0 I_1}{4\pi} \ln \frac{h^2 + (a + b)^2}{h^2 + a^2}$$

This is the flux due to I_1 linking I_2 per unit length along the wires. Thus the required mutual inductance per unit length of the wires is given by

$$\mathcal{L}_{21} = \frac{\psi_{21}}{I_1} = \frac{\mu_0}{4\pi} \ln \frac{h^2 + (a + b)^2}{h^2 + a^2} \text{ henrys/m} \quad \blacksquare$$

6.5   Magnetic Circuits

Let us consider the two structures shown in Figs. 6.20(a) and (b). The structure of Fig. 6.20(a) is a toroidal conductor of uniform conductivity σ and having a cross-sectional area A and mean circumference l. There is an infinitesimal gap a–b across which a potential difference of V_0 volts is maintained

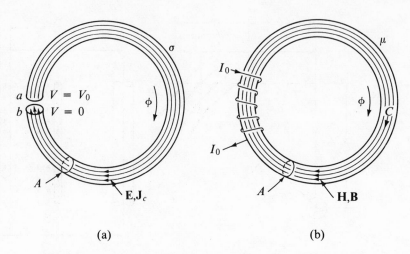

(a)   (b)

Fig. 6.20. For illustrating the analogy between electric and magnetic circuits.

by connecting an appropriate voltage source. Because of the potential difference, an electric field is established in the toroid and a conduction current results from the higher potential surface *a* to the lower potential surface *b* as shown in the figure. The structure of Fig. 6.20(b) is a toroidal magnetic core of uniform permeability μ and having a cross-sectional area A and mean circumference l. A current I amp is passed through a filamentary wire of N turns wound around the toroid by connecting an appropriate current source. Because of the current through the winding, a magnetic field is established in the toroid and a magnetic flux results in the direction of advance of a right-hand screw as it is turned in the sense of the current.

Since the conduction current cannot leak into the free space surrounding the conductor, it is confined entirely to the conductor. On the other hand, the magnetic flux can leak into the free space surrounding the magnetic core and hence is not confined completely to the core. However, let us consider the case for which $\mu \gg \mu_0$. Applying the boundary conditions at the boundary between a magnetic material of $\mu \gg \mu_0$ and free space as shown in Fig. 6.21, we have

$$B_1 \sin \alpha_1 = B_2 \sin \alpha_2$$

$$H_1 \cos \alpha_1 = H_2 \cos \alpha_2$$

or

$$\frac{B_1}{H_1} \tan \alpha_1 = \frac{B_2}{H_2} \tan \alpha_2$$

$$\frac{\tan \alpha_1}{\tan \alpha_2} = \frac{\mu_2}{\mu_1} \ll 1$$

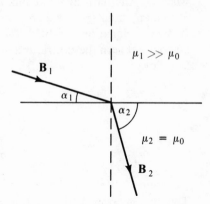

Fig. 6.21. Lines of magnetic flux density at the boundary between free space and a magnetic material of permeability $\mu \gg \mu_0$.

Thus $\alpha_1 \ll \alpha_2$, and

$$\frac{B_2}{B_1} = \frac{\sin \alpha_1}{\sin \alpha_2} \ll 1$$

For example, if the values of μ_1 and α_2 are $1000\mu_0$ and $89°$, respectively, then $\alpha_1 = 3°16'$ and $\sin \alpha_1/\sin \alpha_2 = 0.057$. We can assume for all practical purposes that the magnetic flux is confined entirely to the magnetic core just as the conduction current is confined to the conductor. The structure of Fig. 6.20(b) is then known as a "magnetic circuit" similar to the "electric circuit" of Fig. 6.20(a).

For the structure of Fig. 6.20(a), we have

$$\mathbf{\nabla} \times \mathbf{E} = 0 \tag{6-105a}$$

$$\int_a^b \mathbf{E} \cdot d\mathbf{l} = V_0 \tag{6-105b}$$

$$\mathbf{J}_c = \sigma \mathbf{E} \tag{6-105c}$$

$$I_c = \int_A \mathbf{J}_c \cdot d\mathbf{S} \tag{6-105d}$$

For the structure of Fig. 6.20(b), we have

$$\mathbf{\nabla} \times \mathbf{H} = 0 \tag{6-106a}$$

$$\oint_C \mathbf{H} \cdot d\mathbf{l} = NI_0 \tag{6-106b}$$

$$\mathbf{B} = \mu \mathbf{H} \tag{6-106c}$$

$$\psi = \int_A \mathbf{B} \cdot d\mathbf{S} \tag{6-106d}$$

Equation (6-106a) results from the fact that there are no true currents in the magnetic material. In Eq. (6-106b), the factor N on the right side takes into account the fact that the filamentary wire penetrates a surface bounded by the path C as many times as there are number of turns in the entire winding. Alternatively, if we pull the path C out of the toroid, it will be cut at as many points as there are number of turns in the entire winding, that is, N times. Equations (6-105a)–(6-105d) and (6-106a)–(6-106d) indicate an analogy between the electric and magnetic circuits of Figs. 6.20(a) and 6.20(b) as follows:

$$\mathbf{E} \longleftrightarrow \mathbf{H} \tag{6-107a}$$

$$V_0 \longleftrightarrow NI_0 \tag{6-107b}$$

$$\mathbf{J}_c \longleftrightarrow \mathbf{B} \tag{6-107c}$$

$$\sigma \longleftrightarrow \mu \tag{6-107d}$$

$$I_c \longleftrightarrow \psi \tag{6-107e}$$

This analogy permits the solution of magnetic circuit problems from a knowledge of the solution of electric circuit problems.

The ratio of V_0 to I_c is the resistance R of the electric circuit of Fig. 6.20(a). The analogous quantity for the magnetic circuit of Fig. 6.20(b) is the ratio of NI_0 to ψ. It is known as the reluctance and is denoted by the symbol $\mathfrak{R}$. The resistance is purely a function of the dimensions of the conductor and the conductivity. For a magnetic core of the same dimensions as the conductor, the reluctance can therefore be obtained simply by replacing σ in R by μ. We note, however, that, unlike σ for conductors, μ for magnetic materials used for the cores is a function of the magnetic flux density in the material. This makes the reluctance analogous to a nonlinear resistor. Thus, to complete the analogy, we have

$$R = \frac{V_0}{I_c} \longleftrightarrow \mathfrak{R} = \frac{NI_0}{\psi} \tag{6-107f}$$

For the structure of Fig. 6.20(a), an exact expression for the resistance can be obtained by taking into account the variation of $\mathbf{E}$ over the cross section of the toroid. However, an approximate expression sufficient for practical purposes can be obtained by assuming that $\mathbf{E}$ is uniform over the cross-sectional area and equal to its value at the mean radius of the toroid, especially if the radius of cross-section is small compared to the mean radius of the toroid. Thus

$$lE_\phi = V_0$$

$$J_c = \sigma E_\phi = \frac{\sigma V_0}{l}$$

$$I_c = J_c A = \frac{\sigma V_0 A}{l}$$

$$R = \frac{V_0}{I_c} = \frac{l}{\sigma A} \tag{6-108}$$

It follows from the analogy that the reluctance of the structure of Fig.

6.20(b) is

$$\Re = \frac{NI_0}{\psi} = \frac{l}{\mu A} \tag{6-109}$$

In fact, if we assume that the magnetic field intensity **H** is uniform over the cross-sectional area and equal to its value at the mean radius of the toroid, we have

$$lH_\phi = NI_0$$

$$B_\phi = \mu H_\phi = \frac{\mu NI_0}{l}$$

$$\psi = B_\phi A = \frac{\mu NI_0 A}{l}$$

$$\Re = \frac{NI_0}{\psi} = \frac{l}{\mu A}$$

which agrees with (6-109). The equivalent circuit representations of (6-108) and (6-109) are shown in Figs. 6.22(a) and (b).

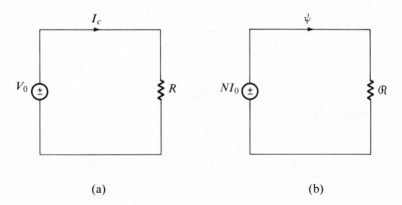

(a)   (b)

Fig. 6.22. Equivalent circuit representations for the structures of Figs. 6.20(a) and (b).

EXAMPLE 6-15. The structure shown in Fig. 6.23(a) is that of a magnetic circuit containing three legs with an air gap in the center leg. A filamentary wire of N turns carrying current I is wound around the center leg. The core material is annealed sheet steel, for which the B versus H curve is shown in Fig. 6.23(b). The dimensions of the magnetic circuit are

$$A_2 = 5 \text{ cm}^2 \qquad A_1 = A_3 = 3 \text{ cm}^2$$

$$l_2 = 10 \text{ cm} \qquad l_1 = l_3 = 20 \text{ cm}, \quad l_g = 0.1 \text{ cm}$$

We wish to obtain the equivalent circuit and find NI required to establish a magnetic flux of 8×10^{-4} Wb in the air gap.

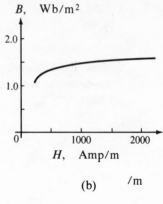

(a)

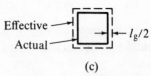

(b)

(c)

Fig. 6.23. (a) A magnetic circuit. (b) *B-H* curve for annealed sheet steel. (c) Effective and actual cross sections for the air gap of the magnetic circuit of (a).

The current in the winding establishes a magnetic flux in leg 2 which divides between legs 1 and 3. In the air gap, fringing of the flux occurs. This is taken into account by using an effective cross section which is greater than the actual cross section, as shown in Fig. 6.23(c). Using subscripts 1, 2, 3, and g for the fields and permeabilities associated with the three legs and the air gap, respectively, we can write the following equations:

$$NI = H_2 l_2 + H_g l_g + H_1 l_1$$

$$= \psi_2 \frac{l_2}{\mu_2 A_2} + \psi_g \frac{l_g}{\mu_g A_g} + \psi_1 \frac{l_1}{\mu_1 A_1} \qquad (6\text{-}110)$$

$$= \psi_2 \mathcal{R}_2 + \psi_g \mathcal{R}_g + \psi_1 \mathcal{R}_1$$

$$0 = H_3 l_3 - H_1 l_1$$

$$= \psi_3 \frac{l_3}{\mu_3 A_3} - \psi_1 \frac{l_1}{\mu_1 A_1} \qquad (6\text{-}111)$$

$$= \psi_3 \mathcal{R}_3 - \psi_1 \mathcal{R}_1$$

The equivalent circuit corresponding to Eqs. (6-110) and (6-111) can be drawn as shown in Fig. 6.24, taking into account the fact that $\psi_g = \psi_2$. To determine the required NI, we note that

$$B_2 = \frac{\psi_2}{A_2} = \frac{8 \times 10^{-4}}{5 \times 10^{-4}} = 1.6 \text{ Wb/m}^2$$

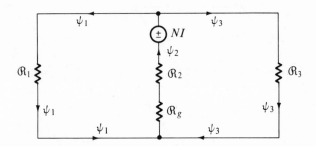

Fig. 6.24. Equivalent circuit for analyzing the magnetic circuit of Fig. 6.23(a).

From Fig. 6.23(b), the value of H_2 is 2200 amp/m. Since legs 1 and 3 are identical, their reluctances are equal so that the flux ψ_2 divides equally between the two legs. Thus $\psi_1 = \psi_3 = \psi_2/2 = 4 \times 10^{-4}$ Wb/m². Then

$$B_1 = \frac{\psi_1}{A_1} = \frac{4 \times 10^{-4}}{3 \times 10^{-4}} = 1.333 \text{ Wb/m}^2$$

From Fig. 6.23(b), the value of H_1 is 475 amp/m. The effective cross section of the air gap is $(\sqrt{5} + l_g)^2 = 2.34$ cm². The flux density in the air gap is

$$B_g = \frac{\psi_g}{A_g} = \frac{8 \times 10^{-4}}{2.34^2 \times 10^{-4}} = 1.46 \text{ Wb/m}^2$$

The magnetic field intensity in the air gap is

$$H_g = \frac{B_g}{\mu_g} = \frac{B_g}{\mu_0} = \frac{1.46}{4\pi \times 10^{-7}} = 0.1162 \times 10^7 \text{ Wb/m}^2$$

From (6-110), we then have

$$NI = H_2 l_2 + H_g l_g + H_1 l_1$$
$$= 2200 \times 0.10 + 0.1162 \times 10^7 \times 10^{-3} + 475 \times 0.20$$
$$= 1477 \text{ amp-turns}$$

We note that a large part of the ampere-turns is due to high reluctance of the air gap. ∎

6.6 Quasistatics; The Field Basis of Low-Frequency Circuit Theory

In Section 6.4 we considered three structures, shown in Figs. 6.13(a), (b), and (c), from the point of view of static fields. Let us now consider the three structures driven by time-varying sources. The resulting fields are then time varying. From Maxwell's equations for time-varying fields, we know that a time-varying electric field is accompanied by a time-varying magnetic field and vice versa. Thus, for example, a time-varying voltage source applied to the structure of Fig. 6.13(b) results in a time-varying electric field which has associated with it a time-varying magnetic field as shown in Fig. 6.25(a).

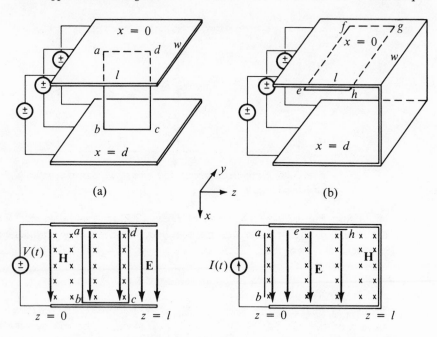

Fig. 6.25. For illustrating the behavior of the structures of Figs. 6.13(b) and (c) for time-varying sources.

A certain amount of magnetic energy is then associated with the structure in addition to the electric energy. We can no longer say that the structure behaves like a single capacitor as in the case of static fields. Furthermore, applying Faraday's law to a rectangular path *abcda* as shown in Fig. 6.25(a), we have

$$\int_a^b \mathbf{E} \cdot d\mathbf{l} - \int_d^c \mathbf{E} \cdot d\mathbf{l} = \frac{d}{dt} \int_{\substack{\text{area} \\ abcd}} \mu H_y \, dS \tag{6-112}$$

It follows from (6-112) that the voltage between *a* and *b* is not necessarily equal to the voltage between *d* and *c* because of the time-varying magnetic field. The voltage along the structure is dependent on *z*. However, under certain conditions, the time-varying magnetic field is negligible so that the electric field distribution at any time can be approximated by the static field distribution resulting from a constant voltage source between the plates having a value equal to that of the source voltage at that time. Such approximations are known as quasistatic approximations and the corresponding fields are known as quasistatic fields. Thus, for the structure of Fig. 6.13(b) under the quasistatic approximation, $\partial \mathbf{B}/\partial t$ is negligible so that

$$\nabla \times \mathbf{E} \approx 0 \tag{6-113}$$

$$E_x(t) \approx \frac{V(t)}{d} \tag{6-114}$$

The magnitude of the resulting time-varying charge on either plate is

$$Q(t) = (lw)\epsilon E_x(t) = \frac{\epsilon wl}{d}V(t) = CV(t) \qquad (6\text{-}115)$$

where $C = \epsilon wl/d$ is the same as the capacitance obtained for the direct voltage source. Differentiating both sides of (6-115) with respect to time, we have

$$\frac{dQ}{dt} = \frac{d}{dt}(CV) \qquad (6\text{-}116)$$

But, according to the law of conservation of charge, dQ/dt must be equal to the current I flowing into the plate from the voltage source. Thus Eq. (6-116) becomes

$$I = \frac{d}{dt}(CV) \qquad (6\text{-}117)$$

which is the familiar voltage-to-current relationship used in circuit theory for a capacitor. For the sinusoidally time-varying case, Eq. (6-117) reduces to

$$\bar{I} = j\omega C \bar{V} \qquad (6\text{-}118)$$

where $\bar{I}$ and $\bar{V}$ are the phasor current and phasor voltage, respectively, and ω is the radian frequency of the voltage source.

Similarly, a time-varying current source applied to the structure of Fig. 6.13(c) results in a time-varying magnetic field which has associated with it a time-varying electric field as shown in Fig. 6.25(b). A certain amount of electric energy is then associated with the structure in addition to the magnetic energy. We can no longer say that the structure behaves like a single inductor as in the case of static fields. Furthermore, applying the integral form of Maxwell's curl equation for **H** to a rectangular path *efghe* as shown in Fig. 6.25(b), we have

$$\int_e^f \mathbf{H} \cdot d\mathbf{l} - \int_h^g \mathbf{H} \cdot d\mathbf{l} = \frac{d}{dt} \int_{\substack{\text{area}\\efgh}} \epsilon E_x \, dS \qquad (6\text{-}119)$$

Since H is zero outside the structure, it follows from (6-119) that the current crossing the line *ef* is not necessarily equal to the current crossing the line *hg* because of the time-varying electric field. The current flowing along the structure is dependent on z. However, under the quasistatic approximation, $\partial\mathbf{D}/\partial t$ is negligible so that the magnetic field distribution at any time can be approximated by the static magnetic field distribution resulting from the flow of a direct current having a value equal to that of the source current at that time. Thus

$$\mathbf{V} \times \mathbf{H} \approx 0 \qquad (6\text{-}120)$$

$$H_y(t) \approx \frac{I(t)}{w} \qquad (6\text{-}121)$$

The resulting time-varying magnetic flux linking the current is

$$\psi(t) = (dl)\mu H_y(t) = \frac{\mu dl}{w} I(t) = LI(t) \tag{6-122}$$

where $L = \mu dl/w$ is the same as the inductance obtained for the direct current source. Differentiating both sides of (6-122) with respect to time, we have

$$\frac{d\psi}{dt} = \frac{d}{dt}(LI) \tag{6-123}$$

However, applying Faraday's law to the rectangular contour bounding the magnetic flux linking the current and noting that the contribution to $\oint \mathbf{E} \cdot d\mathbf{l}$ is entirely from the path ab shown in Fig. 6.25(b), we have

$$\int_a^b \mathbf{E} \cdot d\mathbf{l} = \frac{d\psi}{dt} \tag{6-124}$$

The left side of (6-124) is the voltage $V(t)$ across the current source. Thus Eq. (6-123) becomes

$$V = \frac{d}{dt}(LI) \tag{6-125}$$

which is the familiar voltage-to-current relationship used in circuit theory for an inductor. For the sinusoidally time-varying case, Eq. (6-125) reduces to

$$\bar{V} = j\omega L\bar{I} \tag{6-126}$$

where $\bar{V}$ and $\bar{I}$ are the phasor voltage and phasor current, respectively, and ω is the radian frequency of the current source.

Finally, for the structure of Fig. 6.13(a) under the quasistatic approximation, both $\partial \mathbf{B}/\partial t$ and $\partial \mathbf{D}/\partial t$ are negligible so that

$$\nabla \times \mathbf{E} \approx 0 \tag{6-127a}$$

$$\nabla \times \mathbf{H} \approx \mathbf{J}_c \tag{6-127b}$$

In view of (6-127a), we have

$$E_x(t) = \frac{V(t)}{d} \tag{6-128}$$

The conduction current flowing from the upper plate to the lower plate is

$$I_c(t) = (lw)\sigma E_x(t) = \frac{\sigma wl}{d} V(t) \tag{6-129}$$

In view of (6-127b), $\oint \mathbf{H} \cdot d\mathbf{l}$ around a rectangular path surrounding the conductor in the cross-sectional plane is equal to the conduction current I_c. But the same quantity is also equal to the current I drawn from the voltage source. Thus

$$I(t) = \frac{\sigma wl}{d} V(t) = GV(t) \tag{6-130a}$$

or

$$V(t) = \frac{d}{\sigma wl} I(t) = RI(t) \tag{6-130b}$$

where $G = \sigma wl/d$ and $R = d/\sigma wl$ are the same as the conductance and resistance, respectively, obtained for the direct voltage source. Equations (6-130a) and (6-130b) are the familiar voltage-to-current relationships used in circuit theory for conductance and resistance, respectively. For the sinusoidally time-varying case, we have

$$\bar{I} = G\bar{V} \qquad (6\text{-}131a)$$

and

$$\bar{V} = R\bar{I} \qquad (6\text{-}131b)$$

where $\bar{I}$ and $\bar{V}$ are the phasor current and phasor voltage, respectively.

To summarize what we have learned thus far in this section, the voltage-to-current relationships used in circuit theory for a capacitor, inductor, and resistor given by (6-117), (6-125), and (6-130b), respectively, are valid only under the quasistatic approximation. For the quasistatic approximation to hold, $\partial \mathbf{B}/\partial t$ must be negligible for the case of the capacitor, $\partial \mathbf{D}/\partial t$ must be negligible for the case of the inductor, and both $\partial \mathbf{B}/\partial t$ and $\partial \mathbf{D}/\partial t$ must be negligible for the case of the resistor. To illustrate a method for determining the quantitative condition for the quasistatic approximation to hold in a particular case, we consider the structure of Fig. 6.25(b) in detail for the sinusoidally time-varying case in the following example.

EXAMPLE 6-16.   The parallel plate structure of Fig. 6.25(b) is driven by a sinusoidally time-varying current source. It is desired to show that the quasistatic approximation holds, that is, that the structure behaves like a single inductor as viewed by the current source for the condition

$$f \ll \frac{1}{2\pi l \sqrt{\mu \epsilon}}$$

where f is the frequency of the current source and μ and ϵ are the permeability and permittivity, respectively, of the medium between the plates.

Under the quasistatic approximation, the time-varying magnetic field distribution at any particular time must be approximately the same as that of the static magnetic field resulting from a direct current equal to the value of the source current at that time. Thus, denoting the phasor corresponding to this magnetic field by $\bar{H}_y^q$, we have

$$\bar{H}_y^q = \frac{\bar{I}_0}{w} \qquad (6\text{-}132)$$

where $\bar{I}_0 = [\bar{I}]_{z=0}$ is the phasor corresponding to the source current. This time-varying magnetic field induces a time-varying electric field in the x direction in accordance with Maxwell's curl equation for $\mathbf{E}$, given in phasor form by

$$\nabla \times \bar{\mathbf{E}} = -j\omega\bar{\mathbf{B}}$$

Denoting the phasor corresponding to this electric field by $\bar{E}'_x$, we have

$$\frac{\partial \bar{E}'_x}{\partial z} = -j\omega \bar{B}^q_y = -j\omega \mu \bar{H}^q_y = -j\omega \mu \frac{\bar{I}_0}{w} \tag{6-133}$$

Integrating (6-133) with respect to z, we obtain

$$\bar{E}'_x = -j\omega \mu \frac{\bar{I}_0}{w} (z - l) \tag{6-134}$$

where we have evaluated the arbitrary constant of integration by using the boundary condition that $[\bar{E}'_x]_{z=l} = 0$. If $\partial D/\partial t$ is not negligible, the time-varying electric field corresponding to the phasor $\bar{E}'_x$ produces a time-varying magnetic field in the y direction in accordance with Maxwell's curl equation for **H**, given in phasor form by

$$\nabla \times \bar{\mathbf{H}} = j\omega \bar{\mathbf{D}}$$

Denoting the phasor corresponding to this induced magnetic field by $\bar{H}'_y$, we have

$$-\frac{\partial \bar{H}'_y}{\partial z} = j\omega \bar{D}'_x = j\omega \epsilon \bar{E}'_x = \omega^2 \mu \epsilon \frac{\bar{I}_0}{w} (z - l) \tag{6-135}$$

Integrating (6-135) with respect to z, we obtain

$$\bar{H}'_y = -\omega^2 \mu \epsilon \frac{\bar{I}_0}{w} \left[\frac{(z-l)^2}{2} - \frac{l^2}{2} \right] \tag{6-136}$$

where we have evaluated the arbitrary constant of integration by using the boundary condition that $[\bar{H}'_y]_{z=0} = 0$ since the condition that the current at $z = 0$, as determined by the tangential magnetic field intensity at $z = 0$, must be equal to the source current is satisfied by (6-132) alone.

Now, the time-varying magnetic field corresponding to the phasor given by (6-136) induces a time-varying electric field. Denoting the phasor corresponding to this induced electric field by $\bar{E}''_x$, we have

$$\frac{\partial \bar{E}''_x}{\partial z} = -j\omega \mu \bar{H}'_y = j\omega^3 \mu^2 \epsilon \frac{\bar{I}_0}{w} \left[\frac{(z-l)^2}{2} - \frac{l^2}{2} \right] \tag{6-137}$$

Integrating (6-137) with respect to z, we obtain

$$\bar{E}''_x = j\omega^3 \mu^2 \epsilon \frac{\bar{I}_0}{w} \left[\frac{(z-l)^3}{6} - \frac{l^2(z-l)}{2} \right] \tag{6-138}$$

where we have again evaluated the arbitrary constant of integration by using the boundary condition that $[\bar{E}''_x]_{z=l} = 0$. Continuing in this manner, we obtain the successively induced magnetic and electric fields as

$$\bar{H}''_y = \omega^4 \mu^2 \epsilon^2 \frac{\bar{I}_0}{w} \left[\frac{(z-l)^4}{24} - \frac{l^2(z-l)^2}{4} + \frac{5l^4}{24} \right] \tag{6-139}$$

$$\bar{E}'''_x = -j\omega^5 \mu^3 \epsilon^2 \frac{\bar{I}_0}{w} \left[\frac{(z-l)^5}{120} - \frac{l^2(z-l)^3}{12} + \frac{5l^4(z-l)}{24} \right] \tag{6-140}$$

$$\bar{H}_y''' = -\omega^6 \mu^3 \epsilon^3 \frac{\bar{I}_0}{w} \left[\frac{(z-l)^6}{720} - \frac{l^2(z-l)^4}{48} + \frac{5l^4(z-l)^2}{48} - \frac{61l}{720} \right] \quad (6\text{-}141)$$

$$\bar{E}_x''' = \cdots$$

$$\bar{H}_y'''' = \cdots$$

The total electric field is given by

$$
\begin{aligned}
\bar{E}_x &= \bar{E}_x' + \bar{E}_x'' + \bar{E}_x''' + \cdots \\
&= -j\omega\mu \frac{\bar{I}_0}{w}(z-l) + j\omega^3\mu^2\epsilon \frac{\bar{I}_0}{w}\left[\frac{(z-l)^3}{6} - \frac{l^2(z-l)}{2} \right] \\
&\quad - j\omega^5\mu^3\epsilon^2 \frac{\bar{I}_0}{w}\left[\frac{(z-l)^5}{120} - \frac{l^2(z-l)^3}{12} + \frac{5l^4(z-l)}{24} \right] + \cdots \\
&= -j\sqrt{\frac{\mu}{\epsilon}} \frac{\bar{I}_0}{w}\left(1 + \frac{\omega^2\mu\epsilon l^2}{2} + \frac{5\omega^4\mu^2\epsilon^2 l^4}{24} + \cdots \right) \\
&\quad \times \left[\omega\sqrt{\mu\epsilon}(z-l) - (\omega\sqrt{\mu\epsilon})^3\frac{(z-l)^3}{3!} + (\omega\sqrt{\mu\epsilon})^5\frac{(z-l)^5}{5!} - \cdots \right] \\
&= -j\sqrt{\frac{\mu}{\epsilon}} \frac{\bar{I}_0}{w} \frac{\sin \omega\sqrt{\mu\epsilon}(z-l)}{\cos \omega\sqrt{\mu\epsilon}l}
\end{aligned}
\quad (6\text{-}142)
$$

The total electric field at $z=0$ is given by

$$[\bar{E}_x]_{z=0} = j\sqrt{\frac{\mu}{\epsilon}} \frac{\bar{I}_0}{w} \tan \omega\sqrt{\mu\epsilon}l \quad (6\text{-}143)$$

This result could have been obtained simply by adding $[\bar{E}_x']_{z=0}$, $[\bar{E}_x'']_{z=0}$, $[\bar{E}_x''']_{z=0}$, and so on. However, Eq. (6-142) was derived to point out that the electric field and hence the voltage along the structure varies sinusoidally with distance. Similarly, if we add $\bar{H}_y^q$, $\bar{H}_y'$, $\bar{H}_y''$, $\bar{H}_y'''$, and so on, we obtain the total magnetic field as

$$\bar{H}_y = \frac{\bar{I}_0}{w} \frac{\cos \omega\sqrt{\mu\epsilon}(z-l)}{\cos \omega\sqrt{\mu\epsilon}l} \quad (6\text{-}144)$$

indicating that the magnetic field and hence the current along the structure varies cosinusoidally with distance.

The phasor voltage across the current source is given by

$$
\begin{aligned}
\bar{V}_0 &= [\bar{V}]_{z=0} = \int_a^b [\bar{E}_x]_{z=0}\, dl \\
&= j\sqrt{\frac{\mu}{\epsilon}} \frac{\bar{I}_0 d}{w} \tan \omega\sqrt{\mu\epsilon}l \\
&= j\omega \frac{\mu dl}{w} \bar{I}_0 \frac{\tan \omega\sqrt{\mu\epsilon}l}{\omega\sqrt{\mu\epsilon}l} \\
&= j\omega L \bar{I}_0 \frac{\tan \omega\sqrt{\mu\epsilon}l}{\omega\sqrt{\mu\epsilon}l}
\end{aligned}
\quad (6\text{-}145)
$$

where $L = \mu dl/w$ is the inductance of the structure computed from static field considerations. Equation (6-145) represents the voltage-to-current relationship at the source end of the structure under the condition for which $\partial \mathbf{D}/\partial t$ is not negligible. For $\omega\sqrt{\mu\epsilon}l \ll 1$, $\tan \omega\sqrt{\mu\epsilon}l \approx \omega\sqrt{\mu\epsilon}l$ and Eq. (6-145) reduces to

$$\bar{V}_0 = j\omega L \bar{I}_0$$

which is the voltage-to-current relationship for a single inductor. Thus, for the quasistatic approximation to hold, the condition to be satisfied is

$$f \ll \frac{1}{2\pi l\sqrt{\mu\epsilon}} \tag{6-146}$$

As a numerical example, for $l = 0.1$ m, $\mu = \mu_0$, and $\epsilon = \epsilon_0$, the value of $1/2\pi l\sqrt{\mu\epsilon}$ is $(1500/\pi) \times 10^6$. For a value of $1/10$ for $\omega\sqrt{\mu\epsilon}l$, the frequency must be less than $150/\pi$ MHz for the structure to behave essentially like a single inductor. ∎

EXAMPLE 6-17.   In Example 6-16 we showed that the quasistatic approximation holds for the structure of Fig. 6.25(b) for the condition $f \ll 1/2\pi l\sqrt{\mu\epsilon}$. The structure then behaves like a single inductor as shown in Fig. 6.26(a). It is desired to examine the behavior of the structure as viewed from the source end for frequencies beyond the value for which the quasistatic approximation holds.

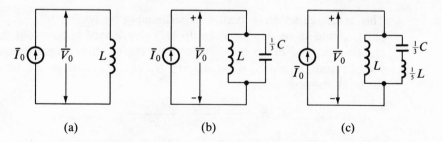

(a)   (b)   (c)

Fig. 6.26. (a) Equivalent circuit for the input behavior of the structure of Fig. 6.25(b) under quasistatic approximation. (b) and (c) Same as (a) except for frequencies higher and higher than those for which the quasistatic approximation is valid. The values of L and C are $\mu dl/w$ and $\epsilon wl/d$, respectively.

Expressing $\tan \omega\sqrt{\mu\epsilon}l$ as a sum of infinite series in powers of $\omega\sqrt{\mu\epsilon}l$, Eq. (6-145) can be written as

$$\bar{V}_0 = j\omega L \bar{I}_0 \left(1 + \frac{1}{3}\omega^2 \mu\epsilon l^2 + \frac{2}{15}\omega^4 \mu^2\epsilon^2 l^4 + \cdots\right) \tag{6-147}$$

For the quasistatic approximation, we neglect all the terms involving powers of $\omega\sqrt{\mu\epsilon}l$ in comparison with 1 in the series on the right side of (6-147). For a frequency slightly higher than the value for which condition (6-146) is acceptable, we have to include the second term in the series. Thus we have

$$\bar{V}_0 \approx j\omega L \bar{I}_0 \left(1 + \frac{1}{3}\omega^2 \mu\epsilon l^2\right)$$

$$= j\omega L \bar{I}_0 \left(1 + \frac{1}{3}\omega^2 LC\right) \tag{6-148}$$

where $C = \epsilon wl/d$ is the capacitance computed from static field considerations if the structure were open circuited at $z = l$. Rearranging (6-148), we get

$$\bar{I}_0 = \frac{\bar{V}_0}{j\omega L(1 + \frac{1}{3}\omega^2 LC)} \approx \frac{\bar{V}_0}{j\omega L}\left(1 - \frac{1}{3}\omega^2 LC\right)$$

$$= \bar{V}_0 \left(\frac{1}{j\omega L} + j\omega \frac{C}{3}\right) \tag{6-149}$$

The voltage-to-current relationship given by (6-149) corresponds to that of an inductor of value L in parallel with a capacitor of value $\frac{1}{3}C$ as shown in Fig. 6.26(b). Thus the same structure which behaves almost like a single inductor at low frequencies governed by (6-146) acts like an inductor in parallel with a capacitor as the frequency is increased. For still higher frequencies, we have to include one more term in the series on the right side of (6-147), giving us

$$\bar{V}_0 \approx j\omega L \bar{I}_0 \left(1 + \frac{1}{3}\omega^2 \mu\epsilon l^2 + \frac{2}{15}\omega^4 \mu^2\epsilon^2 l^4\right)$$

$$= j\omega L \bar{I}_0 \left(1 + \frac{1}{3}\omega^2 LC + \frac{2}{15}\omega^4 L^2 C^2\right)$$

or

$$\bar{I}_0 = \frac{\bar{V}_0}{j\omega L}\left(1 + \frac{1}{3}\omega^2 LC + \frac{2}{15}\omega^4 L^2 C^2\right)^{-1}$$

$$= \frac{\bar{V}_0}{j\omega L}\left(1 - \frac{1}{3}\omega^2 LC - \frac{1}{45}\omega^4 L^2 C^2 + \text{higher-order terms}\right)$$

$$\approx \frac{\bar{V}_0}{j\omega L}\left(1 - \frac{1}{3}\omega^2 LC - \frac{1}{45}\omega^4 L^2 C^2\right)$$

$$= \frac{\bar{V}_0}{j\omega L} + \bar{V}_0 \left(j\frac{\omega C}{3} + j\frac{\omega^3 LC^2}{45}\right) \tag{6-150}$$

$$= \frac{\bar{V}_0}{j\omega L} + \frac{\bar{V}_0}{1/[(j\omega C/3)(1 + \omega^2 LC/15)]}$$

$$\approx \frac{\bar{V}_0}{j\omega L} + \frac{\bar{V}_0}{(3/j\omega C)(1 - \omega^2 LC/15)}$$

$$= \frac{\bar{V}_0}{j\omega L} + \frac{\bar{V}_0}{(3/j\omega C) + (j\omega L/5)}$$

The equivalent circuit corresponding to (6-150) is shown in Fig. 6.26(c). It is now evident that as the frequency is increased, more and more elements are added to the equivalent circuit. For an arbitrarily high frequency, we must include all terms in the series. We then have, from (6-145),

$$\frac{\bar{V}_0}{\bar{I}_0} = j\omega L \frac{\tan \omega\sqrt{\mu\epsilon}l}{\omega\sqrt{\mu\epsilon}l}$$

$$= j\sqrt{\frac{L}{C}} \tan \omega\sqrt{LC} \tag{6-151}$$

Since $\tan \omega\sqrt{LC}$ can be negative if $\omega\sqrt{LC}$ falls in the second or third quadrant, the reactance viewed by the current source can even be capacitive! ∎

6.7 Transmission-Line Equations; The Distributed Circuit Concept

In Example 6-16 we obtained the fields between the parallel plates of the structure shown in Fig. 6.25(b) by starting with the quasistatic magnetic field and using successively the two curl equations

$$\nabla \times \bar{\mathbf{E}} = -j\omega\bar{\mathbf{B}} = -j\omega\mu\bar{\mathbf{H}} \tag{6-152a}$$

$$\nabla \times \bar{\mathbf{H}} = j\omega\bar{\mathbf{D}} = j\omega\epsilon\bar{\mathbf{E}} \tag{6-152b}$$

to find the successively induced electric and magnetic fields. The total electric field is the sum of all the electric fields found successively and the total magnetic field is the sum of all the magnetic fields found successively. These two total fields must satisfy the two curl equations simultaneously. Thus, denoting the total electric field by $\bar{\mathbf{E}}(z) = \bar{E}_x(z)\mathbf{i}_x$ and the total magnetic field by $\bar{\mathbf{H}}(z) = \bar{H}_y(z)\mathbf{i}_y$, we have, from (6-152a) and (6-152b), respectively,

$$\frac{\partial \bar{E}_x}{\partial z} = -j\omega\mu\bar{H}_y \tag{6-153a}$$

$$-\frac{\partial \bar{H}_y}{\partial z} = j\omega\epsilon\bar{E}_x \tag{6-153b}$$

However, the voltage between the two conductors in any plane normal to the z direction is given by

$$\bar{V}(z) = \int_{x=0}^{d} \bar{E}_x(z)\,dx = \bar{E}_x d \tag{6-154a}$$

The current along the conductors crossing any plane normal to the z direction is given by

$$\bar{I}(z) = \int_{y=0}^{w} \bar{H}_y(z)\,dy = \bar{H}_y w \tag{6-154b}$$

Substituting for $\bar{E}_x$ and $\bar{H}_y$ in (6-153a) and (6-153b) from (6-154a) and (6-154b), respectively, we get

$$\frac{\partial \bar{V}}{\partial z} = -j\omega\frac{\mu d}{w}\bar{I} = -j\omega\mathscr{L}\bar{I} \tag{6-155}$$

$$\frac{\partial \bar{I}}{\partial z} = -j\omega\frac{\epsilon w}{d}\bar{V} = -j\omega\mathscr{C}\bar{V} \tag{6-156}$$

where $\mathcal{L} = \mu d/w$ and $\mathcal{C} = \epsilon w/d$ are the inductance and capacitance, respectively, per unit length of the structure computed from static fields.

Equations (6-155) and (6-156) relate the time-varying voltage distribution along the z direction to the time-varying current distribution along the z direction. While we have obtained these equations for the particular case of a structure consisting of two parallel plane conductors, they are general and hold for any structure consisting of two parallel, infinitely long, perfect conductors having arbitrary but uniform cross sections. To prove this, let us consider such a structure having the cross section shown in Fig. 6.27.

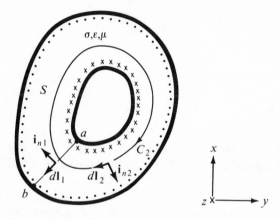

Fig. 6.27. For deriving the transmission-line equations.

For the sake of generality, we consider the dielectric to be imperfect with uniform conductivity σ and also work with arbitrarily time-varying fields instead of sinusoidally time-varying fields. Thus the electric and magnetic fields between the conductors are given by

$$\begin{aligned} \mathbf{E}(x, y, z, t) &= E_x(x, y, z, t)\mathbf{i}_x + E_y(x, y, z, t)\mathbf{i}_y \\ &= \mathbf{E}_{xy}(x, y, z, t) \end{aligned} \tag{6-157a}$$

$$\begin{aligned} \mathbf{H}(x, y, z, t) &= H_x(x, y, z, t)\mathbf{i}_x + H_y(x, y, z, t)\mathbf{i}_y \\ &= \mathbf{H}_{xy}(x, y, z, t) \end{aligned} \tag{6-157b}$$

Substituting (6-157a) and (6-157b) in

$$\nabla \times \mathbf{E} = -\frac{\partial \mathbf{B}}{\partial t}$$

we have

$$-\frac{\partial E_y}{\partial z}\mathbf{i}_x + \frac{\partial E_x}{\partial z}\mathbf{i}_y = -\frac{\partial B_x}{\partial t}\mathbf{i}_x - \frac{\partial B_y}{\partial t}\mathbf{i}_y \tag{6-158}$$

Taking the cross product of both sides of (6-158) with the unit vector $\mathbf{i}_z$, we get

$$-\frac{\partial E_y}{\partial z}\mathbf{i}_y - \frac{\partial E_x}{\partial z}\mathbf{i}_x = -\frac{\partial}{\partial t}(\mathbf{i}_z \times \mathbf{B}_{xy})$$

or

$$\frac{\partial \mathbf{E}_{xy}}{\partial z} = \frac{\partial}{\partial t}(\mathbf{i}_z \times \mathbf{B}_{xy}) \qquad (6\text{-}159)$$

Performing line integration of both sides of (6-159) from point *a* on the inner conductor to point *b* on the outer conductor, we have

$$\int_a^b \frac{\partial \mathbf{E}_{xy}}{\partial z} \cdot d\mathbf{l}_1 = \int_a^b \frac{\partial}{\partial t}(\mathbf{i}_z \times \mathbf{B}_{xy}) \cdot d\mathbf{l}_1$$

or

$$\frac{\partial}{\partial z}\int_a^b \mathbf{E}_{xy} \cdot d\mathbf{l}_1 = -\frac{\partial}{\partial t}\int_a^b \mathbf{B}_{xy} \cdot (\mathbf{i}_z \times d\mathbf{l}_1)$$

$$= -\frac{\partial}{\partial t}\int_a^b \mathbf{B}_{xy} \cdot \mathbf{i}_{n1} \, dl_1 \qquad (6\text{-}160)$$

where $\mathbf{i}_{n1}$ is the unit vector normal to $d\mathbf{l}_1$ as shown in Fig. 6.27. The integral on the left side of (6-160) is simply the voltage V between the conductors in the plane in which the line integral is evaluated since the magnetic field has no z component. The integral on the right side of (6-160) is the magnetic flux per unit length in the z direction, linking the inner conductor if the conductors are carrying a direct current equal to the current I crossing the plane containing the path *ab*. It is therefore equal to $\mathcal{L}I$, where $\mathcal{L}$ is the inductance per unit length of the structure computed from static field considerations. Thus we have

$$\frac{\partial V(z, t)}{\partial z} = -\frac{\partial}{\partial t}[\mathcal{L}I(z, t)] = -\mathcal{L}\frac{\partial I(z, t)}{\partial t} \qquad (6\text{-}161)$$

Similarly, substituting (6-157a) and (6-157b) in

$$\mathbf{\nabla} \times \mathbf{H} = \mathbf{J}_c + \frac{\partial \mathbf{D}}{\partial t} = \sigma \mathbf{E} + \frac{\partial \mathbf{D}}{\partial t}$$

we have

$$-\frac{\partial H_y}{\partial z}\mathbf{i}_x + \frac{\partial H_x}{\partial z}\mathbf{i}_y = \sigma E_x \mathbf{i}_x + \sigma E_y \mathbf{i}_y + \frac{\partial D_x}{\partial t}\mathbf{i}_x + \frac{\partial D_y}{\partial t}\mathbf{i}_y \qquad (6\text{-}162)$$

Taking the cross product of both sides of (6-162) with the unit vector $\mathbf{i}_z$, we get

$$-\frac{\partial H_y}{\partial z}\mathbf{i}_y - \frac{\partial H_x}{\partial z}\mathbf{i}_x = \sigma(\mathbf{i}_z \times \mathbf{E}_{xy}) + \frac{\partial}{\partial t}(\mathbf{i}_z \times \mathbf{D}_{xy})$$

or

$$\frac{\partial \mathbf{H}_{xy}}{\partial z} = -\sigma(\mathbf{i}_z \times \mathbf{E}_{xy}) - \frac{\partial}{\partial t}(\mathbf{i}_z \times \mathbf{D}_{xy}) \qquad (6\text{-}163)$$

Performing line integration of both sides of (6-163) around the closed path C_2 surrounding the inner conductor, we have

$$\oint_{C_2} \frac{\partial \mathbf{H}_{xy}}{\partial z} \cdot d\mathbf{l}_2 = -\oint_{C_2} \sigma(\mathbf{i}_z \times \mathbf{E}_{xy}) \cdot d\mathbf{l}_2 - \oint_{C_2} \frac{\partial}{\partial t}(\mathbf{i}_z \times \mathbf{D}_{xy}) \cdot d\mathbf{l}_2$$

or

$$\frac{\partial}{\partial z} \oint_{C_2} \mathbf{H}_{xy} \cdot d\mathbf{l}_2 = -\oint_{C_2} \sigma \mathbf{E}_{xy} \cdot (d\mathbf{l}_2 \times \mathbf{i}_z) - \frac{\partial}{\partial t} \oint_{C_2} \mathbf{D}_{xy} \cdot (d\mathbf{l}_2 \times \mathbf{i}_z)$$

$$= -\oint_{C_2} \sigma \mathbf{E}_{xy} \cdot \mathbf{i}_{n2} \, dl_2 - \frac{\partial}{\partial t} \oint_{C_2} \mathbf{D}_{xy} \cdot \mathbf{i}_{n2} \, dl_2 \qquad (6\text{-}164)$$

where $\mathbf{i}_{n2}$ is the unit vector normal to $d\mathbf{l}_2$ on C_2 as shown in Fig. 6.27. The integral on the left side of (6-164) is the current I in the positive z direction on the inner conductor (or the current in the negative z direction on the outer conductor) crossing the plane in which the closed path C_2 lies since the electric field has no z component. The first integral on the right side of (6-164) is the conduction current per unit length in the z direction, flowing from the inner to the outer conductor if the voltage between the two conductors is a direct voltage equal to the voltage V in the plane containing the path C_2. This current is equal to $\mathcal{G}V$, where $\mathcal{G}$ is the conductance per unit length computed from static field considerations. The second integral on the right side of (6-164) is the displacement flux, per unit length in the z direction, from the inner to the outer conductor if the voltage between the two conductors is a direct voltage equal to the voltage V in the plane containing the path C_2. This flux is equal to the magnitude of the charge per unit length on either conductor, which in turn is equal to $\mathcal{C}V$, where $\mathcal{C}$ is the capacitance per unit length computed from static field considerations. Thus we have

$$\frac{\partial I(z, t)}{\partial z} = -\mathcal{G}V(z, t) - \frac{\partial}{\partial t}[\mathcal{C}V(z, t)]$$

$$= -\mathcal{G}V(z, t) - \mathcal{C}\frac{\partial V(z, t)}{\partial t} \qquad (6\text{-}165)$$

Equations (6-161) and (6-165) describe the behavior of the voltage and current as functions of distance along the structure and of time. The structure itself is known as a transmission line since electromagnetic energy transmission occurs along the structure due to the time-varying fields, as we will learn later. Equations (6-161) and (6-165) are therefore known as the transmission-line equations. To obtain the circuit equivalent of the transmission-line equations, let us consider a section of infinitesimal length Δz along the line between z and $z + \Delta z$. From (6-161) and (6-165), we then have

$$\underset{\Delta z \to 0}{\text{Lim}} \frac{V(z + \Delta z, t) - V(z, t)}{\Delta z} = -\mathcal{L}\frac{\partial I(z, t)}{\partial t}$$

$$\underset{\Delta z \to 0}{\text{Lim}} \frac{I(z + \Delta z, t) - I(z, t)}{\Delta z} = \underset{\Delta z \to 0}{\text{Lim}}\left[-\mathcal{G}V(z + \Delta z, t) - \mathcal{C}\frac{\partial V(z + \Delta z, t)}{\partial t}\right]$$

or, for $\Delta z \rightarrow 0$,

$$V(z + \Delta z, t) - V(z, t) = -\mathcal{L}\,\Delta z\,\frac{\partial I(z, t)}{\partial t} \tag{6-166a}$$

$$I(z + \Delta z, t) - I(z, t) = -\mathcal{G}\,\Delta z\,V(z + \Delta z, t) - \mathcal{C}\,\Delta z\,\frac{\partial V(z + \Delta z, t)}{\partial t} \tag{6-166b}$$

The circuit theory equivalent of Eqs. (6-166a) and (6-166b) can be drawn as shown in Fig. 6.28 by recognizing that Eq. (6-166a) is Kirchhoff's voltage

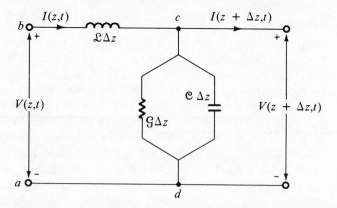

Fig. 6.28. Circuit equivalent for an infinitesimal length Δz of a transmission line.

law written for the loop *abcda* and that Eq. (6-166b) is Kirchhoff's current law written for node *c*. Thus an infinitesimal length Δz of the structure is equivalent to the circuit shown in Fig. 6.28 as $\Delta z \rightarrow 0$. It follows that the circuit representation for a portion of length *l* of the structure consists of infinite number of such sections in cascade as shown in Fig. 6.29. In other

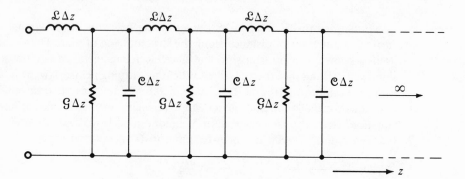

Fig. 6.29. Distributed circuit representation of a transmission line.

words, the structure can no longer be represented by a collection of lumped circuit elements. The conductance, capacitance, and inductance are "distributed" uniformly and overlappingly along the structure, giving rise to the concept of a "distributed circuit." Physically, the electric stored energy, the magnetic stored energy, and the power dissipation due to conduction current flow are distributed uniformly and overlappingly along the line.

Before we conclude this section, we wish to show that the power flow across any cross-sectional plane of the transmission line as computed from surface integration of the Poynting vector is equal to the product of the voltage and current in that plane. To do this, let us again consider the structure of Fig. 6.27. Considering an infinite plane surface (which is a spherical surface of infinite radius and hence a closed surface) in the cross-sectional plane and noting that the fields outside the conductors are zero, the power flow P across any cross-sectional plane is simply the surface integral of the Poynting vector over the cross-sectional surface S between the conductors. Thus

$$
\begin{aligned}
P(z, t) &= \int_S \mathbf{E}_{xy}(z, t) \times \mathbf{H}_{xy}(z, t) \cdot \mathbf{i}_z \, dS \\
&= \int_a^b \oint_{C_2} (\mathbf{E}_{xy} \times \mathbf{H}_{xy}) \cdot (d\mathbf{l}_1 \times d\mathbf{l}_2) \\
&= \int_a^b \oint_{C_2} (\mathbf{E}_{xy} \cdot d\mathbf{l}_1)(\mathbf{H}_{xy} \cdot d\mathbf{l}_2) \\
&\quad - \int_a^b \oint_{C_2} (\mathbf{E}_{xy} \cdot d\mathbf{l}_2)(\mathbf{H}_{xy} \cdot d\mathbf{l}_1)
\end{aligned}
\tag{6-167}
$$

Since we can always choose C_2 such that $d\mathbf{l}_2$ is everywhere normal to $\mathbf{E}_{xy}$ or, alternatively, since we can always choose the path ab such that $d\mathbf{l}_1$ is everywhere normal to $\mathbf{H}_{xy}$, the second integral on the right side of (6-167) is equal to zero. Since $\int_a^b \mathbf{E}_{xy} \cdot d\mathbf{l}_1$ is independent of the path on S chosen from a to b or, alternatively, since $\oint_{C_2} \mathbf{H}_{xy} \cdot d\mathbf{l}_2$ is independent of the contour C_2 on S, Eq. (6-167) simplifies to

$$
\begin{aligned}
P(z, t) &= \left(\int_a^b \mathbf{E}_{xy} \cdot d\mathbf{l}_1 \right)\left(\oint_{C_2} \mathbf{H}_{xy} \cdot d\mathbf{l}_2 \right) \\
&= V(z, t) I(z, t)
\end{aligned}
\tag{6-168}
$$

which is the desired result.

PART II.   Electromagnetic Waves

6.8 The Wave Equation; Uniform Plane Waves and Transmission-Line Waves

In Section 4.11 we stated that time-varying electric and magnetic fields give rise to the phenomenon of electromagnetic wave propagation. In this section we introduce the topic of wave propagation. Let us consider a perfect dielectric region free of charges and for which μ and ϵ are constants. Maxwell's equations for such a medium are

$$\mathbf{V} \cdot \mathbf{D} = 0 \tag{6-169a}$$

$$\mathbf{V} \cdot \mathbf{B} = 0 \tag{6-169b}$$

$$\mathbf{V} \times \mathbf{E} = -\frac{\partial \mathbf{B}}{\partial t} \tag{6-169c}$$

$$\mathbf{V} \times \mathbf{H} = \frac{\partial \mathbf{D}}{\partial t} \tag{6-169d}$$

Taking the curl of both sides of (6-169c), we have

$$\mathbf{V} \times \mathbf{V} \times \mathbf{E} = -\mathbf{V} \times \frac{\partial \mathbf{B}}{\partial t} \tag{6-170}$$

Using the vector identity

$$\mathbf{V} \times \mathbf{V} \times \mathbf{A} \equiv \mathbf{V}(\mathbf{V} \cdot \mathbf{A}) - \mathbf{V}^2 \mathbf{A}$$

for the left side of (6-170) and interchanging $\partial/\partial t$ and the curl operation on the right side since the curl operation has to do with differentiation with respect to space coordinates, we obtain

$$\mathbf{V}(\mathbf{V} \cdot \mathbf{E}) - \mathbf{V}^2\mathbf{E} = -\frac{\partial}{\partial t}(\mathbf{V} \times \mathbf{B}) \tag{6-171}$$

However, from (6-169a), $\mathbf{V} \cdot \mathbf{E} = 0$ and from (6-169d), $\mathbf{V} \times \mathbf{B} = \mu\epsilon \, \partial \mathbf{E}/\partial t$. Thus (6-171) reduces to

$$\mathbf{V}^2\mathbf{E} = \mu\epsilon \frac{\partial^2 \mathbf{E}}{\partial t^2} \tag{6-172}$$

For sinusoidally time-varying fields, we have the phasor form of (6-172) as

$$\mathbf{V}^2\bar{\mathbf{E}} = -\omega^2 \mu\epsilon \bar{\mathbf{E}} \tag{6-173}$$

Note that the left side of (6-172) is the Laplacian of a vector and not of a scalar. Equation (6-172) is known as the vector wave equation. Equating the like components on either side of (6-172), we obtain three scalar wave

408

equations. Thus, in cartesian coordinates, we have

$$\nabla^2 E_x = \mu\epsilon \frac{\partial^2 E_x}{\partial t^2}$$

$$\nabla^2 E_y = \mu\epsilon \frac{\partial^2 E_y}{\partial t^2}$$

$$\nabla^2 E_z = \mu\epsilon \frac{\partial^2 E_z}{\partial t^2}$$

In the most general case, we can have all three components of **E** and each one of these can be dependent on all three space coordinates x, y, and z and on time. But let us assume for simplicity that $E_y = E_z = 0$. Then we have

$$\nabla^2 E_x = \frac{\partial^2 E_x}{\partial x^2} + \frac{\partial^2 E_x}{\partial y^2} + \frac{\partial^2 E_x}{\partial z^2} = \mu\epsilon \frac{\partial^2 E_x}{\partial t^2} \qquad (6\text{-}174)$$

We are still faced with a three-dimensional second-order partial differential equation. Our aim at present is to illustrate that time-varying electric and magnetic fields give rise to electromagnetic wave propagation. Hence let us simplify the problem further by assuming that E_x is independent of x and y. Thus

$$\mathbf{E} = E_x(z, t)\mathbf{i}_x \qquad (6\text{-}175)$$

and Eq. (6-174) simplifies to

$$\frac{\partial^2 E_x}{\partial z^2} = \mu\epsilon \frac{\partial^2 E_x}{\partial t^2} \qquad (6\text{-}176)$$

Equation (6-176) is the one-dimensional scalar wave equation. Its solution can be found by using the Laplace transform technique or the separation of variables technique. However, we will here write down the solution and show that it indeed satisfies the equation. Thus let us consider

$$E_x(z, t) = A\, f(t - \sqrt{\mu\epsilon}z) + B\, g(t + \sqrt{\mu\epsilon}z) \qquad (6\text{-}177)$$

where f and g are any functions of the respective arguments and A and B are arbitrary constants. Then

$$\frac{\partial E_x}{\partial z} = -A\sqrt{\mu\epsilon}\, f'(t - \sqrt{\mu\epsilon}z) + B\, g'(t + \sqrt{\mu\epsilon}z)$$

$$\frac{\partial^2 E_x}{\partial z^2} = A\mu\epsilon\, f''(t - \sqrt{\mu\epsilon}z) + B\mu\epsilon\, g''(t + \sqrt{\mu\epsilon}z) \qquad (6\text{-}178a)$$

$$\frac{\partial E_x}{\partial t} = A\, f'(t - \sqrt{\mu\epsilon}z) + B\, g'(t + \sqrt{\mu\epsilon}z)$$

$$\frac{\partial^2 E_x}{\partial t^2} = A\, f''(t - \sqrt{\mu\epsilon}z) + B\, g''(t + \sqrt{\mu\epsilon}z) \qquad (6\text{-}178b)$$

where the primes denote differentiation with respect to the respective arguments. From (6-178a) and (6-178b), we note that (6-177) satisfies (6-176) and hence is the solution for (6-176). The forms of the functions f and g depend upon the particular problem under consideration. Some examples are $\cos \omega(t - \sqrt{\mu\epsilon}z)$, $e^{-(t-\sqrt{\mu\epsilon}z)^2}$, and $(t + \sqrt{\mu\epsilon}z) \sin(t + \sqrt{\mu\epsilon}z)$. In the general case, the solution can be a superposition of several different functions of $(t - \sqrt{\mu\epsilon}z)$ and $(t + \sqrt{\mu\epsilon}z)$.

To discuss the meaning of the functions f and g in the solution for E_x, let us consider a specific example

$$f(t - \sqrt{\mu\epsilon}z) = e^{-(t-\sqrt{\mu\epsilon}z)}u(t - \sqrt{\mu\epsilon}z)$$

Assigning one value for t at a time, we can obtain a series of functions of z. The time history of these functions can be illustrated conveniently by a three-dimensional plot in which the three axes represent time t, distance z, and the value of the function f. Such a plot for the function under consideration is shown in Fig. 6.30. We note from Fig. 6.30 that the function of z at

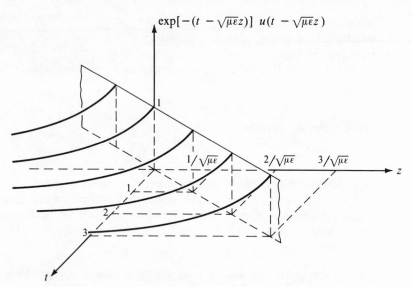

Fig. 6.30. Three-dimensional representation of the function $e^{-(t-\sqrt{\mu\epsilon}z)}u(t - \sqrt{\mu\epsilon}z)$ for illustrating the concept of a traveling wave.

any value of time is exactly the same as the function of z at a preceding value of time but shifted towards the direction of increasing z. For example, by following the peak in the function, we note that from time $t = 0$ to time $t = 1$, the peak shifts from $z = 0$ to $z = 1/\sqrt{\mu\epsilon}$. Thus the function $f(t - \sqrt{\mu\epsilon}z)$ represents a waveform traveling in the positive z direction with a velocity $1/\sqrt{\mu\epsilon}$. The solution is said to correspond to a traveling wave in the positive z direction, or a $(+)$ wave. The word "wave" is used here in the sense that it represents any arbitrary function of z and not necessarily a sinusoidally varying function of z. The fact that the velocity of propagation is $1/\sqrt{\mu\epsilon}$ can be proved in general by following any particular point of the function and noting down its positions z_1 and z_2 for two times t_1 and t_2. Obviously then,

$$t_1 - \sqrt{\mu\epsilon}\,z_1 = t_2 - \sqrt{\mu\epsilon}\,z_2$$

or the velocity of propagation is

$$v = \frac{z_2 - z_1}{t_2 - t_1} = \frac{1}{\sqrt{\mu\epsilon}}$$

Note that the units of $1/\sqrt{\mu\epsilon}$ are

$$\left[\frac{\text{newtons}}{(\text{ampere})^2} \times \frac{(\text{coulomb})^2}{(\text{newton})(\text{meter})^2}\right]^{-1/2} = \frac{\text{ampere-meters}}{\text{coulomb}} = \frac{\text{meters}}{\text{second}}$$

For free space, $1/\sqrt{\mu\epsilon} = 1/\sqrt{\mu_0\epsilon_0} = 3 \times 10^8$ m/sec, which is the velocity of light.

We now suspect that the function $g(t + \sqrt{\mu\epsilon}z)$ represents wave motion in the direction of decreasing values of z, that is, in the negative z direction. To check if this is true, we note that, to follow a particular point associated with the function, an observer has to move in space and time such that

$$t + \sqrt{\mu\epsilon}z = \text{constant}$$

or

$$dt + \sqrt{\mu\epsilon}\,dz = 0$$

or the velocity with which the observer has to move in the positive z direction is

$$v = \frac{dz}{dt} = -\frac{1}{\sqrt{\mu\epsilon}}$$

The negative sign for the velocity signifies that the observer must actually move in the negative z direction with a velocity of $1/\sqrt{\mu\epsilon}$. Thus the function $g(t + \sqrt{\mu\epsilon}z)$ does indeed represent a traveling wave in the negative z direction, or a $(-)$ wave.

Now, the solution for the magnetic field associated with the electric field can be obtained by substituting (6-177) into (6-169c). Thus we obtain

$$-\frac{\partial \mathbf{B}}{\partial t} = \frac{\partial E_x}{\partial z}\mathbf{i}_y = [-A\sqrt{\mu\epsilon}\,f'(t - \sqrt{\mu\epsilon}z) + B\sqrt{\mu\epsilon}\,g'(t + \sqrt{\mu\epsilon}z)]\mathbf{i}_y$$

or

$$\mathbf{H} = H_y(z, t)\mathbf{i}_y = \left[A\sqrt{\frac{\epsilon}{\mu}}\,f(t - \sqrt{\mu\epsilon}z) - B\sqrt{\frac{\epsilon}{\mu}}\,g(t + \sqrt{\mu\epsilon}z)\right]\mathbf{i}_y$$

Defining

$$\eta = \sqrt{\frac{\mu}{\epsilon}} \tag{6-179}$$

we have

$$H_y(z, t) = \frac{1}{\eta}[A\,f(t - \sqrt{\mu\epsilon}z) - B\,g(t + \sqrt{\mu\epsilon}z)] \tag{6-180}$$

The quantity η is known as the intrinsic or characteristic impedance of the medium. Note that the units of η are

$$\left[\frac{\text{newtons}}{(\text{ampere})^2} \div \frac{(\text{coulomb})^2}{(\text{newton})(\text{meter})^2} \right]^{1/2} = \frac{\text{newton-meters}}{\text{coulomb-ampere}} = \frac{\text{volts}}{\text{ampere}} = \text{ohms}$$

For free space, $\eta = \eta_0 = \sqrt{\mu_0/\epsilon_0} = 120\pi$ or 377 ohms.

Denoting the electric fields in the $(+)$ and $(-)$ waves as E_x^+ and E_x^-, respectively, and the magnetic fields in the $(+)$ and $(-)$ waves as H_y^+ and H_y^-, respectively, we have

$$E_x = E_x^+ + E_x^- \qquad (6\text{-}181\text{a})$$

$$H_y = H_y^+ + H_y^- \qquad (6\text{-}181\text{b})$$

Comparing (6-181a) and (6-181b) with (6-177) and (6-180), respectively, we note that

$$H_y^+ = \frac{E_x^+}{\eta} \qquad \text{and} \qquad H_y^- = -\frac{E_x^-}{\eta} \qquad (6\text{-}182)$$

The Poynting vectors associated with the $(+)$ and $(-)$ waves are

$$\mathbf{P}^+ = E_x^+ \mathbf{i}_x \times H_y^+ \mathbf{i}_y = E_x^+ \mathbf{i}_x \times \frac{E_x^+}{\eta} \mathbf{i}_x = \frac{(E_x^+)^2}{\eta} \mathbf{i}_z \qquad (6\text{-}183\text{a})$$

$$\mathbf{P}^- = E_x^- \mathbf{i}_x \times H_y^- \mathbf{i}_y = (E_x^- \mathbf{i}_x) \times \left(-\frac{E_x^-}{\eta} \mathbf{i}_y \right) = -\frac{(E_x^-)^2}{\eta} \mathbf{i}_z \qquad (6\text{-}183\text{b})$$

Equations (6-183a) and (6-183b) indicate that the power flow associated with the $(+)$ wave is indeed in the positive z direction and the power flow associated with the $(-)$ wave is indeed in the negative z direction.

Summarizing what we have learned thus far in this section, time-varying electric and magnetic fields give rise to electromagnetic wave propagation. A simple solution consists of waves traveling in the positive and negative z directions and having electric fields entirely in the x direction and magnetic fields entirely in the y direction. Furthermore, the fields are uniform in the planes transverse to the direction of propagation, that is, in the planes $z = $ constant. For this reason they are known as uniform plane waves. In reality, uniform plane waves do not exist. However, at distances far from a radiating antenna and the ground, the radiated waves are approximately uniform plane waves. The uniform plane waves are a very important building block in the study of electromagnetic waves.

EXAMPLE 6-18. A uniform plane wave traveling in the positive z direction in free space has its electric field entirely along the x direction. The time variation of the electric field intensity in the $z = 0$ plane is shown in Fig. 6.31(a). Find and sketch the variation with z of the magnetic field intensity at $t = 2$ μsec.

The magnetic field intensity is entirely in the y direction and its value for any (z, t) is equal to the value of the corresponding electric field intensity

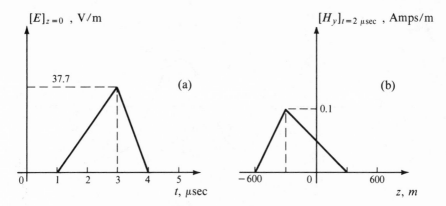

Fig. 6.31. (a) Electric field intensity in the $z = 0$ plane versus time. (b) Magnetic field intensity at $t = 2$ μsec versus z for the uniform plane wave of Example 6.18.

divided by 377 ohms. The velocity of propagation of the wave in free space is 3×10^8 m/sec. Since the wave is propagating in the positive z direction, an amplitude which exists in the $z = 0$ plane at any time t must exist in the plane $z = (2 \times 10^{-6} - t) \times 3 \times 10^8$ m at $t = 2$ μsec. Hence the variation with z of the magnetic field intensity at $t = 2$ μsec is as shown in Fig. 6.31(b). ∎

We now direct our attention to the (+) wave to define certain important parameters for the sinusoidally time-varying case and to develop expressions for the fields in a uniform plane wave traveling in an arbitrary direction with reference to a coordinate system. Let us consider a uniform plane wave characterized by sinusoidally time-varying electric and magnetic fields given by

$$\mathbf{E}(z, t) = E_0 \cos [\omega(t - \sqrt{\mu\epsilon}z) + \phi_0] \mathbf{i}_x \qquad (6\text{-}184\text{a})$$

$$\mathbf{H}(z, t) = \frac{E_0}{\sqrt{\mu/\epsilon}} \cos [\omega(t - \sqrt{\mu\epsilon}z) + \phi_0] \mathbf{i}_y \qquad (6\text{-}184\text{b})$$

where E_0 and ϕ_0 are constants. At any particular value of z, say z_0, the fields vary with time in the manner shown in Fig. 6.32(a). Any particular value of the field repeats in time at intervals of $2\pi/\omega$. The number of times the value repeats in 1 sec is equal to $\omega/2\pi$ or f, which is the well-known parameter frequency. At any particular value of time, say $t = t_0$, the fields vary with distance z in the manner shown in Fig. 6.32(b). Any particular value of the field repeats in distance at intervals of $2\pi/\omega\sqrt{\mu\epsilon}$. This interval is known as the wavelength, λ. Thus

$$\lambda = \frac{2\pi}{\omega\sqrt{\mu\epsilon}} = \frac{1}{f\sqrt{\mu\epsilon}} \qquad (6\text{-}185)$$

But $1/\sqrt{\mu\epsilon}$ is the velocity of propagation of the wave. In this case, it is known

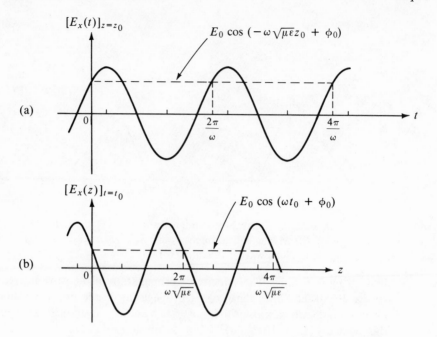

Fig. 6.32. (a) Electric field intensity in a $z = $ constant plane versus time. (b) Electric field intensity at a fixed time versus z, for a uniform plane wave in the sinusoidal steady state and traveling in the z direction.

as the phase velocity since the argument of the cosine function is known as the phase and an observer has to travel with a velocity $1/\sqrt{\mu\epsilon}$ in the z direction to follow a constant phase of the field, that is, to stay on a particular constant phase surface. The constant phase surfaces are the planes $z = $ constant. Denoting the phase velocity by v_p, we have

$$v_p = \frac{1}{\sqrt{\mu\epsilon}} \tag{6-186}$$

Substituting (6-186) into (6-185), we get

$$v_p = \lambda f \tag{6-187}$$

Equation (6-187) is an important relationship which relates the space and time variations of the fields in an electromagnetic wave. For free space, Eq. (6-187) gives a simple formula

(wavelength in meters) $\times$ (frequency in megahertz) $= 300$

The quantity $\omega\sqrt{\mu\epsilon}$ is the rate at which the phase changes with distance z at any particular time. It is known as the phase constant and is denoted by β. Thus

$$\beta = \omega\sqrt{\mu\epsilon} \tag{6-188}$$

and

$$\lambda = \frac{2\pi}{\beta} \tag{6-189}$$

$$v_p = \frac{\omega}{\beta} \tag{6-190}$$

The units of β are (radians/second)(seconds/meter) or radians per meter.

For a wave traveling in the z direction, the phase changes most rapidly in the z direction since, looking in any other direction, the distance between any two particular constant phase surfaces is longer than the distance between the same two constant phase surfaces as seen looking in the z direction, as shown in Fig. 6.33. Thus, if we choose the coordinate system such

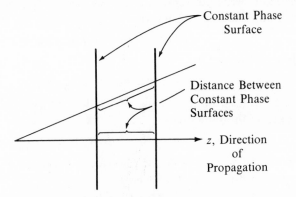

Fig. 6.33. Distances between two constant phase surfaces for a uniform plane wave as seen along different directions.

that the wave is traveling in an arbitrary direction with reference to the coordinate system, the rates at which the phase changes along the coordinate axes are all less than the rate at which the phase changes along the direction of propagation which is normal to the constant phase surfaces. Denoting the phase constants along the x, y, and z directions by β_x, β_y, and β_z, respectively, and the phase at the origin at $t = 0$ by ϕ_0, we note that the phase at any point (x, y, z) is $\omega t - (\beta_x x + \beta_y y + \beta_z z) + \phi_0$. The constant phase surfaces are the planes given by

$$\beta_x x + \beta_y y + \beta_z z = \text{constant} \tag{6-191}$$

The direction of the gradient of the scalar function $\beta_x x + \beta_y y + \beta_z z$ is the direction of the normal to the constant phase surfaces and hence is the direction of propagation whereas the magnitude of the gradient gives the rate of change of phase with distance or the phase constant β along the normal and hence along the direction of propagation. Thus, noting that

$$\mathbf{V}(\beta_x x + \beta_y y + \beta_z z) = \beta_x \mathbf{i}_x + \beta_y \mathbf{i}_y + \beta_z \mathbf{i}_z$$

the direction of propagation is along the vector $\beta_x \mathbf{i}_x + \beta_y \mathbf{i}_y + \beta_z \mathbf{i}_z$ and the phase constant along the direction of propagation is

$$\beta = (\beta_x^2 + \beta_y^2 + \beta_z^2)^{1/2} \tag{6-192}$$

We can combine these two facts by defining vector $\boldsymbol{\beta}$ as

$$\boldsymbol{\beta} = \beta_x \mathbf{i}_x + \beta_y \mathbf{i}_y + \beta_z \mathbf{i}_z \tag{6-193}$$

so that the direction of $\boldsymbol{\beta}$ is the direction of propagation and the magnitude of $\boldsymbol{\beta}$ is the phase constant. Hence $\boldsymbol{\beta}$ is known as the propagation vector. The phase at any point (x, y, z) can then be written as $\omega t - \boldsymbol{\beta} \cdot \mathbf{r} + \phi_0$, where $\mathbf{r}$ is the position vector $x\mathbf{i}_x + y\mathbf{i}_y + z\mathbf{i}_z$.

Denoting the electric field intensity in the plane of zero phase as $\mathbf{E}_0$, we can now write the expression for the electric field intensity vector associated with a uniform plane wave propagating along the direction of $\boldsymbol{\beta}$ as

$$\mathbf{E} = \mathbf{E}_0 \cos(\omega t - \boldsymbol{\beta} \cdot \mathbf{r} + \phi_0) \tag{6-194a}$$

or the complex vector as

$$\bar{\mathbf{E}} = \mathbf{E}_0 e^{j\phi_0} e^{-j\boldsymbol{\beta}\cdot\mathbf{r}} = \bar{\mathbf{E}}_0 e^{-j\boldsymbol{\beta}\cdot\mathbf{r}} \tag{6-194b}$$

where $\bar{\mathbf{E}}_0 = \mathbf{E}_0 e^{j\phi_0}$. Since $\mathbf{E}_0$ must be entirely transverse to the direction of propagation, it follows that

$$\boldsymbol{\beta} \cdot \mathbf{E}_0 = 0 \qquad \text{or} \qquad \boldsymbol{\beta} \cdot \bar{\mathbf{E}}_0 = 0 \tag{6-195}$$

Similarly, the magnetic field intensity vector associated with the wave which is in phase with $\mathbf{E}$ can be written as

$$\mathbf{H} = \mathbf{H}_0 \cos(\omega t - \boldsymbol{\beta} \cdot \mathbf{r} + \phi_0) \tag{6-196a}$$

or the complex vector as

$$\bar{\mathbf{H}} = \mathbf{H}_0 e^{j\phi_0} e^{-j\boldsymbol{\beta}\cdot\mathbf{r}} = \bar{\mathbf{H}}_0 e^{-j\boldsymbol{\beta}\cdot\mathbf{r}} \tag{6-196b}$$

where $\bar{\mathbf{H}}_0 = \mathbf{H}_0 e^{j\phi_0}$. Since $\mathbf{H}_0$ must be entirely transverse to the direction of propagation, it follows that

$$\boldsymbol{\beta} \cdot \mathbf{H}_0 = 0 \qquad \text{or} \qquad \boldsymbol{\beta} \cdot \bar{\mathbf{H}}_0 = 0 \tag{6-197}$$

Furthermore $\mathbf{E}_0$ and $\mathbf{H}_0$ must be normal to each other with their cross product (Poynting vector) pointing in the direction of propagation and with the ratio of their magnitudes given by

$$\frac{E_0}{H_0} = \sqrt{\frac{\mu}{\epsilon}} = \frac{\omega\mu}{\omega\sqrt{\mu\epsilon}} = \frac{\omega\mu}{\beta} \tag{6-198}$$

In vector notation, we express the preceding statement as

$$\mathbf{H}_0 = \frac{1}{\omega\mu}\boldsymbol{\beta} \times \mathbf{E}_0 \tag{6-199}$$

and hence

$$\bar{\mathbf{H}} = \frac{1}{\omega\mu}\boldsymbol{\beta} \times \bar{\mathbf{E}} \tag{6-200}$$

These properties associated with a uniform plane wave are illustrated in Fig. 6.34.

The wavelength λ along the direction of propagation is given by

$$\lambda = \frac{2\pi}{\beta} \tag{6-201}$$

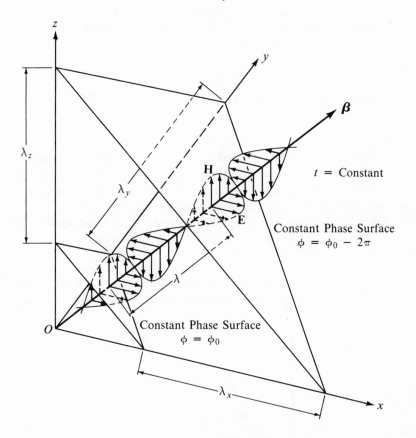

Fig. 6.34. For illustrating the various concepts associated with a uniform plane wave traveling in an arbitrary direction.

The apparent wavelengths λ_x, λ_y, and λ_z along the coordinate axes x, y, and z, respectively, as shown in Fig. 6.34 are given by

$$\lambda_x = \frac{2\pi}{\beta_x} \qquad \lambda_y = \frac{2\pi}{\beta_y} \qquad \lambda_z = \frac{2\pi}{\beta_z} \tag{6-202}$$

Substituting (6-201) and (6-202) into (6-192), we have

$$\frac{1}{\lambda^2} = \frac{1}{\lambda_x^2} + \frac{1}{\lambda_y^2} + \frac{1}{\lambda_z^2} \tag{6-203}$$

The phase velocity along the direction of propagation is given by

$$v_p = \frac{\omega}{\beta} \qquad (6\text{-}204)$$

For an observer moving along the x axis, y and z are constants. Hence the observer has to travel with a velocity equal to ω/β_x to remain on a particular constant phase surface. This velocity is known as the apparent phase velocity in the x direction. Thus the apparent phase velocities v_{px}, v_{py}, and v_{pz} in the x, y, and z directions are

$$v_{px} = \frac{\omega}{\beta_x} \qquad v_{py} = \frac{\omega}{\beta_y} \qquad v_{pz} = \frac{\omega}{\beta_z} \qquad (6\text{-}205)$$

Substituting (6-204) and (6-205) into (6-192), we have

$$\frac{1}{v_p^2} = \frac{1}{v_{px}^2} + \frac{1}{v_{py}^2} + \frac{1}{v_{pz}^2} \qquad (6\text{-}206)$$

Note that the apparent wavelengths and phase velocities along the coordinate axes are greater than the wavelength and the phase velocity, respectively, along the direction of propagation, since the phase changes less rapidly with distance along the coordinate axes than along the direction of propagation. We will now consider an example to consolidate our understanding of the uniform plane wave propagating in an arbitrary direction with reference to a set of coordinate axes.

EXAMPLE 6-19.   The orientation of the propagation vector $\boldsymbol{\beta}$ for a uniform plane wave of 12 MHz propagating in free space is as shown in Fig. 6.35. It makes

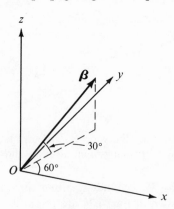

Fig. 6.35. Orientation of the propagation vector $\boldsymbol{\beta}$ for the uniform plane wave of Example 6-19.

an angle of 30° upwards with the horizontal (xy) plane and its projection on the xy plane makes an angle of 60° with the x axis. The electric field intensity has no upward (z) component and its magnitude as a function of time at $x = 0$, $y = 0$ and $z = 0$ is $10 \cos(\omega t - 30°)$ volts/m, where ω is the angular frequency. It is desired to find the expressions for the complex field vectors $\bar{\mathbf{E}}$ and $\bar{\mathbf{H}}$.

Since the medium is free space, the phase velocity along the propagation vector is $1/\sqrt{\mu_0 \epsilon_0} = 3 \times 10^8$ m/sec. From (6-204), we have

$$\beta = \frac{\omega}{v_p} = \frac{2\pi f}{v_p} = \frac{2\pi \times 12 \times 10^6}{3 \times 10^8} = 0.08\pi$$

From the given orientation of the propagation vector, we have

$$\boldsymbol{\beta} = 0.08\pi(\cos 30° \cos 60° \, \mathbf{i}_x + \cos 30° \sin 60° \, \mathbf{i}_y + \sin 30° \, \mathbf{i}_z)$$
$$= 0.02\pi(\sqrt{3} \, \mathbf{i}_x + 3\mathbf{i}_y + 2\mathbf{i}_z)$$

The solution for $\bar{\mathbf{E}}$ is of the form $\bar{\mathbf{E}}_0 \, e^{-j\boldsymbol{\beta}\cdot\mathbf{r}}$. Since $\mathbf{E}$ has no z component, we can write

$$\bar{\mathbf{E}}_0 = \bar{E}_{x0}\mathbf{i}_x + \bar{E}_{y0}\mathbf{i}_y$$

From (6-195), we have

$$\boldsymbol{\beta} \cdot \bar{\mathbf{E}}_0 = \beta_x \bar{E}_{x0} + \beta_y \bar{E}_{y0} = 0$$

Since β_x and β_y are both real, $\bar{E}_{x0}$ and $\bar{E}_{y0}$ must be either in phase or in phase opposition for the above equation to be true. Hence let

$$\bar{E}_{x0} = E_{x0}e^{j\alpha} \qquad \text{and} \qquad \bar{E}_{y0} = E_{y0}e^{j\alpha}$$

so that

$$\bar{\mathbf{E}}_0 = (E_{x0}\mathbf{i}_x + E_{y0}\mathbf{i}_y)e^{j\alpha}$$

and

$$\beta_x E_{x0} + \beta_y E_{y0} = 0.02\pi(\sqrt{3} \, E_{x0} + 3E_{y0}) = 0$$

or

$$E_{x0} = -\sqrt{3} \, E_{y0}$$

From the given function of time for the electric field intensity magnitude at $x = 0$, $y = 0$ and $z = 0$, that is, $\mathbf{r} = 0$, we have

$$|E_{x0}\mathbf{i}_x + E_{y0}\mathbf{i}_y|e^{j\alpha} = 10e^{-j30°}$$

or

$$|E_{x0}\mathbf{i}_x + E_{y0}\mathbf{i}_y| = [E_{x0}^2 + E_{y0}^2]^{1/2} = 10 \qquad \text{and} \qquad \alpha = -\frac{\pi}{6}$$

Substituting $E_{x0} = -\sqrt{3} \, E_{y0}$ in the above equation, we obtain $4E_{y0}^2 = 100$ or $E_{y0} = 5$ and $E_{x0} = -5\sqrt{3}$. Thus

$$\bar{\mathbf{E}}_0 = (-5\sqrt{3} \, e^{-j\pi/6}\mathbf{i}_x + 5e^{-j\pi/6}\mathbf{i}_y)$$

The required expression for $\mathbf{E}$ is then given by

$$\bar{\mathbf{E}} = 5(-\sqrt{3} \, \mathbf{i}_x + \mathbf{i}_y)e^{-j\pi/6}e^{-j0.02\pi(\sqrt{3}x+3y+2z)}$$

The corresponding expression for $\bar{\mathbf{H}}$ can be obtained by using (6-200) as follows:

$$\bar{H} = \frac{1}{\omega\mu}\boldsymbol{\beta} \times \bar{E}$$

$$= \frac{1}{96\pi}\begin{vmatrix} \mathbf{i}_x & \mathbf{i}_y & \mathbf{i}_z \\ \sqrt{3} & 3 & 2 \\ -\sqrt{3} & 1 & 0 \end{vmatrix} e^{-j\pi/6}e^{-j0.02\pi(\sqrt{3}x+3y+2z)}$$

$$= \frac{1}{48\pi}(-\mathbf{i}_x - \sqrt{3}\,\mathbf{i}_y + 2\sqrt{3}\,\mathbf{i}_z)e^{-j\pi/6}e^{-j0.02\pi(\sqrt{3}x+3y+2z)}$$

We can also find the wavelength along the direction of propagation and the apparent wavelengths and velocities of propagation along the x, y, and z axes. Thus

$$\lambda = \frac{2\pi}{\beta} = \frac{2\pi}{0.08\pi} = 25 \text{ m}$$

$$\lambda_x = \frac{2\pi}{\beta_x} = \frac{2\pi}{0.02\sqrt{3}\,\pi} = 57.7 \text{ m}$$

$$\lambda_y = \frac{2\pi}{\beta_y} = \frac{2\pi}{0.06\pi} = 33.3 \text{ m}$$

$$\lambda_z = \frac{2\pi}{\beta_z} = \frac{2\pi}{0.04\pi} = 50 \text{ m}$$

$$v_{px} = \frac{\omega}{\beta_x} = \frac{24\pi \times 10^6}{0.02\sqrt{3}\,\pi} = 6.928 \times 10^8 \text{ m/sec}$$

$$v_{py} = \frac{\omega}{\beta_y} = \frac{24\pi \times 10^6}{0.06\pi} = 4 \times 10^8 \text{ m/sec}$$

$$v_{pz} = \frac{\omega}{\beta_z} = \frac{24\pi \times 10^6}{0.04\pi} = 6 \times 10^8 \text{ m/sec}$$

Note that

$$\frac{1}{57.7^2} + \frac{1}{33.3^2} + \frac{1}{50^2} = \frac{1}{25^2}$$

and

$$\frac{1}{6.928^2} + \frac{1}{4^2} + \frac{1}{6^2} = \frac{1}{3^2}$$

in agreement with (6-203) and (6-206), respectively. ∎

In Section 4.9 we discussed polarization of vector fields. The fields we found in the preceding example are linearly polarized. We then say that the wave is linearly polarized. If we combine two linearly polarized uniform plane waves propagating in the same direction and having electric field vectors equal in magnitude but oriented perpendicular to each other and differing in phase by $\pi/2$, we obtain a circularly polarized uniform plane wave. For example, a uniform plane wave characterized by the electric field intensity vector

$$\bar{\mathbf{E}} = 2.5(-\mathbf{i}_x - \sqrt{3}\,\mathbf{i}_y + 2\sqrt{3}\,\mathbf{i}_z)e^{j\pi/3}e^{-j0.02\pi(\sqrt{3}x+3y+2z)}$$

when superimposed with the uniform plane wave of Example 6-19, would result in a circularly polarized uniform plane wave. In general, two linearly polarized uniform plane waves propagating in the same direction result in an elliptically polarized uniform plane wave.

We have introduced the topic of electromagnetic wave propagation by considering uniform plane waves. The uniform plane waves are a special class of waves known as transverse electromagnetic waves, abbreviated TEM waves, so named because the electric and magnetic fields are entirely transverse to the direction of propagation, that is, components of **E** and **H** along the direction of propagation are zero. For a general TEM wave, the fields are not uniform but are functions of position in the transverse plane. The electromagnetic field between the conductors of a transmission line made up of perfect conductors is entirely transverse to the line axis and is in general nonuniform in the cross-sectional plane. In fact, we considered such a field [Eqs. (6-157a) and (6-157b)] in Section 6.7 and, by substituting into Maxwell's curl equations, we obtained the transmission-line equations given by (6-161) and (6-165). For a perfect dielectric medium between the conductors, that is, for $\sigma = 0$, these equations are

$$\frac{\partial V(z, t)}{\partial z} = -\mathcal{L}\frac{\partial I(z, t)}{\partial t} \tag{6-207}$$

and

$$\frac{\partial I(z, t)}{\partial z} = -\mathcal{C}\frac{\partial V(z, t)}{\partial t} \tag{6-208}$$

where, with reference to Fig. 6.27,

$$V(z, t) = \int_a^b \mathbf{E}(x, y, z, t) \cdot d\mathbf{l}_1 \tag{6-209a}$$

and

$$I(z, t) = \oint_{C_2} \mathbf{H}(x, y, z, t) \cdot d\mathbf{l}_2 \tag{6-209b}$$

are, respectively, the voltage between the conductors and the current along the conductors for any (z, t).

Eliminating I from (6-207) and (6-208), we obtain a differential equation for V alone as

$$\frac{\partial^2 V(z, t)}{\partial z^2} = \mathcal{L}\mathcal{C}\frac{\partial^2 V(z, t)}{\partial t^2} \tag{6-210}$$

This equation is completely analogous to Eq. (6-176). It is the wave equation for the TEM wave propagation guided by the conductors of the transmission line except that it is written in terms of the voltage between the conductors instead of the electric field in the medium between the conductors. We can write the solution for (6-210) from our experience with the solution of (6-176). The solution is

$$V(z, t) = V^+(t - \sqrt{\mathcal{L}\mathcal{C}}z) + V^-(t + \sqrt{\mathcal{L}\mathcal{C}}z) \qquad (6\text{-}211)$$

where the subscripts $+$ and $-$ indicate $(+)$ and $(-)$ waves. The corresponding solution for the line current I can be obtained by substituting (6-211) into (6-207) or (6-208). This gives

$$I(z, t) = \sqrt{\frac{\mathcal{C}}{\mathcal{L}}}[V^+(t - \sqrt{\mathcal{L}\mathcal{C}}z) - V^-(t + \sqrt{\mathcal{L}\mathcal{C}}z)]$$

Defining

$$Z_0 = \sqrt{\frac{\mathcal{L}}{\mathcal{C}}} \qquad (6\text{-}212)$$

we have

$$I(z, t) = \frac{1}{Z_0}[V^+(t - \sqrt{\mathcal{L}\mathcal{C}}z) - V^-(t + \sqrt{\mathcal{L}\mathcal{C}}z)] \qquad (6\text{-}213)$$

The quantity Z_0 is the characteristic impedance of the transmission line analogous to the intrinsic impedance of the dielectric medium.

Thus the general solutions for the voltage and current along a transmission line are superpositons of $(+)$ and $(-)$ traveling waves along the line with velocities equal to $1/\sqrt{\mathcal{L}\mathcal{C}}$ in the respective directions. We will refer to these voltage and current waves as "transmission-line waves." They are completely analogous to the uniform plane waves with the analogy as follows:

$$V \longleftrightarrow E_x$$
$$I \longleftrightarrow H_y$$
$$\mathcal{L} \longleftrightarrow \mu$$
$$\mathcal{C} \longleftrightarrow \epsilon \qquad (6\text{-}214)$$
$$\frac{1}{\sqrt{\mathcal{L}\mathcal{C}}} \longleftrightarrow \frac{1}{\sqrt{\mu\epsilon}}$$
$$\sqrt{\frac{\mathcal{L}}{\mathcal{C}}} \longleftrightarrow \sqrt{\frac{\mu}{\epsilon}}$$

We should, however, keep in mind that the phenomenon is one of transverse *electromagnetic* waves guided by the conductors of the transmission line. It is not necessary to work with the fields since, because of the transverse electromagnetic nature of the fields, we are able to define uniquely the voltage and current for any transverse plane. In other words, if we consider two points a and b in the same transverse plane on the two conductors, the voltage between these two points is uniquely defined by the electric field in that plane since a closed path lying in that plane and passing through a and b does not enclose any magnetic flux. Similarly, the current flowing across a transverse plane in one direction along the inner conductor and returning in the opposite direction along the outer conductor is uniquely defined by the magnetic field in that plane since a closed path surrounding the inner conductor and lying in that plane does not enclose any electric flux. One or both of these properties are not satisfied if one or both of the fields have axial (or longitudinal) components. This is the case for TE or transverse electric waves which contain a magnetic field component along the guide axis and for TM or transverse magnetic waves which contain an electric field component along the guide axis. We will discuss such waves in Section 6.12.

Returning to the solutions for the voltage and current for the transmission-line waves given by (6-211) and (6-213), we write them concisely as

$$V = V^+ + V^- \tag{6-215a}$$

$$I = I^+ + I^- \tag{6-215b}$$

In writing (6-215a) and (6-215b), we follow the notation that both I^+ and I^- flow in the positive z direction along one conductor (say, a) and return in the negative z-direction along the other conductor (say, b) and that both V^+ and V^- have the same polarities with conductor a positive with respect to conductor b, as shown in Fig. 6.36. These notations are consistent with

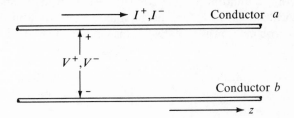

Fig. 6.36. Polarities for voltages and currents associated with $(+)$ and $(-)$ transmission-line waves.

the corresponding notations for the z-directed uniform plane waves which consider the electric fields for both $(+)$ and $(-)$ waves to be in the x direction and the magnetic fields for both $(+)$ and $(-)$ waves to be in the y direction. Comparing (6-215a) and (6-215b) with (6-211) and (6-213), respectively, we have

$$I^+ = \frac{V^+}{Z_0} \tag{6-216a}$$

and

$$I^- = -\frac{V^-}{Z_0} \tag{6-216b}$$

The power flow in the z direction associated with the $(+)$ wave is

$$P^+ = V^+ I^+ = V^+\left(\frac{V^+}{Z_0}\right) = \frac{(V^+)^2}{Z_0} \tag{6-217a}$$

The power flow in the z direction associated with the $(-)$ wave is

$$P^- = V^-I^- = V^-\left(-\frac{V^-}{Z_0}\right) = -\frac{(V^-)^2}{Z_0} \qquad (6\text{-}217\text{b})$$

where the minus sign on the right side signifies that the power flow is indeed in the negative z direction. Thus, if the notation of Fig. 6.36 is followed, together with Eqs. (6-216a) and (6-216b), the different directions of power flow for the $(+)$ and $(-)$ waves are taken care of automatically.

6.9 Traveling Waves in Time Domain

In the previous section we introduced uniform plane waves and transmission-line waves and discussed the analogy between them. We are now ready to consider simultaneously uniform plane waves incident normally on plane boundaries between different dielectric media and transmission-line waves. In this section we will discuss transient waves. To do this, let us consider the case of two semiinfinite perfect dielectric media characterized by ϵ_1, μ_1 and ϵ_2, μ_2, respectively, and separated by the plane $z = 0$ as shown in Fig. 6.37(a). A uniform plane wave with electric field E_x^+ and magnetic field H_y^+ is incident normally on the boundary. The transmission-line analogy of this problem consists of two transmission lines of different characteristic

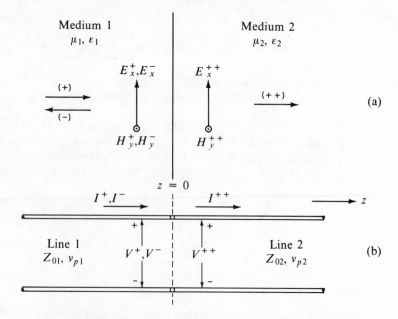

Fig. 6.37. (a) Normal incidence of a uniform plane wave on a plane boundary between two semiinfinite dielectric media. (b) Transmission-line analog of (a).

impedances Z_{01} and Z_{02} and velocities of propagation v_{p1} and v_{p2}, respectively, connected in cascade as shown in Fig. 6.37(b). The specification of Z_0 and v_p for a transmission line is equivalent to the specification of $\mathcal{L}$ and $\mathcal{C}$ since $Z_0 = \sqrt{\mathcal{L}/\mathcal{C}}$ and $v_p = 1/\sqrt{\mathcal{L}\mathcal{C}}$. A $(+)$ wave characterized by voltage V^+ and current I^+ is incident on the junction $z = 0$. We are not interested in the time variation of the incident waves at present. We merely wish to determine the transmission and reflection properties at the boundary. Obviously, there is no need to write equations for both the plane wave and transmission line cases because of the analogy. Hence we will simply write the equations in terms of the transmission-line parameters V, I, and Z_0 with the understanding that they can be replaced by E_x, H_y, and η, respectively.

The relationship between V^+ and I^+ is given by

$$I^+ = \frac{V^+}{Z_{01}} \tag{6-218}$$

The incident wave cannot be transmitted into line 2 as it is, since the voltage-to-current ratio in line 2 must be equal to Z_{02}. Thus, let the transmitted wave voltage and current be V^{++} and I^{++}, respectively. The incident and transmitted waves alone cannot satisfy the boundary conditions at the junction, which require that the voltages on either side of the junction be equal and the currents on either side of the junction be equal. These conditions are analogous to the boundary conditions for the fields, which state that the tangential electric fields (E_x) must be continuous and that the tangential magnetic fields (H_y) must be continuous (in the absence of a surface current) at the boundary between the dielectrics. To satisfy the boundary conditions, there is only one possibility. This is setting up a $(-)$ wave in line 1 which reflects part of the incident power into line 1. Let the voltage and current in this reflected wave be V^- and I^-, respectively. The voltage-to-current relationships for the transmitted and reflected waves are

$$I^{++} = \frac{V^{++}}{Z_{02}} \tag{6-219}$$

and

$$I^- = -\frac{V^-}{Z_{01}} \tag{6-220}$$

The boundary conditions at $z = 0$ are

$$V^+ + V^- = V^{++} \tag{6-221a}$$

$$I^+ + I^- = I^{++} \tag{6-221b}$$

Substituting (6-218), (6-219), and (6-220) into (6-221b), we have

$$\frac{V^+}{Z_{01}} - \frac{V^-}{Z_{01}} = \frac{V^{++}}{Z_{02}} \tag{6-222}$$

Solving (6-221a) and (6-222) for V^-, we get

$$V^- = V^+ \frac{Z_{02} - Z_{01}}{Z_{02} + Z_{01}} \tag{6-223}$$

We now define a quantity Γ, known as the voltage reflection coefficient, as the ratio of the $(-)$ wave or reflected wave voltage to the $(+)$ wave or incident wave voltage. From (6-223), the voltage reflection coefficient is given by

$$\Gamma = \frac{V^-}{V^+} = \frac{Z_{02} - Z_{01}}{Z_{02} + Z_{01}} \tag{6-224}$$

We then note that the current reflection coefficient is

$$\frac{I^-}{I^+} = \frac{-V^-/Z_{01}}{V^+/Z_{01}} = -\frac{V^-}{V^+} = -\Gamma \tag{6-225}$$

We also define a quantity τ_V, known as the voltage transmission coefficient, as the ratio of the $(++)$ wave or transmitted wave voltage to the $(+)$ wave or incident wave voltage. Thus

$$\tau_V = \frac{V^{++}}{V^+} = \frac{V^+ + V^-}{V^+} = 1 + \frac{V^-}{V^+}$$
$$= 1 + \Gamma = \frac{2Z_{02}}{Z_{02} + Z_{01}} \tag{6-226}$$

The current transmission coefficient τ_I is given by

$$\tau_I = \frac{I^{++}}{I^+} = \frac{I^+ + I^-}{I^+} = 1 + \frac{I^-}{I^+}$$
$$= 1 - \Gamma = \frac{2Z_{01}}{Z_{02} + Z_{01}} \tag{6-227}$$

At this point, we may be surprised to note that the transmitted voltage or the transmitted current can be greater than the incident voltage or the incident current, respectively, depending upon whether Γ is positive or negative, that is, $Z_{02} > Z_{01}$ or $Z_{02} < Z_{01}$. However, this is not of concern since it is the power balance that must be satisfied. To check this, we note that

$$P^+, \text{ incident power} = V^+ I^+$$
$$P^-, \text{ reflected power} = V^- I^- = (\Gamma V^+)(-\Gamma I^+)$$
$$= -\Gamma^2 V^+ I^+ = -\Gamma^2 P^+$$

where the minus sign signifies that the actual power flow is in the negative z direction, and

$$P^{++}, \text{ transmitted power} = V^{++} I^{++} = [(1 + \Gamma)V^+][(1 - \Gamma)I^+]$$
$$= (1 - \Gamma^2)V^+ I^+ = (1 - \Gamma^2)P^+$$

Thus $P^{++} = P^+ + P^-$, which verifies the power balance at the junction. The fact is that if the transmitted voltage is greater than the incident voltage, the transmitted current is less than the incident current and vice versa so that the transmitted power is less than the incident power. We will now consider an example to illustrate the application of the formulas for the reflection and transmission coefficients.

EXAMPLE 6-20. In Fig. 6.38, the region $z < 0$ is free space, the region $0 < z < 1.5$ m is a perfect dielectric of permittivity $4\epsilon_0$, and the region $z > 1.5$ m is a perfect dielectric of permittivity $9\epsilon_0$. The leading edge of a uniform plane wave of $0.01\,\mu$sec duration and having E_x equal to 1 volt/m is incident on the $z = -3$ m plane at $t = 0$. It is desired to find and sketch E_x in the planes $z = -3$ m and $z = 2.5$ m as functions of time for $t \geq 0$.

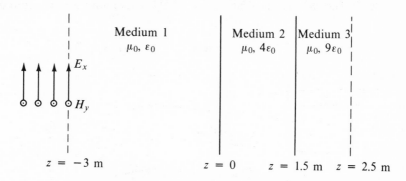

Fig. 6.38. Three dielectric media for Example 6-20.

The intrinsic impedances for the three media are η_0, $\eta_0/2$, and $\eta_0/3$, respectively, where $\eta_0\ (= \sqrt{\mu_0/\epsilon_0})$ is 377 ohms. The velocities of propagation in the three media are c, $c/2$, and $c/3$, respectively, where $c\ (= 1/\sqrt{\mu_0\epsilon_0})$ is 3×10^8 m/sec. The leading edge of the uniform plane wave strikes the interface $z = 0$ at $t = 3/(3 \times 10^8)$ sec $= 0.01\,\mu$sec. The reflection coefficient for this wave at $z = 0$ is $(\eta_0/2 - \eta_0)/(\eta_0/2 + \eta_0) = -\frac{1}{3}$ and the transmission coefficient is $1 + (-\frac{1}{3}) = \frac{2}{3}$. Hence the reflected wave E_x has a value $-\frac{1}{3}$ that of the incident wave E_x and its leading edge reaches the $z = -3$ m plane at $t = 0.02\,\mu$sec. The transmitted wave E_x has a value $\frac{2}{3}$ that of the incident wave E_x and its leading edge strikes the interface $z = 1.5$ m at $t = [10^{-8} + 1.5/(1.5 \times 10^8)]$ sec $= 0.02\,\mu$sec. The reflection coefficient for this wave at $z = 1.5$ m is $(\eta_0/3 - \eta_0/2)/(\eta_0/3 + \eta_0/2) = -\frac{1}{5}$ and the transmission coefficient is $1 + (-\frac{1}{5}) = \frac{4}{5}$. Thus the transmitted wave E_x in medium 3 has a value of $(\frac{4}{5} \times \frac{2}{3})$ or $\frac{8}{15}$ that of the incident wave E_x in medium 1. Its leading edge reaches the $z = 2.5$ m plane at $t = (2 \times 10^{-8} + 1/10^8) = 0.03\,\mu$sec. Now, the reflected wave at the interface $z = 1.5$ m travels towards the interface $z = 0$ and strikes it at $t = 0.03\,\mu$sec. It then violates the boundary conditions at $z = 0$, which have thus far been satisfied by the incident and reflected waves in medium 1 and the transmitted wave in medium 2 if they still exist at the interface. In any case, to satisfy the boundary conditions, it sets up a reflected wave into medium 2 and a transmitted wave into medium 1. By superposition, the reflection and transmission coefficients are the same as if this wave alone were incident on the interface. Hence the reflection coefficient for this wave at $z = 0$ is $(\eta_0 - \eta_0/2)/(\eta_0 + \eta_0/2) = \frac{1}{3}$ and the transmission coefficient is $1 + \frac{1}{3} = \frac{4}{3}$. The transmitted wave travels towards the interface $z = -3$ m. The reflected wave travels towards the interface $z = 1.5$ m and sets up reflected and transmitted waves. This process continues indefinitely.

To keep track of the bouncing back and forth of the transient waves between the interfaces, we resort to a "bounce diagram" as shown in Fig. 6.39. The bounce diagram is essentially a two-dimensional representation

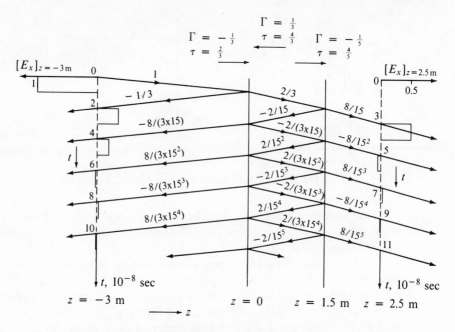

Fig. 6.39. Bounce diagram for keeping track of transient waves in the dielectric media of Fig. 6.38.

of transient waves bouncing back and forth. Distance along the direction of propagation is represented horizontally and time is represented vertically. Reflection and transmission coefficient values at the interfaces are written at the top of the diagram for quick reference, with appropriate arrows indicating directions of incidence. Criss-cross lines are drawn as shown on the diagram to indicate the progress of the waves as functions of z and t, with the numerical value of E_x for each leg of travel shown beside the line corresponding to that leg. The time functions of E_x representing the waves for each leg are drawn along the time axes in the planes of interest as shown on the bounce diagram. The bounce diagram of Fig. 6.39 is for E_x (or V). Similar bounce diagrams can be drawn for H_y (or I), taking note that the reflection coefficient for H_y (or I) is the negative of the reflection coefficient for E_x (or V). From Fig. 6.39, we can now draw the required sketches of E_x versus t in the planes

$z = -3$ m and $z = 2.5$ m. These are shown in Figs. 6.40(a) and (b), respectively. ∎

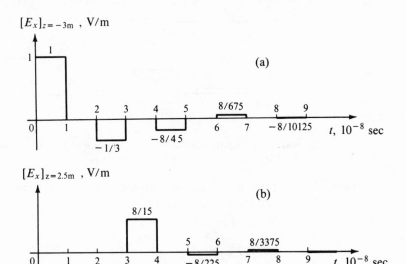

Fig. 6.40. E_x in the planes $z = -3$ m and $z = 2.5$ m versus time from the bounce diagram of Fig. 6.39.

6.10 Traveling Waves in Sinusoidal Steady State; Standing Waves

In the preceding section we discussed transient traveling waves. In this section we consider traveling waves in sinusoidal steady state. Once again, we deal simultaneously with uniform plane waves at normal incidence and transmission-line waves, keeping in mind the analogy between the two. From (6-211) and (6-213), the general solutions for the line voltage and line current in the sinusoidal steady state are

$$V(z, t) = V^+ \cos\left[\omega(t - \sqrt{\mathcal{LC}}z) + \phi^+\right] + V^- \cos\left[\omega(t + \sqrt{\mathcal{LC}}z) + \phi^-\right]$$

$$I(z, t) = \frac{1}{Z_0}\{V^+ \cos\left[\omega(t - \sqrt{\mathcal{LC}}z) + \phi^+\right] - V^- \cos\left[\omega(t + \sqrt{\mathcal{LC}}z) + \phi^-\right]\}$$

The corresponding expressions for the phasor line voltage and phasor line current are

$$\bar{V}(z) = \bar{V}^+ e^{-j\beta z} + \bar{V}^- e^{j\beta z} \tag{6-228a}$$

$$\bar{I}(z) = \frac{1}{Z_0}(\bar{V}^+ e^{-j\beta z} - \bar{V}^- e^{j\beta z}) \tag{6-228b}$$

where we have substituted β for $\omega\sqrt{\mathcal{LC}}$. For sinusoidal steady-state problems, it is convenient to use a distance variable d which is in opposition to z, that is, a variable which increases as we go away from the load and towards the generator as shown in Fig. 6.41. The wave which progresses away from the generator is still denoted as the (+) wave and the wave which progresses

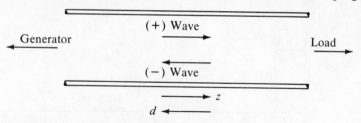

Fig. 6.41. For illustrating the distance variable *d* used for sinusoidal steady-state analysis of traveling waves.

towards the generator is still denoted as the (−) wave. In terms of *d*, the solutions for $\bar{V}$ and $\bar{I}$ are then given by

$$\bar{V}(d) = \bar{V}^+ e^{j\beta d} + \bar{V}^- e^{-j\beta d} \tag{6-229a}$$

$$\bar{I}(d) = \frac{1}{Z_0}(\bar{V}^+ e^{j\beta d} - \bar{V}^- e^{-j\beta d}) \tag{6-229b}$$

We will be working with these equations for the remainder of this section.

Let us now consider a semiinfinite perfect dielectric medium characterized by ϵ and μ and bounded by a perfect conductor in the plane $d = 0$ as shown in Fig. 6.42(a). The corresponding transmission-line equivalent is a

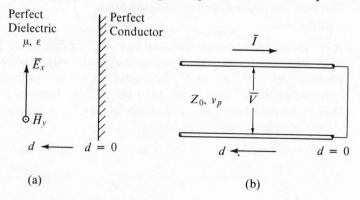

(a)   (b)

Fig. 6.42. (a) Normal incidence of a uniform plane wave on a plane perfect conductor. (b) Transmission-line analog of (a).

line short circuited at $d = 0$ as shown in Fig. 6.42(b). Let us assume that sinusoidally time-varying traveling waves exist in the medium due to a source which is not shown in the figure and that conditions have reached steady state. We wish to determine the characteristics of the waves satisfying the boundary condition at the perfect conductor (or short circuit). This boundary condition is

$$[\bar{E}_x]_{d=0} = 0 \qquad \text{or} \qquad [\bar{V}]_{d=0} = 0$$

Applying this boundary condition to the general solution for $\bar{V}(d)$ given by (6-229a), we obtain

$$0 = \bar{V}^+ + \bar{V}^- \qquad \text{or} \qquad \bar{V}^- = -\bar{V}^+$$

The particular solutions for the voltage and current are then given by

$$\bar{V}(d) = \bar{V}^+ e^{j\beta d} - \bar{V}^+ e^{-j\beta d} = 2j\bar{V}^+ \sin \beta d \qquad (6\text{-}230\text{a})$$

$$\bar{I}(d) = \frac{1}{Z_0}(\bar{V}^+ e^{j\beta d} + \bar{V}^+ e^{-j\beta d}) = 2\frac{\bar{V}^+}{Z_0} \cos \beta d \qquad (6\text{-}230\text{b})$$

The instantaneous voltage and current are given by

$$\begin{aligned} V(d, t) &= \mathcal{Re}[\bar{V}(d)e^{j\omega t}] \\ &= \mathcal{Re}(2e^{j\pi/2}\,|\,\bar{V}^+\,|\,e^{j\theta} \sin \beta d\, e^{j\omega t}) \\ &= -2\,|\bar{V}^+|\sin \beta d \sin (\omega t + \theta) \end{aligned} \qquad (6\text{-}231\text{a})$$

$$\begin{aligned} I(d, t) &= \mathcal{Re}[\bar{I}(d)\,e^{j\omega t}] \\ &= \mathcal{Re}\left(2\frac{|\,\bar{V}^+\,|}{Z_0}e^{j\theta} \cos \beta d\, e^{j\omega t}\right) \\ &= 2\frac{|\,\bar{V}^+\,|}{Z_0} \cos \beta d \cos (\omega t + \theta) \end{aligned} \qquad (6\text{-}231\text{b})$$

where θ is the phase angle of $\bar{V}^+$. The instantaneous line voltage and line current given by (6-231a) and (6-231b), respectively, are sketched in Fig. 6.43 as functions of d for various values of t. The following characteristics can be inferred from these sketches:

(a) The line voltage is zero at $d = 0,\ \pi/\beta,\ 2\pi/\beta, \ldots = 0,\ \lambda/2,\ \lambda, \ldots$ for all values of time. Hence there is no power flow across these planes for all values of time. If we short circuit the line (or place perfect conductors) at these values of d, there will be no effect on the voltage and current (fields) at any other value of d.

(b) The line current is zero at $d = \pi/2\beta,\ 3\pi/2\beta, \ldots = \lambda/4,\ 3\lambda/4, \ldots$ for all values of time. Hence there is no power flow across these planes for all values of time. If we open circuit the line (or place imaginary magnetic conductors) at these values of d, there will be no change in the voltage and current (fields) at any other value of d.

(c) *Wherever* the line voltage has maximum amplitude, the line current has zero amplitude and vice versa. Thus the line voltage and line current are out of phase in distance by $\pi/2\beta$ or $\lambda/4$.

(d) *Whenever* the line voltage is maximum at all values of d, the line current is zero at all values of d and vice versa. Thus the line voltage and line current are out of phase in time by $\pi/2\omega$ or $T/4$, where T is the period corresponding to ω.

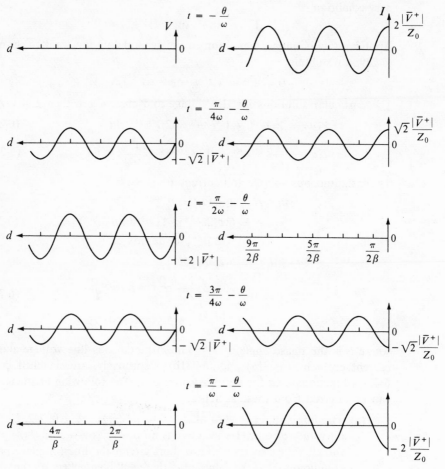

Fig. 6.43. Voltage and current versus distance for various values of time for the short-circuited line of Fig. 6.42(b).

We conclude from these characteristics that the situation for a short-circuited line consists of voltage and current waves which stand still and only increase and decrease in amplitude in each section of $\lambda/2$ in length between the voltage nodes (zeros) and between the current nodes (zeros), respectively, similar to the oscillations executed by a string tied down at one end and vibrated at a point half a wavelength from the tie-down point. These waves are therefore known as "complete standing waves." Complete standing waves are the result of $(+)$ and $(-)$ traveling waves of equal magnitude. Whatever power is incident on the short circuit by the $(+)$ wave is reflected entirely in the form of the $(-)$ wave since the short circuit cannot absorb any power. While there is instantaneous power flow at values of d between the voltage and current nodes, there is no time-average power flow for any value of d, as can be seen from

$$\langle P \rangle = \frac{1}{2} \Re e [\bar{V}(d) \bar{I}^*(d)]$$

$$= \frac{1}{2} \Re e \left[(2j\bar{V}^+ \sin \beta d) \left(2 \frac{\bar{V}^{+*}}{Z_0} \cos \beta d \right) \right]$$

$$= \frac{1}{2} \Re e \left(2j \frac{|\bar{V}^+|^2}{Z_0} \sin 2\beta d \right) = 0$$

The amplitudes of the sinusoidally time-varying line voltage and line current as functions of d are

$$|\bar{V}(d)| = 2|j||\bar{V}^+||\sin \beta d| = 2|\bar{V}^+||\sin \beta d| \qquad (6\text{-}232a)$$

$$|\bar{I}(d)| = 2 \frac{|\bar{V}^+|}{Z_0} |\cos \beta d| \qquad (6\text{-}232b)$$

These amplitudes are sketched in Fig. 6.44. The patterns of Fig. 6.44 are known as "standing wave patterns." Standing wave patterns are easily measured in the laboratory with the aid of moving probes which sample the electric field.

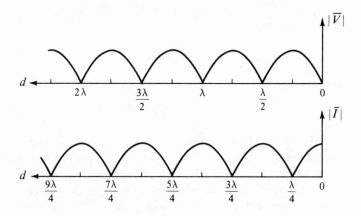

Fig. 6.44. Standing wave patterns for voltage and current along a short-circuited line.

EXAMPLE 6-21. A transmission line of length l and short circuited at both ends has certain energy stored in it. From the preceding discussion, this energy must exist in the form of complete standing waves on the line. What are the possible standing wave patterns and the corresponding frequencies?

The voltage must be zero at both ends of the line since it is short circuited at both ends. It follows from the standing wave patterns of Fig. 6.44 that the current must be maximum at both ends. Thus the possible voltage and current standing wave patterns are as shown in Fig. 6.45. They must consist of integral numbers of half-sinusoidal variations over the length

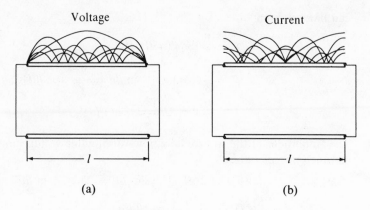

Fig. 6.45. Standing wave patterns for (a) voltage and (b) current along a line short circuited at both ends.

of the line; that is, the wavelengths λ_n corresponding to these standing wave patterns must be such that

$$l = n\frac{\lambda_n}{2} \qquad n = 1, 2, 3, \ldots$$

or

$$\lambda_n = \frac{2l}{n} \qquad n = 1, 2, 3, \ldots$$

The corresponding frequencies are

$$f_n = \frac{v_p}{\lambda_n} = \frac{n v_p}{2l} \qquad n = 1, 2, 3, \ldots$$

where v_p is the phase velocity. These frequencies are known as the "natural frequencies of oscillation." The standing wave patterns are said to correspond to the different "natural modes of oscillation." The lowest frequency (corresponding to the longest wavelength) is known as the "fundamental" frequency of oscillation and the corresponding mode is known as the fundamental mode. The quantity n is called the mode number. ∎

Returning now to the expressions for the phasor line voltage and the phasor line current given by (6-230a) and (6-230b), respectively, we define the ratio of these two quantities as the line (or wave) impedance $\bar{Z}(d)$ at that point seen looking towards the short circuit. Thus

$$\bar{Z}(d) = \frac{\bar{V}(d)}{\bar{I}(d)} = \frac{2j\bar{V}^+ \sin \beta d}{2(\bar{V}^+/Z_0) \cos \beta d} = jZ_0 \tan \beta d \qquad (6\text{-}233)$$

In particular, the input impedance $\bar{Z}_{\text{in}}$ of a short-circuited line of length l is given by

$$\bar{Z}_{\text{in}} = jZ_0 \tan \beta l = jZ_0 \tan \frac{2\pi f}{v_p} l \qquad (6\text{-}234)$$

This expression is the same as the expression (6-151) derived by the step-by-step solution of Maxwell's curl equations for the fields in the parallel-plate structure of Fig. 6.25(b). We now note that the condition for quasistatic approximation given by (6-146) can be stated alternatively as

$$\beta l \ll 1 \qquad \text{or} \qquad l \ll \lambda$$

For a fixed l, $\tan(2\pi f/v_p)l$ becomes alternatively positive and negative as f increases and hence the input reactance alternates between inductive and capacitive as illustrated in Fig. 6.46. It can be seen that frequencies at which

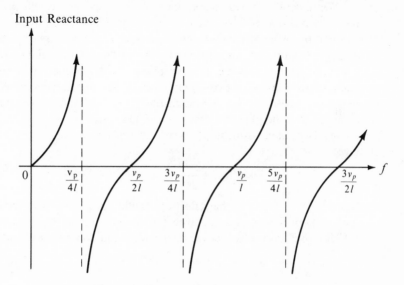

Fig. 6.46. Variation of the input reactance of a short-circuited line of length l with frequency.

the input reactance is zero are the same as the natural frequencies of oscillation if the input were short circuited. Likewise, the frequencies at which the input reactance is infinity are the same as the natural frequencies of oscillation if the input were open circuited. These properties of short-circuited line sections permit them to be used as inductive and capacitive elements and resonant circuits at high frequencies. We will illustrate an application by means of the following example.

EXAMPLE 6-22.   To determine the location of a short circuit in an air-insulated parallel wire line, a voltage generator of variable frequency is connected at its input. The generator frequency is varied continuously from a value of 100 MHz upwards and the current drawn from the generator is monitored. It is found that the current reaches a minimum at 100.02 MHz and then a maximum at 100.05 MHz. How far is the location of the short circuit from the generator?

The current minimum occurs at a frequency for which the input impedance is infinity. The current maximum occurs at a frequency for which the input impedance is zero. From Fig. 6.46, the difference between adjacent frequencies for which the input reactances of a short-circuited line are infinity and zero is equal to $v_p/4l$. Hence, for this problem, $v_p/4l$ is $(100.05 - 100.02)$ or 0.03 MHz. Since the line is air insulated, the velocity of propagation is 3×10^8 m/sec. Hence $l = (3 \times 10^8)/(4 \times 0.03 \times 10^6) = 2500$ m. Thus the location of the short circuit is 2.5 km away from the generator. ∎

We have thus far discussed complete standing waves which result from the superposition of $(+)$ and $(-)$ waves of equal magnitudes. Let us now consider the general case of the superposition of $(+)$ and $(-)$ waves of unequal magnitudes, thereby giving rise to "partial standing waves." Such a situation can arise when uniform plane waves are incident normally on a plane interface between two different dielectrics or interfaces between several dielectrics in cascade. We first define the generalized reflection coefficient $\bar{\Gamma}(d)$ as the ratio of the phasor voltage associated with the $(-)$ wave to the phasor voltage associated with the $(+)$ wave at a given d. Thus, from (6-229a), we have

$$\bar{\Gamma}(d) = \frac{\bar{V}^- e^{-j\beta d}}{\bar{V}^+ e^{j\beta d}} = \frac{\bar{V}^-}{\bar{V}^+} e^{-j2\beta d} = \bar{\Gamma}(0)e^{-j2\beta d} \tag{6-235}$$

where $\bar{\Gamma}(0) = \bar{V}^-/\bar{V}^+$ is the reflection coefficient at $d = 0$. We note that the magnitude of $\bar{\Gamma}(d)$ is constant whereas the phase angle changes linearly with d. Using (6-235), we can write the general solutions for $\bar{V}(d)$ and $\bar{I}(d)$ as

$$\bar{V}(d) = \bar{V}^+ e^{j\beta d}[1 + \bar{\Gamma}(d)] \tag{6-236a}$$

$$\bar{I}(d) = \frac{\bar{V}^+}{Z_0} e^{j\beta d}[1 - \bar{\Gamma}(d)] \tag{6-236b}$$

The line (or wave) impedance $\bar{Z}(d)$ is given by

$$\bar{Z}(d) = \frac{\bar{V}(d)}{\bar{I}(d)} = Z_0 \frac{1 + \bar{\Gamma}(d)}{1 - \bar{\Gamma}(d)} \tag{6-237}$$

Conversely,

$$\bar{\Gamma}(d) = \frac{\bar{Z}(d) - Z_0}{\bar{Z}(d) + Z_0} \tag{6-238}$$

To study the standing wave patterns corresponding to (6-236a) and (6-236b), we look at the magnitudes of $\bar{V}(d)$ and $\bar{I}(d)$. These are given by

$$|\bar{V}(d)| = |\bar{V}^+||e^{j\beta d}||1 + \bar{\Gamma}(d)|$$
$$= |\bar{V}^+||1 + \bar{\Gamma}(0)e^{-j2\beta d}| \tag{6-239a}$$

$$|\bar{I}(d)| = \frac{|\bar{V}^+|}{Z_0}|e^{j\beta d}||1 - \bar{\Gamma}(d)|$$
$$= \frac{|\bar{V}^+|}{Z_0}|1 - \bar{\Gamma}(0)e^{-j2\beta d}| \tag{6-239b}$$

To sketch $|\bar{V}(d)|$ and $|\bar{I}(d)|$, it is sufficient if we consider the quantities $|1 + \bar{\Gamma}(0)e^{-j2\beta d}|$ and $|1 - \bar{\Gamma}(0)e^{-j2\beta d}|$ since $|\bar{V}^{+}|$ is simply a constant, dependent upon the source of the waves. Each of these quantities consists of two complex numbers one of which is a constant equal to $(1 + j0)$ and the other of which has a constant magnitude $|\bar{\Gamma}(0)|$ but a variable phase angle $\theta - 2\beta d$, where θ is the phase angle of $\bar{\Gamma}(0)$. To evaluate $|1 + \bar{\Gamma}(0)e^{-j2\beta d}|$ and $|1 - \bar{\Gamma}(0)e^{-j2\beta d}|$, we make use of the constructions in the complex $\bar{\Gamma}$ plane as shown in Figs. 6.47(a) and (b), respectively. In both

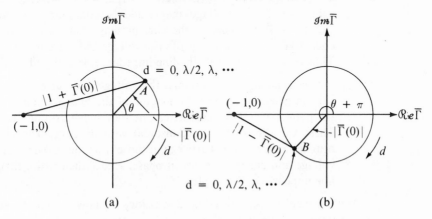

Fig. 6.47. $\bar{\Gamma}$-plane diagrams for sketching the voltage and current standing wave patterns for a partial standing wave.

diagrams, we draw circles with centers at the origin and having radii equal to $|\bar{\Gamma}(0)|$. For $d = 0$, the complex number $\bar{\Gamma}(0)e^{-j2\beta d}$ is equal to $\bar{\Gamma}(0)$ or $|\bar{\Gamma}(0)|e^{j\theta}$, which is represented by point A in Fig. 6.47(a). To add $(1 + j0)$ and $\bar{\Gamma}(0)$, we simply draw a line from the point $(-1, 0)$ to the point A. The length of this line gives $|1 + \bar{\Gamma}(0)|$, which is proportional to the amplitude of the voltage standing wave at $d = 0$. As d increases, point A, representing $\bar{\Gamma}(0)e^{-j2\beta d}$, moves around the circle in the clockwise direction. The line joining $(-1, 0)$ to the point A whose length is $|1 + \bar{\Gamma}(0)e^{-j2\beta d}|$ executes the motion of a "crank." To subtract $\bar{\Gamma}(0)$ from $(1 + j0)$ we locate point B in Fig. 6.47(b), which is diametrically opposite to point A in Fig. 6.47(a), and draw a line from $(-1, 0)$ to point B. The length of this line gives $|1 - \bar{\Gamma}(0)|$, which is proportional to the amplitude of the current standing wave at $d = 0$. As d increases, B moves around the circle in the clockwise direction following the movement of A in Fig. 6.47(a). The line joining $(-1, 0)$ to the point B whose length is $|1 - \bar{\Gamma}(0)e^{-j2\beta d}|$ executes the motion of a "crank." From these constructions, we note the following facts:

(a) Point A lies along the positive real axis and point B lies along the negative real axis for $\theta - 2\beta d = 0, -2\pi, -4\pi, -6\pi, \ldots$ or $d = (\lambda/4\pi)(\theta + 2n\pi)$, where $n = 0, 1, 2, 3, \ldots$. Hence, at these values of d, the voltage magnitude is maximum and equal to $|\bar{V}^+|[1 + |\bar{\Gamma}(0)|]$ whereas the current magnitude is minimum and equal to $(|\bar{V}^+|/Z_0)[1 - |\bar{\Gamma}(0)|]$. The voltage and current are in phase. Thus their ratio, that is, the line impedance, is real and maximum.

(b) Point A lies along the negative real axis and point B lies along the positive real axis for $\theta - 2\beta d = -\pi, -3\pi, -5\pi, -7\pi, \ldots$ or $d = (\lambda/4\pi)[\theta + (2n - 1)\pi]$, where $n = 1, 2, 3, 4, \ldots$. Hence, at these values of d, the voltage magnitude is minimum and equal to $|\bar{V}^+|[1 - |\bar{\Gamma}(0)|]$ whereas the current magnitude is maximum and equal to $(|\bar{V}^+|/Z_0)[1 + |\bar{\Gamma}(0)|]$. The voltage and current are in phase. Thus their ratio, that is, the line impedance, is real and minimum.

(c) Between maxima and minima, the voltage and current magnitudes vary in accordance with the lengths of the lines joining $(-1, 0)$ to the points A and B, respectively, as they move around the circles. These variations are not sinusoidal with distance. The variations near the minima are sharper than those near the maxima. Also, the voltage and current are not in phase. Hence their ratio, that is, the line impedance, is complex.

With the aid of the preceding discussion, we now sketch the standing wave patterns for the line voltage and line current as shown in Fig. 6.48. The standing wave patterns should not be misinterpreted as the voltage and current remaining constant with time at a given point. On the other hand, at every point on the line, the voltage and current vary sinusoidally with time as shown in the insets of Fig. 6.48, with the amplitudes of these sinusoidal variations equal to the magnitudes indicated by the standing wave patterns. We now define a quantity known as the voltage standing wave ratio, abbreviated as VSWR, as the ratio of the maximum voltage $V_{\max}$ to the minimum voltage $V_{\min}$ in the standing wave patterns. Thus

$$\text{VSWR} = \frac{V_{\max}}{V_{\min}} = \frac{|\bar{V}^+|[1 + |\bar{\Gamma}(0)|]}{|\bar{V}^+|[1 - |\bar{\Gamma}(0)|]} = \frac{1 + |\bar{\Gamma}(0)|}{1 - |\bar{\Gamma}(0)|} \qquad (6\text{-}240)$$

The VSWR is a measure of the standing waves on the line. It is an easily measurable parameter. We note the following special cases:

(a) For a pure traveling wave, that is, for a $(+)$ wave alone, $\bar{V}^- = 0$, $\bar{\Gamma} = 0$, and hence VSWR $= 1$; that is, the standing wave pattern is simply a line representing constant magnitude. This is the case if the line is infinitely long or if it is terminated by its characteristic impedance.

(b) For a complete standing wave, $\bar{V}^-$ and $\bar{V}^+$ have equal magnitudes; $|\bar{\Gamma}| = 1$ and hence VSWR $= \infty$. This is indeed the case with the standing wave pattern of Fig. 6.44.

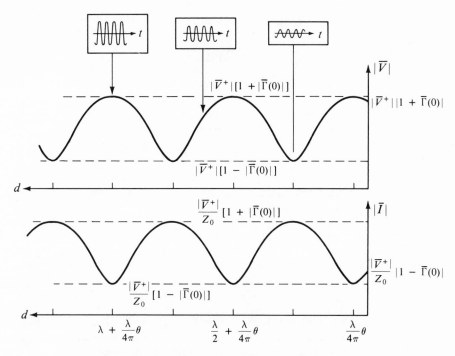

Fig. 6.48. Voltage and current standing wave patterns for a partial standing wave. The insets show time variations of voltage at points along the line.

EXAMPLE 6-23.   In Fig. 6.49(a), a plane dielectric slab of thickness 3.75 cm and permittivity $4\epsilon_0$ is sandwiched between two semiinfinite media 1 and 3. Medium 1 is free space and medium 3 is a perfect dielectric of permittivity $9\epsilon_0$. A uniform plane wave of frequency 3000 MHz is incident normally on the slab from medium 1 and sinusoidal steady-state conditions are established in all media. It is desired to find and sketch the standing wave patterns for the fields in all media.

The intrinsic impedances of the three media are η_0, $\eta_0/2$, and $\eta_0/3$, respectively. The velocities of propagation in the three media are c, $c/2$, and $c/3$, respectively, where $c = 3 \times 10^8$ m/sec. The wavelength in medium 2 for 3000 MHz is 5 cm. Hence the electrical length of medium 2 is $3\lambda/4$. The transmission-line analog of the problem is shown in Fig. 6.49(b). We solve this problem in a step-by-step manner as follows:

(a) Line 3 has only a $(+)$ wave since it extends to infinity. Hence VSWR for that line is equal to 1. The line impedance is independent of distance and equal to the characteristic impedance $\eta_0/3$.

(b) At the junction between two lines, the boundary conditions dictate that $\bar{V}$ and $\bar{I}$ be continuous (analogous to $\bar{E}_x$ and $\bar{H}_y$ being continuous at

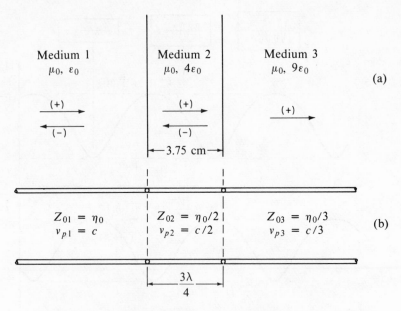

Fig. 6.49. (a) A plane dielectric slab sandwiched between two semiinfinite dielectric media. (b) Transmission-line analog of (a) for uniform plane waves at normal incidence.

the interface between the two dielectric media). Hence the ratio of these two quantities must be continuous. The line impedance at the right end of line 2 is therefore equal to the line impedance at the left end of line 3, which is equal to $\eta_0/3$.

(c) From (6-238), the reflection coefficient at the right end of line 2 is

$$\bar{\Gamma}_1 = \frac{\eta_0/3 - Z_{02}}{\eta_0/3 + Z_{02}} = \frac{\eta_0/3 - \eta_0/2}{\eta_0/3 + \eta_0/2} = -\frac{1}{5}$$

(d) From (6-240), VSWR for line 2 is

$$\frac{1 + |\bar{\Gamma}_1|}{1 - |\bar{\Gamma}_1|} = \frac{1 + \frac{1}{5}}{1 - \frac{1}{5}} = 1.5$$

Also, since Γ_1 is purely real and negative, the voltage magnitude is minimum at the right end of line 2, as can be seen from the construction of Fig. 6.47(a).

(e) From (6-235), the reflection coefficient at the left end of line 2 is

$$\bar{\Gamma}_2 = \bar{\Gamma}_1 e^{-j2\beta_2(3\lambda_2/4)} = -\frac{1}{5}e^{-j3\pi} = \frac{1}{5}$$

(f) From (6-237), the line impedance at the left end of line 2 is

$$Z_{02}\frac{1 + \bar{\Gamma}_2}{1 - \bar{\Gamma}_2} = \frac{\eta_0}{2}\frac{1 + \frac{1}{5}}{1 - \frac{1}{5}} = \frac{3}{4}\eta_0$$

(g) Since the line impedance at a junction between two lines has to be continuous from the discussion in (b) above, the line impedance at the right end of line 1 is $\frac{3}{4}\eta_0$. From (6-238), the reflection coefficient at the right end of line 1 is

$$\bar{\Gamma}_3 = \frac{3\eta_0/4 - Z_{01}}{3\eta_0/4 + Z_{01}} = \frac{3\eta_0/4 - \eta_0}{3\eta_0/4 + \eta_0} = -\frac{1}{7}$$

(h) From (6-240), VSWR for line 1 is

$$\frac{1 + |\bar{\Gamma}_3|}{1 - |\bar{\Gamma}_3|} = \frac{1 + \frac{1}{7}}{1 - \frac{1}{7}} = \frac{4}{3}$$

Also, since $\bar{\Gamma}_3$ is purely real and negative, the line voltage is a minimum at the right end of line 1.

From the above results and noting that the wavelength in medium 1 for 3000 MHz is 10 cm, we now sketch the standing wave pattern for the electric field intensity (based on a magnitude of unity in medium 3) as shown by the solid curves in Fig. 6.50. The standing wave pattern for the magnetic

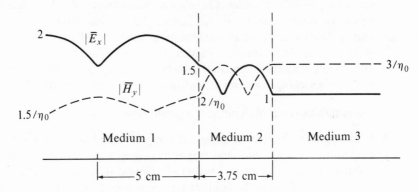

Fig. 6.50. Standing wave patterns for the fields in the three media of Fig. 6.49.

field intensity follows from the fact that $|\bar{E}_x|/|\bar{H}_y| = \eta_0/3$ in medium 3 and by noting that wherever $|\bar{E}_x|$ is maximum, $|\bar{H}_y|$ is minimum and vice versa. It is shown by the dashed curves in Fig. 6.50. ∎

EXAMPLE 6-24.   One important type of problem is that of matching between two dielectric media of different permittivities. For example, in Fig. 6.49(a), we can choose the thickness and permittivity of medium 2 so that reflected wave is eliminated in medium 1. Then all the power incident on the interface between media 1 and 2 is transmitted into medium 3 (although standing waves exist in medium 2). Let us determine the minimum thickness and permittivity of medium 2 required to achieve such a match.

To determine the required quantities, we note that, for a particular line of characteristic impedance Z_0, the product of the line impedances at two values of d separated by an odd multiple of $\lambda/4$ is given by

$$\{\bar{Z}[d]\} \left\{\bar{Z}\left[d + (2n - 1)\frac{\lambda}{4}\right]\right\}$$

$$= \left\{Z_0\frac{1 + \bar{\Gamma}(d)}{1 - \bar{\Gamma}(d)}\right\} \left\{Z_0\frac{1 + \bar{\Gamma}[d + (2n - 1)\lambda/4]}{1 - \bar{\Gamma}[d + (2n - 1)\lambda/4]}\right\}$$

$$= Z_0^2\left[\frac{1 + \bar{\Gamma}(d)}{1 - \bar{\Gamma}(d)}\right]\left[\frac{1 + \bar{\Gamma}(d)e^{-j2\beta(2n-1)\lambda/4}}{1 - \bar{\Gamma}(d)e^{-j2\beta(2n-1)\lambda/4}}\right] \qquad (6\text{-}241)$$

$$= Z_0^2\left[\frac{1 + \bar{\Gamma}(d)}{1 - \bar{\Gamma}(d)}\right]\left[\frac{1 + \bar{\Gamma}(d)e^{-j(2n-1)\pi}}{1 - \bar{\Gamma}(d)e^{-j(2n-1)\pi}}\right]$$

$$= Z_0^2\left[\frac{1 + \bar{\Gamma}(d)}{1 - \bar{\Gamma}(d)}\right]\left[\frac{1 - \bar{\Gamma}(d)}{1 + \bar{\Gamma}(d)}\right] = Z_0^2$$

where n can take any integer value. For eliminating standing waves in line 1, the impedance seen at the right end of line 1 must be equal to $Z_{01} = \eta_0$. Hence the line impedance at the left end of line 2 must be η_0. However, the impedance seen at the right end of line 2 is equal to $Z_{03} = \eta_0/3$. Hence, according to (6-241), we must have a minimum length of $\lambda/4$ and a characteristic impedance equal to $\sqrt{\eta_0(\eta_0/3)}$ or $\eta_0/\sqrt{3}$ for line 2 to achieve the required match. For the intrinsic impedance of medium 2 to be $\eta_0/\sqrt{3}$, its permittivity must be $3\epsilon_0$. Since the wavelength for 3000 MHz in medium 2 is then $10/\sqrt{3}$ cm, the minimum required thickness is $2.5/\sqrt{3}$ or 1.4434 cm. This technique of matching is known as matching by "quarter-wave transformer." ∎

6.11 Transmission-Line Matching; the Smith Chart

In the previous section we discussed complete standing waves resulting from $(+)$ and $(-)$ waves of equal magnitudes, and then partial standing waves resulting from $(+)$ and $(-)$ waves of unequal magnitudes. While standing waves are useful from the point of view of energy storage, they are unwanted from the point of view of energy transmission. To elaborate upon this, we note that the time-average power flow down the line is given by

$$\langle P \rangle = \frac{1}{2}\mathfrak{Re}(\bar{V}\bar{I}^*)$$

$$= \frac{1}{2}\mathfrak{Re}\left\{\bar{V}^+e^{j\beta d}[1 + \bar{\Gamma}(d)]\frac{\bar{V}^{+*}}{Z_0}e^{-j\beta d}[1 - \bar{\Gamma}^*(d)]\right\}$$

$$= \frac{1}{2}\mathfrak{Re}\left\{\frac{|\bar{V}^+|^2}{Z_0}[1 - |\bar{\Gamma}(d)|^2 + \bar{\Gamma}(d) - \bar{\Gamma}^*(d)]\right\}$$

$$= \frac{|\bar{V}^+|^2}{2Z_0}[1 - |\bar{\Gamma}(d)^2|] = \frac{|\bar{V}^+|^2}{2Z_0}[1 - |\bar{\Gamma}(0)|^2] \qquad (6\text{-}242)$$

$$= \left\{\frac{|\bar{V}^+|}{2}[1 + |\bar{\Gamma}(0)|]\right\}\left\{\frac{|\bar{V}^+|}{Z_0}[1 - |\bar{\Gamma}(0)|]\right\}$$

$$= \frac{V_{max}V_{min}}{2Z_0} = \frac{I_{max}I_{min}Z_0}{2}$$

$$= \frac{V_{max}^2}{2(\text{VSWR})Z_0} = \frac{I_{max}^2}{2(\text{VSWR})}Z_0$$

where V_{max}, V_{min}, I_{max}, and I_{min} are the maximum and minimum magnitudes of voltage and current, respectively, in the standing wave patterns. From (6-242), the limitations imposed by standing waves on power transfer down a line are evident. For a particular line, there is an upper limit for the electric field which the dielectric can withstand and hence there is a breakdown voltage. For a particular value of the breakdown voltage, the power that can be transmitted down the line is inversely proportional to the VSWR, according to (6-242). Similarly, there can be an upper limit for the current that can be carried by the conductors of the line without overheating them. Again, (6-242) states that, for a particular value of this current, the power that can be transmitted down the line is inversely proportional to the VSWR.

Another and perhaps more serious limitation imposed by standing waves concerns the input impedance of the line. In the presence of standing waves, that is, when the load impedance is not equal to the characteristic impedance, it follows from (6-237) that the input impedance of the line will vary with frequency since the electrical length of the line and hence $\bar{\Gamma}(d) = \bar{\Gamma}(0)e^{-j2\beta d}$ changes. This sensitivity to frequency increases with the electrical length of the line. To show this, let the length of the line be $l = n\lambda$. If the frequency is changed by an amount Δf, then the change in n is given by

$$\Delta n = \Delta\left(\frac{l}{\lambda}\right) = \Delta\left(\frac{lf}{v_p}\right) = \frac{l}{v_p}\Delta f = \frac{n\lambda}{v_p}\Delta f = n\frac{\Delta f}{f}$$

Thus Δn, the change in the number of wavelengths corresponding to the line length, is proportional to n.

For these reasons, it is necessary to eliminate standing waves on the line by connecting a "matching" device near the load such that the line views an effective impedance equal to its own characteristic impedance on the generator side of the matching device. The matching device should not at the same time absorb any power. Small sections of short-circuited lines known as stubs connected in parallel with the line at appropriate distances from the load are used for this purpose since their input impedance is purely reactive and hence they do not absorb any power. Indeed we are making use of standing waves (on the stub) to eliminate standing waves (on the line between the generator and the stub)! This technique of matching is known as stub matching. We now illustrate the principle behind the stub matching technique by means of an example.

EXAMPLE 6-25.   A lossless transmission line having a characteristic impedance of 50 ohms is terminated by a load impedance $\bar{Z}_R$ equal to $(30 - j40)$ ohms. It is desired to find the location and the length of a lossless, short-circuited stub connected in parallel with the line so that a match is obtained between the generator driving this line and the load, assuming that the characteristic impedance of the stub is 50 ohms.

The principle behind the stub matching technique consists of finding the location nearest to the load at which the real part of the line admittance (reciprocal of the line impedance) is equal to the line characteristic admittance

Y_0 (reciprocal of the line characteristic impedance Z_0). The imaginary part of the line admittance is then cancelled by placing in parallel with the line a short-circuited stub of appropriate length so that its input susceptance is equal to the negative of the imaginary part of the line admittance to the right of the stub as shown in Fig. 6.51. The line admittance seen from the left of

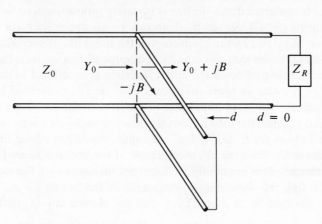

Fig. 6.51. Transmission-line matching by means of a stub.

the stub is then equal to $(Y_0 + jB) + (-jB) = Y_0$. The line impedance seen from the left of the stub into the junction of the line and the stub is therefore equal to Z_0 and a match is achieved. To find the required parameters, we proceed in a step-by-step manner as follows:

(a) Find the reflection coefficient at the load.

$$\bar{\Gamma}(0) = \frac{\bar{Z}_R - Z_0}{\bar{Z}_R + Z_0} = \frac{(30 - j40) - 50}{(30 - j40) + 50} = 0.5e^{-j\pi/2}$$

(b) Find the reflection coefficient as a function of d.

$$\bar{\Gamma}(d) = \bar{\Gamma}(0)e^{-j2\beta d} = 0.5e^{-j(2\beta d + \pi/2)}$$

(c) Find the line admittance as a function of d.

$$\bar{Y}(d) = \frac{1}{\bar{Z}(d)} = \frac{1}{Z_0}\left[\frac{1 - \bar{\Gamma}(d)}{1 + \bar{\Gamma}(d)}\right]$$

$$= \frac{1}{50}\left[\frac{1 - 0.5e^{-j(2\beta d + \pi/2)}}{1 + 0.5e^{-j(2\beta d + \pi/2)}}\right]$$

$$= 0.02\frac{0.75 + j\cos 2\beta d}{1.25 - \sin 2\beta d}$$

(d) Set the real part of $\bar{Y}(d)$ equal to Y_0 and solve for d.

$$0.02\frac{0.75}{1.25 - \sin 2\beta d} = 0.02$$

or

$$\sin 2\beta d = 0.5$$

$$2\beta d = \frac{\pi}{6} \quad \text{or} \quad \frac{5\pi}{6}$$

$$d = \frac{\lambda}{24} \quad \text{or} \quad \frac{5\lambda}{24}$$

Thus the stub must be located at a distance $\lambda/24$ or $5\lambda/24$ from the load.

(e) To find the length of the stub, we note that the imaginary part of $\bar{Y}(d)$ is $(0.02 \cos 2\beta d)/(1.25 - \sin 2\beta d)$. Its value at the stub location is

$$B = \begin{cases} 0.02 \times 1.15 & \text{for } d = \dfrac{\lambda}{24} \\[2mm] 0.02 \times (-1.15) & \text{for } d = \dfrac{5\lambda}{24} \end{cases}$$

(f) The input impedance of a short-circuited line of length l is given by (6-234). The input admittance is

$$\bar{Y}_{in} = \frac{1}{\bar{Z}_{in}} = \frac{1}{jZ_0 \tan \beta l} = -jY_0 \cot \beta l$$

Thus the stub length l must be such that

$$-jY_0 \cot \beta l = \begin{cases} -j0.02 \times 1.15 & \text{for } d = \dfrac{\lambda}{24} \\[2mm] j0.02 \times 1.15 & \text{for } d = \dfrac{5\lambda}{24} \end{cases}$$

or

$$l = \begin{cases} 0.113\lambda & \text{for } d = \dfrac{\lambda}{24} \\[2mm] 0.387\lambda & \text{for } d = \dfrac{5\lambda}{24} \end{cases} \quad \blacksquare$$

The steps involved in the analytical solution of the stub matching problem in the preceding example consist of conversion from line impedance to reflection coefficient, then going along the constant $|\bar{\Gamma}|$ circle in the complex-plane diagram of Fig. 6.47 to find $\bar{\Gamma}(d)$ and then converting back to impedance. This process of conversion and reconversion from one quantity to the other can be eliminated by constructing a chart which associates with each point in the complex $\bar{\Gamma}$ plane the corresponding impedance or admittance. One such chart is known as the Smith chart. To discuss the basis of Smith chart construction, we define the normalized line impedance, $\bar{z}(d)$, as the ratio of the line impedance $\bar{Z}(d)$ to the characteristic impedance Z_0. Thus

$$\bar{z}(d) = \frac{\bar{Z}(d)}{Z_0} = \frac{1 + \bar{\Gamma}(d)}{1 - \bar{\Gamma}(d)} \tag{6-243}$$

Conversely,

$$\bar{\Gamma}(d) = \frac{\bar{z}(d) - 1}{\bar{z}(d) + 1} \qquad (6\text{-}244)$$

Letting $\bar{z}(d) = r + jx$, we have

$$\bar{\Gamma}(d) = \frac{r + jx - 1}{r + jx + 1} = \frac{(r - 1) + jx}{(r + 1) + jx}$$

and

$$|\bar{\Gamma}(d)| = \left[\frac{(r - 1)^2 + x^2}{(r + 1)^2 + x^2}\right]^{1/2} \leq 1$$

for positive values of r. Thus, for passive line impedances, the reflection coefficient lies inside or on the circle of unit radius in the $\bar{\Gamma}$ plane. We will hereafter call this circle the unit circle. Conversely, each point inside or on the unit circle represents a possible value of reflection coefficient corresponding to a unique value of passive normalized line impedance in view of (6-243). Hence all possible values of passive normalized line impedances can be mapped onto the region bounded by the unit circle.

To determine how the normalized line impedance values are mapped onto the region bounded by the unit circle, we note that

$$\bar{\Gamma} = \frac{r + jx - 1}{r + jx + 1} = \frac{r^2 - 1 + x^2}{(r + 1)^2 + x^2} + j\frac{2x}{(r + 1)^2 + x^2}$$

so that

$$\mathcal{R}e(\bar{\Gamma}) = \frac{r^2 - 1 + x^2}{(r + 1)^2 + x^2}$$

$$\mathcal{I}m(\bar{\Gamma}) = \frac{2x}{(r + 1)^2 + x^2}$$

Let us now discuss different cases:

(a) $\bar{z}$ is purely real; that is, $x = 0$. Then

$$\mathcal{R}e(\bar{\Gamma}) = \frac{r - 1}{r + 1} \qquad \text{and} \qquad \mathcal{I}m(\bar{\Gamma}) = 0$$

Purely real values of $\bar{z}$ are represented by points on the real axis. For example, $r = 0, \frac{1}{3}, 1, 3,$ and ∞ are represented by $\bar{\Gamma} = -1, -\frac{1}{2}, 0, \frac{1}{2},$ and 1, respectively, as shown in Fig. 6.52(a).

(b) $\bar{z}$ is purely imaginary; that is, $r = 0$. Thus

$$|\bar{\Gamma}| = \left| \frac{x^2 - 1}{x^2 + 1} + j\frac{2x}{x^2 + 1} \right| = 1$$

and

$$\underline{/\bar{\Gamma}} = \tan^{-1}\frac{2x}{x^2 - 1}$$

Purely imaginary values of $\bar{z}$ are represented by points on the unit circle. For example, $x = 0, 1, \infty, -1,$ and $-\infty$ are represented by

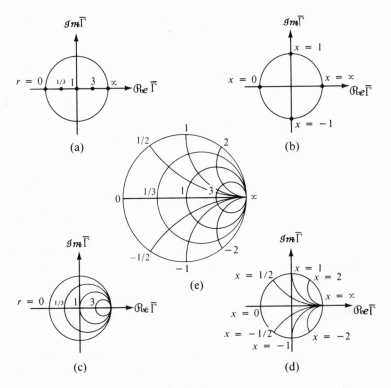

Fig. 6.52. Development of the Smith chart by transformation from $\bar{z}$ to $\bar{\Gamma}$.

$\bar{\Gamma} = 1\underline{/\pi}, 1\underline{/\pi/2}, 1\underline{/0°}, 1\underline{/-\pi/2}$ and $1\underline{/2\pi}$, respectively, as shown in Fig. 6.52(b).

(c) $\bar{z}$ is complex but its real part is constant. Then

$$\left[\mathcal{R}e(\bar{\Gamma}) - \frac{r}{r+1}\right]^2 + [\mathcal{I}m(\bar{\Gamma})]^2$$

$$= \left[\frac{r^2 - 1 + x^2}{(r+1)^2 + x^2} - \frac{r}{r+1}\right]^2 + \left[\frac{2x}{(r+1)^2 + x^2}\right]^2 = \left(\frac{1}{r+1}\right)^2$$

This is the equation of a circle with center at $\mathcal{R}e(\bar{\Gamma}) = r/(r+1)$ and $\mathcal{I}m(\bar{\Gamma}) = 0$ and radius equal to $1/(r+1)$. Thus loci of constant r are circles in the $\bar{\Gamma}$ plane with centers at $[r/(r+1), 0]$ and radii $1/(r+1)$. For example, for $r = 0, \frac{1}{3}, 1, 3,$ and ∞, the centers of the circles are $(0, 0)$, $(\frac{1}{4}, 0)$, $(\frac{1}{2}, 0)$, $(\frac{3}{4}, 0)$, and $(1, 0)$, respectively, and the radii are $1, \frac{3}{4}, \frac{1}{2}, \frac{1}{4}$, and 0, respectively. These circles are shown in Fig. 6.52(c).

(d) $\bar{z}$ is complex but its imaginary part is constant. Then

$$[\mathcal{R}e(\bar{\Gamma}) - 1]^2 + \left[\mathcal{I}m(\bar{\Gamma}) - \frac{1}{x}\right]^2$$

$$= \left[\frac{r^2 - 1 + x^2}{(r+1)^2 + x^2} - 1\right]^2 + \left[\frac{2x}{(r+1)^2 + x^2} - \frac{1}{x}\right]^2 = \left(\frac{1}{x}\right)^2$$

This is the equation of a circle with center at $\mathcal{R}e(\bar{\Gamma}) = 1$ and $\mathcal{I}m(\bar{\Gamma}) = 1/x$ and radius equal to $1/|x|$. Thus locii of constant x are circles in the $\bar{\Gamma}$ plane with centers at $(1, 1/x)$ and radii equal to $1/|x|$. For example, for $x = 0, \pm\frac{1}{2}, \pm1, \pm2,$ and $\pm\infty$, the centers of the circles are $(1, \infty)$, $(1, \pm2)$, $(1, \pm1)$, $(1, \pm\frac{1}{2})$, and $(1, 0)$, respectively, and the radii are $\infty, 2, 1, \frac{1}{2}$, and 0, respectively. Portions of these circles which fall inside the unit circle are shown in Fig. 6.52(d). Portions which fall outside the unit circle represent active impedances.

Combining (c) and (d), we obtain the chart of Fig. 6.52(e). In a commercially available form shown in Fig. 6.53, the Smith chart contains circles of constant r and constant x for very small increments of r and x, respectively, so that interpolation between the contours can be carried out accurately. We now illustrate the application of the Smith chart by means of some examples.

EXAMPLE 6-26.   A transmission line of characteristic impedance 50 ohms is terminated by a load impedance $\bar{Z}_R = (15 - j20)$ ohms. It is desired to find the following quantities by using the Smith chart.

(1) Reflection coefficient at the load.

(2) VSWR on the line.

(3) Distance of the first voltage minimum of the standing wave pattern from the load.

(4) Line impedance at $d = 0.05\lambda$.

(5) Line admittance at $d = 0.05\lambda$.

(6) Location nearest to the load at which the real part of the line admittance is equal to the line characteristic admittance.

We proceed with the solution of the problem in the following step-by-step manner with reference to Fig. 6.54.

(a) Find the normalized load impedance.

$$\bar{z}_R = \frac{\bar{Z}_R}{Z_0} = \frac{15 - j20}{50} = 0.3 - j0.4$$

(b) Locate the normalized load impedance on the Smith chart at the intersection of the 0.3 constant normalized resistance circle and -0.4 constant normalized reactance circle (point A).

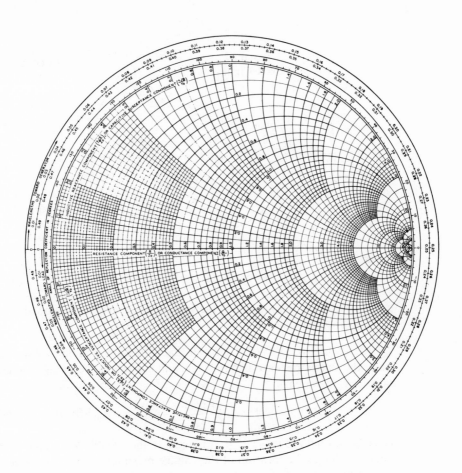

Fig. 6.53. The Smith chart (Copyrighted by and reproduced with the permission of Kay Elemetrics Corp., Pine Brook, N.J.).

(c) Locating point A actually amounts to computing the reflection coefficient at the load since the Smith chart is a transformation in the $\bar{\Gamma}$ plane. The magnitude of the reflection coefficient is the distance from the center (O) of the Smith chart (origin of the $\bar{\Gamma}$ plane) to the point A based on a radius of unity for the outermost circle. For this example, $|\bar{\Gamma}(0)| = 0.6$. The phase angle of $\bar{\Gamma}(0)$ is the angle measured from the horizontal axis to the right of O (positive real axis in the $\bar{\Gamma}$ plane) to the line OA in the counterclockwise direction. This angle is indicated on the chart along its circumference. For this example, $\underline{/\bar{\Gamma}(0)} = 227°$. Thus

$$\bar{\Gamma}(0) = 0.6e^{j227°}$$

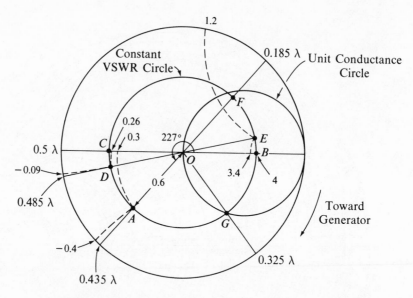

Fig. 6.54. For illustrating the various procedures to be followed in using the Smith chart.

(d) To find the VSWR, we recall that at the location of a voltage maximum, the line impedance is purely real and maximum. Denoting this impedance as R_{max}, we have

$$R_{max} = \frac{V_{max}}{I_{min}} = \frac{|\bar{V}^+|(1 + |\bar{\Gamma}|)}{(|\bar{V}^+|/Z_0)(1 - |\bar{\Gamma}|)} = Z_0(\text{VSWR}) \qquad (6\text{-}245)$$

Thus the normalized value of R_{max} is equal to the VSWR. We therefore move along the line to the location of the voltage maximum, which involves going around the constant $|\bar{\Gamma}|$ circle to the point on the positive real axis. To do this on the Smith chart, we draw a circle passing through A and with center at O. This circle is known as the "constant VSWR circle" since for points on this circle, $|\bar{\Gamma}|$ and hence $\text{VSWR} = (1 + |\bar{\Gamma}|)/(1 - |\bar{\Gamma}|)$ is a constant. Impedance values along this circle are normalized line impedances as seen moving along the line. In particular, since point B (the intersection of the constant VSWR circle with the horizontal axis to the right of O) corresponds to voltage maximum, the normalized impedance value at point B which is purely real and maximum, is equal to the VSWR. Thus, for this example, $\text{VSWR} = 4$.

(e) Just as point B represents the position of a voltage maximum on the line, point C (intersection of the constant VSWR circle with the horizontal axis to the left of O, i.e., the negative real axis of the $\bar{\Gamma}$ plane) represents the location of a voltage minimum. Hence, to find the distance of the first voltage minimum from the load, we move along the constant VSWR circle starting at point A (load impedance) towards the generator (clockwise direction on the chart) to reach point C. Distance moved along the constant VSWR circle in this process can be determined by recognizing that one complete revolution around the chart ($\bar{\Gamma}$-plane diagram) constitutes movement on the line by 0.5λ. However, it is not necessary to compute in this manner since distance scales in terms of λ are provided along the periphery of the chart for movement in both directions. For this example, the distance from the load to the first voltage minimum $= (0.5 - 0.435)\lambda = 0.065\lambda$. Conversely, if the VSWR and the location of the voltage minimum are specified, we can find the load impedance following the above procedures in reverse.

(f) To find the line impedance at $d = 0.05\lambda$, we start at point A and move along the constant VSWR circle towards the generator (in the clockwise direction) by a distance of 0.05λ to reach point D. This step is equivalent to finding the reflection coefficient at $d = 0.05\lambda$ knowing the reflection coefficient at $d = 0$ and then computing the normalized line impedance by using (6-243). Thus, from the coordinates corresponding to point D, the normalized line impedance at $d = 0.05\lambda$ is $(0.26 - j0.09)$ and hence the line impedance at $d = 0.05\lambda$ is $50(0.26 - j0.09)$ or $(13 - j4.5)$ ohms.

(g) To find the line admittance at $d = 0.05\lambda$, we recall that

$$[\bar{Z}(d)]\left[\bar{Z}\left(d + \frac{\lambda}{4}\right)\right] = Z_0^2$$

so that

$$[\bar{z}(d)]\left[\bar{z}\left(d + \frac{\lambda}{4}\right)\right] = 1$$

or

$$\bar{y}(d) = \bar{z}\left(d + \frac{\lambda}{4}\right) \tag{6-246}$$

Thus the normalized line admittance at a point D is the same as the normalized line impedance at a distance $\lambda/4$ from it. Hence, to find $\bar{y}(0.05\lambda)$, we start at point D and move along the constant VSWR circle by a distance $\lambda/4$ to reach point E (we note that this point is diametrically opposite to point D) and read its coordinates. This gives $\bar{y}(0.05\lambda) = (3.4 + j1.2)$. We then have $\bar{Y}(0.05\lambda) = \bar{y}(0.05\lambda) \times Y_0 = (3.4 + j1.2) \times 1/50 = (0.068 + j0.024)$ mhos.

(h) Relationship (6-246) permits us to use the Smith chart as an admittance chart instead of an impedance chart. In other words, if we want to find the normalized line admittance $\bar{y}(Q)$ at a point Q on the line, knowing the normalized line admittance $\bar{y}(P)$ at another point P on the line, we can simply locate $\bar{y}(P)$ by entering the chart at coordinates equal to its real and imaginary parts and then moving along the constant VSWR circle by the amount of the distance from P to Q in the proper direction to obtain the coordinates equal to the real and imaginary parts of $\bar{y}(Q)$. Thus it is not necessary first to locate $\bar{z}(P)$ diametrically opposite to $\bar{y}(P)$ on the constant VSWR circle, then move along the constant VSWR circle to locate $\bar{z}(Q)$, and then find $\bar{y}(Q)$ diametrically opposite to $\bar{z}(Q)$. To find the location nearest to the load at which the real part of the line admittance is equal to the line characteristic admittance, we first locate $\bar{y}(0)$ at point F diametrically opposite to point A which corresponds to $\bar{z}(0)$. We then move along the constant VSWR circle towards the generator to reach point G on the circle corresponding to constant real part equal to unity (we call this circle the "unit conductance circle"). Distance moved from F to G is read off the chart as $(0.325 - 0.185)\lambda = 0.14\lambda$. This is the distance closest to the load at which the real part of the normalized line admittance is equal to unity and hence the real part of the line admittance is equal to the line characteristic admittance. ∎

EXAMPLE 6-27. It is desired to solve the stub matching problem of Example 6-25 by using the Smith chart.

We make use of the principle of stub matching illustrated in Example 6-25 and the procedures learned in Example 6-26 to solve this problem in the following step-by-step manner with reference to Fig. 6.55.

(a) Find the normalized load impedance.

$$\bar{z}_R = \frac{\bar{Z}_R}{Z_0} = \frac{30 - j40}{50} = 0.6 - j0.8$$

Locate the normalized load impedance on the Smith chart at point A.

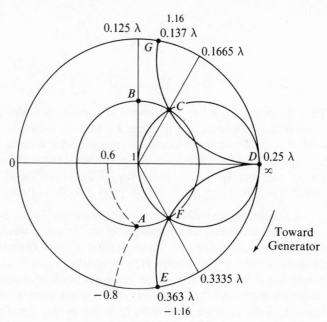

Fig. 6.55. Solution of transmission-line matching problem by using the Smith chart.

(b) Draw the constant VSWR circle passing through point *A*. This is the locus of the normalized line impedance as well as the normalized line admittance. Starting at point *A*, go around the constant VSWR circle by half a revolution to reach point *B* diametrically opposite to point *A*. Point *B* corresponds to the normalized load admittance.

(c) Starting at point *B*, go around the constant VSWR circle towards the generator until point *C* on the unit conductance circle is reached. This point corresponds to the normalized line admittance having the real part equal to unity and hence it corresponds to the location of the stub. The distance moved from point *B* to point *C* (not from point *A* to point *C*) is equal to the distance from the load at which the stub must be located. Thus the location of the stub from the load $= (0.1665 - 0.125)\lambda = 0.0415\lambda$.

(d) Read off the Smith chart the normalized susceptance value corresponding to point *C*. This value is 1.16 and it is the imaginary part of the normalized line admittance at the location of the stub. The imaginary part of the line admittance is equal to $1.16 \times Y_0 = (1.16/50)$ mhos. The input susceptance of the stub must therefore be equal to $-(1.16/50)$ mhos.

(e) This step consists of finding the length of a short-circuited stub having an input susceptance equal to $-(1.16/50)$ mhos. We can use the Smith chart for this purpose since this simply consists of finding the distance between two points on a line (the stub in this case) at which the admittances (purely imaginary in this case) are known. Thus, since the short circuit corresponds to a susceptance of infinity, we start at point *D* and move towards the generator along the constant VSWR circle through *D* (the outermost circle) to reach point *E* corresponding to $-j1.16$, which is the input admittance of the stub normalized with respect to its own characteristic admittance. The distance moved from *D* to *E* is the required length of the stub. Thus length of the short-circuited stub $= (0.363 - 0.25)\lambda = 0.113\lambda$.

(f) The results obtained for the location and the length of the stub agree with one of the solutions found analytically in Example 6-25. The second solution can be obtained by noting that in step (c) above, we can go around the constant VSWR circle from point *B* until point *F* on the unit conductance circle is reached instead of stopping at point *C*. The stub location for this solution is $(0.3335 - 0.125)\lambda = 0.2085\lambda$. The required input susceptance of the stub is $(1.16/50)$ mhos. The length of the stub is the distance from point *D* to point *G* in the clockwise direction. This is $(0.137 + 0.25)\lambda = 0.387\lambda$. These values are the same as the second solution obtained in Example 6-25. ∎

We have illustrated the use of the Smith chart by considering the transmission-line matching problem. However, from the procedures learned in Example 6.26, it can be seen that the Smith chart can be used for all transmission-line and analogous plane-wave problems involving reflection, transmission, and matching. As a further illustration of the applications of the Smith chart, we will learn in the following section that waveguide problems can be treated by using transmission-line equivalents. Thus the Smith chart can be used for solving these and many other problems.

6.12 Waveguides; Dispersion and Group Velocity

In Section 6.8 we obtained the solution for the one-dimensional wave equation as $(+)$ and $(-)$ uniform plane waves traveling along that dimension and then deduced the expressions for the fields in a uniform plane wave traveling in an arbitrary direction with reference to a coordinate system. We now make use of these expressions to discuss uniform plane waves incident obliquely on a perfect conductor and then introduce the concept of waveguides. Since an arbitrarily polarized uniform plane wave can be decomposed into linearly polarized uniform plane waves, we consider linearly polarized uniform plane waves only for this discussion. Let us consider a perfect conductor occupying the $x = 0$ plane and upon which is incident a uniform plane wave having the electric field vector

$$
\begin{aligned}
\bar{\mathbf{E}}_i &= \bar{E}_0 e^{-j\boldsymbol{\beta}_i \cdot \mathbf{r}} \mathbf{i}_y \\
&= \bar{E}_0 e^{-j(\beta \cos \theta_i\, \mathbf{i}_x + \beta \sin \theta_i\, \mathbf{i}_z) \cdot \mathbf{r}} \mathbf{i}_y \\
&= \bar{E}_0 e^{-j(\beta x \cos \theta_i + \beta z \sin \theta_i)} \mathbf{i}_y
\end{aligned}
\tag{6-247a}
$$

where $\bar{E}_0$ is a constant, $\beta = \omega\sqrt{\mu\epsilon}$, and θ_i is the angle between the propagation vector $\boldsymbol{\beta}_i$ and the normal to the conductor as shown in Fig. 6.56. The expression for the corresponding magnetic field vector can be obtained by using (6-200) as follows:

$$
\begin{aligned}
\bar{\mathbf{H}}_i &= \frac{1}{\omega\mu}\boldsymbol{\beta}_i \times \bar{\mathbf{E}}_i \\
&= \sqrt{\frac{\epsilon}{\mu}}(-\bar{E}_0 \sin \theta_i\, \mathbf{i}_x + \bar{E}_0 \cos \theta_i\, \mathbf{i}_z)e^{-j(\beta x \cos \theta_i + \beta z \sin \theta_i)}
\end{aligned}
\tag{6-247b}
$$

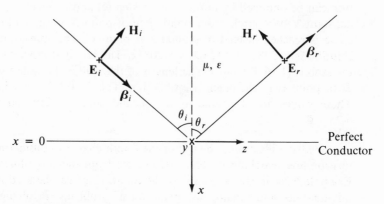

Fig. 6.56. Oblique incidence of a uniform plane wave on a perfect conductor.

Since the boundary condition at a perfect conductor surface dictates that the tangential component of the electric field be zero, a reflected wave must exist which cancels completely the tangential component (which is the only component in this case) of the electric field vector of the incident wave at the surface of the conductor. Such cancellation is possible only if the tangential component of the electric field in the reflected wave at the surface of the conductor is entirely in the y direction, that is, the same as the direction of the tangential component of the electric field vector of the incident wave. Furthermore, since we are dealing with linearly polarized uniform plane waves, the electric field in the reflected wave must everywhere be in the same direction. Hence it must have a y component only everywhere. Thus the electric and magnetic fields of the reflected wave can be written as

$$\begin{aligned}
\bar{\mathbf{E}}_r &= \bar{E}_0' e^{-j\boldsymbol{\beta}_r \cdot \mathbf{r}} \mathbf{i}_y \\
&= \bar{E}_0' e^{-j(-\beta \cos \theta_r \, \mathbf{i}_x + \beta \sin \theta_r \, \mathbf{i}_z) \cdot \mathbf{r}} \mathbf{i}_y \\
&= \bar{E}_0' e^{j(\beta x \cos \theta_r - \beta z \sin \theta_r)} \mathbf{i}_y
\end{aligned} \tag{6-248a}$$

$$\begin{aligned}
\bar{\mathbf{H}}_r &= \frac{1}{\omega \mu} \boldsymbol{\beta}_r \times \bar{\mathbf{E}}_r \\
&= \sqrt{\frac{\epsilon}{\mu}} (-\bar{E}_0' \sin \theta_r \, \mathbf{i}_x - \bar{E}_0' \cos \theta_r \, \mathbf{i}_z) e^{j(\beta x \cos \theta_r - \beta z \sin \theta_r)}
\end{aligned} \tag{6-248b}$$

where $\bar{E}_0'$ is a constant, $\beta = \omega \sqrt{\mu\epsilon}$, and θ_r is the angle between the propagation vector $\boldsymbol{\beta}_r$ and the normal to the conductor as shown in Fig. 6.56.

Adding the incident and reflected fields, we obtain the components of the total electric and magnetic fields as

$$\bar{E}_y = \bar{E}_0 e^{-j(\beta x \cos \theta_i + \beta z \sin \theta_i)} + \bar{E}_0' e^{j(\beta x \cos \theta_r - \beta z \sin \theta_r)} \tag{6-249a}$$

$$\begin{aligned}
\bar{H}_x = \sqrt{\frac{\epsilon}{\mu}} [&-\bar{E}_0 \sin \theta_i \, e^{-j(\beta x \cos \theta_i + \beta z \sin \theta_i)} \\
&- \bar{E}_0' \sin \theta_r \, e^{j(\beta x \cos \theta_r - \beta z \sin \theta_r)}]
\end{aligned} \tag{6-249b}$$

$$\begin{aligned}
\bar{H}_z = \sqrt{\frac{\epsilon}{\mu}} [&\bar{E}_0 \cos \theta_i \, e^{-j(\beta x \cos \theta_i + \beta z \sin \theta_i)} \\
&- \bar{E}_0' \cos \theta_r \, e^{j(\beta x \cos \theta_r - \beta z \sin \theta_r)}]
\end{aligned} \tag{6-249c}$$

Applying the boundary condition at the surface of the conductor, we have

$$[\bar{E}_y]_{x=0} = \bar{E}_0 e^{-j\beta z \sin \theta_i} + \bar{E}_0' e^{-j\beta z \sin \theta_r} = 0 \qquad \text{for all } z \tag{6-250}$$

Equation (6-250) can be satisfied only if the exponential factors are equal for all z. Thus we obtain the result

$$\theta_r = \theta_i \tag{6-251}$$

that is, the angle of reflection is equal to the angle of incidence, which is the familiar law of reflection in optics. Substituting (6-251) into (6-250), we have

$$\bar{E}_0' = -\bar{E}_0 \tag{6-252}$$

Substituting (6-251) and (6-252) into (6-249a)–(6-249c), we obtain the following expressions for the components of the total fields:

$$\bar{E}_y = -2j\bar{E}_0 \sin(\beta x \cos\theta_i) e^{-j\beta z \sin\theta_i} \tag{6-253a}$$

$$\bar{H}_x = \sqrt{\frac{\epsilon}{\mu}} 2j\bar{E}_0 \sin\theta_i \sin(\beta x \cos\theta_i) e^{-j\beta z \sin\theta_i} \tag{6-253b}$$

$$\bar{H}_z = \sqrt{\frac{\epsilon}{\mu}} 2\bar{E}_0 \cos\theta_i \cos(\beta x \cos\theta_i) e^{-j\beta z \sin\theta_i} \tag{6-253c}$$

The exponential factor in (6-253a)–(6-253c) lends a pure traveling wave character in the z direction to the fields whereas the sine and cosine factors involving x lend a complete standing wave character in the x direction. In fact, the complex Poynting vector is given by

$$\bar{\mathbf{P}} = \frac{1}{2}\bar{\mathbf{E}} \times \bar{\mathbf{H}}^* = \frac{1}{2}[\bar{E}_y\bar{H}_z^*\mathbf{i}_x + (-\bar{E}_y)(\bar{H}_x^*)\mathbf{i}_z]$$

$$= \frac{1}{2}\sqrt{\frac{\epsilon}{\mu}}[-2j\,|\bar{E}_0|^2 \cos\theta_i \sin(2\beta x \cos\theta_i)\,\mathbf{i}_x \tag{6-254}$$

$$+ 4\,|\bar{E}_0|^2 \sin\theta_i \sin^2(\beta x \cos\theta_i)\,\mathbf{i}_z]$$

Thus the time-average power flow is entirely in the z direction whereas the reactive power flow is associated entirely with the x direction. The situation can therefore be described as one of complete standing waves in the x direction traveling as a whole in the z direction.

We note from (6-253a) that $\bar{E}_y$ is equal to zero not only at the surface of the conductor ($x = 0$), but also in other planes given by

$$\sin(\beta x \cos\theta_i) = 0$$

or

$$\beta x \cos\theta_i = -m\pi \qquad m = 1, 2, 3, \ldots$$

or

$$x = -\frac{m\pi}{\beta \cos\theta_i} = -\frac{m\lambda}{2\cos\theta_i} \qquad m = 1, 2, 3, \ldots \tag{6-255}$$

where $\lambda = 2\pi/\beta$ is the wavelength along the direction of incidence (or reflection). Introduction of perfect conductors in planes parallel to the conductor surface and at distances of integral multiples of $\lambda/(2\cos\theta_i)$ from it does not alter in any way the total field, once it is established. Let us introduce a perfectly conducting plate in the plane $x = -m\lambda/(2\cos\theta_i)$ as shown in Fig. 6.57, where m can take any integer value. The two conductors support standing waves in the x direction while permitting traveling waves in the z direction. The phenomenon is actually one of uniform plane waves bouncing obliquely between the two plane conductors as shown in Fig. 6.57. The structure is known as a parallel-plate "waveguide." The total magnetic field has a component in the z direction, which is the direction of time-average power flow whereas the electric field is entirely transverse to the z direction. For this reason, the waves are known as "transverse electric" or TE waves.

Let us now fix the spacing between the parallel plates as a and discuss the

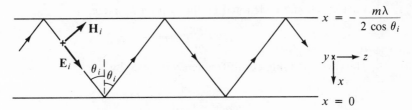

$$x = -\frac{m\lambda}{2\cos\theta_i}$$

$$x = 0$$

Fig. 6.57. Bouncing of a uniform plane wave obliquely along a parallel-plate waveguide.

behavior of the guided waves as the frequency of the source exciting these waves is varied. Setting $m\lambda/(2\cos\theta_i)$ equal to a, we have

$$\cos\theta_i = \frac{m\lambda}{2a} \tag{6-256}$$

From (6-256), we note that, for very high frequencies, $\lambda \approx 0$, $\cos\theta_i \approx 0$, $\theta_i \approx 90°$, and the waves slide between the plates almost like a TEM wave. As the frequency is decreased, λ increases, $\cos\theta_i$ increases, θ_i decreases, and the waves bounce obliquely between the plates, progressing in the z direction until, for $\lambda = 2a/m$, $\cos\theta_i = 1$, $\theta_i = 0°$, and the waves bounce back and forth between the plates and normal to them so that there is no progress in the z direction. These different cases are illustrated in Figs. 6.58(a)–(d).

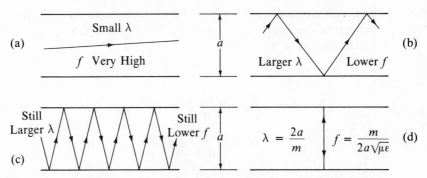

Fig. 6.58. Bouncing of uniform plane waves of different frequencies between parallel plane conductors of fixed spacing for illustrating the "cutoff" phenomenon.

For $\lambda > 2a/m$, $\cos\theta_i > 1$, $\sin\theta_i = \sqrt{1 - \cos^2\theta_i}$ becomes imaginary, the exponents in the expressions for the total fields become real, and the situation no longer corresponds to one of wave propagation; the fields diminish in magnitude along z. Thus there is a wavelength below which propagation occurs and above which there is no propagation. This is known as the cutoff wavelength and is denoted by the symbol λ_c. Here,

$$\lambda_c = \frac{2a}{m} \qquad m = 1, 2, 3, \ldots \tag{6-257}$$

The corresponding cutoff frequency is given by

$$f_c = \frac{v_p}{\lambda_c} = \frac{m}{2a\sqrt{\mu\epsilon}} \qquad m = 1, 2, 3, \ldots \tag{6-258}$$

where $v_p = 1/\sqrt{\mu\epsilon}$ is the phase velocity along the direction of incidence (or reflection). For $f > f_c$, propagation occurs and for $f < f_c$, there is no propagation.

Substituting λ_c for $2a/m$ in (6-256), we have

$$\cos\theta_i = \frac{\lambda}{\lambda_c}, \ \sin\theta_i = \sqrt{1 - \left(\frac{\lambda}{\lambda_c}\right)^2}$$

$$\beta\cos\theta_i = \frac{2\pi}{\lambda}\frac{\lambda}{\lambda_c} = \frac{2\pi}{\lambda_c} = \frac{m\pi}{a}$$

$$\beta\sin\theta_i = \frac{2\pi}{\lambda}\sqrt{1 - \left(\frac{\lambda}{\lambda_c}\right)^2}$$

But $\beta\sin\theta_i$ is the component of the propagation vector $\boldsymbol{\beta}_i$ in the z direction, that is, along the guide axis. Hence the wavelength in the z direction, which we call the guide wavelength λ_g, is given by

$$\lambda_g = \frac{2\pi}{\beta\sin\theta_i} = \frac{\lambda}{\sqrt{1 - (\lambda/\lambda_c)^2}} = \frac{\lambda}{\sqrt{1 - (f_c/f)^2}} \tag{6-259}$$

Now, substituting for $\beta\cos\theta_i$ and $\beta\sin\theta_i$ in the expressions for the components of the total fields given by (6-253a)–(6-253c), we obtain expressions independent of θ_i as

$$\bar{E}_y = -2j\bar{E}_0 \sin\left(\frac{m\pi x}{a}\right) e^{-j(2\pi/\lambda_g)z} \tag{6-260a}$$

$$\bar{H}_x = 2j\frac{\bar{E}_0}{\eta}\frac{\lambda}{\lambda_g} \sin\left(\frac{m\pi x}{a}\right) e^{-j(2\pi/\lambda_g)z} \tag{6-260b}$$

$$\bar{H}_z = 2\frac{\bar{E}_0}{\eta}\frac{\lambda}{\lambda_c} \cos\left(\frac{m\pi x}{a}\right) e^{-j(2\pi/\lambda_g)z} \tag{6-260c}$$

where $\eta = \sqrt{\mu/\epsilon}$ and λ_g and λ_c are given by (6-259) and (6-257), respectively. The solution for the fields corresponding to each value of m is called a mode. The x dependence of the fields is sinusoidal with m half-sine variations between the plates. The fields are independent of the y coordinate; that is, they have zero half-sine variations along the y direction. The solutions are therefore said to correspond to $\text{TE}_{m,0}$ modes, where the first and second subscripts represent the number of half-sine variations of the fields in the x and y directions, respectively. The cutoff wavelength is smaller and the cutoff frequency is higher, the larger the value of m. For any particular wave frequency, all modes for which the cutoff frequencies are less than the wave frequency can propagate down the guide. The mode which has the lowest cutoff frequency is known as the dominant mode. Here, the $TE_{1,0}$ mode is the dominant mode.

From the expressions for the fields, we note that the constant phase surfaces are the planes $z = $ constant. The rate of change of phase with distance along z, that is, along the normal to the constant phase surfaces, is $2\pi/\lambda_g$. Hence the phase velocity in the z direction, which we denote as v_{pz}, is given by

$$v_{pz} = \frac{\omega}{(2\pi/\lambda_g)} = \frac{\omega}{\beta \sin \theta_i} = \frac{v_p}{\sqrt{1 - (\lambda/\lambda_c)^2}} = \frac{v_p}{\sqrt{1 - (f_c/f)^2}} \qquad (6\text{-}261)$$

where $v_p = 1/\sqrt{\mu\epsilon}$. We note that v_{pz} is simply the apparent phase velocity of the obliquely bouncing waves along the z direction. We also note that v_{pz} is a function of frequency f, the consequence of which we will discuss later in this section. The constant amplitude surfaces are given by $x = $ constant. Thus, for the total fields, the amplitude is not constant over the constant phase surfaces.

From the point of view of time-average power flow, the field components of interest are $-\bar{E}_y$ and $\bar{H}_x$, as can be seen from (6-254). The wave impedance obtained by taking the ratio of these two components is known as the guide impedance and is denoted by the symbol η_g. Thus

$$\eta_g = \frac{-\bar{E}_y}{\bar{H}_x} = \eta \frac{\lambda_g}{\lambda} = \frac{\eta}{\sqrt{1 - (\lambda/\lambda_c)^2}} = \frac{\eta}{\sqrt{1 - (f_c/f)^2}} \qquad (6\text{-}262)$$

Now, using the analogy

$$
\begin{aligned}
-\bar{E}_y &\longleftrightarrow \bar{V} \\
\bar{H}_x &\longleftrightarrow \bar{I} \\
\lambda_g &\longleftrightarrow \lambda \\
v_{pz} &\longleftrightarrow v_p \\
\eta_g &\longleftrightarrow \eta
\end{aligned}
\qquad (6\text{-}263)
$$

we can develop a transmission-line equivalent as shown in Fig. 6.59 which is valid for power flow in the z direction. Employing the transmission-line techniques discussed in Sections 6.9, 6.10, and 6.11 in conjunction with this equivalent, we can solve reflection, transmission, and matching problems

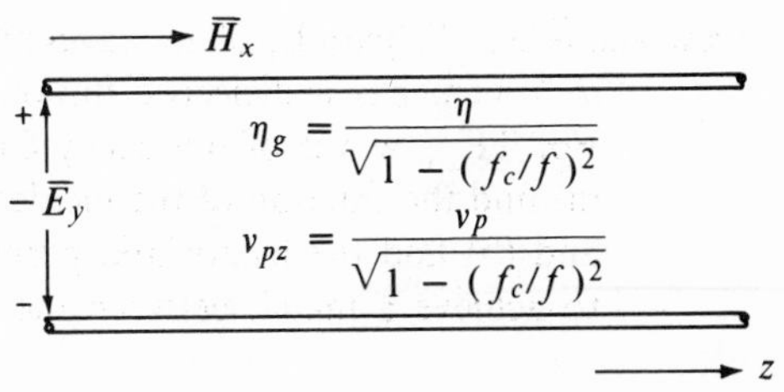

Fig. 6.59. Transmission-line equivalent for power flow along the guide for TE waves in a parallel-plate waveguide.

$$\eta_g = \frac{\eta}{\sqrt{1 - (f_c/f)^2}}$$

$$v_{pz} = \frac{v_p}{\sqrt{1 - (f_c/f)^2}}$$

involving TE modes in waveguides. The proof is left as an exercise (Problem 6.68) for the student. We will now consider some examples, to consolidate what we have learned thus far in this section.

EXAMPLE 6-28. The dimension a of a parallel-plate waveguide is 5.0 cm. Determine the propagating $\text{TE}_{m,0}$ modes for a wave frequency of 10,000 MHz, assuming free space between the plates. For each propagating mode, find (a) the cutoff frequency f_c, (b) the angle θ_i at which the wave bounces obliquely between the conductors, (c) the guide wavelength λ_g, (d) the phase velocity v_{pz}, and (e) the guide impedance η_g.

From (6-257), the cutoff wavelengths are $\lambda_c = 2a/m = 10/m$ cm. The wave frequency of 10,000 MHz corresponds to a wavelength λ of 3 cm in free space. Hence the propagating $\text{TE}_{m,0}$ modes are $\text{TE}_{1,0}(\lambda_c = 10 \text{ cm})$, $\text{TE}_{2,0}(\lambda_c = 5 \text{ cm})$, and $\text{TE}_{3,0}(\lambda_c = 10/3 \text{ cm})$. For each propagating mode, the quantities f_c, θ_i, λ_g, v_{pz}, and η_g can be computed by using the following formulas:

$$f_c = \frac{v_p}{\lambda_c} = \frac{1}{\lambda_c \sqrt{\mu_0 \epsilon_0}}$$

$$\theta_i = \cos^{-1} \frac{\lambda}{\lambda_c}$$

$$\lambda_g = \frac{\lambda}{\sqrt{1 - (\lambda/\lambda_c)^2}}$$

$$v_{pz} = \frac{v_p}{\sqrt{1 - (\lambda/\lambda_c)^2}} \qquad \text{where } v_p = \frac{1}{\sqrt{\mu_0 \epsilon_0}}$$

$$\eta_g = \frac{\eta}{\sqrt{1 - (\lambda/\lambda_c)^2}} \qquad \text{where } \eta = \sqrt{\frac{\mu_0}{\epsilon_0}}$$

The computed values are as follows:

Mode	$TE_{1,0}$	$TE_{2,0}$	$TE_{3,0}$
f_c, MHz	3000	6000	9000
θ_i, deg	72.55	53.13	25.15
λ_g, cm	3.145	3.75	6.883
v_{pz}, m/sec	3.145×10^8	3.75×10^8	6.883×10^8
η_g, ohms	395.2	471.2	864.9

EXAMPLE 6-29. A parallel-plate waveguide extending in the z direction and having $a = 3$ cm has a dielectric discontinuity at $z = 0$ as shown in Fig. 6.60(a). For $\text{TE}_{1,0}$ waves of frequency 6,000 MHz incident from the free-space side, (a) find the fraction of the incident power transmitted into the region $z > 0$, and (b) find the length and permittivity of a quarter-wave section required to achieve a match between the two media.

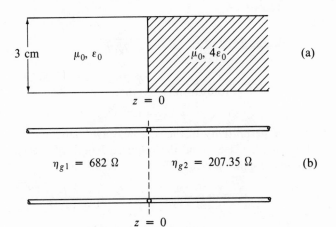

Fig. 6.60. (a) Dielectric discontinuity in a parallel-plate waveguide. (b) Transmission-line equivalent for power flow across
the discontinuity for the $TE_{1,0}$ mode.

Since the discontinuity exists over the entire transverse section of the
waveguide, we can use the transmission-line equivalent of Fig. 6.59 for each
section of the guide. For the $TE_{1,0}$ mode, $\lambda_c = 2a = 6$ cm. For $f = 6000$
MHz, the wavelength in free space is $\lambda_1 = 5$ cm and the wavelength in a
dielectric of permittivity $4\epsilon_0$ is $\lambda_2 = 2.5$ cm. Since λ_1 and λ_2 are both less
than λ_c, the $TE_{1,0}$ mode can propagate in both sections. Denoting the guide
parameters associated with sections 1 and 2 by subscripts 1 and 2, respectively,
we have

$$\eta_{g1} = \frac{\eta_1}{\sqrt{1 - (\lambda_1/\lambda_c)^2}} = \frac{377}{\sqrt{1 - (5/6)^2}} = 682 \text{ ohms}$$

$$\eta_{g2} = \frac{\eta_2}{\sqrt{1 - (\lambda_2/\lambda_c)^2}} = \frac{188.5}{\sqrt{1 - (2.5/6)^2}} = 207.35 \text{ ohms}$$

The transmission-line equivalent for power flow in the z direction is shown
in Fig. 6.60(b). The reflection coefficient at $z = 0$ is then given by

$$\bar{\Gamma} = \frac{\eta_{g2} - \eta_{g1}}{\eta_{g2} + \eta_{g1}} = \frac{207.35 - 682}{207.35 + 682} = -0.5337$$

Thus the fraction of incident power transmitted into the region $z > 0$ is
$1 - |\bar{\Gamma}|^2 = 1 - 0.5337^2 = 0.715$. The characteristic impedance of a quarter-
wave section required to achieve a match between line 1 and line 2 must be
equal to $\sqrt{\eta_{g1}\eta_{g2}}$. Denoting the parameters associated with the quarter-
wave section by subscript 3, we have

$$\eta_{g3} = \frac{\eta_3}{\sqrt{1 - (\lambda_3/\lambda_c)^2}} = \sqrt{\eta_{g1}\eta_{g2}}$$

or

$$\frac{\eta_1\sqrt{\epsilon_1/\epsilon_3}}{\sqrt{1-(\lambda_1/\lambda_c)^2(\epsilon_1/\epsilon_3)}} = \sqrt{\eta_{g1}\eta_{g2}}$$

$$\frac{\epsilon_1/\epsilon_3}{1-(5/6)^2(\epsilon_1/\epsilon_3)} = \frac{\eta_{g1}\eta_{g2}}{\eta_1^2} = \frac{682 \times 207.35}{377^2} = 0.995$$

$$\epsilon_3 = 1.6995\epsilon_0$$

Hence the permittivity of the quarter-wave matching section must be equal to $1.6995\epsilon_0$. To determine the required length of the matching section, we compute the guide wavelength in the section as

$$\lambda_{g3} = \frac{\lambda_3}{\sqrt{1-(\lambda_3/\lambda_c)^2}} = \frac{\lambda_1/\sqrt{1.6995}}{\sqrt{1-(\lambda_1/\lambda_c)^2(1/1.6995)}}$$

$$= \frac{3.8355}{\sqrt{1-0.4086}} = 4.9874 \text{ cm}$$

Hence the required length $= \lambda_g/4 = 1.24685$ cm. ∎

We have merely introduced the concept of a waveguide by considering $\text{TE}_{m,0}$ modes in a parallel-plate guide. Since the electric field is entirely along the y direction, that is, tangential to the plates, introduction of two more conductors in two $y = $ constant planes, say $y = 0$ and $y = b$, does not in any way alter the field configuration of the $\text{TE}_{m,0}$ mode. We then have a metallic pipe with rectangular cross section in the xy plane as shown in Fig. 6.61. Such a structure is known as a "rectangular waveguide." The fields

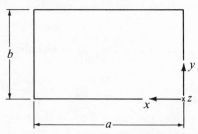

Fig. 6.61. Cross section of a rectangular waveguide.

in the $\text{TE}_{m,0}$ modes have m half-sinusoidal variations in the x direction and no variations in the y direction. They are due to uniform plane waves having electric field in the y direction only and bouncing obliquely between the walls $x = 0$ and $x = a$. In a similar manner, we can have uniform plane waves having electric field in the x direction only and bouncing obliquely between the walls $y = 0$ and $y = b$, resulting in $\text{TE}_{0,n}$ modes. The cutoff wavelengths and frequencies for these modes can be obtained by substituting b for a and n for m in (6-257) and (6-258), respectively. We can even have $\text{TE}_{m,n}$ modes due to uniform plane waves having both x and y components of electric field and bouncing between all four walls, satisfying the boundary condition that the tangential electric fields at the walls are zero. We can repeat the entire discussion by starting with uniform plane waves incident obliquely on a perfect conductor with their magnetic field entirely parallel to the plane of the conductor, leading to transverse magnetic or TM modes. We should, however, note that $TM_{m,0}$ and $TM_{0,n}$ modes are not possible in rectangular waveguides. To see why this is so, we note, for example, that $TM_{m,0}$ modes in parallel-plate waveguides contain x components of electric fields and it is not possible to place conductors in $y =$ constant planes without creating half-sine variations of E_x in the y direction. For a particular frequency, all modes for which the cutoff frequencies are less than that frequency can propagate along the guide. However, in practice, waveguides are designed to transmit only the dominant mode, that is, the TE_{10} mode by a suitable choice of the dimensions a and b.

We will now discuss the consequence of v_{pz}, the phase velocity along the guide axis, being a function of frequency. Let us consider a wave which is made up of a group of waves of different frequencies. If the phase velocity is independent of frequency, the different frequency components maintain the same phase relationships at each and every point along the direction of propagation, thereby preserving the waveshape as it travels. We can then say that the group as a whole travels with the phase velocity. If, on the other hand, the phase velocity is dependent on frequency, the different frequency components do not maintain the same phase relationships at points along the direction of propagation, thereby changing the waveshape. This phenomenon is known as "dispersion," so termed after the phenomenon of dispersion of colors by a prism. In the presence of dispersion, we cannot say that the group as a whole travels with any one of the phase velocities of its components. However, we can attribute a velocity known as the "group velocity," denoted by v_g for the group travel under certain conditions.

To discuss the concept of group velocity, let us consider a group of two waves of frequencies ω_1 and ω_2 $(> \omega_1)$. Let the associated phase constants be β_1 and β_2. Then the phase velocities associated with ω_1 and ω_2 are $v_{p1} = \omega_1/\beta_1$ and $v_{p2} = \omega_2/\beta_2$, respectively. Let us consider an instant of time, say $t = 0$, at which the variations of the two waveforms with distance are as shown in Fig. 6.62(a), in which there is a coincidence of the two waveforms at the point designated A_1, A_2. For the parallel-plate waveguide, Eq. (6-261) indicates that the phase velocity decreases as frequency is increased. Hence, as the two waves travel along z, the waveform for ω_2 slides backwards relative to the waveform for ω_1. Thus, while the points B_1 and B_2 of Fig. 6.62(a) both move in the positive z direction as time progresses, the spacing between them decreases continuously until, at a time Δt, the two points coincide as shown in Fig. 6.62(b). The variation with distance of one waveform relative to the other is then exactly the same as in Fig. 6.62(a). For an observer, the group as a whole appears to be shifted in

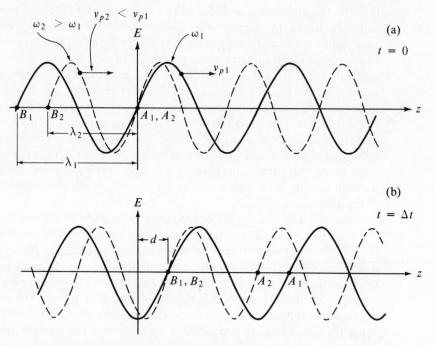

Fig. 6.62. For illustrating the concept of group velocity and for deriving an expression for the group velocity.

distance by d in time Δt. Hence the group velocity is

$$v_g = \frac{d}{\Delta t}$$

But

the distance moved by B_1 in time $\Delta t = \lambda_1 + d$

the distance moved by B_2 in time $\Delta t = \lambda_2 + d$

where λ_1 and λ_2 are the wavelengths corresponding to ω_1 and ω_2. From the phase velocities associated with ω_1 and ω_2, we then have

$$\lambda_1 + d = \frac{\omega_1}{\beta_1}\Delta t$$

$$\lambda_2 + d = \frac{\omega_2}{\beta_2}\Delta t$$

These two equations can be solved to obtain Δt and d as

$$\Delta t = \frac{\lambda_1 - \lambda_2}{(\omega_1/\beta_1) - (\omega_2/\beta_2)} = 2\pi\frac{\beta_2 - \beta_1}{\omega_1\beta_2 - \omega_2\beta_1}$$

$$d = \frac{(\omega_2/\beta_2)\lambda_1 - (\omega_1/\beta_1)\lambda_2}{(\omega_1/\beta_1) - (\omega_2/\beta_2)} = 2\pi\frac{\omega_2 - \omega_1}{\omega_1\beta_2 - \omega_2\beta_1}$$

so that

$$v_g = \frac{\omega_2 - \omega_1}{\beta_1 - \beta_1} \tag{6-264}$$

Between times zero and Δt, the distance variation of one waveform relative to the other is obviously such that the group is not identical to a displaced version of the group at $t = 0$. However, let us look at the waveform obtained by adding the two signals. This is given by

$$E = E_0 \cos(\omega_1 t - \beta_1 z) + E_0 \cos(\omega_2 t - \beta_2 z)$$

$$= E_0 \cos\left[\left(\frac{\omega_1 + \omega_2}{2}t - \frac{\beta_1 + \beta_2}{2}z\right) - \left(\frac{\omega_2 - \omega_1}{2}t - \frac{\beta_2 - \beta_1}{2}z\right)\right]$$

$$+ E_0 \cos\left[\left(\frac{\omega_1 + \omega_2}{2}t - \frac{\beta_1 + \beta_2}{2}z\right) + \left(\frac{\omega_2 - \omega_1}{2}t - \frac{\beta_2 - \beta_1}{2}z\right)\right]$$

$$= 2E_0 \cos\left(\frac{\omega_2 - \omega_1}{2}t - \frac{\beta_2 - \beta_1}{2}z\right) \cos\left(\frac{\omega_1 + \omega_2}{2}t - \frac{\beta_1 + \beta_2}{2}z\right)$$

$$\tag{6-265}$$

The right side of (6-265) represents a wave of frequency $(\omega_1 + \omega_2)/2$ traveling with a phase velocity $(\omega_1 + \omega_2)/(\beta_1 + \beta_2)$ and with its amplitude modulated in accordance with another wave of frequency $(\omega_2 - \omega_1)/2$ traveling with a phase velocity $(\omega_2 - \omega_1)/(\beta_2 - \beta_1)$, as shown in Fig. 6.63. Thus,

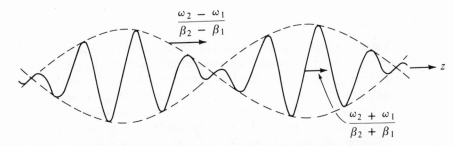

Fig. 6.63. For illustrating that the envelope of the superposition of two waves of frequencies ω_1 and ω_2 and phase constants β_1 and β_2, respectively, moves with the group velocity $(\omega_2 - \omega_1)/(\beta_2 - \beta_1)$.

although the waveform for $(\omega_1 + \omega_2)/2$ is changing in phase in accordance with the phase velocity $(\omega_1 + \omega_2)/(\beta_1 + \beta_2)$, its envelope is moving with the velocity $(\omega_2 - \omega_1)/(\beta_2 - \beta_1)$. As far as the amplitude is concerned, the entire group appears to be moving with the velocity $(\omega_2 - \omega_1)/(\beta_2 - \beta_1)$.

For the parallel-plate waveguide, the phase constant corresponding to v_{pz} is

$$\beta_z = \frac{\omega}{v_{pz}} = \frac{\omega}{v_p}\sqrt{1 - \left(\frac{f_c}{f}\right)^2} = \beta\sqrt{1 - \left(\frac{\omega_c}{\omega}\right)^2} \tag{6-266}$$

The variation of β_z with ω is shown in Fig. 6.64. A diagram of this kind is known as the $\omega - \beta_z$ diagram or the dispersion diagram. The phase velocity corresponding to any particular frequency is given by the slope of the line

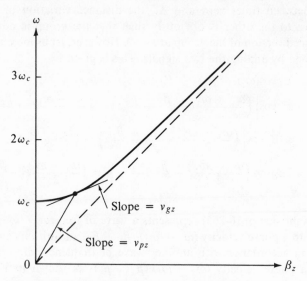

Fig. 6.64. β_z versus ω for the parallel-plate wave-guide.

drawn from the origin to the point on the curve corresponding to that frequency. The group velocity corresponding to any two frequencies ω_1 and ω_2 is given by the slope of the line joining the two points on the curve corresponding to those two frequencies. If we have a band of frequencies, we can find group velocities for each pair of these frequencies in this manner. We can attribute a group velocity to the entire group only if all these group velocities are equal. From Fig. 6.64, we see that this is not possible for a wide band of frequencies because of the nonlinear dependence of β_z upon ω. Hence it is not meaningful to talk of a group velocity for a group of waves comprising a wide band of frequencies. If, on the other hand, the frequencies are contained in a narrow band about a predominant frequency ω, then we can approximate the nonlinear $\omega - \beta_z$ curve in that narrow band by a straight line having the slope equal to that of the actual curve at ω so that it is meaningful to attribute a velocity to that group. This group velocity is given by

$$v_{gz} = \frac{d\omega}{d\beta_z} \qquad (6\text{-}267)$$

For β_z given by (6-266),

$$\frac{d\beta_z}{d\omega} = \beta\left[1 - \left(\frac{\omega_c}{\omega}\right)^2\right]^{-1/2}\frac{\omega_c^2}{\omega^3} + \frac{d\beta}{d\omega}\left[1 - \left(\frac{\omega_c}{\omega}\right)^2\right]^{1/2}$$

$$= \frac{\omega_c^2}{v_p\omega^2}\left[1 - \left(\frac{\omega_c}{\omega}\right)^2\right]^{-1/2} + \frac{1}{v_p}\left[1 - \left(\frac{\omega_c}{\omega}\right)^2\right]^{1/2}$$

$$= \frac{1}{v_p}\left[1 - \left(\frac{\omega_c}{\omega}\right)^2\right]^{-1/2}$$

and

$$v_{gz} = \frac{d\omega}{d\beta_z} = v_p\sqrt{1 - \left(\frac{\omega_c}{\omega}\right)^2}$$

$$= v_p\sqrt{1 - \left(\frac{f_c}{f}\right)^2} = v_p\sqrt{1 - \left(\frac{\lambda}{\lambda_c}\right)^2} < v_p \qquad (6\text{-}268)$$

Substituting for $\sqrt{1 - (\lambda/\lambda_c)^2}$ in terms of θ_i, we have

$$v_{gz} = v_p \sin\theta_i$$

Thus the group velocity is the component of v_p along the z axis. It is the distance between two constant z planes divided by the time taken by a point on the obliquely bouncing wavefront to pass from one plane to the other as shown in Fig. 6.65. This gives the physical interpretation for the group

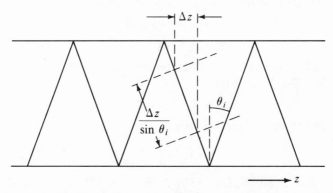

Fig. 6.65. For illustrating that the group velocity is the velocity with which energy propagates along the guide axis.

velocity as the velocity with which energy propagates along the guide axis. In fact, this physical interpretation is valid not only in this case but, in general, whenever a meaningful velocity can be attributed to the group.

Finally, we note a simple relationship between v_{pz}, v_{gz}, and v_p as

$$v_{pz}v_{gz} = v_p^2 \qquad (6\text{-}269)$$

Because the dispersion which we have discussed here is caused by the geometry associated with the bouncing of the waves between the walls, it is known as "geometric dispersion." There are other types of dispersion as we will learn in later sections. The relationship (6-267) also holds for these other types of dispersion since its derivation is independent of the mechanism causing the dispersion.

6.13 Waves in Imperfect Dielectrics and Conductors; Attenuation and the Skin Effect

Thus far we have been concerned with wave propagation in perfect dielectric media ($\sigma = 0$). In this section we will discuss wave propagation in lossy media, especially in good conductors. We restrict our discussion to sinusoidal steady state. For a medium characterized by conductivity σ, permittivity ϵ, and permeability μ, we recall that Maxwell's curl equations are given by

$$\mathbf{V} \times \mathbf{E} = -\frac{\partial \mathbf{B}}{\partial t} = -\mu \frac{\partial \mathbf{H}}{\partial t}$$

$$\mathbf{V} \times \mathbf{H} = \mathbf{J} + \frac{\partial \mathbf{D}}{\partial t} = \sigma \mathbf{E} + \epsilon \frac{\partial \mathbf{E}}{\partial t}$$

For sinusoidally time-varying fields, we have

$$\mathbf{V} \times \bar{\mathbf{E}} = -j\omega\mu\bar{\mathbf{H}} \tag{6-270a}$$

$$\mathbf{V} \times \bar{\mathbf{H}} = \sigma\bar{\mathbf{E}} + j\omega\epsilon\bar{\mathbf{E}} = (\sigma + j\omega\epsilon)\bar{\mathbf{E}} \tag{6-270b}$$

Taking the curl of (6-270a) on both sides and using the vector identity for $\mathbf{V} \times \mathbf{V} \times \bar{\mathbf{E}}$, we obtain

$$\mathbf{V}(\mathbf{V} \cdot \bar{\mathbf{E}}) - \mathbf{V}^2\bar{\mathbf{E}} = -j\omega\mu\mathbf{V} \times \bar{\mathbf{H}} \tag{6-271}$$

But from (6-270b), we have

$$\mathbf{V} \cdot \bar{\mathbf{E}} = \frac{1}{\sigma + j\omega\epsilon}\mathbf{V} \cdot \mathbf{V} \times \bar{\mathbf{H}} = 0 \tag{6-272}$$

Substituting (6-272) and (6-270b) into (6-271), we obtain the vector wave equation for the electric field as

$$\mathbf{V}^2\bar{\mathbf{E}} = j\omega\mu(\sigma + j\omega\epsilon)\bar{\mathbf{E}} \tag{6-273}$$

Defining a complex quantity $\bar{\gamma}$ as

$$\bar{\gamma}^2 = j\omega\mu(\sigma + j\omega\epsilon) \tag{6-274}$$

we write (6-273) as

$$\mathbf{V}^2\bar{\mathbf{E}} = \bar{\gamma}^2\bar{\mathbf{E}} \tag{6-275}$$

Assuming that the electric field has only an x component, which is dependent on the z coordinate only, that is,

$$\bar{\mathbf{E}} = \bar{E}_x(z)\mathbf{i}_x$$

Eq. (6-275) reduces to

$$\frac{\partial^2 \bar{E}_x}{\partial z^2} = \bar{\gamma}^2 \bar{E}_x \tag{6-276}$$

The solution for (6-276) is given by

$$\bar{E}_x(z) = \bar{A}e^{-\bar{\gamma}z} + \bar{B}e^{\bar{\gamma}z} \tag{6-277}$$

where $\bar{A}$ and $\bar{B}$ are arbitrary constants. Since $\bar{\gamma}$ is a complex number, we can write

$$\bar{\gamma} = \alpha + j\beta \tag{6-278}$$

where α and β are the real and imaginary parts of $\bar{\gamma}$, respectively. Substituting (6-278) into (6-277) and also writing $\bar{A} = Ae^{j\theta}$ and $\bar{B} = Be^{j\phi}$, we have

$$\bar{E}_x(z) = Ae^{-\alpha z}e^{-j\beta z}e^{j\theta} + Be^{\alpha z}e^{j\beta z}e^{j\phi} \tag{6-279}$$

and

$$\begin{aligned} E_x(z, t) &= \Re e[\bar{E}_x(z)e^{j\omega t}] \\ &= Ae^{-\alpha z}\cos(\omega t - \beta z + \theta) + Be^{\alpha z}\cos(\omega t + \beta z + \phi) \end{aligned} \tag{6-280}$$

Ignoring the factors $e^{-\alpha z}$ and $e^{\alpha z}$ on the right side of (6-280) for a moment, we note that the first and second terms represent the (+) and (−) waves, respectively. The factor $e^{-\alpha z}$ decreases in value as z increases, thereby resulting in attenuation of the (+) wave as it progresses in the positive z direction. Similarly, the factor $e^{\alpha z}$ decreases in value as z decreases, thereby resulting in the attenuation of the (−) wave as it progresses in the negative z direction. The factor α is therefore known as the "attenuation constant." The units of α are nepers per meter. The word "neper" is a variation of the spelling of the name Napier. The factor β is, of course, the "phase constant" associated with the traveling waves. Since α and β together characterize the propagation of the wave, the factor $\bar{\gamma}$ is known as the "propagation constant." Since we have identified the two terms on the right side of (6-277) as representing (+) and (−) waves, respectively, we can replace $\bar{A}$ and $\bar{B}$ by $\bar{E}_x^+$ and $\bar{E}_x^-$, respectively, and write

$$\bar{E}_x(z) = \bar{E}_x^+ e^{-\bar{\gamma}z} + \bar{E}_x^- e^{\bar{\gamma}z} \tag{6-281a}$$

The corresponding solution for $\bar{\mathbf{H}}$ contains a y component only which can be obtained by substituting (6-281a) into (6-270a). Thus

$$\bar{H}_y(z) = \frac{1}{\bar{\eta}}(\bar{E}_x^+ e^{-\gamma z} - \bar{E}_x^- e^{\gamma z}) \tag{6-281b}$$

where

$$\bar{\eta} = \frac{j\omega\mu}{\bar{\gamma}} = \sqrt{\frac{j\omega\mu}{\sigma + j\omega\epsilon}} \tag{6-282}$$

is the intrinsic impedance of the medium, which is now complex. Equations (6-281a) and (6-281b) together represent uniform plane-wave solution for the lossy medium since, in the planes of constant phase, the amplitudes of the fields are uniform although there is attenuation from one plane to another.

To obtain the expressions for α and β, we substitute (6-278) into (6-274) and equate the real and imaginary parts on both sides of the resulting equation. Thus we have

$$\alpha^2 - \beta^2 = -\omega^2 \mu\epsilon \quad \text{and} \quad 2\alpha\beta = \omega\mu\sigma$$

Solving these two equations for α, we get

$$\alpha^2 = \frac{\omega^2 \mu\epsilon}{2}\left(\pm\sqrt{1 + \frac{\sigma^2}{\omega^2\epsilon^2}} - 1\right)$$

The minus sign associated with the square root in the above equation makes α imaginary. Hence we ignore it to obtain

$$\alpha = \omega\left[\frac{\mu\epsilon}{2}\left(\sqrt{1 + \frac{\sigma^2}{\omega^2\epsilon^2}} - 1\right)\right]^{1/2} \tag{6-283}$$

and

$$\beta = \sqrt{\alpha^2 + \omega^2\mu\epsilon} = \omega\left[\frac{\mu\epsilon}{2}\left(\sqrt{1 + \frac{\sigma^2}{\omega^2\epsilon^2}} + 1\right)\right]^{1/2} \tag{6-284}$$

Note that if $\sigma = 0$, Eqs. (6-283), (6-284), and (6-282) give $\alpha = 0$, $\beta = \omega\sqrt{\mu\epsilon}$, and $\bar\eta = \sqrt{\mu/\epsilon}$, which correspond to a perfect dielectric medium. Since β given by (6-284) is not a linear function of ω, the wave propagation in the lossy medium is characterized by dispersion. This type of dispersion is known as "conductive" dispersion since it is due to the conductivity of the medium.

The expressions for α and β given by (6-283) and (6-284), respectively, are very complicated. They can, however, be reduced to simple expressions for two special cases. We now consider these two special cases:

(a) *Good dielectrics:* $\sigma \ll \omega\epsilon$; that is, conduction current is very small compared to displacement current. We can then write

$$\sqrt{1 + \left(\frac{\sigma}{\omega\epsilon}\right)^2} \approx 1 + \frac{\sigma^2}{2\omega^2\epsilon^2} - \frac{\sigma^4}{8\omega^4\epsilon^4}$$

The simplified expressions for α, β, and $\bar\eta$ are

$$\alpha \approx \omega\sqrt{\frac{\mu\epsilon}{2}\left(\frac{\sigma^2}{2\omega^2\epsilon^2} - \frac{\sigma^4}{8\omega^4\epsilon^4}\right)} \approx \frac{1}{2}\sigma\sqrt{\frac{\mu}{\epsilon}}\left(1 - \frac{\sigma^2}{8\omega^2\epsilon^2}\right) \tag{6-285a}$$

$$\beta \approx \omega\sqrt{\frac{\mu\epsilon}{2}\left(2 + \frac{\sigma^2}{2\omega^2\epsilon^2}\right)} \approx \omega\sqrt{\mu\epsilon}\left(1 + \frac{\sigma^2}{8\omega^2\epsilon^2}\right) \tag{6-285b}$$

$$\bar\eta = \sqrt{\frac{j\omega\mu}{j\omega\epsilon(1 + \sigma/j\omega\epsilon)}} = \sqrt{\frac{\mu}{\epsilon}}\left(1 - j\frac{\sigma}{\omega\epsilon}\right)^{-1/2}$$

$$\approx \sqrt{\frac{\mu}{\epsilon}}\left[\left(1 - \frac{3}{8}\frac{\sigma^2}{\omega^2\epsilon^2}\right) + j\frac{\sigma}{2\omega\epsilon}\right] \tag{6-285c}$$

where we have retained all terms up to and including the second power in $\sigma/\omega\epsilon$. Although the first-term approximation of the attenuation constant given by (6-285a) seems to be independent of frequency, σ and ϵ are, in general, functions of frequency as stated in Section 5.10. In fact, the quantity $\sigma/\omega\epsilon$ is very nearly constant for several dielectrics over wide frequency ranges.

(b) *Good conductors:* $\sigma \gg \omega\epsilon$; that is, conduction current is very large compared to displacement current. We can then write

$$\sqrt{1 + \left(\frac{\sigma}{\omega\epsilon}\right)^2} \approx \frac{\sigma}{\omega\epsilon}$$

The simplified expressions for α, β, and $\bar{\eta}$ are

$$\alpha \approx \omega\sqrt{\frac{\mu\epsilon}{2}\left(\frac{\sigma}{\omega\epsilon} - 1\right)} \approx \omega\sqrt{\frac{\mu\sigma}{2\omega}} = \sqrt{\frac{\omega\mu\sigma}{2}} = \sqrt{\pi f \mu\sigma} \qquad (6\text{-}286a)$$

$$\beta \approx \omega\sqrt{\frac{\mu\epsilon}{2}\left(\frac{\sigma}{\omega\epsilon} + 1\right)} \approx \omega\sqrt{\frac{\mu\sigma}{2\omega}} = \sqrt{\frac{\omega\mu\sigma}{2}} = \sqrt{\pi f \mu\sigma} \qquad (6\text{-}286b)$$

$$\bar{\eta} \approx \sqrt{\frac{j\omega\mu}{\sigma}} = (1 + j)\sqrt{\frac{\omega\mu}{2\sigma}} = (1 + j)\sqrt{\frac{\pi f \mu}{\sigma}} \qquad (6\text{-}286c)$$

EXAMPLE 6-30. For uniform plane waves in sea water ($\sigma = 4$ mhos/m, $\epsilon = 80\epsilon_0$, $\mu = \mu_0$), find α, β, $\bar{\eta}$, and λ for two frequencies: (a) 10,000 MHz and (b) 25 kHz.

The frequency at which $\sigma = \omega\epsilon$ is equal to $4/(2\pi \times 80 \times 10^{-9}/36\pi)$ or 900 MHz. Hence, for 10,000 MHz, $\sigma \ll \omega\epsilon$, sea water is a good dielectric and for 25 kHz, $\sigma \gg \omega\epsilon$, sea water is a good conductor.

Thus, for 10,000 MHz, we have

$$\alpha \approx \frac{1}{2}\sigma\sqrt{\frac{\mu}{\epsilon}}\left(1 - \frac{\sigma^2}{8\omega^2\epsilon^2}\right) \approx \frac{1}{2}\sigma\sqrt{\frac{\mu}{\epsilon}} = \frac{1}{2} \times 4 \times \sqrt{\frac{\mu_0}{80\epsilon_0}} = 2 \times \frac{377}{\sqrt{80}}$$

$$= 84.3 \text{ nepers/m}$$

$$\beta \approx \omega\sqrt{\mu\epsilon}\left(1 + \frac{\sigma^2}{8\omega^2\epsilon^2}\right) \approx \omega\sqrt{\mu\epsilon} = \frac{2\pi \times 10 \times 10^9 \times \sqrt{80}}{3 \times 10^8}$$

$$= 1873 \text{ rad/m}$$

$$\bar{\eta} \approx \sqrt{\frac{\mu}{\epsilon}}\left[\left(1 - \frac{3}{8}\frac{\sigma^2}{\omega^2\epsilon^2}\right) + j\frac{\sigma}{2\omega\epsilon}\right] \approx \sqrt{\frac{\mu}{\epsilon}} = \frac{377}{\sqrt{80}} = 42.15 \text{ ohms}$$

$$\lambda = \frac{2\pi}{\beta} = \frac{2\pi}{1873} = 3.353 \times 10^{-3} \text{ m} = 3.353 \text{ mm}$$

as compared to 30 mm in free space.
For 25 kHz, we have

$$\alpha \approx \sqrt{\pi f \mu\sigma} = \sqrt{\pi \times 25 \times 10^3 \times 4\pi \times 10^{-7} \times 4} = 0.2\pi \text{ nepers/m}$$

$$\beta \approx \sqrt{\pi f \mu\sigma} = \alpha = 0.2\pi \text{ rad/m}$$

$$\bar{\eta} \approx (1 + j)\sqrt{\frac{\pi f \mu}{\sigma}} = (1 + j)\sqrt{\frac{\pi \times 25 \times 10^3 \times 4\pi \times 10^{-7}}{4}}$$

$$= 0.05\pi(1 + j) \text{ ohms}$$

$$\lambda = \frac{2\pi}{\beta} = \frac{2\pi}{0.2\pi} = 10 \text{ m}$$

as compared to 12 km in free space.

From these values, we conclude that low-frequency waves are more suitable for communication with underwater objects. We should, however, note that since the wavelength in free space is large for low frequencies, long antennas are required in air. ▊

Let us now consider a uniform plane wave incident normally on a semiinfinite plane slab of good conductor as shown in Fig. 6.66. Since the

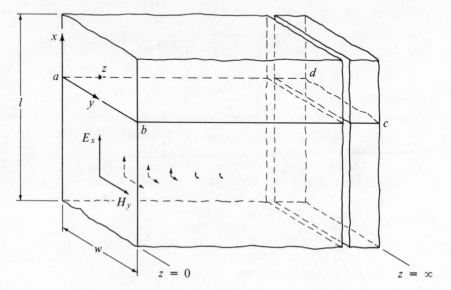

Fig. 6.66. Normal incidence of a uniform plane wave on a semiinfinite plane slab of good conductor, for illustrating the concept of skin depth.

conductor is of infinite depth, only a $(+)$ wave can exist inside the conductor. The fields inside the conductor are therefore given by

$$\bar{E}_x(z) = \bar{E}_{x_0} e^{-\bar{\gamma} z} = \bar{E}_{x_0} e^{-\sqrt{\pi f \mu \sigma} z} e^{-j\sqrt{\pi f \mu \sigma} z} \tag{6-287a}$$

$$\bar{H}_y(z) = \frac{\bar{E}_{x_0}}{\bar{\eta}} e^{-\bar{\gamma} z} = \frac{\bar{E}_{x_0}}{\bar{\eta}} e^{-\sqrt{\pi f \mu \sigma} z} e^{-j\sqrt{\pi f \mu \sigma} z} \tag{6-287b}$$

where $\bar{E}_{x_0}$ is the value of $\bar{E}_x$ at the surface $z = 0$ of the conductor. To obtain an idea of how rapidly the wave is attenuated, we use the concept of "skin depth" or depth of penetration applied to plane conductors. The skin depth denoted by the symbol δ is defined as the depth or distance from the surface of the plane conductor at which the magnitude of the field is e^{-1} times its value at the surface. From (6-287a), we note that

$$\delta = \frac{1}{\sqrt{\pi f \mu \sigma}} \tag{6-288}$$

Thus the skin depth is inversely proportional to σ, μ, and f. For copper, the conductivity is 5.8×10^7 mhos/m so that the skin depth is given by

$$\delta_{\text{copper}} = \frac{1}{\sqrt{\pi f \times 4\pi \times 10^{-7} \times 5.8 \times 10^7}} = \frac{0.066}{\sqrt{f}} \text{ m} \qquad (6\text{-}289)$$

For a frequency of 10^6 Hz, the skin depth of copper is 0.066 mm. Thus the fields are attenuated to e^{-1} times their values at the surface in a distance of 0.066 mm even at the low frequency of 1 MHz. In a distance of one wavelength in the conductor, the attenuation is equal to $e^{\alpha\lambda} = e^{\alpha(2\pi/\beta)} = e^{2\pi}$ nepers/m since β is equal to α. In terms of decibels, this attenuation is $20 \log_{10} e^{2\pi} = 20 \log_e e^{2\pi}/\log_e 10 = 40\pi/2.3026 = 54.5$. In view of the rapid attenuation of the fields inside the conductor, the fields and the current given by density $\mathbf{J} = \sigma\mathbf{E}$ are concentrated close to the surface of the conductor. This phenomenon is known as the "skin effect."

Because of the skin effect, a conductor of finite thickness equal to a few skin depths can for all practical purposes be considered as a conductor of infinite depth. Hence, if a wave is incident upon it from one side, its effect is not felt on the other side, thereby "shielding" one side of the conductor from the other. Furthermore, since there is no reflected wave inside the conductor, we can compute the power flow into the conductor by surface integration of the Poynting vector corresponding to $\bar{E}_x$ and $\bar{H}_y$ given by (6-287a) and (6-287b). Thus, noting that

$$\bar{E}_x(0) = \bar{E}_{x_0} \qquad \text{and} \qquad \bar{H}_y(0) = \frac{\bar{E}_{x_0}}{\bar{\eta}}$$

we obtain the complex Poynting vector at the conductor surface as

$$\bar{\mathbf{P}} = \frac{1}{2}[\bar{E}_x(0)\mathbf{i}_x \times \bar{H}_y^*(0)\mathbf{i}_y] = \frac{1}{2}\bar{E}_{x_0}\frac{\bar{E}_{x_0}^*}{\bar{\eta}^*}\mathbf{i}_z$$

$$= \frac{1}{2}\frac{|\bar{E}_{x_0}|^2}{\bar{\eta}^*}\mathbf{i}_z = \frac{1}{2}\frac{|\bar{H}_y(0)|^2\,|\bar{\eta}|^2}{\bar{\eta}^*}\mathbf{i}_z$$

$$= \frac{1}{2}\bar{\eta}\,|\bar{H}_y(0)|^2\mathbf{i}_z$$

For a surface S of length l in the x direction and width w in the y direction, the complex power flow into the conductor is

$$\bar{P}_{\text{in}} = \int_S \bar{\mathbf{P}} \cdot d\mathbf{S} = \bar{P}_z(lw) = \frac{1}{2}lw\bar{\eta}\,|\bar{H}_y(0)|^2 \qquad (6\text{-}290)$$

However, applying Maxwell's curl equation for $\bar{\mathbf{H}}$ in integral form to the closed path *abcda* shown in Fig. 6.66, we have

$$\oint_{abcda} \bar{\mathbf{H}} \cdot d\mathbf{l} = \int_{\substack{\text{area} \\ abcd}} (\sigma\bar{\mathbf{E}} + j\omega\epsilon\bar{\mathbf{E}}) \cdot d\mathbf{S} \qquad (6\text{-}291)$$

The left side of (6-291) has a contribution from *ab* only, since along *bc* and *da*, $\bar{\mathbf{H}}$ is perpendicular to the path and along *cd*, $\bar{\mathbf{H}}$ is zero. Along *ab*, $\bar{\mathbf{H}} = \bar{H}_y(0)\mathbf{i}_y$ so that the integral is $w\bar{H}_y(0)$. On the right side of (6-291), the second term in the integrand can be ignored since $\sigma \gg \omega\epsilon$. Hence the integral is simply the conduction current flowing in the conductor. Denoting this by $\bar{I}_x$, we have

$$w\bar{H}_y(0) = \bar{I}_x \qquad \text{or} \qquad \bar{H}_y(0) = \frac{\bar{I}_x}{w} \tag{6-292}$$

Substituting (6-292) into (6-290), we get

$$\bar{P}_{\text{in}} = \frac{1}{2}\bar{\eta}\frac{l}{w}|\bar{I}_x|^2 \tag{6-293}$$

Substituting for $\bar{\eta}$ in (6-293) from (6-286c), we have

$$\bar{P}_{\text{in}} = \frac{1}{2}(1+j)\sqrt{\frac{\pi f\mu}{\sigma}}\frac{l}{w}|\bar{I}_x|^2$$

$$= \frac{1}{2}(1+j)\frac{l}{\sigma\delta w}|\bar{I}_x|^2 \tag{6-294}$$

$$= \frac{1}{2}\frac{l}{\sigma\delta w}|\bar{I}_x|^2 + j\frac{1}{2}\frac{l}{\sigma\delta w}|\bar{I}_x|^2$$

From (5-206), the real part on the right side of (6-294) is the time-average power dissipated in the conductor. It is also exactly the result that would be obtained by computing the time-average power dissipated under quasistatic conditions in a conductor of length l, width w, thickness δ, and conductivity σ if the current $\bar{I}_x$ were distributed uniformly over the cross section of the conductor. This gives an alternative significance for the skin depth δ. We will denote the resistance $l/\sigma\delta w$ by the symbol R_s. From (5-206), the imaginary part on the right side of (6-294) is 2ω times the time-average magnetic stored energy in the conductor since the time-average electric stored energy is negligible in view of $\sigma \gg \omega\epsilon$. In fact, a volume integration of $\frac{1}{4}\mu|\bar{H}_y|^2$ gives exactly the imaginary part of the right side of (6-294) divided by 2ω, that is,

$$\frac{1}{4}\frac{l}{\omega\sigma\delta w}|\bar{I}_x|^2$$

This energy is the same as the time-average magnetic energy stored under quasistatic conditions in an inductor of value $l/\omega\sigma\delta w$ if the current $|\bar{I}_x|$ were flowing in it. This inductance is the internal inductance of the conductor which we denote as L_i. Thus the impedance offered by a portion of the conductor of length l and width w to the current flowing in it is given by

$$\bar{Z}_i = R_s + j\omega L_i = \frac{l}{\sigma\delta w} + j\frac{l}{\sigma\delta w} \tag{6-295}$$

This impedance is known as the "internal impedance." We may emphasize that the formulas for skin depth and internal impedance developed here are strictly valid for plane conductors only. However, if the radius of a cylindrical conductor is very large compared to the skin depth for the material of the conductor, these formulas can be used with negligible error.

EXAMPLE 6-31.   Figure 6.67 shows the cross section of a hollow cylindrical conductor of radius a and thickness $d \ll a$, in which current flows axially. It is desired to find the approximate expression for the internal impedance of the conductor per unit length in the axial direction if the skin depth δ for the material is $\ll d$.

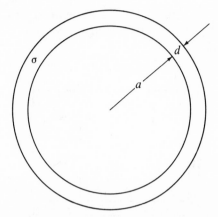

Fig. 6.67. Cross section of a hollow cylindrical conductor of thickness small compared to its radius.

   Since d is $\ll a$, we can assume that the required internal impedance is approximately equal to the internal impedance of a plane conductor of appropriate width. If d is not $\ll a$, we cannot use this approximation and the problem must be solved in cylindrical coordinates. If δ is $\ll d$, it is actually immaterial whether the conductor is hollow or not since the current does not penetrate much below the surface and hence the depth can be assumed to be infinity for the purpose of computing the internal impedance. Thus the required internal impedance is approximately the same as the internal impedance of a plane conductor of infinite depth and width equal to $2\pi a$. From (6-295), this is equal to $(1 + j)/2\pi a\sigma\delta$ per unit length in the axial direction. ∎

   In Sections 6.8, 6.9, and 6.10, we considered transmission-line waves between perfect conductors with the medium between them as a perfect dielectric. These waves are exactly TEM since the perfect conductors ($\sigma = \infty$) do not require any axial electric field to maintain a current flow along them. If the dielectric is now made imperfect, the waves are still exactly TEM except that attenuation takes place as they propagate down the line. In fact, the transmission-line equivalent circuit in Section 6.7 was derived by considering the dielectric to be imperfect. On the other hand, if the conductors are imperfect, the finite conductivity requires an axial electric field for the current to flow along the conductors. This axial electric field in the conductors is accompanied by an axial electric field in the dielectric since the boundary condition at the interface between the dielectric and the conductor requires that the tangential electric field be continuous. Thus the electric field between the conductors is no longer entirely transverse and hence the waves are no longer exactly TEM waves. However, if the conductors are good conductors, as is the case in practice, the axial electric field is very small compared to the transverse electric field and the waves between the conductors are almost TEM waves. In the conductors, the axial component of the electric field dominates so that power flow is almost normal to the dielectric-conductor interface. The situation as compared to the perfect conductor case is illustrated in Fig. 6.68. Thus, as the wave propagates, it gets attenuated

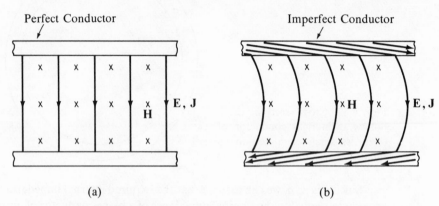

Perfect Conductor   Imperfect Conductor

(a)   (b)

Fig. 6.68. Fields for a transmission line employing (a) perfect conductors, and (b) imperfect conductors.

partly due to power dissipation in the lossy dielectric and partly due to energy leakage into the conductors which is dissipated in the conductors. The power dissipation in the lossy dielectric is accounted for in the distributed equivalent circuit by the conductance in parallel with the capacitor. The power dissipation in the conductors can be accounted for by introducing into the series branch an impedance which is offered by the conductors to the current flow. Since the current flow is almost parallel to the conductor surface, this impedance is approximately the same as the internal impedance given by (6-295) per unit length. Thus we obtain the distributed equivalent circuit for a lossy transmission line as shown in Fig. 6.69, where the factor 2 takes into account the two conductors and

$\mathcal{R}_s$ = resistance per unit length of the conductor due to skin effect,
$\mathcal{L}_i$ = internal inductance per unit length of the conductor due to skin effect,
$\mathcal{G}, \mathcal{L}, \mathcal{C}$ = conductance, inductance, and capacitance per unit length if the conductors were perfect.

The circuit of Fig. 6.69 forms the basis for lossy transmission-line theory which follows along lines similar to lossless transmission-line theory but is characterized by attenuation and dispersion.

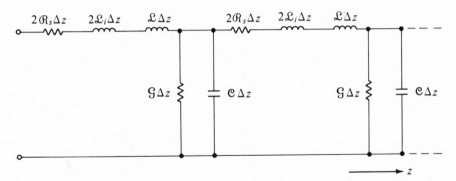

Fig. 6.69. Distributed equivalent circuit for a lossy transmission line.

The concept of internal impedance is useful not only for a lossy transmission line but for any system involving waves between imperfect but good conductors, since it permits the estimation of the power loss in the conductors from the solutions for the fields in the corresponding lossless case. This is because, for good conductors, it is reasonable to assume that the fields between the conductors differ very little from the lossless case so that the current flowing in the conductors can be obtained from the tangential magnetic fields. Then, since these currents flow very nearly parallel to the conductor surface, power loss can be computed by using $\frac{1}{2}|\bar{I}|^2 R_s$. We will use this technique in the following section for deriving the Q factor for a parallel-plate resonator employing imperfect but good conductors.

6.14   Resonators; Laser Oscillation

In Section 6.10 we discussed complete standing waves resulting from the superposition of $(+)$ and $(-)$ waves of equal magnitudes. For a short-circuited line (or a semiinfinite dielectric medium terminated by a perfect conductor), we found that the line voltage (or the electric field) is zero at distances of integral multiples of $\lambda/2$ from the short circuit (or the perfect conductor). Hence, if we short circuit the line (or place a perfect conductor) at these points, there will be no effect on the voltage and current (or fields) at any other point. Alternatively, if we have a line of length l which is short circuited at both ends (or a dielectric medium between two parallel, perfectly conducting plates) and containing some stored energy, this energy must exist in the form of complete standing waves having wavelengths such that $l = n\lambda_n/2$, that is, $\lambda_n = 2l/n$, or $\beta_n = 2\pi/\lambda_n = n\pi/l$, where $n = 1, 2, 3, \ldots$ as discussed in Example 6-21. The corresponding frequencies are given by $\omega_n = n\pi v_p/l$.

Thus let us consider a system of two infinite, parallel, perfectly conducting plates as shown in Fig. 6.70, between which the medium is a perfect

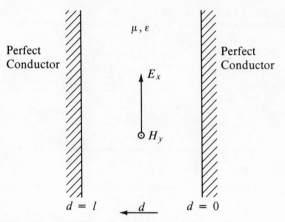

Fig. 6.70. Parallel-plate resonator consisting of two infinite-plane, perfectly conducting plates.

dielectric and energy is stored in the form of standing waves having field components E_x and H_y. From (6-231a) and (6-231b), the expressions for these fields that satisfy the boundary condition at $d = 0$ can be written as

$$E_x(d, t) = -2E_0 \sin \beta d \sin \omega t \qquad (6\text{-}296a)$$

$$H_y(d, t) = 2\frac{E_0}{\eta} \cos \beta d \cos \omega t \qquad (6\text{-}296b)$$

where E_0 is a constant, the value of which we need not know, and $\eta = \sqrt{\mu/\epsilon}$. Substituting $\beta = n\pi/l$ and $\omega = n\pi v_p/l = n\pi/l\sqrt{\mu\epsilon}$ in (6-296a) and (6-296b), we obtain

$$E_x(d, t) = -2E_0 \sin \frac{n\pi d}{l} \sin \frac{n\pi t}{l\sqrt{\mu\epsilon}} \qquad (6\text{-}297a)$$

$$H_y(d, t) = 2\sqrt{\frac{\epsilon}{\mu}} E_0 \cos \frac{n\pi d}{l} \cos \frac{n\pi t}{l\sqrt{\mu\epsilon}} \qquad (6\text{-}297b)$$

which satisfy the boundary condition at $d = l$ for all t. The instantaneous electric and magnetic stored energy densities associated with these fields are

$$w_e(d, t) = \frac{1}{2}\epsilon E_x^2 = 2\epsilon E_0^2 \sin^2 \frac{n\pi d}{l} \sin^2 \frac{n\pi t}{l\sqrt{\mu\epsilon}} \qquad (6\text{-}298a)$$

$$w_m(d, t) = \frac{1}{2}\mu H_y^2 = 2\epsilon E_0^2 \cos^2 \frac{n\pi d}{l} \cos^2 \frac{n\pi t}{l\sqrt{\mu\epsilon}} \qquad (6\text{-}298b)$$

Let us for simplicity consider the case $n = 1$, that is, for which the standing waves have one-half wavelength between the plates, and sketch the energy densities as functions of d for different values of t, as shown in Fig. 6.71. We note from Fig. 6.71 and from Eqs. (6-298a) and (6-298b) that the

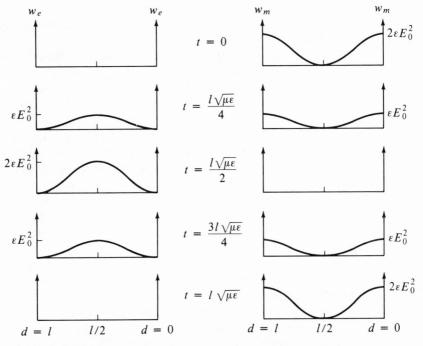

Fig. 6.71. Electric and magnetic energy densities versus d for various values of t for the parallel-plate resonator of Fig. 6.70.

stored energy density at all points is entirely magnetic at certain times ($t = 0$, $l\sqrt{\mu\epsilon}$, . . .) and entirely electric at certain other times ($t = l\sqrt{\mu\epsilon}/2$, $3l\sqrt{\mu\epsilon}/2$, . . .) with the bulk of the magnetic energy stored close to the conducting plates and the bulk of the electric energy stored close to half way between the plates for all times. The total energy density in the two fields from $d = 0$ to $d = l$ must be constant with respect to time. To show that this is indeed true, we write

$$
\begin{aligned}
w(t) &= \int_{d=0}^{l} w_e(d, t)\, dd + \int_{d=0}^{l} w_m(d, t)\, dd \\
&= \int_{d=0}^{l} 2\epsilon E_0^2 \sin^2 \frac{\pi d}{l} \sin^2 \frac{\pi t}{l\sqrt{\mu\epsilon}}\, dd \\
&\quad + \int_{d=0}^{l} 2\epsilon E_0^2 \cos^2 \frac{\pi d}{l} \cos^2 \frac{\pi t}{l\sqrt{\mu\epsilon}}\, dd \\
&= 2\epsilon E_0^2 \frac{l}{2} \left(\sin^2 \frac{\pi t}{l\sqrt{\mu\epsilon}} + \cos^2 \frac{\pi t}{l\sqrt{\mu\epsilon}} \right) = \epsilon_0 E_0^2 l
\end{aligned}
\tag{6-299}
$$

The same result holds for any value of n. This process of exchange of energy stored between the plates from one field to the other is the phenomenon of resonance. The parallel-plate structure itself is known as a resonator, the distributed counterpart of a lumped parameter resonant circuit. The frequencies $f_n = n/2l\sqrt{\mu\epsilon}$ are the resonant frequencies or the natural frequencies of oscillation of the parallel-plate resonator.

The same concept can be extended to waveguides, discussed in Section 6.12. For example, by superimposing two $TE_{m,0}$ waves of equal amplitudes propagating in positive and negative z directions in a parallel-plate waveguide, we can obtain complete standing $TE_{m,0}$ waves in the guide, having nodes (zeros) of E_y at intervals of integer multiples of $\lambda_g/2$ in z. By placing perfect conductors in these planes, we do not alter the fields in any other plane. Conversely, by placing perfect conductors in two transverse planes of a parallel-plate waveguide, separated by a distance d, we create a resonator which supports standing waves of guide wavelengths $\lambda_{gn} = 2d/l$, where $l = 1, 2, 3, \ldots$. The corresponding modes are designated as $TE_{m,0,l}$ modes where l stands for the number of half-wavelengths in the z direction. Proceeding in this manner to rectangular waveguides leads to resonators which are enclosed by perfect conductors on all sides. These are known as cavity resonators although the term "cavity" is also used for partially enclosed resonators. We will, however, not pursue these ideas any further, but consider the effect of conductor losses.

If the conductors of a resonator are imperfect, some of the energy is dissipated in them as it oscillates from one field to the other. We then associate a Q or "quality factor" to the resonator. The quality factor is defined as

$$Q = 2\pi \frac{\text{energy stored}}{\text{energy dissipated per cycle}}$$

$$= 2\pi \frac{\text{energy stored}}{\text{energy dissipated per second/number of cycles per second}} \qquad (6\text{-}300)$$

$$= 2\pi f \frac{\text{energy stored}}{\text{time-average power dissipated}}$$

If the conductors are good conductors, the losses are small. The stored energy and power dissipated are then computed by assuming that the fields in the resonator are the same as in the lossless case, that is, perfect conductor case. We will use this technique to find the Q of a parallel-plate resonator in the following example.

EXAMPLE 6-32.   For the parallel-plate resonator of Fig. 6.70, it is desired to find the Q, assuming that the plates are made up of imperfect conductors of conductivity σ and having thickness of several skin depths for the frequencies of interest.

From (6-299), the energy stored in the resonator per unit area of the plates is given by

$$w \approx \epsilon E_0^2 l \qquad (6\text{-}301)$$

To find the time-average power dissipated, we note that the current flowing in the conductor per unit width in the y direction is equal to the tangential magnetic field at the surface of the conductor in accordance with (6-292), since the thickness of the conductor is several skin depths. Thus, for the conductor at $d = 0$,

$$I_x = H_y(0) = 2\sqrt{\frac{\epsilon}{\mu}}\, E_0 \cos\frac{n\pi}{l\sqrt{\mu\epsilon}}t$$

or

$$\bar{I}_x = 2\sqrt{\frac{\epsilon}{\mu}}\, E_0$$

From (6-295), the resistance offered by the conductor per unit length in the x direction and unit width in the y direction is

$$R_s = \frac{1}{\sigma\delta} = \sqrt{\frac{\pi f \mu}{\sigma}}$$

where δ is the skin depth for the frequency of interest. Thus the time-average power dissipated in the conductor per unit surface area is

$$P_d = \frac{1}{2}|\bar{I}_x|^2 R_s = 2\frac{\epsilon}{\mu} E_0^2 \sqrt{\frac{\pi f \mu}{\sigma}} = 2\epsilon E_0^2 \sqrt{\frac{\pi f}{\mu\sigma}} \qquad (6\text{-}302)$$

Similarly, the time-average power dissipated in the conductor at $d = l$ can be found to be the same as given by (6-302). Thus, from (6-300), (6-301), and (6-302), we have

$$Q = 2\pi f \frac{\epsilon E_0^2 l}{4\epsilon E_0^2 \sqrt{\pi f/\mu\sigma}} = \frac{l}{2}\sqrt{\pi f \mu\sigma} = \frac{l}{2\delta} \qquad (6\text{-}303)$$

As a numerical example, we note that, for $l = 1$ cm and free space between the plates, the wavelength corresponding to the fundamental frequency of oscillation, that is, for $n = 1$, is 2 cm and hence the frequency is 15,000 MHz. For plates made of copper, the skin depth at 15,000 MHz is $0.066/(\sqrt{15 \times 10^9})$ m or 5.38×10^{-5} cm. Hence, from (6-303), the value of Q is $1/(2 \times 5.38 \times 10^{-5})$ or 9280, which is very large compared to values encountered in circuit theory. It is left as an exercise (Problem 6.82) for the student to show that for a particular mode of operation, that is, for a fixed value of n, Q is inversely proportional to $\sqrt{f}$. The above formula for Q takes into account only the losses in the conductors. In practice, there are other losses, for example, losses in the dielectric and losses due to radiation. ∎

The bouncing of $(+)$ and $(-)$ waves between two parallel plates which results in resonance as we discussed for the parallel-plate resonator is employed at optical frequencies in the Fabry–Perot resonator for laser amplification and oscillation. The Fabry–Perot resonator consists of two plane reflecting surfaces between which is an optically active medium characterized by a propagation constant $\bar{\gamma} = \alpha + j\beta$, where α is negative. To determine the condition for oscillation, let us consider a normally incident uniform plane wave passing through the surface $z = 0$ and setting up incident and reflected waves in the active medium as shown in Fig. 6.72. The steady-state situation in the medium can be thought of as a superposition of an

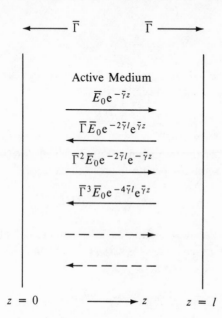

Fig. **6.72.** Bouncing of $(+)$ and $(-)$ waves in an active medium between two parallel plates.

infinite number of $(+)$ and $(-)$ waves due to reflections and rereflections at the plates $z = 0$ and $z = l$. Thus, denoting the electric field in the initial $(+)$ wave in the medium (i.e., the wave which would exist if the medium extended to $z = \infty$) to be $\bar{E}_0 e^{-\bar{\gamma}z}$, we obtain the field at $z = l$ in the reflected or $(-)$ wave due to it as $\bar{\Gamma}\bar{E}_0 e^{-\bar{\gamma}l}$, where $\bar{\Gamma}$ is the reflection coefficient at $z = l$. Since this $(-)$ wave is propagating towards $z = 0$, its field at any value of z is $\bar{\Gamma}\bar{E}_0 e^{-\bar{\gamma}l}e^{-\bar{\gamma}(l-z)}$ or $\bar{\Gamma}\bar{E}_0 e^{-2\bar{\gamma}l}e^{\bar{\gamma}z}$. Thus the $(-)$ wave field at $z = 0$ is $\bar{\Gamma}\bar{E}_0 e^{-2\bar{\gamma}l}$. Then the field at $z = 0$ in the rereflected or $(-+)$ wave due to the reflection of the $(-)$ wave at $z = 0$ is $(\bar{\Gamma})(\bar{\Gamma}\bar{E}_0 e^{-2\bar{\gamma}l})$, where we assume that the reflecting surfaces are identical and hence the reflection coefficient for the $(-)$ wave at $z = 0$ is the same as the reflection coefficient for the $(+)$ wave at $z = l$. Since the $(-+)$ wave is propagating towards $z = l$, its field at any value of z is $\bar{\Gamma}^2\bar{E}_0 e^{-2\bar{\gamma}l}e^{-\bar{\gamma}z}$. We can continue in this manner to obtain an infinite number of $(+)$ and $(-)$ waves in the active medium as shown in Fig. 6.72. The total field in the medium is the superposition of the fields in all these waves. Thus it is given by

$$
\begin{aligned}
\bar{E}(z) &= \bar{E}_0 e^{-\bar{\gamma}z} + \bar{\Gamma}\bar{E}_0 e^{-2\bar{\gamma}l}e^{\bar{\gamma}z} \\
&\quad + \bar{\Gamma}^2\bar{E}_0 e^{-2\bar{\gamma}l}e^{-\bar{\gamma}z} + \bar{\Gamma}^3\bar{E}_0 e^{-4\bar{\gamma}l}e^{\bar{\gamma}z} \\
&\quad + \bar{\Gamma}^4\bar{E}_0 e^{-4\bar{\gamma}l}e^{-\bar{\gamma}z} + \cdots \\
&= \bar{E}_0[e^{-\bar{\gamma}z}(1 + \bar{\Gamma}^2 e^{-2\bar{\gamma}l} + \bar{\Gamma}^4 e^{-4\bar{\gamma}l} + \cdots) \\
&\quad + \bar{\Gamma}e^{\bar{\gamma}z}e^{-2\bar{\gamma}l}(1 + \bar{\Gamma}^2 e^{-2\bar{\gamma}l} + \cdots)] \\
&= \bar{E}_0 \frac{e^{-\bar{\gamma}z} + \bar{\Gamma}e^{-2\bar{\gamma}l}e^{\bar{\gamma}z}}{1 - \bar{\Gamma}^2 e^{-2\bar{\gamma}l}}
\end{aligned} \tag{6-304}
$$

From (6-304), we note that the condition for oscillation, that is, for a field to be set up in the medium for zero $\bar{E}_0$, is

$$ 1 - \bar{\Gamma}^2 e^{-2\bar{\gamma}l} = 0 $$

or

$$ \bar{\Gamma}e^{-\bar{\gamma}l} = \pm 1 \tag{6-305} $$

Denoting $\bar{\Gamma} = |\bar{\Gamma}|e^{j\theta}$ and substituting for $\bar{\gamma}$ in terms of α and β, we write (6-305) as

$$ |\bar{\Gamma}|e^{j\theta}e^{-\alpha l}e^{-j\beta l} = 1e^{\pm jn\pi} \qquad n = 0, 1, 2, 3, \ldots $$

or

$$ |\bar{\Gamma}|e^{-\alpha l} = 1 \quad \text{and} \quad \theta - \beta l = \pm n\pi $$

$$ \alpha = \frac{1}{l}\ln|\bar{\Gamma}| \quad \text{and} \quad \beta l = \theta \pm n\pi, \; n = 0, 1, 2, 3, \ldots \tag{6-306} $$

where we choose only those values of n for which βl is greater than zero. While the condition $\beta l = \theta + n\pi$ can be satisfied for several frequencies for a given l, the condition $\alpha = (1/l)\ln|\bar{\Gamma}|$ is satisfied by a particular active medium only for a narrow range of frequencies, so that oscillation occurs only in that narrow range of frequencies. Note that for $\bar{\Gamma} = -1$ as is the case for perfectly conducting plates, the condition for oscillation is

$$ \alpha = \frac{1}{l}\ln 1 = 0 \quad \text{and} \quad \beta l = n\pi, \; n = 1, 2, 3, \ldots $$

which agrees with the result for the parallel-plate resonator.

6.15   Waves in Plasma; Ionospheric Propagation

Thus far we have discussed wave propagation in free space and perfect dielectrics and then in lossy dielectrics and good conductors. In free space and perfect dielectrics, the conduction current is zero so that the current is entirely of the displacement type. In lossy dielectrics, we have both conduction and displacement currents but the conduction current is small compared to the displacement current. In good conductors, the displacement current is negligible compared to the conduction current. In this section we will discuss wave propagation in plasma. Plasma is a gaseous medium in which the atoms are ionized to produce positive ions and electrons, which are free to move under the influence of the electric and magnetic fields of a wave incident upon the medium. The positive ions are, however, heavy compared to electrons so that they are relatively immobile. The electron motion produces a current which influences the wave propagation. This current is different from the conduction current in metallic conductors, which is due to electron drift with an average velocity owing to the frictional mechanism provided by their collisions with the atomic lattice. The electrons in the plasma, on the other hand, are accelerated by the electric field although losing some of the energy due to their collisions with the heavy particles and other electrons. We will, however, neglect the effect of these collisions as well as the influence on the motion of an electron by the neighboring electrons. In addition, since the magnetic field of the incident wave has negligible influence on the electron motion, its effect will be ignored.

Thus the equation of motion of an electron is given by

$$\frac{d}{dt}(m\mathbf{v}) = e\mathbf{E} \tag{6-307}$$

where e and m are the charge and mass of the electron, $\mathbf{v}$ is its velocity, and $\mathbf{E}$ is the electric field of the wave. If N is the number density of the electrons in the plasma, the current density resulting from their motion is given by

$$\mathbf{J} = Ne\mathbf{v} \tag{6-308}$$

Combining (6-307) and (6-308), we get

$$\frac{\partial \mathbf{J}}{\partial t} = Ne\frac{d\mathbf{v}}{dt} = \frac{Ne^2}{m}\mathbf{E} \tag{6-309}$$

For sinusoidally time-varying fields of radian frequency ω, we have

$$j\omega\bar{\mathbf{J}} = \frac{Ne^2}{m}\bar{\mathbf{E}}$$

or

$$\bar{\mathbf{J}} = -j\frac{Ne^2}{m\omega}\bar{\mathbf{E}} \tag{6-310}$$

Equation (6-310) gives the expression for the current density which we have to use for $\bar{\mathbf{J}}$ in Maxwell's equation for $\nabla \times \bar{\mathbf{H}}$ to discuss wave propagation in plasma. Thus we have

$$\nabla \times \bar{\mathbf{E}} = -j\omega\mu_0\bar{\mathbf{H}} \tag{6-311a}$$

$$\nabla \times \bar{\mathbf{H}} = \bar{\mathbf{J}} + j\omega\epsilon_0\bar{\mathbf{E}}$$

$$= -j\frac{Ne^2}{m\omega}\bar{\mathbf{E}} + j\omega\epsilon_0\bar{\mathbf{E}} \tag{6-311b}$$

$$= j\omega\epsilon_0\left(1 - \frac{Ne^2}{m\omega^2\epsilon_0}\right)\bar{\mathbf{E}}$$

Since the free electrons and heavy positive particles are distributed with statistical uniformity in the ionized region, the net space charge is zero so that

$$\mathbf{V} \cdot \bar{\mathbf{E}} = \frac{\bar{\rho}}{\epsilon_0} = 0 \tag{6-312}$$

Taking the curl of both sides of (6-311a) and making use of (6-311b) and (6-312), we obtain

$$\nabla^2 \bar{\mathbf{E}} = -\omega^2 \mu_0 \epsilon_0 \left(1 - \frac{Ne^2}{m\omega^2 \epsilon_0}\right) \bar{\mathbf{E}} \tag{6-313}$$

Equation (6-313) is the wave equation for a plasma medium. Comparing it with (6-173), we note that it is similar to the wave equation for a perfect dielectric medium with the permittivity ϵ replaced by $\epsilon_0(1 - Ne^2/m\omega^2\epsilon_0)$. We may therefore call the quantity $\epsilon_0(1 - Ne^2/m\omega^2\epsilon_0)$ the effective permittivity of a plasma medium.

We now define a quantity known as the plasma frequency, f_N, as

$$f_N = \frac{1}{2\pi}\sqrt{\frac{Ne^2}{m\epsilon_0}} = \sqrt{80.6N} \tag{6-314}$$

where f_N is in hertz and N is in electrons per cubic meter. The plasma frequency is simply another way of specifying the electron density in the plasma. Substituting (6-314) into (6-313), we have

$$\nabla^2 \bar{\mathbf{E}} = -\omega^2 \mu_0 \epsilon_0 \left(1 - \frac{f_N^2}{f^2}\right) \bar{\mathbf{E}} = \bar{\gamma}^2 \bar{\mathbf{E}}$$

where the propagation constant $\bar{\gamma}$ is given by

$$\bar{\gamma} = j\omega \sqrt{\mu_0 \epsilon_0 \left(1 - \frac{f_N^2}{f^2}\right)} \tag{6-315}$$

Thus wave propagation in plasma is characterized by the propagation constant given by (6-315). We note that for $f > f_N$, $(1 - f_N^2/f^2) > 0$, $\bar{\gamma}$ is purely imaginary, and the wave is propagated. For $f < f_N$, $(1 - f_N^2/f^2) < 0$, $\bar{\gamma}$ is purely real, and the fields are attenuated. For the propagating range of frequencies, the phase constant is

$$\beta = \omega \sqrt{\mu_0 \epsilon_0 \left(1 - \frac{f_N^2}{f^2}\right)} \tag{6-316}$$

and the phase velocity v_p is given by

$$v_p = \frac{\omega}{\beta} = \frac{1}{\sqrt{\mu_0\epsilon_0}\sqrt{1 - f_N^2/f^2}} = \frac{c}{\sqrt{1 - f_N^2/f^2}} \tag{6-317}$$

where c is the velocity of light in free space. In view of the dependence of v_p on the wave frequency, wave propagation in plasma is characterized by dispersion. This dispersion is known as parametric dispersion from the point of view that it is a consequence of the frequency dependence of the effective permittivity of the medium. The group velocity is given by

$$v_g = \frac{d\omega}{d\beta} = c\sqrt{1 - \frac{f_N^2}{f^2}}$$

Note that

$$v_p v_g = c^2$$

EXAMPLE 6-33.   An important example of plasma is the ionosphere, which is a region of the upper atmosphere extending from about 50 km to more than 1000 km above the earth. In this region the constituent gases are ionized, mostly due to ultraviolet radiation from the sun. The electron density in the ionosphere exists in several layers known as D, E, and F layers in which the ionization changes with the hour of the day, the season, and the sunspot cycle. For the purpose of our discussion, we will assume that the electron density increases continuously from zero at the lower boundary, reaching a peak at some height, typically lying between 250 and 350 km, and then decreases continuously as shown in Fig. 6.73(a). We will assume that it is uniform geographically, which is not the case in reality, and that the geometry is plane instead of spherical. Furthermore, wave propagation in the ionosphere is complicated by the presence of the earth's magnetic field. We will here ignore the effect of the earth's magnetic field. Let us consider a uniform plane wave of frequency f incident obliquely at the lower boundary of such a plane ionosphere at an angle θ_0 with the normal to the boundary, as shown in Fig. 6.73(b).

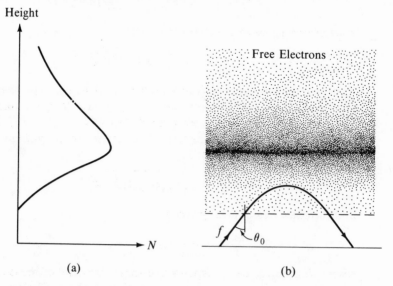

(a)   (b)

Fig. 6.73. (a) Variation of electron density versus height for a simplified ionosphere. (b) Path of a wave incident obliquely on the ionosphere.

We wish to investigate the path of the wave as it propagates in the ionized medium.

We divide the region into several infinitesimal slabs, in each of which the electron density can be considered to be uniform with height. Let us consider the boundary between the free space and the first slab, for which we will denote the plasma frequency as $f_{N,1}$. From (6-317), the phase velocity along the direction of propagation, that is, normal to the constant phase surfaces in this slab, is given by

$$v_{p,1} = \frac{c}{\sqrt{1 - f_{N,1}^2/f^2}}$$

For the waves in the free space and in the slab to be in step at the boundary, their apparent phase velocities along the boundary must be equal. This is the same as saying that the apparent wavelengths along the boundary must be equal. Since $v_{p,1} > c$, this is possible only if the direction of travel of the wave is bent away from the normal to the boundary as shown in Fig. 6.74.

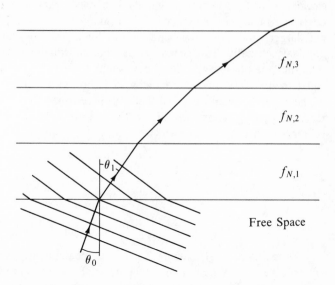

Fig. 6.74. For illustrating the bending of the path of a wave as it propagates in the ionosphere.

Thus, denoting the angle between the normal to the boundary and direction of travel in the slab by θ_1, we have

$$\frac{c}{\sin \theta_0} = \frac{v_{p,1}}{\sin \theta_1}$$

or

$$\frac{c}{v_{p,1}} \sin \theta_1 = \sin \theta_0$$

Applying the same argument from one slab to the next, we obtain, for the ith and $(i - 1)$th slabs,

$$\frac{c}{v_{p,i}} \sin \theta_i = \frac{c}{v_{p,i-1}} \sin \theta_{i-1}$$

The quantity c/v_p is known as the phase refractive index. It is denoted by the symbol μ, which is not to be confused with permittivity. Thus we have

$$\mu_i \sin \theta_i = \mu_{i-1} \sin \theta_{i-1}$$

which is known as Snell's law. For the series of slabs, we then have

$$\mu_i \sin \theta_i = \mu_{i-1} \sin \theta_{i-1} = \mu_{i-2} \sin \theta_{i-2} = \cdots$$
$$= \mu_2 \sin \theta_2 = \mu_1 \sin \theta_1 = \sin \theta_0$$

As the number of slabs is increased indefinitely, we approach the limiting case in which the path of the wave is no longer a series of straight lines but a continuous curve. As the wave penetrates into regions of higher and higher electron density, the phase velocity becomes larger and larger, the phase refractive index becomes smaller and smaller, the angle θ becomes larger and larger, and the path bends gradually away from the normal to the boundary. Finally, a level may be reached at which the electron density is such that the phase refractive index is equal to $\sin \theta_0$, so that $\sin \theta$ becomes equal to unity, $\theta = 90°$, and the path is horizontal. Due to the curvature of the path, it is bent over and the wave is returned to the ground by a symmetrical path as shown in Fig. 6.73(b). For the level at which the path becomes horizontal, we have

$$\mu = \sin \theta_0$$

or

$$\sqrt{1 - \frac{f_N^2}{f^2}} = \sin \theta_0 \qquad \qquad (6\text{-}318)$$

$$f_N = f \cos \theta_0$$

Thus a wave of frequency f which is incident obliquely at an angle θ_0 with the normal to the boundary is reflected from a level at which the plasma frequency is equal to $f \cos \theta_0$. For the special case of normal incidence on the ionosphere, $\theta_0 = 0$ and the condition for reflection is

$$f_N = f$$

The wave is then reflected from a level at which the plasma frequency is equal to the wave frequency. Hence vertically incident waves of frequencies less than the maximum plasma frequency, typically about 10 MHz (but varying with time of the day, season, sunspot cycle, and geographic location) are reflected. Vertically incident waves of frequencies greater than the maximum plasma frequency are transmitted. The same is, of course, true if the transmitter is above the ionosphere. As the angle of incidence is made oblique, waves of larger frequencies are reflected in accordance with (6-318). The earth's curvature, however, sets a limit for the highest frequency which can be reflected. ∎

6.16   Radiation of Electromagnetic Waves

Thus far we have assumed that electromagnetic fields exist in a medium and then discussed their characteristics based on Maxwell's equations. In this section we will discuss how these fields are produced and are "radiated" away from the sources. To do this, we consider Maxwell's equations including the source terms and solve them simultaneously. The Maxwell's equations are

$$\mathbf{V} \cdot \mathbf{D} = \rho \qquad (6\text{-}319a)$$

$$\mathbf{V} \cdot \mathbf{B} = 0 \qquad (6\text{-}319b)$$

$$\mathbf{V} \times \mathbf{E} = -\frac{\partial \mathbf{B}}{\partial t} \qquad (6\text{-}319c)$$

$$\mathbf{V} \times \mathbf{H} = \mathbf{J} + \frac{\partial \mathbf{D}}{\partial t} \qquad (6\text{-}319d)$$

where ρ and $\mathbf{J}$ are the source charge and current densities, respectively. To solve (6-319a)–(6-319d) simultaneously, we recall from Chapters 3 and 4 the following: In view of (6-319b), we can express $\mathbf{B}$ as the curl of a vector potential $\mathbf{A}$; that is,

$$\mathbf{B} = \mathbf{V} \times \mathbf{A} \qquad (6\text{-}320)$$

Then, substituting (6-320) into (6-319c) and rearranging, we have

$$\mathbf{V} \times \left(\mathbf{E} + \frac{\partial \mathbf{A}}{\partial t} \right) = 0$$

so that $\mathbf{E} + \partial \mathbf{A}/\partial t$ can be expressed as the gradient of a scalar potential. Thus $\mathbf{E} + \partial \mathbf{A}/\partial t = -\mathbf{V}V$, or

$$\mathbf{E} = -\mathbf{V}V - \frac{\partial \mathbf{A}}{\partial t} \qquad (6\text{-}321)$$

We now substitute (6-320) and (6-321) into (6-319a) and (6-319d) to obtain a pair of coupled equations in V and $\mathbf{A}$. These are

$$\mathbf{V} \cdot \left(-\mathbf{V}V - \frac{\partial \mathbf{A}}{\partial t} \right) = \frac{\rho}{\epsilon}$$

$$\mathbf{V} \times \mathbf{V} \times \mathbf{A} - \mu\epsilon \frac{\partial}{\partial t}\left(-\mathbf{V}V - \frac{\partial \mathbf{A}}{\partial t} \right) = \mu \mathbf{J}$$

or

$$\nabla^2 V + \frac{\partial}{\partial t}(\mathbf{V} \cdot \mathbf{A}) = -\frac{\rho}{\epsilon} \qquad (6\text{-}322a)$$

$$\nabla^2 \mathbf{A} - \mathbf{V}\left(\mathbf{V} \cdot \mathbf{A} + \mu\epsilon \frac{\partial V}{\partial t} \right) - \mu\epsilon \frac{\partial^2 \mathbf{A}}{\partial t^2} = -\mu \mathbf{J} \qquad (6\text{-}322b)$$

Equations (6-322a) and (6-322b) seem to be very complicated. However, a vector is uniquely defined only if both its curl and divergence are specified. While the curl of **A** is given by (6-320), we have not yet specified the divergence of **A**. We now do this by setting

$$\nabla \cdot \mathbf{A} = -\mu\epsilon \frac{\partial V}{\partial t} \tag{6-323}$$

which is known as the Lorentz condition. This uncouples the equations (6-322a) and (6-322b) to give us

$$\nabla^2 V - \mu\epsilon \frac{\partial^2 V}{\partial t^2} = -\frac{\rho}{\epsilon} \tag{6-324}$$

$$\nabla^2 \mathbf{A} - \mu\epsilon \frac{\partial^2 \mathbf{A}}{\partial t^2} = -\mu \mathbf{J} \tag{6-325}$$

If we can solve these two equations for given charge and current distributions of densities ρ and **J**, respectively, we can then find the fields by using (6-321) and (6-320).

Before we discuss the solution of (6-324) and (6-325), we will show that the continuity equation is implied by the Lorentz condition. To do this, we take the Laplacian of both sides of (6-323). We then have

$$\nabla^2 (\nabla \cdot \mathbf{A}) = -\mu\epsilon \nabla^2 \frac{\partial V}{\partial t}$$

or

$$\nabla \cdot \nabla^2 \mathbf{A} = -\mu\epsilon \frac{\partial}{\partial t} \nabla^2 V \tag{6-326}$$

Substituting for $\nabla^2 \mathbf{A}$ and $\nabla^2 V$ in (6-326) from (6-325) and (6-324), respectively, we get

$$\nabla \cdot \left(\mu\epsilon \frac{\partial^2 \mathbf{A}}{\partial t^2} - \mu \mathbf{J} \right) = -\mu\epsilon \frac{\partial}{\partial t} \left(\mu\epsilon \frac{\partial^2 V}{\partial t^2} - \frac{\rho}{\epsilon} \right)$$

or

$$\mu\epsilon \frac{\partial^2}{\partial t^2} \left(\nabla \cdot \mathbf{A} + \mu\epsilon \frac{\partial V}{\partial t} \right) = \mu \left(\nabla \cdot \mathbf{J} + \frac{\partial \rho}{\partial t} \right)$$

Thus, by assuming the Lorentz condition, we imply $\nabla \cdot \mathbf{J} + \partial\rho/\partial t = 0$, which is the continuity equation. Since the continuity equation must be satisfied by physical charge and current distributions, it is appropriate to use the Lorentz condition to uncouple (6-322a) and (6-322b).

Returning now to Eqs. (6-324) and (6-325), we note that their forms are familiar. They are wave equations with source terms on the right sides. Hence they are inhomogeneous wave equations. We will discuss the solutions to these equations from our knowledge of static fields and our experience with the homogeneous wave equations. It is sufficient if we discuss the solution for one of the two equations. The solution for the second equation follows from similarity. Let us therefore consider Eq. (6-324). For static fields, this equation reduces to

$$\nabla^2 V = -\frac{\rho}{\epsilon}$$

which is Poisson's equation for the electrostatic potential. Let us consider a point charge Q_0 at the origin. The electrostatic potential due to this point charge is given by

$$V(r) = \frac{Q_0}{4\pi\epsilon r}$$

For the time-varying case, we know that electromagnetic effects propagate with a finite velocity v which for the homogeneous wave equation corresponding to (6-324) is $1/\sqrt{\mu\epsilon}$. Hence, if the point charge at the origin is varying with time (due to current flowing into and/or away from the origin), its effect is felt at a distance r from the origin after a time delay of r/v. Conversely, the effect felt at a distance r from the origin at time t is due to the value of the charge which existed at the origin at an earlier time $t - r/v$. Thus, if the point charge at the origin is varying in the manner $Q_0 \sin \omega t$, we expect the time-varying electric potential due to it to be

$$V(r, t) = \frac{Q_0 \sin \omega(t - r/v)}{4\pi\epsilon r} \tag{6-327}$$

To verify if our reasoning is correct, we note that

$$\nabla^2 V = \nabla^2 \left[\frac{Q_0 \sin \omega(t - r/v)}{4\pi\epsilon r} \right]$$

$$= \frac{Q_0}{4\pi\epsilon} \left\{ \left[\sin \omega \left(t - \frac{r}{v} \right) \right] \nabla^2 \frac{1}{r} \right.$$

$$+ 2\nabla \sin \omega \left(t - \frac{r}{v} \right) \cdot \nabla \frac{1}{r} + \frac{1}{r} \nabla^2 \sin \omega \left(t - \frac{r}{v} \right) \right\}$$

$$= -\frac{Q_0 \delta(\mathbf{r}) \sin \omega t}{\epsilon} - \frac{\omega^2 Q_0 \sin \omega(t - r/v)}{4\pi\epsilon r v^2} \tag{6-328a}$$

where we have used the vector identity

$$\nabla^2(\phi\psi) = \phi \nabla^2 \psi + 2 \nabla\phi \cdot \nabla\psi + \psi \nabla^2\phi$$

and the relation (see Problem 2-58)

$$\nabla^2 \frac{1}{r} = -4\pi\delta(\mathbf{r})$$

We also note that

$$\mu\epsilon \frac{\partial^2 V}{\partial t^2} = -\frac{\omega^2 Q_0 \sin \omega(t - r/v)}{4\pi\epsilon r v^2} \tag{6-328b}$$

From (6-328a) and (6-328b), we have

$$\nabla^2 V - \mu\epsilon \frac{\partial^2 V}{\partial t^2} = -\frac{Q_0 \delta(\mathbf{r}) \sin \omega t}{\epsilon}$$

which agrees with (6-324) for a point charge $Q_0 \sin \omega t$ at the origin.

It follows from (6-327) that, for a time-varying volume charge of density $\rho(\mathbf{r}', t)$ in an infinitesimal volume dv' at a point $P(\mathbf{r}')$, the time-varying electric potential at a point $Q(\mathbf{r})$ is given by

$$dV(\mathbf{r}, t) = \frac{\rho(\mathbf{r}', t - |\mathbf{r} - \mathbf{r}'|/v)}{4\pi\epsilon \, |\mathbf{r} - \mathbf{r}'|} \, dv' \qquad (6\text{-}329\text{a})$$

Similarly, from Eq. (6-325), the time-varying magnetic vector potential at a point $Q(\mathbf{r})$ due to a time-varying volume current of density $\mathbf{J}(\mathbf{r}', t)$ in an infinitesimal volume dv' at a point $P(\mathbf{r}')$ is given by

$$d\mathbf{A}(\mathbf{r}, t) = \frac{\mu \mathbf{J}(\mathbf{r}', t - |\mathbf{r} - \mathbf{r}'|/v)}{4\pi \, |\mathbf{r} - \mathbf{r}'|} \, dv' \qquad (6\text{-}329\text{b})$$

Equations (6-329a) and (6-329b) tell us that, to find the time-varying electromagnetic potentials at a point $Q(\mathbf{r})$ at a time t due to a volume charge $\rho \, dv'$ and a volume current $\mathbf{J} \, dv'$ at a point $P(\mathbf{r}')$, we can make use of the expressions for V and $\mathbf{A}$ for the static case except that we have to use those values of ρ and $\mathbf{J}$ which existed at P at a time $t - |\mathbf{r} - \mathbf{r}'|/v$. For this reason, these potentials are known as the "retarded potentials." The retarded potentials for volume charge and current distributions in an extended volume V' are given by the integrals of (6-329a) and (6-329b). These are

$$V(\mathbf{r}, t) = \int_{V'} \frac{\rho(\mathbf{r}', t - |\mathbf{r} - \mathbf{r}'|/v)}{4\pi\epsilon \, |\mathbf{r} - \mathbf{r}'|} \, dv' \qquad (6\text{-}330)$$

$$\mathbf{A}(\mathbf{r}, t) = \int_{V'} \frac{\mu \mathbf{J}(\mathbf{r}', t - |\mathbf{r} - \mathbf{r}'|/v)}{4\pi \, |\mathbf{r} - \mathbf{r}'|} \, dv' \qquad (6\text{-}331)$$

We will now evaluate the retarded potentials and then the fields for a simple but a very useful source known as the Hertzian dipole. We will find that the field expressions we will obtain are quite complicated even for this simplest case. The Hertzian dipole is an oscillating version of the static electric dipole. It consists of two equal and opposite time-varying charges $Q_1(t) = Q_0 \sin \omega t$ and $Q_2(t) = -Q_0 \sin \omega t$ separated by an infinitesimal distance dl. We will place the dipole at the origin and orient it along the z axis. The dipole moment is then given by $d\mathbf{p} = Q_0 \, dl \sin \omega t \, \mathbf{i}_z$. To satisfy the continuity equation, we connect the two charges by a filamentary wire so that the current flowing in the wire from Q_2 to Q_1 is

$$I(t) = \frac{dQ_1}{dt} = -\frac{dQ_2}{dt} = \omega Q_0 \cos \omega t = I_0 \cos \omega t$$

where $I_0 = \omega Q_0$. The Hertzian dipole and the time variations of Q_1, Q_2, and I are shown in Fig. 6.75.

With reference to the notation of Fig. 6.75(a), the time-varying electric

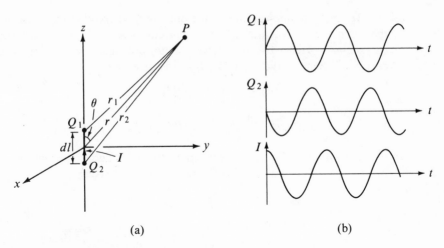

Fig. 6.75. (a) Hertzian dipole. (b) Time variations of Q_1, Q_2, and I for the Hertzian dipole.

potential at point P is given by

$$V = \frac{Q_1(t - r_1/v)}{4\pi\epsilon r_1} + \frac{Q_2(t - r_2/v)}{4\pi\epsilon r_2}$$

$$= \frac{Q_0 \sin \omega(t - r_1/v)}{4\pi\epsilon r_1} - \frac{Q_0 \sin \omega(t - r_2/v)}{4\pi\epsilon r_2}$$

We will let $dl \rightarrow 0$, keeping the product $Q_0\, dl$ constant and thereby obtaining a point dipole. We then have

$$V \approx \frac{Q_0 \sin \omega\{t - [r - (dl/2)\cos\theta]/v\}}{4\pi\epsilon[r - (dl/2)\cos\theta]} - \frac{Q_0 \sin \omega\{t - [r + (dl/2)\cos\theta]/v\}}{4\pi\epsilon[r + (dl/2)\cos\theta]}$$

$$= \frac{Q_0}{4\pi\epsilon r}\Bigg[\Big(1 + \frac{dl}{2r}\cos\theta\Big)\sin\omega\Big(t - \frac{r}{v} + \frac{dl}{2v}\cos\theta\Big)$$

$$-\Big(1 - \frac{dl}{2r}\cos\theta\Big)\sin\omega\Big(t - \frac{r}{v} - \frac{dl}{2v}\cos\theta\Big)\Bigg]$$

$$= \frac{Q_0}{4\pi\epsilon r}\Bigg[2\sin\Big(\frac{\omega\, dl}{2v}\cos\theta\Big)\cos\omega\Big(t - \frac{r}{v}\Big)$$ (6-332)

$$+ \frac{dl\cos\theta}{r}\cos\Big(\frac{\omega\, dl}{2v}\cos\theta\Big)\sin\omega\Big(t - \frac{r}{v}\Big)\Bigg]$$

$$\rightarrow \frac{Q_0 dl\cos\theta}{4\pi\epsilon r}\Bigg[\frac{\omega\cos\omega(t - r/v)}{v} + \frac{\sin\omega(t - r/v)}{r}\Bigg]$$

We note from (6-332) that the time-varying electric potential due to the dipole is not simply equal to the electrostatic potential $(Q_0\, dl \cos\theta)/4\pi\epsilon r^2$ times the retardation factor $\sin\omega(t - r/v)$, but has an additional term. This arises because of the phase difference between the time-varying potentials associated with the individual point charges of the dipole.

To find the time-varying vector potential, we recall from Chapter 3 that the static vector potential due to a current element $I_0 \, dl \, \mathbf{i}_z$ at the origin is $(\mu I_0 \, dl/4\pi r)\mathbf{i}_z$. Hence the time-varying vector potential due to the time-varying current element of the Hertzian dipole is given by

$$
\begin{aligned}
\mathbf{A} &= \frac{\mu I_0 \, dl \cos \omega(t - r/v)}{4\pi r} \mathbf{i}_z \\
&= \frac{\mu I_0 \, dl \cos \omega(t - r/v)}{4\pi r} (\cos \theta \, \mathbf{i}_r - \sin \theta \, \mathbf{i}_\theta)
\end{aligned}
\tag{6-333}
$$

We will now obtain the electromagnetic fields due to the Hertzian dipole by using (6-321) and (6-320). From (6-321), we have

$$
\begin{aligned}
\mathbf{E} &= -\nabla V - \frac{\partial \mathbf{A}}{\partial t} \\
&= \left(-\frac{\partial V}{\partial r} - \frac{\partial A_r}{\partial t} \right) \mathbf{i}_r + \left(-\frac{1}{r}\frac{\partial V}{\partial \theta} - \frac{\partial A_\theta}{\partial t} \right) \mathbf{i}_\theta
\end{aligned}
\tag{6-334a}
$$

From (6-320), we have

$$
\begin{aligned}
\mathbf{H} &= \frac{\mathbf{B}}{\mu} = \frac{1}{\mu}\nabla \times \mathbf{A} \\
&= \frac{1}{\mu r}\left[\frac{\partial}{\partial r}(r A_\theta) - \frac{\partial A_r}{\partial \theta} \right] \mathbf{i}_\phi
\end{aligned}
\tag{6-334b}
$$

Thus the field components are given by

$$
\begin{aligned}
E_r &= -\frac{\partial V}{\partial r} - \frac{\partial A_r}{\partial t} \\
&= \frac{2\omega \, Q_0 \, dl \cos \theta}{4\pi \epsilon}\left[\frac{\sin \omega(t - r/v)}{\omega r^3} + \frac{\cos \omega(t - r/v)}{v r^2} \right]
\end{aligned}
\tag{6-335a}
$$

$$
\begin{aligned}
E_\theta &= -\frac{1}{r}\frac{\partial V}{\partial \theta} - \frac{\partial A_\theta}{\partial t} \\
&= \frac{\omega Q_0 \, dl \sin \theta}{4\pi \epsilon}\left[\frac{\sin \omega(t - r/v)}{\omega r^3} + \frac{\cos \omega(t - r/v)}{v r^2} \right. \\
&\qquad\qquad \left. - \frac{\omega \sin \omega(t - r/v)}{v^2 r} \right]
\end{aligned}
\tag{6-335b}
$$

$$
\begin{aligned}
H_\phi &= \frac{1}{\mu r}\left[\frac{\partial}{\partial r}(r A_\theta) - \frac{\partial A_r}{\partial \theta} \right] \\
&= \frac{I_0 \, dl \sin \theta}{4\pi}\left[\frac{\cos \omega(t - r/v)}{r^2} - \frac{\omega \sin \omega(t - r/v)}{v r} \right]
\end{aligned}
\tag{6-335c}
$$

Alternatively, E_r and E_θ can be obtained from H_ϕ by using Maxwell's curl equation for $\mathbf{H}$ in which case it is not necessary to determine V. Writing the field expressions in phasor form, we have

$$
\begin{aligned}
\bar{E}_r &= \frac{2\bar{I}_0 \, dl \cos \theta}{4\pi \epsilon}\left(-\frac{j}{\omega r^3} + \frac{1}{v r^2} \right) e^{-j\omega r/v} \\
&= -\frac{2\beta^2 \eta \bar{I}_0 \, dl \cos \theta}{4\pi}\left[\frac{1}{(j\beta r)^3} + \frac{1}{(j\beta r)^2} \right] e^{-j\beta r}
\end{aligned}
\tag{6-336}
$$

$$\bar{E}_\theta = \frac{\bar{I}_0 \, dl \sin \theta}{4\pi\epsilon} \left(-\frac{j}{\omega r^3} + \frac{1}{vr^2} + \frac{j\omega}{v^2 r} \right) e^{-j\omega r/v}$$

$$= -\frac{\beta^2 \eta \bar{I}_0 \, dl \sin \theta}{4\pi} \left[\frac{1}{(j\beta r)^3} + \frac{1}{(j\beta r)^2} + \frac{1}{j\beta r} \right] e^{-j\beta r} \qquad (6\text{-}337)$$

$$\bar{H}_\phi = \frac{\bar{I}_0 \, dl \sin \theta}{4\pi} \left(\frac{1}{r^2} + \frac{j\omega}{vr} \right) e^{-j\omega r/v}$$

$$= -\frac{\beta^2 \bar{I}_0 \, dl \sin \theta}{4\pi} \left[\frac{1}{(j\beta r)^2} + \frac{1}{j\beta r} \right] e^{-j\beta r} \qquad (6\text{-}338)$$

where $\beta = \omega/v$, $\eta = \sqrt{\mu/\epsilon} = 1/\epsilon v$, and $\bar{I}_0 = I_0 = \omega Q_0$.

We note from (6-335a)–(6-335c) or (6-336)–(6-338) that the field expressions contain terms involving $1/r^3$, $1/r^2$, and $1/r$. Very close to the dipole, the $1/r^3$ and $1/r^2$ terms dominate the $1/r$ terms. Far from the dipole, the $1/r^3$ and $1/r^2$ terms are negligible and the fields are determined by the $1/r$ terms. To see how far from the dipole, let us first consider the H_ϕ component. The magnitudes of the two terms are equal for $r = v/\omega = 1/\beta = \lambda/2\pi \approx 0.16\lambda$. For the E_θ component the combined magnitude of the $1/r^3$ and $1/r^2$ terms is equal to the $1/r$ term for

$$\left(\frac{1}{\omega r^3} \right)^2 + \left(\frac{1}{vr^2} \right)^2 = \left(\frac{\omega}{v^2 r} \right)^2$$

or

$$r^4 - \left(\frac{\lambda}{2\pi} \right)^2 r^2 - \left(\frac{\lambda}{2\pi} \right)^4 = 0$$

$$r = \sqrt{\frac{1 + \sqrt{5}}{2}} \frac{\lambda}{2\pi} \approx 0.2\lambda$$

Thus, even in a distance of few wavelengths from the dipole, we can neglect the $1/r^3$ and $1/r^2$ terms in comparison with the $1/r$ terms. The field expressions then reduce to

$$\bar{E}_r = 0$$

$$\bar{E}_\theta = \frac{j\omega \bar{I}_0 \, dl \sin \theta}{4\pi\epsilon v^2 r} e^{-j\omega r/v} = \frac{j\beta\eta \bar{I}_0 \, dl \sin \theta}{4\pi r} e^{-j\beta r} \qquad (6\text{-}339)$$

$$\bar{H}_\phi = \frac{j\omega \bar{I}_0 \, dl \sin \theta}{4\pi vr} e^{-j\omega r/v} = \frac{j\beta \bar{I}_0 \, dl \sin \theta}{4\pi r} e^{-j\beta r} \qquad (6\text{-}340)$$

These fields are known as the "radiation fields" because they are the components which contribute to radiation of electromagnetic waves away from the dipole. In fact, we will learn later that the $1/r^3$ and $1/r^2$ terms do not contribute to the time-average power flow even near the dipole. We note that the ratio of E_θ to H_ϕ given by (6-339) and (6-340) is equal to $\eta = \sqrt{\mu/\epsilon}$ as for the case of the fields associated with a uniform plane wave, although the constant phase surfaces are $r = $ constant and the constant amplitude surfaces are $(\sin\theta)/r = $ constant. However, let us consider a spherical surface of large radius and centered at the dipole and divide it into small regions, in each of which $\sin \theta$ may be considered to be constant. Then each small region is approximately a plane surface on which the phase as well as magnitude are constants. Thus, over each small region, the fields are almost like uniform plane waves, with the amplitude differing from one region to the other. This is what we meant by the statement in Section 6.8 that, far from a radiating antenna, the radiated waves are approximately uniform plane waves.

Returning now to the field expressions given by (6-336)–(6-338), we obtain the complex Poynting vector as

$$
\begin{aligned}
\bar{\mathbf{P}} &= \frac{1}{2}\bar{\mathbf{E}} \times \bar{\mathbf{H}}^* \\
&= \frac{1}{2}(\bar{E}_\theta \bar{H}_\phi^* \mathbf{i}_r - \bar{E}_r \bar{H}_\phi^* \mathbf{i}_\theta) \\
&= \frac{|\bar{I}_0|^2 \, (dl)^2 \sin^2 \theta}{32\pi^2 \epsilon}\left(\frac{-j}{\omega r^3} + \frac{1}{vr^2} + \frac{j\omega}{v^2 r}\right)\left(\frac{1}{r^2} - \frac{j\omega}{vr}\right)\mathbf{i}_r \\
&\quad - \frac{|\bar{I}_0|^2 \, (dl)^2 \sin 2\theta}{32\pi^2 \epsilon}\left(\frac{-j}{\omega r^3} + \frac{1}{vr^2}\right)\left(\frac{1}{r^2} - \frac{j\omega}{vr}\right)\mathbf{i}_\theta \\
&= \frac{|\bar{I}_0|^2 \, (dl)^2 \sin^2 \theta}{32\pi^2 \epsilon}\left(\frac{\omega^2}{v^3 r^2} - j\frac{1}{\omega r^5}\right)\mathbf{i}_r \\
&\quad + j\frac{|\bar{I}_0|^2 \, (dl)^2 \sin 2\theta}{32\pi^2 \epsilon}\left(\frac{\omega}{v^2 r^3} + \frac{1}{\omega r^5}\right)\mathbf{i}_\theta
\end{aligned}
\tag{6-341}
$$

The time-average Poynting vector is given by

$$
\begin{aligned}
\langle \mathbf{P} \rangle &= \mathcal{R}e[\bar{\mathbf{P}}] \\
&= \frac{|\bar{I}_0|^2 \, (dl)^2 \sin^2 \theta}{32\pi^2 \epsilon}\frac{\omega^2}{v^3 r^2}\mathbf{i}_r
\end{aligned}
\tag{6-342}
$$

which is exactly the same as the time-average Poynting vector due to the radiation fields given by (6-339) and (6-340). Thus the near fields, that is, the $1/r^3$ and $1/r^2$ terms, do not contribute to the time-average power flow even near the dipole. They contribute only to the reactive power, which is entirely due to them since the reactive power associated with the radiation fields is zero.

By integrating the time-average Poynting vector given by (6-342) over a surface of radius r centered at the dipole, we obtain the time-average power radiated by the dipole as

$$
\begin{aligned}
\langle P_{\text{rad}} \rangle &= \int_{\theta=0}^{\pi} \int_{\phi=0}^{2\pi} \langle \mathbf{P} \rangle \cdot r^2 \sin \theta \, d\theta \, d\phi \, \mathbf{i}_r \\
&= \int_{\theta=0}^{\pi} \int_{\phi=0}^{2\pi} \frac{\omega^2 |\bar{I}_0|^2 \, (dl)^2}{32\pi^2 \epsilon v^3} \sin^3 \theta \, d\theta \, d\phi \\
&= \frac{\omega^2 |\bar{I}_0|^2 \, (dl)^2}{32\pi^2 \epsilon v^3}\frac{8\pi}{3} = \frac{\omega^2 |\bar{I}_0|^2 \, (dl)^2}{12\pi \epsilon v^3} \\
&= \frac{\eta \beta^2 |\bar{I}_0|^2 \, (dl)^2}{12\pi} = \frac{\pi \eta |\bar{I}_0|^2}{3}\left(\frac{dl}{\lambda}\right)^2
\end{aligned}
\tag{6-343}
$$

We now see why the near fields cannot contribute to time-average power flow. The reason is that, from conservation of energy, the time-average power flow across a spherical surface of one radius must be equal to the time-average power flow across a spherical surface of a different radius, that is, it must be independent of r as indicated by (6-343). Since the surface area of the sphere varies as r^2, only those components of **E** and **H** which vary as $1/r$ can satisfy this condition.

Rewriting (6-343) as

$$\langle P_{\text{rad}} \rangle = \frac{1}{2} |\bar{I}_0|^2 \left[\frac{2\pi\eta}{3} \left(\frac{dl}{\lambda} \right)^2 \right]$$

we note that the power radiated by the dipole is the same as the time-average power dissipated in a resistance of value $[(2\pi\eta/3)(dl/\lambda)^2]$ when a current $I_0 \cos \omega t$ is passed through it. This is known as the "radiation resistance" and is denoted by the symbol R_{rad}. Thus, for the Hertzian dipole,

$$R_{\text{rad}} = \frac{2\pi\eta}{3} \left(\frac{dl}{\lambda} \right)^2 \text{ ohms}$$

For $\eta = \eta_0 = 120\pi$, that is, for the dipole in free space, we have

$$R_{\text{rad}} = 80\pi^2 \left(\frac{dl}{\lambda} \right)^2 \text{ ohms} \qquad (6\text{-}344)$$

As a numerical example, for dl/λ equal to 0.01, R_{rad} is equal to 0.08 ohms. This value is too small to make a Hertzian dipole of dl/λ equal to 0.01 an effective radiator. This is why a practical dipole must be an appreciable fraction of a wavelength long. But then, Eq. (6-344) is no longer correct for the radiation resistance since the variation of current along the length of the dipole must be taken into account in obtaining the radiation fields and hence the radiated power. This can be done by considering the dipole as a series of Hertzian dipoles connected end to end and then using superposition. We will illustrate this by means of an example.

EXAMPLE 6-34.   A practical short dipole is a center-fed straight wire antenna, having a length that is short compared to a wavelength. The current distribution along the wire can be approximated as shown in Fig. 6.76(a) in which the magnitude decreases uniformly from a maximum at the center to zero at the ends. It is desired to find the radiation resistance of the short dipole.

With reference to Fig. 6.76(a), the current distribution along the dipole can be written as

$$\bar{I}(z) = \begin{cases} \bar{I}_0 \left(1 - \dfrac{2z}{L} \right) & \text{for } 0 < z < \dfrac{L}{2} \\[2mm] \bar{I}_0 \left(1 + \dfrac{2z}{L} \right) & \text{for } -\dfrac{L}{2} < z < 0 \end{cases} \qquad (6\text{-}345)$$

where $\bar{I}_0$ is a constant. To determine the radiation fields, we can represent

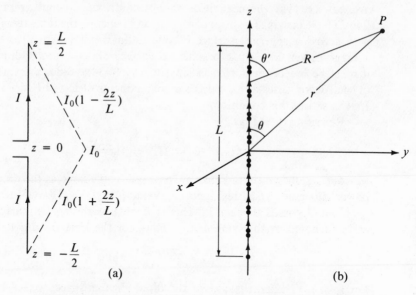

Fig. 6.76. (a) Current distribution along a short dipole. (b) Representation of the short dipole as a series of Hertzian dipoles for computing the radiation fields and the radiation resistance.

the short dipole as a series of Hertzian dipoles of infinitesimal lengths dz as shown in Fig. 6.76(b). From (6-339) and (6-340) and from superposition, the radiation fields for the short dipole are then given by

$$\bar{E}_\theta = \int_{z=-L/2}^{L/2} \frac{j\beta\eta\bar{I}(z)\sin\theta'}{4\pi R} e^{-j\beta R}\, dz \tag{6-346a}$$

$$\bar{H}_\phi = \int_{z=-L/2}^{L/2} \frac{j\beta\bar{I}(z)\sin\theta'}{4\pi R} e^{-j\beta R}\, dz \tag{6-346b}$$

where R and θ' are as shown in Fig. 6.76(b). For $R \gg L$, as is the case for radiation fields, we can set $\theta' \approx \theta$ and $R \approx r$ in the numerators and denominators of the integrands on the right sides of (6-346a) and (6-346b). For the R in the exponential factors, however, we substitute $(r - z\cos\theta)$ because, depending on the value of β, $e^{-j\beta R}$ can vary appreciably for $-L/2 < z < L/2$. Considering (6-346a), we then have

$$\bar{E}_\theta = \int_{z=-L/2}^{L/2} \frac{j\beta\eta\bar{I}(z)\sin\theta}{4\pi r} e^{-j\beta r} e^{j\beta z\cos\theta}\, dz$$

$$= \frac{j\beta\eta\sin\theta}{4\pi r} e^{-j\beta r} \int_{z=-L/2}^{L/2} \bar{I}(z) e^{j\beta z\cos\theta}\, dz \tag{6-347}$$

Substituting (6-345) into (6-347), we obtain

$$\bar{E}_\theta = \frac{j\beta\eta\,\bar{I}_0\sin\theta}{4\pi r}e^{-j\beta r}\left[\int_{z=0}^{L/2}\left(1-\frac{2z}{L}\right)e^{j\beta z\cos\theta}\,dz\right.$$

$$\left.+\int_{z=-L/2}^{0}\left(1+\frac{2z}{L}\right)e^{j\beta z\cos\theta}\,dz\right]$$

$$= \frac{j\beta\eta\bar{I}_0\sin\theta}{4\pi r}e^{-j\beta r}\int_{z=0}^{L/2}\left(1-\frac{2z}{L}\right)(e^{j\beta z\cos\theta}+e^{-j\beta z\cos\theta})\,dz$$

$$= \frac{j\beta\eta\bar{I}_0\sin\theta}{2\pi r}e^{-j\beta r}\int_{z=0}^{L/2}\left(1-\frac{2z}{L}\right)\cos(\beta z\cos\theta)\,dz$$

(6-348)

However, for $L\ll\lambda$, $\beta L=2\pi L/\lambda\ll 1$, and

$$\cos(\beta z\cos\theta)=1-\frac{(\beta z\cos\theta)^2}{2}+\cdots\approx 1\text{ for }-\frac{L}{2}<z<\frac{L}{2}$$

so that (6-348) simplfies to

$$\bar{E}_\theta = \frac{j\beta\eta\bar{I}_0\sin\theta}{2\pi r}e^{-j\beta r}\int_{z=0}^{L/2}\left(1-\frac{2z}{L}\right)dz$$

$$= \frac{j\beta\eta L\bar{I}_0\sin\theta}{8\pi r}e^{-j\beta r}$$

Likewise,

$$\bar{H}_\phi = \frac{j\beta L\bar{I}_0\sin\theta}{8\pi r}e^{-j\beta r}$$

The time-average radiated power is then given by

$$\langle P_{\text{rad}}\rangle = \int_{\theta=0}^{\pi}\int_{\phi=0}^{2\pi}\frac{1}{2}(\bar{E}_\theta\bar{H}_\phi^*)r^2\sin\theta\,d\theta\,d\phi$$

$$= \int_{\theta=0}^{\pi}\int_{\phi=0}^{2\pi}\frac{\beta^2\eta L^2|\bar{I}_0|^2}{128\pi^2}\sin^3\theta\,d\theta\,d\phi$$

$$= \frac{1}{2}|\bar{I}_0|^2\left[\frac{\pi\eta}{6}\left(\frac{L}{\lambda}\right)^2\right]$$

Thus, for $\eta=\eta_0=120\pi$, the radiation resistance of a short dipole of length L is given by

$$R_{\text{rad}}\text{ (short dipole)} = 20\pi^2\left(\frac{L}{\lambda}\right)^2 \qquad (6\text{-}349)$$

As a numerical example, for $L/\lambda=0.1$, $R_{\text{rad}}\approx 2$ ohms. ∎

We will conclude this section with a brief discussion of the directional properties of the Hertzian and short dipoles. In this connection, we define the radiation intensity U of an antenna in a given direction as the power radiated per unit solid angle in that direction. Since the surface area of a sphere of radius r is $4\pi r^2$ and the solid angle subtended by it at its center is 4π, the surface area per unit solid angle is r^2. Thus the radiation intensity is given by

$$U = \langle \mathbf{P} \rangle \cdot r^2 \mathbf{i}_r, \text{ watts/steradian}$$

From (6-342) the radiation intensity for the Hertzian dipole is

$$U = \frac{|\bar{I}_0|^2 (dl)^2 \omega^2}{32\pi^2 \epsilon v^3} \sin^2 \theta$$

The quantity

$$\frac{|\bar{I}_0|^2 (dl)^2 \omega^2}{32\pi^2 \epsilon v^3}$$

is a constant for a particular frequency and hence, by dividing U by this quantity, we obtain the normalized radiation intensity U_n, as

$$U_n = \sin^2 \theta \tag{6-350}$$

The same result holds for the short dipole of Example 6-34 since the power radiated by it is also proportional to $\sin^2 \theta$. A plot of U_n given by (6-350) versus θ is shown in Fig. 6.77. This plot illustrates the directional properties

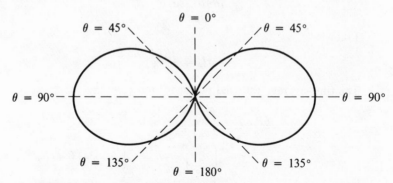

Fig. 6.77. Normalized radiation intensity versus θ for Hertzian and short dipoles.

of the Hertzian and short dipoles. Their radiation intensities are maximum for $\theta = 90°$, that is, broadside to the dipole and zero for $\theta = 0$ and $180°$, that is, along the dipole. The directivity D of an antenna is defined as the ratio of the maximum radiation intensity to the average radiation intensity. Thus

$$D = \frac{[U_n]_{\max}}{(1/4\pi) \int U_n \, d\Omega} = \frac{4\pi [U_n]_{\max}}{\int_{\theta=0}^{\pi} \int_{\phi=0}^{2\pi} U_n(\theta, \phi) \sin \theta \, d\theta \, d\phi}$$

where Ω denotes the solid angle. For the Hertzian and short dipoles,

$$D = \frac{4\pi [\sin^2 \theta]_{\max}}{\int_{\theta=0}^{\pi} \int_{\phi=0}^{2\pi} \sin^3 \theta \, d\theta \, d\phi} = \frac{4\pi}{8\pi/3} = \frac{3}{2}$$

PROBLEMS

6.1.   For the following charge distributions, find the electrostatic potential everywhere using Poisson's and Laplace's equations.

(a) $\rho = \begin{cases} z & \text{for } |z| < a \\ 0 & \text{for } |z| > a \end{cases}$ cartesian coordinates

(b) $\rho = \begin{cases} \rho_0 & \text{for } r < a \\ 0 & \text{for } r > a \end{cases}$ cylindrical coordinates

(c) $\rho = \begin{cases} 0 & \text{for } r < a \\ \rho_0 & \text{for } a < r < b \\ 0 & \text{for } r > b \end{cases}$ spherical coordinates

(d) $\rho = \begin{cases} \rho_0\left(1 - \dfrac{r^2}{a^2}\right) & \text{for } r < a \\ 0 & \text{for } r > a \end{cases}$ spherical coordinates

6.2.   Show that the equation of motion of an electron in the space-charge limited vacuum diode of Example 6-2 is given by

$$\frac{d^3x}{dt^3} = \frac{eJ_0}{m\epsilon_0}$$

where e and m are the charge and mass of the electron, respectively, and J_0 is the current density. For an electron leaving the cathode at $t = 0$ and subject to the conditions stated in Example 6-2, obtain the solution for $x(t)$ by solving the equation of motion. Then find the solution for V, which should agree with (6-22).

6.3.   Verify the general solutions for the one-dimensional Laplace's equations and the particular solutions for the particular sets of boundary conditions listed in Table 6.1.

6.4.   Two conductors occupying the surfaces $r = a$ and $r = b$ in cylindrical coordinates are kept at potentials $V = V_0$ and $V = 0$, respectively. The region $a < r < c (< b)$ is a perfect dielectric of permittivity ϵ_1 and the region $c < r < b$ is a perfect dielectric of permittivity ϵ_2. Find the solutions for the potentials in the two regions and the potential at the boundary $r = c$.

6.5.   Two parallel conducting plates occupying the planes $x = 0$ and $x = d$ are kept at potentials $V = 0$ and $V = V_0$, respectively. The medium between the two plates is a perfect dielectric of nonuniform permittivity given by

$$\epsilon = \epsilon_1 + (\epsilon_2 - \epsilon_1)\frac{x}{d}$$

where ϵ_1 and ϵ_2 are constants. Find the solutions for the potential and the electric field intensity between the plates.

6.6.   The region $0 < x < d$ is occupied by a medium characterized by the magnetization vector $\mathbf{M} = M_0(d - x)\mathbf{i}_x$, where M_0 is a constant. By solving the analogous electrostatic problem, obtain $\mathbf{H}$ and $\mathbf{B}$ both inside and outside the region $0 < x < d$.

6.7. The region $r < a$ in spherical coordinates is occupied by a medium characterized by the magnetization vector $\mathbf{M} = M_0 \mathbf{i}_z$, where M_0 is a constant. (a) Set up the analogous electrostatic problem for obtaining $\mathbf{H}$ and $\mathbf{B}$ both inside and outside the region $r < a$. (b) Find the electric field intensity for this electrostatic problem from the answer to part (*d*) of Problem 5.11. (c) Find $\mathbf{H}$ and $\mathbf{B}$ both inside and outside the region $r < a$.

6.8. A conductor occupying the surfaces $x > 0$, $y = 0$ and $y > 0$, $x = 0$ is kept at zero potential. A second conductor occupying the surface $xy = 2$ is kept at a potential of 100 volts, making sure that the edges where the two conductors touch are insulated. The medium between the conductors is charge free. Find the solutions for the potential and the electric field intensity between the conductors. Find the surface charge densities on the conductors.

6.9. The potential distribution at the mouth of the slot of Fig. 6.6 is given by

$$V = V_1 \sin \frac{\pi y}{b} + V_2 \sin \frac{3\pi y}{b} \qquad \text{for } x = a, 0 < y < b$$

where V_1 and V_2 are constants. Find the solution for the potential distribution in the slot. Repeat the problem for

$$V = V_1 \sin^3 \frac{\pi y}{b} \qquad \text{for } x = a, 0 < y < b$$

6.10. Two conductors occupying the planes $x = 0$ and $x = a$ are kept at zero potentials. A third conductor occupying the surface $y = 0, 0 < x < a$ is kept at a constant potential V_0, making sure that the edges are insulated. Find the solutions for the potential in the region $0 < x < a$ for both $y > 0$ and $y < 0$. Show that the potential at large values of $|y|$ varies with x approximately as $\sin (\pi x/a)$.

6.11. A thin rectangular slab of uniform conductivity σ_0 mhos/m, shown in Fig. 6.78, has its edges coated with perfectly conducting material, making sure that the

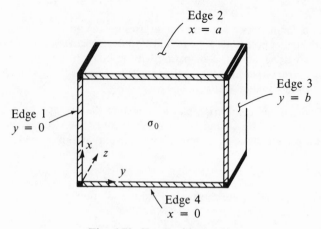

Fig. 6.78. For Problem 6.11.

corners are insulated. For each of the following cases, find the solution for the potential and hence for the current density in the conductor:

(a) Edges 1 and 3 kept at zero potential; edge 2 kept at potential V_0 and edge 4 kept at potential $-V_0$.

(b) Edges 1 and 3 kept at zero potential; edges 2 and 4 kept at potential V_0.

(c) Edges 1 and 4 kept at zero potential; edge 2 kept at potential V_1 and edge 3 kept at potential V_2.

6.12. For the triangular box of Fig. 6.10, assume that the longer side is kept at zero potential and the shorter sides are kept at a potential of 100 volts. Find the potentials at points a, b, and c.

6.13. An infinitely long line charge of uniform density ρ_{L0} C/m is situated parallel to and at a distance d from a grounded infinite plane conductor. Obtain the image charge and show that the induced surface charge on the conductor per unit length parallel to the line charge is equal to $-\rho_{L0}$.

6.14. For each of the arrangements shown in Fig. 6.79, find the image charges required to determine the electric field on the side of the actual charges. For case (a), find the electric field intensity everywhere on the conductor surface and show that the total induced charge is $-Q$.

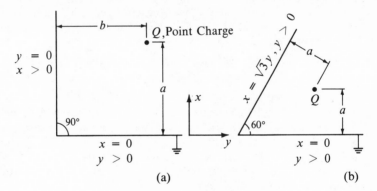

Fig. 6.79. For Problem 6.14.

6.15. For the infinitely long line charge of uniform density ρ_{L0} C/m parallel to an infinitely long grounded conducting cylinder in Example 6-11, show that the induced surface charge per unit length of the cylinder is $-\rho_{L0}$.

6.16. A point charge Q is situated at a distance d from the center of a grounded spherical conductor of radius a $(< d)$. Show that the image charge required for computing the field outside the spherical conductor is a point charge of value $-Qa/d$, lying at a distance a^2/d from the center of the conductor along the line joining the center to the charge Q and on the side of Q. What is the induced charge on the surface of the conductor?

6.17. For the problem of Example 6-4:

(a) Find the electric field intensities in the two regions $0 < x < t$ and $t < x < d$.

(b) Find the surface charge densities on the plates $x = 0$ and $x = d$.

(c) Find the capacitance C per unit area of the plates and show that

$$\frac{1}{C} = \frac{1}{\epsilon_1/t} + \frac{1}{\epsilon_2/(d - t)}$$

6.18. For the parallel-plate arrangement of Problem 6.5, find the capacitance per unit area of the plates in three ways:

(a) From the definition $C = Q/V_0$, where Q is the magnitude of the charge per unit area on either plate.

(b) By evaluating the electric stored energy in the dielectric per unit area of the plates and using (6-81).

(c) By dividing the dielectric into several slabs, each having an infinitesimal thickness and using the result of Problem 6.17.

6.19. Derive the expressions for the conductance, capacitance, and inductance per unit length of the two-conductor configuration of Fig. 6.15(b).

6.20. For the two-conductor configuration of Fig. 6.15(d), find the locations of a pair of equal and opposite, infinitely long, uniform line charges parallel to the conductors such that two of the equipotential surfaces corresponding to the pair of line charges are the surfaces occupied by the conductors. Then find the expressions for conductance, capacitance, and inductance per unit length of the conductor system. Let d be equal to zero and show that these expressions reduce to those for the configuration of Fig. 6.15(b).

6.21. A current I amp flows with nonuniform volume density given by

$$\mathbf{J} = J_0 \frac{r}{a} \mathbf{i}_z$$

along an infinitely long cylindrical conductor of radius a having the z axis as its axis. The current returns with uniform surface density in the opposite direction along the surface of an infinitely long perfectly conducting cylinder of radius $b\ (> a)$ and coaxial with the inner conductor. Find the internal inductance per unit length of the inner conductor by using the method of flux linkages. Verify your answer by using the energy method.

6.22. A filamentary wire carrying a current I amp is closely wound around a toroidal magnetic core of rectangular cross section as shown in Fig. 6.80. The mean radius

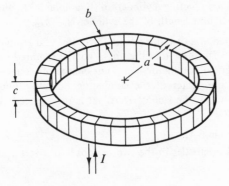

Fig. 6.80. For Problem 6.22.

of the toroidal core is a and the number of turns per unit length along the mean circumference of the toroid is N. Find the inductance of the toroid.

6.23. An infinitely long, uniformly wound solenoid of radius a and having N turns per unit length carries a current I amp. Find the inductance per unit length of the solenoid.

6.24. Show that $L_{21} = L_{12}$.

6.25. An infinitely long, uniformly wound solenoid of radius a and having N_1 turns per unit length is coaxial with another infinitely long, uniformly wound solenoid of radius $b \, (> a)$ and having N_2 turns per unit length. Find the mutual inductance per unit length of the solenoids.

6.26. A cylindrical slab of material lying between plane surfaces $z = 0$ and $z = d$ and having a cross-sectional area $A = \pi a^2$ is characterized by nonuniform conductivity

$$\sigma = \frac{\sigma_0}{1 + z/d}$$

permittivity $\epsilon = 4\epsilon_0$, and permeability $\mu = 2\mu_0$, where σ_0 is a constant. The surfaces $z = 0$ and $z = d$ are perfectly conducting. A current flows through perfectly conducting filamentary wires into the center of the plane surface $z = d$ and out of the center of the plane surface $z = 0$. Assume that this current is established by appropriate connection of a battery of voltage V_0 which is far away from the material so that the magnetic field outside the slab may be considered to be the same as that due to an infinitely long wire along the axis of the slab (z axis). Find the following quantities:

(a) The electric field intensity, the conduction current density, and the displacement flux density in the material.

(b) The surface charge densities on the perfectly conducting surfaces $z = 0$ and $z = d$.

(c) The true charge density in the material.

(d) The polarization vector and the polarization charge distribution in the material.

(e) The magnetic field intensity and the magnetic flux density in the material.

(f) The current drawn from the battery and the magnetic field intensity outside the material.

(g) The surface current density on the perfectly conducting surfaces $z = 0$ and $z = d$.

(h) The magnetization vector and the magnetization current distribution in the material.

(i) The power dissipation density and the power dissipated in the material and the conductance of the configuration.

(j) The electric stored energy density and the electric stored energy in the material and the capacitance of the configuration.

(k) The magnetic stored energy density and the magnetic stored energy in the material and the internal inductance of the configuration.

(l) The power flow into the material evaluated by surface integration of the Poynting vector.

6.27. A toroidal magnetic core of circular cross section and with an air gap has the following dimensions:

$$\text{area of cross section} = 2 \text{ cm}^2$$
$$\text{mean circumference} = 20 \text{ cm}$$
$$\text{air gap width} = 0.1 \text{ cm}$$

Find the ampere turns required to establish a magnetic flux of 3×10^{-4} Wb in the air gap if the core is made of annealed sheet steel. The effective area of the air gap is that of a circle whose radius exceeds the actual radius by half the width of the air gap.

6.28. For the magnetic circuit of Fig. 6.23, assume that there is no air gap. If NI is equal to 150 amp-turns, find the magnetic flux density in leg 2.

6.29. For the structure of Fig. 6.25(a), show that, under quasistatic conditions, the rate at which energy flows into the volume of the structure as obtained by surface integration of the Poynting vector over the surface bounding the volume is equal to

$$\frac{d}{dt}\left[\frac{1}{2}CV^2(t)\right]$$

6.30. For the structure of Fig. 6.25(b), show that, under quasistatic conditions, the rate at which energy flows into the volume of the structure as obtained by surface integration of the Poynting vector over the surface bounding the volume is equal to

$$\frac{d}{dt}\left[\frac{1}{2}LI^2(t)\right]$$

6.31. By proceeding in a manner similar to that in Example 6-16, show that the quasistatic approximation holds for the parallel-plate structure of Fig. 6.25(a), that is, the structure behaves like a single capacitor, for the condition

$$f \ll \frac{1}{2\pi l\sqrt{\mu\epsilon}}$$

Examine the input behavior of the structure for frequencies beyond the value for which the quasistatic approximation holds.

6.32. A time-varying voltage source drives the structure of Fig. 6.13(a). Assume that the conductor is a good conductor so that the displacement current can be neglected compared to the conduction current. Show that the quasistatic approximation holds, that is, the structure behaves essentially like a single resistor, for the condition

$$f \ll \frac{1}{\pi\mu\sigma l^2}$$

Investigate the approximation quantitatively for copper. Examine the input behavior of the structure for frequencies slightly beyond the value for which the quasistatic approximation holds and also for frequencies for which $f \gg 1/\pi\mu\sigma l^2$.

6.33. The structure shown in Fig. 6.81 is an arrangement of two parallel perfectly conducting plates connected at one end by a third perfectly conducting plate. A current source $I(t) = 1 \cos 2\pi ft$ amp is connected between the plates at the other end so that it supplies a z-directed current uniformly distributed in the y direction to the structure. The medium between the plates is free space. For the purpose of this

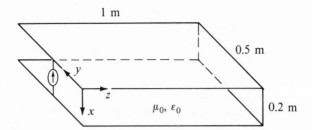

Fig. 6.81. For Problem 6.33.

problem, the arrangement can be assumed to be part of a structure infinite in extent in the y direction. The dimensions of the structure are indicated in the figure.

(a) Find the voltage developed across the current source in the steady state if $f = 150$ Hz.

(b) Repeat part (a) if $f = 150$ MHz.

6.34. Derive the transmission-line equations by considering the special case of two infinitely long, coaxial cylindrical conductors. Also show that the power flow along the conductor system is equal to the product of the voltage between the conductors and current along the conductors.

6.35. Show that two alternative representations of the circuit equivalent of the transmission-line equations (6-161) and (6-165) are as shown in Figs. 6.82(a) and (b).

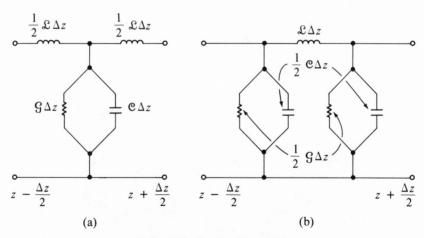

Fig. 6.82. For Problem 6.35.

6.36. Starting with the curl of both sides of (6-169d), derive the wave equation for **H** given by

$$\nabla^2 \mathbf{H} = \mu\epsilon \frac{\partial^2 \mathbf{H}}{\partial t^2}$$

6.37. Solve Eq. (6-176) by using the separation of variables technique.

6.38. Draw three-dimensional sketches similar to that of Fig. 6.30 for the following functions:

(a) $e^{-|t-z|}$

(b) $e^{-|t+z|}$

(c) $(z-t)^2[u(z-t) - u(z-t-2)]$

6.39. A spherical balloon of uniform surface charge density and having its center at the origin possesses a constant total charge Q. Its radius c is made to vary sinusoidally between a minimum of $(a-b)$ and a maximum of $(a+b)$ in the manner

$$c = a + b\cos 2\pi t.$$

(a) Describe and sketch how the electric field intensity vector **E** varies with time in three regions

$$0 < r < (a-b) \qquad (a-b) < r < (a+b) \qquad (a+b) < r < \infty$$

Assume uniform surface charge density for all t.

(b) From your answer to part (a) for the region $(a+b) < r < \infty$, what can you infer about wave propagation due to the fluctuating balloon? Explain.

6.40. A uniform plane wave traveling in the negative z direction in free space has its electric field entirely along the x direction. The space variation of the electric field intensity at time $t = 0$ is shown in Fig. 6.83. Find and sketch the time variation of the magnetic field intensity in the $z = 200$ m plane.

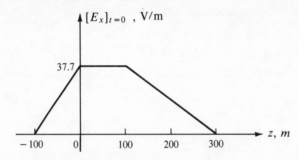

Fig. 6.83. For Problem 6.40.

6.41. The electric field intensity associated with a uniform plane wave traveling in a perfect dielectric medium is given by

$$E_x(z, t) = 10 \cos (2\pi \times 10^7 t - 0.1\pi z) \text{ volts/m}$$

(a) Sketch E_x versus t for two values of z, $z = 0$ and $z = 5$ m. What is the frequency of the wave?

(b) Sketch E_x versus z for two values of t, $t = 0$ and $t = \frac{1}{4} \times 10^{-7}$ sec. What is the wavelength?

(c) What is the velocity of propagation?

(d) Write the expression for the magnetic field intensity associated with the wave if $\mu = \mu_0$.

6.42. The complex electric field vector of a uniform plane wave propagating in free space is given by

$$\bar{\mathbf{E}} = (-\mathbf{i}_x - 2\sqrt{3}\,\mathbf{i}_y + \sqrt{3}\,\mathbf{i}_z)e^{-j0.04\pi(\sqrt{3}x - 2y - 3z)} \quad \text{volts/m}$$

(a) What is the direction of propagation of the wave?

(b) Find the wavelength along the direction of propagation.

(c) Find the frequency of the wave.

(d) Find the apparent wavelengths and the apparent phase velocities along the x, y, and z axes.

(e) Discuss the polarization of the wave.

(f) Obtain the expression for the complex magnetic field vector of the wave.

6.43.   A complex electric field vector is given by

$$\bar{\mathbf{E}} = \left[\left(-\sqrt{3} - j\frac{1}{2}\right)\mathbf{i}_x + \left(1 - j\frac{\sqrt{3}}{2}\right)\mathbf{i}_y + j\sqrt{3}\,\mathbf{i}_z\right]e^{-j0.02\pi(\sqrt{3}x + 3y + 2z)} \quad \text{volts/m}$$

(a) Perform the necessary tests and determine if the given $\bar{\mathbf{E}}$ represents the electric field of a uniform plane wave.

(b) If your answer to part (a) is "yes," repeat Problem 6.42 for the electric field vector of this problem.

6.44.   The complex electric and magnetic field vectors in a perfect dielectric medium are given by

$$\bar{\mathbf{E}} = (-j\mathbf{i}_x - 2\mathbf{i}_y + j\sqrt{3}\,\mathbf{i}_z)e^{-j0.05\pi(\sqrt{3}x + z)} \quad \text{volts/m}$$

$$\bar{\mathbf{H}} = \frac{1}{60\pi}(\mathbf{i}_x - j2\mathbf{i}_y - \sqrt{3}\,\mathbf{i}_z)e^{-j0.05\pi(\sqrt{3}x + z)} \quad \text{amp/m}$$

(a) Perform the necessary tests and determine if these vectors represent the fields associated with a uniform plane wave.

(b) If your answer to part (a) is "yes," find the direction of propagation, the wavelength along the direction of propagation, the velocity along the direction of propagation, and the frequency. Also, discuss the polarization of the wave.

6.45.   Show that the units of $1/\sqrt{\mathscr{L}\mathscr{C}}$ are meters per second and the units of $\sqrt{\mathscr{L}/\mathscr{C}}$ are ohms.

6.46.   The plane $z = 0$ is occupied by a perfect conductor. The medium $z < 0$ is free space. The leading edge of a uniform plane wave traveling in the positive z direction and having $E_x(z)$ as shown in Fig. 6.84 is incident on the plane $z = -150$ m at

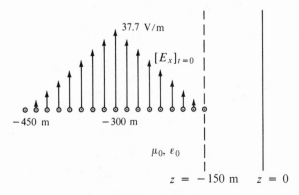

Fig. 6.84. For Problem 6.46.

$t = 0$. Find and sketch E_x and H_y versus z for t equal to $\frac{1}{4}$ μsec, $\frac{3}{4}$ μsec, 1 μsec, $1\frac{1}{4}$ μsec, and 2 μsec. Also sketch E_x and H_y versus t in the plane $z = -150$ m.

6.47. For the problem of Example 6-20:

(a) Sketch E_x versus z for $t = 0.015$ μsec and 0.035 μsec.

(b) Draw the bounce diagram for H_y and sketch H_y in the planes $z = -3$ m and $z = 2.5$ m as functions of time for $t \geq 0$. Also sketch H_y versus z for values of t equal to 0.015 μsec and 0.035 μsec.

6.48. In the transmission-line system shown in Fig. 6.85, the switch S is closed at $t = 0$.

(a) Show that, for $0 < t < l/v$, a (+) wave of voltage

$$V^+(z, t) = \frac{Z_0}{R_g + Z_0} V_g\left(t - \frac{z}{v}\right)$$

exists on the line. What is the current associated with the (+) wave?

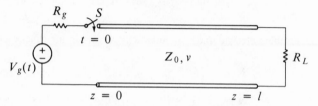

Fig. 6.85. For Problem 6.48.

(b) Show that for $l/v < t < 2l/v$, a (−) wave of voltage

$$V^-(z, t) = \frac{Z_0}{R_g + Z_0} \Gamma_R V_g\left(t - \frac{2l}{v} + \frac{z}{v}\right) \qquad \text{where } \Gamma_R = \frac{R_L - Z_0}{R_L + Z_0}$$

exists on the line in addition to the (+) wave specified in part (a). What is the current associated with the (−) wave?

(c) Show that for $2l/v < t < 3l/v$, a (−+) wave of voltage

$$V^{-+}(z, t) = \frac{Z_0}{R_g + Z_0} \Gamma_R \Gamma_g V_g\left(t - \frac{2l}{v} - \frac{z}{v}\right) \qquad \text{where } \Gamma_g = \frac{R_g - Z_0}{R_g + Z_0}$$

exists on the line in addition to the (+) and (−) waves specified in parts (a) and (b), respectively. What is the current associated with the (−+) wave?

(d) Show that the line voltage and line current at $t = \infty$ are given by the expressions

$$V_{SS}(z, t) = \frac{Z_0}{R_g + Z_0}\left[\sum_{n=0}^{\infty} (\Gamma_R \Gamma_g)^n V_g\left(t - \frac{2nl}{v} - \frac{z}{v}\right)\right.$$
$$\left. + \Gamma_R \sum_{n=0}^{\infty} (\Gamma_R \Gamma_g)^n V_g\left(t - \frac{2nl}{v} + \frac{z}{v} - \frac{2l}{v}\right)\right]$$

$$I_{SS}(z, t) = \frac{1}{R_g + Z_0}\left[\sum_{n=0}^{\infty} (\Gamma_R \Gamma_g)^n V_g\left(t - \frac{2nl}{v} - \frac{z}{v}\right)\right.$$
$$\left. - \Gamma_R \sum_{n=0}^{\infty} (\Gamma_R \Gamma_g)^n V_g\left(t - \frac{2nl}{v} + \frac{z}{v} - \frac{2l}{v}\right)\right]$$

(e) Obtain closed-form expressions for $V_{SS}(z, t)$ and $I_{SS}(z, t)$ for two cases: (i) $V_g(t) = V_0$, a constant and (ii) $V_g(t) = V_0 \cos \omega t$.

6.49. A transmission-line of characteristic impedance Z_0 is terminated by an inductor of value L henries. A $(+)$ wave of constant voltage V_0 is incident on the termination at $t = 0$. Show that the resulting $(-)$ wave voltage at the termination is given by

$$V^-(t) = -V_0 + 2V_0 e^{-(Z_0/L)t}$$

6.50. In Fig. 6.86, a transmission-line of characteristic impedance 50 ohms is terminated by a passive nonlinear element having the volt-ampere characteristic indicated in the figure. If a $(+)$ wave of constant voltage 10 volts is incident on the termination, find the resulting $(-)$ wave voltage.

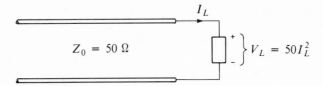

Fig. 6.86. For Problem 6.50.

6.51. Draw sketches of V and I given by Eqs. (6-231a) and (6-231b), respectively, versus t for values of d equal to 0, $\lambda/8$, $\lambda/4$, $3\lambda/8$, and $\lambda/2$. Consider $\theta = 0$ for simplicity.

6.52. A transmission-line of length l is short circuited at one end and open circuited at the other end. What are the natural frequencies of oscillation? Sketch the voltage and current standing wave patterns for the first few modes. Repeat for a line of length l which is open circuited at both ends.

6.53. The transmission-line system shown in Fig. 6.87 is in sinusoidal steady state. The voltage source $V_g(t)$ is equal to $10 \cos 1000\pi t + 5 \cos 2000\pi t$ volts.

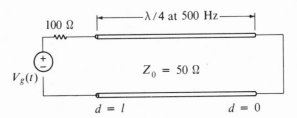

Fig. 6.87. For Problem 6.53.

(a) What is the impedance seen looking into the input terminals of the line for $f = 500$ Hz? Sketch the voltage and current standing wave patterns for $f = 500$ Hz.

(b) What is the impedance seen looking into the input terminals of the line for $f = 1000$ Hz? Sketch the voltage and current standing wave patterns for $f = 1000$ Hz.

(c) From the standing wave patterns of parts (a) and (b), compute the values of root-mean-square line voltages and line currents at $d = 0$, $d = l/2$, and $d = l$.

6.54. Find the two lowest frequencies (zero excluded) for which a transmission-line of length l short circuited at its far end behaves at its input as an inductor of value equal to its inductance computed from static field considerations.

6.55. Show that the minima in the standing wave patterns of Fig. 6.48 are sharper than the maxima.

6.56. Show that the line impedance at a voltage maximum is $Z_0(\text{VSWR})$ and the line impedance at a voltage minimum is $Z_0/(\text{VSWR})$.

6.57. Repeat Example 6-23 for frequency of the uniform plane wave equal to 6000 MHz. Find the fraction of the incident power transmitted into medium 3.

6.58. Repeat Example 6-23 for frequency of the uniform plane wave equal to 1500 MHz. Find the fraction of the incident power transmitted into medium 3. Also find the wave impedance in medium 1 at a distance of 4 cm from the interface between media 1 and 2.

6.59. Find the thickness and permittivity of a quarter-wave dielectric coating which will eliminate reflections of uniform plane waves of frequency 1500 MHz incident normally from free space onto a dielectric of permittivity $16\epsilon_0$. Assume all media to have $\mu = \mu_0$.

6.60. A transmission line of characteristic impedance 50 ohms is terminated by an unknown load impedance $\bar{Z}_R$. Standing wave measurements indicate VSWR equal to 3.0. Distance between successive voltage minima is 20 cm and distance between load and first voltage minimum is 15 cm.

(a) Find $\bar{Z}_R$.

(b) Find the location nearest to the load and the characteristic impedance of a quarter-wave section required to achieve a match between the line and the load.

6.61. A transmission line of characteristic impedance 50 ohms is terminated by a certain load impedance. It is found that the VSWR on the line is equal to 5.0. The first voltage minimum is located to be at 0.1λ from the load. Determine analytically the location and the length of a short-circuited stub connected in parallel with the line so that a match is obtained between the line and the load. Assume the characteristic impedance of the stub to be 50 ohms. Repeat the problem for characteristic impedance of stub equal to 100 ohms.

6.62. A transmission line of characteristic impedance 50 ohms is terminated by a certain load impedance. It is found that the VSWR on the line is equal to 3.0. The first voltage minimum is located at 5.80 cm from the load and the next voltage minimum at 25.80 cm from the load. Find analytically the value of the minimum VSWR that can be achieved on the line by placing a stub in parallel with the line at the load.

6.63. A transmission line of characteristic impedance 100 ohms is terminated by a load impedance $(80 + j200)$ ohms. Using the Smith chart, find the following quantities:

(a) The reflection coefficient at the load.

(b) VSWR on the line.

(c) The distance of the first voltage minimum of the standing wave pattern from the load.

(d) The line impedance at $d = 0.1\lambda$.

(e) The line admittance at $d = 0.1\lambda$.

(f) The location nearest to the load at which the real part of the line admittance is equal to the line characteristic admittance.

6.64. Solve Problem 6.58 using the Smith chart.

6.65. Solve Problem 6.61 using the Smith chart.

6.66. Solve Problem 6.62 using the Smith chart.

6.67. The dimension a of a parallel-plate waveguide filled with a dielectric of permittivity $\epsilon = 4\epsilon_0$ is 4.0 cm. Determine the propagating $TE_{m,0}$ modes for a wave frequency of 6000 MHz. For each propagating mode, find (a) the cutoff frequency, (b) the angle θ_i at which the wave bounces obliquely between the conductors, (c) the guide wavelength λ_g, (d) the phase velocity v_{pz}, and (e) the guide impedance η_g.

6.68. Consider a parallel-plate waveguide extending in the z direction with a dielectric discontinuity at $z = 0$. A $TE_{m,0}$ wave is incident on the discontinuity from the side $z < 0$. By making use of the boundary conditions at the discontinuity, show that each section of the guide can be replaced by the corresponding transmission-line equivalent shown in Fig. 6.59, for the purpose of solving reflection, transmission, and matching problems involving power flow in the z direction.

6.69. In Section 6.12 we introduced transverse electric or TE waves by considering oblique incidence of a linearly polarized uniform plane wave on a perfect conductor with its electric field entirely parallel to the plane of the conductor. To investigate transverse magnetic or TM waves, consider a linearly polarized, uniform plane wave having its magnetic field entirely along the y direction and incident obliquely upon a perfect conductor occupying the $x = 0$ plane as shown in Fig. 6.88.

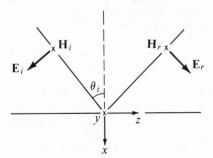

Fig. 6.88. For Problem 6.69.

(a) Obtain the expressions for the total fields.

(b) Show that $\bar{E}_z = 0$ at the surface of the conductor as well as in planes $x = -m\lambda/(2 \cos \theta_i)$, $m = 1, 2, 3, \ldots$.

(c) For a parallel-plate guide of spacing a between the plates, find the expressions for the cutoff wavelengths, cutoff frequencies, guide wavelengths, and the phase velocities in the z direction for the $TM_{m,0}$ modes.

(d) Write expressions for the total fields in the guide independent of θ_i.

(e) Define guide impedance and obtain the transmission-line equivalent for power flow along the guide.

6.70.   Using the transmission-line equivalent determined in Problem 6.69, repeat Example 6-29 for $TM_{1,0}$ waves of frequency 6000 MHz.

6.71.   Show that

$$v_{gz} = \frac{v_{pz}}{1 - (\omega/v_{pz})(dv_{pz}/d\omega)}$$

6.72.   For the parallel-plate waveguide of Problem 6.67, obtain the group velocities for the propagating modes for a wave of frequency 6000 MHz.

6.73.   For a tapered transmission line, the inductance and capacitance per unit length are functions of position z along the line.

(a) Show that the line voltage and line current in the sinusoidal steady state satisfy the equations

$$\frac{\partial^2 \bar{V}}{\partial z^2} - \frac{1}{\mathcal{L}}\left(\frac{\partial \mathcal{L}}{\partial z}\right)\left(\frac{\partial \bar{V}}{\partial z}\right) + \omega^2 \mathcal{L}\mathcal{C}\bar{V} = 0$$

$$\frac{\partial^2 \bar{I}}{\partial z^2} - \frac{1}{\mathcal{C}}\left(\frac{\partial \mathcal{C}}{\partial z}\right)\left(\frac{\partial \bar{I}}{\partial z}\right) + \omega^2 \mathcal{L}\mathcal{C}\bar{I} = 0$$

(b) If $\mathcal{L}(z)$ and $\mathcal{C}(z)$ for a particular tapered transmission line are given by

$$\mathcal{L}(z) = \mathcal{L}_0 e^{-az} \quad \text{and} \quad \mathcal{C} = \mathcal{C}_0 e^{az}$$

where $\mathcal{L}_0$, $\mathcal{C}_0$, and a are constants, find the solutions for $\bar{V}$ and $\bar{I}$ and show that there exists a cutoff frequency below which wave propagation does not occur.

6.74.   Obtain the expression for the attenuation constant per wavelength in a lossy medium characterized by σ, μ, and ϵ. Plot the attenuation constant per wavelength versus $\sigma/\omega\epsilon$.

6.75.   For uniform plane waves in fresh lake water ($\sigma = 10^{-3}$ mho/m, $\epsilon = 80\,\epsilon_0$, $\mu = \mu_0$), find α, β, $\bar{\eta}$, and λ for two frequencies: (a) 100 MHz and (b) 10 kHz.

6.76.   A uniform plane wave of frequency f is incident normally from free space onto a plane slab of good conductor of infinite depth and conductivity σ. Obtain the expression for the fraction of the incident power reflected and the fraction of the incident power transmitted into the conductor. Compute numerical values for incidence from free space to copper at 30 MHz.

6.77.   (a) Express Eqs. (6-281a) and (6-281b) in terms of the distance variable d.

(b) Show that the wave impedance $\bar{Z}(d)$ is given by

$$\bar{Z}(d) = \bar{\eta}\frac{1 + \bar{\Gamma}(d)}{1 - \bar{\Gamma}(d)}$$

where $\bar{\Gamma}(d) = \bar{\Gamma}(0)e^{-2\bar{\gamma}d} = \bar{\Gamma}(0)e^{-2\alpha d}e^{-j2\beta d}$.

(c) In Fig. 6.89, a thin slab of good conductor having a thickness t is backed by a perfect dielectric of thickness $\lambda/4$ at the frequency of operation, which in turn is backed by a perfect conductor. Show that, for uniform plane waves incident normally on the good conductor, reflections are eliminated if $\alpha_c t \ll 1$

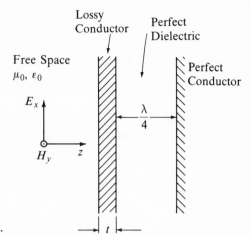

Fig. 6.89. For Problem 6.77.

and $\sigma = 1/\eta_0 t$, where α_c and σ are the attenuation constant and conductivity, respectively, of the good conductor.

6.78. For the semiinfinite plane slab conductor of Fig. 6.66, show that (a) the real part on the right side of (6-294) is the same as the result that would be obtained by a volume integration of the time-average power dissipation density $\frac{1}{2}\sigma|\bar{E}_x|^2$ and (b) the imaginary part on the right side of (6-294) divided by 2ω is the same as the result that would be obtained by a volume integration of the time-average magnetic stored energy density $\frac{1}{4}\mu|\bar{H}_y|^2$.

6.79. For the lossy transmission line of Fig. 6.69,

(a) Write the transmission-line equations.

(b) Find $\bar{\gamma}$ and $\bar{Z}_0$.

(c) Show that for $2\mathcal{R}_i/(2\mathcal{L}_i + \mathcal{L}) = \mathcal{G}/\mathcal{C}$, $\beta = \omega\sqrt{(2\mathcal{L}_i + \mathcal{L})\mathcal{C}}$. What is the attenuation constant for this condition?

6.80. For the parallel-plate resonator of Fig. 6.71, show that the total energy density in the two fields from $d = 0$ to $d = l$ computed by considering the energy density in the electric field at a time at which the magnetic field is zero everywhere between the plates is the same as that given by (6-299).

6.81. The arrangement shown in Fig. 6.90 is that of a parallel-plate resonator made up of two dielectric slabs of thicknesses t and $(l - t)$ and backed by perfect conductors.

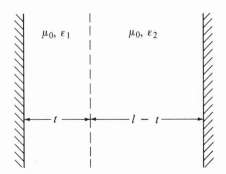

Fig. 6.90. For Problem 6.81.

(a) Show that the resonant frequencies of the system are given by the roots of the equation

$$\tan \omega\sqrt{\mu_0\epsilon_1}t + \sqrt{\frac{\epsilon_1}{\epsilon_2}} \tan \omega\sqrt{\mu_0\epsilon_2}(l - t) = 0$$

(b) Find the three lowest resonant frequencies if $t = l/2$, $l = 5.0$ cm, $\epsilon_1 = \epsilon_0$, and $\epsilon_2 = 4\epsilon_0$.

6.82. For the parallel-plate resonator of Example 6-32, show that, for a particular mode of operation, Q is inversely proportional to $\sqrt{f}$.

6.83. A resonator is formed by placing perfect conductors in two transverse planes $z = 0$ and $z = d$ of a parallel-plate waveguide of spacing a, as shown in Fig. 6.91.

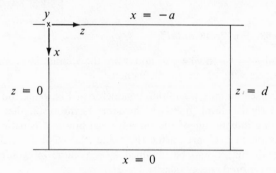

Fig. 6.91. For Problem 6.83.

(a) Show that the resonant frequencies corresponding to the $TE_{m,0,l}$ modes are given by

$$f_{m,0,l} = \frac{1}{2\sqrt{\mu\epsilon}}\sqrt{\left(\frac{m}{a}\right)^2 + \left(\frac{l}{d}\right)^2}$$

Compute the lowest three resonant frequencies if $a = d = 4$ cm. Identify the corresponding mode numbers. Assume free space for the medium between the plates.

(b) Write the expressions for the fields corresponding to the $TE_{1,0,1}$ mode. Derive the expression for the Q of the resonator for the $TE_{1,0,1}$ mode, assuming that the plates are made up of imperfect conductors of conductivity σ and having thicknesses of several skin depths for the frequencies of interest.

6.84. For the parallel-plate resonator of Fig. 6.70, assume that the dielectric is slightly lossy, having a conductivity $\sigma_d \ll \omega\epsilon$.

(a) Assuming the plates to be perfect conductors, show that the Q of the resonator is given by $Q_1 = \omega\epsilon/\sigma_d$.

(b) If, in addition to the slightly lossy dielectric, the plates are made up of slightly lossy conductors, show that the Q of the resonator is given by

$$\frac{1}{Q} = \frac{1}{Q_1} + \frac{1}{Q_2}$$

where Q_1 is as given in part (a) and Q_2 is equal to $l/2\delta$ as derived in Example 6-32.

6.85. Show that the units of $\sqrt{Ne^2/m\epsilon_0}$ are (seconds)$^{-1}$ and that $e^2/4\pi^2m\epsilon_0$ is equal to 80.6.

6.86. The maximum plasma frequency of the plane ionosphere considered in Example 6-33 is 10 MHz.

   (a) What is the minimum value of θ_0 at which a signal of frequency 20 MHz can be incident on the ionosphere in order to get reflected and not to penetrate the ionosphere?

   (b) What is the maximum frequency of a signal incident on the ionosphere at an angle $\theta_0 = 30°$ so that it will be reflected?

6.87. In Fig. 6.92, a satellite signal of frequency $f = 20$ MHz passes through a hypothetical plane slab ionosphere of uniform plasma frequency $f_N = 12$ MHz. The earth's magnetic field and the effect of electron collisions with heavy particles are to be neglected. Find the true elevation angle of the satellite as seen from the receiver.

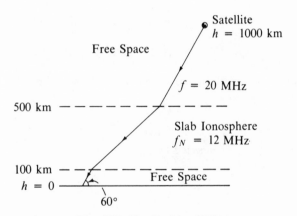

Fig. 6.92. For Problem 6.87.

6.88. A technique of locating the position of an aircraft is by measuring its ranges from a system of satellites of known locations. The apparent range between a satellite and the aircraft is obtained by measuring the time delay of a pulsed continuous wave signal and multiplying it by the velocity of light in free space. The range is an apparent value because the time delay is determined by the group velocity of the signal in the intervening medium which is not free space. For a signal of frequency f much larger than the maximum plasma frequency in the ionosphere and neglecting earth's magnetic field, show that the apparent range is greater than the true range by the amount $(40.3/f^2) \int_A^S N\,ds$, where $\int_A^S N\,ds$ is the integrated electron density in a column of cross section 1 m^2 from the aircraft (A) to the satellite (S) and f is in hertz. For $\int_A^S N\,ds = 10^{18}$ electrons/m^2, find the excess range for two frequencies: (a) 140 MHz, (b) 1,600 MHz.

6.89. Two Hertzian dipoles situated at the origin and carrying currents of the same frequency are oriented along the x and z axes, respectively. The dipoles are of the same length and their currents are equal in magnitude and in phase. Discuss the polarization of the radiation field due to the dipole arrangement at (a) a point along the x axis, (b) a point along the z axis, (c) a point along the y axis, and (d) a point along the line $x = 0$, $y = z$. Repeat for the dipole currents equal in magnitude but differing in phase by $\pi/2$.

6.90. The oscillating version of the static magnetic dipole consists of a circular loop of wire of radius a carrying current varying sinusoidally with time. For circumference of the loop small compared to the wavelength, the current can be considered to be uniform and in phase around the loop so that it is given by $I(t) = I_0 \cos \omega t$. Assume the dipole to be centered at the origin and lying in the xy plane with the current flowing in the ϕ direction.

(a) Find the time-varying magnetic vector potential due to the oscillating magnetic dipole for $r \gg a$.

(b) Obtain the electromagnetic fields due to the oscillating magnetic dipole.

(c) Show that the radiation fields due to the oscillating magnetic dipole are given by

$$\bar{\mathbf{E}} = \frac{\beta^2 \eta \bar{I}_0 \pi a^2}{4\pi r} \sin \theta \, e^{-j\beta r} \mathbf{i}_\phi$$

$$\bar{\mathbf{H}} = -\frac{\beta^2 \bar{I}_0 \pi a^2}{4\pi r} \sin \theta \, e^{-j\beta r} \mathbf{i}_\theta$$

6.91. Fig. 6.93 shows an oscillating electric quadrupole consisting of three time-varying charges given by

$$Q_1(t) = Q_2(t) = Q_0 \sin \omega t$$

$$Q_3(t) = -2Q_0 \sin \omega t$$

The charges are connected by filamentary wires.

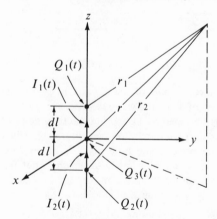

Fig. 6.93. For Problem 6.91.

(a) Write the expressions for the currents $I_1(t)$ and $I_2(t)$ such that the continuity equation is satisfied.

(b) Find the time-varying magnetic vector potential due to the oscillating quadrupole, in the limit that $dl \longrightarrow 0$ keeping $Q_0(dl)^2$ constant.

(c) Find the electromagnetic fields due to the oscillating quadrupole.

(d) Find the radiation fields due to the oscillating quadrupole. Verify by deriving them directly from the radiation fields due to the oscillating dipole given by Eqs. (6-339) and (6-340).

6.92. Find the radiation resistance of a straight copper wire of length 1 cm carrying current of frequency 100 MHz. Compare the radiation resistance with the ohmic resistance of the wire (taking into account skin effect) if it has a cylindrical cross section of radius 1 mm. Repeat for a frequency of 300 MHz.

6.93. A half-wave dipole is a center-fed, straight wire antenna having a length equal to half the wavelength. The current distribution along the half-wave dipole is given by

$$\bar{I}(z) = \bar{I}_0 \cos \frac{\pi z}{L} \qquad \text{for} \qquad -\frac{L}{2} < z < \frac{L}{2}$$

as shown in Fig. 6.94.

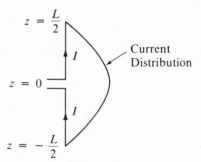

Fig. 6.94. For Problem 6.93.

(a) Show that the radiation fields of the half-wave dipole are

$$\bar{E}_\theta = \frac{j\eta \bar{I}_0 e^{-j(\pi/L)r}}{2\pi r} \frac{\cos [(\pi/2) \cos \theta]}{\sin \theta}$$

$$\bar{H}_\phi = \frac{j\bar{I}_0 e^{-j(\pi/L)r}}{2\pi r} \frac{\cos [(\pi/2) \cos \theta]}{\sin \theta}$$

(b) Show that the radiation resistance of the half-wave dipole in free space is 73 ohms, given that

$$\int_{\theta=0}^{\pi/2} \frac{\cos^2 [(\pi/2) \cos \theta]}{\sin \theta} \, d\theta = 0.609$$

(c) Sketch the normalized radiation intensity pattern.

(d) Show that the directivity of the half-wave dipole is 1.64.

6.94. Two identical short dipoles form an array as shown in Fig. 6.95. Show that the radiation fields due to the array are given by the radiation fields due to one of the dipoles multiplied by the factor $2 \cos [(\beta d \sin \theta \cos \phi)/2]$. Plot the normalized radiation intensity patterns in three planes: (a) $\phi = \pi/2$, (b) $\phi = 0$, and (c) $\theta = \pi/2$ for $d = \lambda/2$.

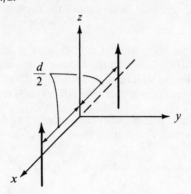

Fig. 6.95. For Problem 6.94.

APPENDIX

UNITS AND DIMENSIONS

In 1960, the International System of Units was given official status at the Eleventh General Conference on weights and measures held in Paris, France. This system of units is an expanded version of the rationalized meter-kilogram-second-ampere (MKSA) system of units and is based on six fundamental or basic units. The six basic units are the units of length, mass, time, current, temperature and luminous intensity.

The international unit of length is the meter. It is exactly 1,650,763.73 times the wavelength in vacuum of the radiation corresponding to the unperturbed transition between the levels $2p_{10}$ and $5d_5$ of the atom of krypton-86, the orange-red line. The international unit of mass is the kilogram. It is the mass of the International Prototype Kilogram which is a particular cylinder of platinum-iridium alloy preserved in a vault at Sèvres, France, by the International Bureau of Weights and Measures. The international unit of time is the second. It is equal to 9,192,631,770 times the period corresponding to the frequency of the transition between the hyperfine levels $F = 4$, $M = 0$ and $F = 3$, $M = 0$ of the fundamental state $^2S_{1/2}$ of the cesium-133 atom unperturbed by external fields.

To present the definition for the international unit of current, we first define the newton, which is the unit of force, derived from the fundamental units meter, kilogram and second in the following manner. Since velocity is rate of change of distance with time, its unit is meter per second. Since acceleration is rate of change of velocity with time, its unit is meter per second per second or meter per second squared. Since force is mass times acceleration, its unit is kilogram-meter per second squared, also known as the newton. Thus, the newton is that force which imparts an acceleration of 1 meter per second squared to a mass of 1 kilogram. The international unit of current, which is the ampere, can now be defined. It is the constant current which when maintained in two straight, infinitely long, parallel conductors of negligible cross section and placed one meter apart in vacuum produces a force of 2×10^{-7} newtons per meter length of the conductors.

The international unit of temperature is the Kelvin degree. It is based on the definition of the thermodynamic scale of temperature by means of the triple-point of water as a fixed fundamental point to which a temperature of exactly 273.16 degrees Kelvin is attributed. The international unit of luminous intensity is the candela. It is defined such that the luminance of a blackbody radiator at the freezing temperature of platinum is 60 candelas per square centimeter.

We have just defined the six basic units of the International System of Units. Two supplementary units are the radian and the steradian for plane angle and solid angle respectively. All other units are derived units. For example, the unit of charge which is the coulomb is the amount of charge transported in 1 second by a current of 1 ampere; the unit of energy which is joule is the work done when the point of application of a force of 1 newton is displaced a distance of 1 meter in the direction of the force; the unit of power which is the watt is the power which gives rise to the production of energy at the rate of 1 joule per second; the unit of electric potential difference which is the volt is the difference of electric potential between two points of a conducting wire carrying constant current of 1 ampere, when the power dissipated between these points is equal to 1 watt; and so on. The units for the various quantities used in this book are listed in Table A.1., together with the symbols of the quantities and their dimensions.

Dimensions are a convenient means of checking the possible validity of a derived equation. The dimension of a given quantity can be expressed as some combination of a set of fundamental dimensions. These fundamental dimensions need not be the same as the quantities corresponding to the basic units. In mechanics, the fundamental dimensions are mass (M), length (L) and time (T). In electromagnetics, it is the usual practice to consider the charge (Q), instead of the current, as the additional fundamental dimension. For the quantities listed in Table A.1., these four dimensions are sufficient. Thus, for example, the dimension of velocity is length (L) divided by time (T), that is LT^{-1}; the dimension of acceleration is length (L) divided by time squared (T^2), that is, LT^{-2}; the dimension of force is mass (M) times acceleration (LT^{-2}), that is, MLT^{-2}; the dimension of ampere is charge (Q) divided by time (T), that is, QT^{-1}; and so on.

To illustrate the application of dimensions for checking the possible validity of a derived equation, let us consider the equation for the velocity of propagation of an electromagnetic wave in free space, given by

$$v = \frac{1}{\sqrt{\mu_0 \epsilon_0}}$$

TABLE A.1. Symbols, Units and Dimensions of Various Quantities

Quantity	Symbol	Unit	Dimensions
Acceleration	$\mathbf{a}$	meter/(second)^2	LT^{-2}
Admittance	$\bar{Y}$	mho	$M^{-1}L^{-2}TQ^2$
Angular velocity	ω	radian/second	T^{-1}
Area	A	square meter	L^2
Attenuation constant	α	neper/meter	L^{-1}
Capacitance	C	farad	$M^{-1}L^{-2}T^2Q^2$
Capacitance per unit length	$\mathfrak{e}$	farad/meter	$M^{-1}L^{-3}T^2Q^2$
Cartesian coordinates	$\begin{cases} x \\ y \\ z \end{cases}$	meter meter meter	L L L
Characteristic admittance	Y_0	mho	$M^{-1}L^{-2}TQ^2$
Characteristic impedance	Z_0	ohm	$ML^2T^{-1}Q^{-2}$
Charge	Q, q	coulomb	Q
Closed path	C	meter	L
Conductance	G	mho	$M^{-1}L^{-2}TQ^2$
Conductance per unit length	$\mathfrak{g}$	mho/meter	$M^{-1}L^{-3}TQ^2$
Conduction current	I_c	ampere	$T^{-1}Q$
Conduction current density	$\mathbf{J}_c$	ampere/square meter	$L^{-2}T^{-1}Q$
Conductivity	σ	mho/meter	$M^{-1}L^{-3}TQ^2$
Current	I	ampere	$T^{-1}Q$
Current transmission coefficient	τ_c	—	—
Cutoff frequency	f_c	hertz	T^{-1}
Cutoff wavelength	λ_c	meter	L
Cylindrical coordinates	$\begin{cases} r, r_c \\ \phi \\ z \end{cases}$	meter radian meter	L — L
Differential length element	dl	meter	L
Differential surface element	dS	square meter	L^2
Differential volume element	dv	cubic meter	L^3
Directivity	D	—	—
Displacement current	I_d	ampere	$T^{-1}Q$
Displacement flux density	$\mathbf{D}$	coulomb/square meter	$L^{-2}Q$
Distance	R, d	meter	L
Drift velocity	$\mathbf{v}_d$	meter/second	LT^{-1}
Electric dipole moment	$\mathbf{p}$	coulomb-meter	LQ
Electric energy	W_e	joule	ML^2T^{-2}
Electric energy density	w_e	joule/cubic meter	$ML^{-1}T^{-2}$
Electric field intensity	$\mathbf{E}$	volt/meter	$MLT^{-2}Q^{-1}$
Electric potential	V	volt	$ML^2T^{-2}Q^{-1}$
Electric susceptibility	χ_e	—	—
Electron density	N	(meter)$^{-3}$	L^{-3}
Electronic charge	e	coulomb	Q
Electronic polarizability	α_e	farad-(meter)^2	$M^{-1}T^2Q^2$
Energy	W	joule	ML^2T^{-2}
Energy density	w	joule/cubic meter	$ML^{-1}T^{-2}$
Force	$\mathbf{F}$	newton	MLT^{-2}
Force per unit volume	$\mathbf{f}$	newton/cubic meter	$ML^{-2}T^{-2}$
Frequency	f	hertz	T^{-1}

TABLE A.1. Cont'd

Quantity	Symbol	Unit	Dimensions
Gravitational field intensity	$\mathbf{g}$	meter/(second)^2	LT^{-2}
Group velocity	v_g	meter/sec	LT^{-1}
Guide impedance	η_g	ohm	$ML^2T^{-1}Q^{-2}$
Guide wavelength	λ_g	meter	L
Impedance	$\bar{Z}$	ohm	$ML^2T^{-1}Q^{-2}$
Incremental relative permeability	μ_{ir}	—	—
Inductance	L	henry	ML^2Q^{-2}
Inductance per unit length	$\mathcal{L}$	henry/meter	MLQ^{-2}
Internal impedance	$\bar{Z}_i$	ohm	$ML^2T^{-1}Q^{-2}$
Internal inductance	L_i, L_{int}	henry	ML^2Q^{-2}
Internal inductance per unit length	$\mathcal{L}_i$	henry/meter	MLQ^{-2}
Intrinsic impedance	η	ohm	$ML^2T^{-1}Q^{-2}$
Intrinsic impedance of free space	η_0	ohm	$ML^2T^{-1}Q^{-2}$
Length	L, l	meter	L
Line charge density	ρ_L	coulomb/meter	$L^{-1}Q$
Linear velocity	$\mathbf{v}$	meter/second	LT^{-1}
Magnetic dipole moment	$\mathbf{m}$	ampere-square meter	$L^2T^{-1}Q$
Magnetic energy	W_m	joule	ML^2T^{-2}
Magnetic energy density	w_m	joule/cubic meter	$ML^{-1}T^{-2}$
Magnetic field intensity	$\mathbf{H}$	ampere/meter	$L^{-1}T^{-1}Q$
Magnetic flux	ψ	weber	$ML^2T^{-1}Q^{-1}$
Magnetic flux density	$\mathbf{B}$	weber/square meter (or tesla)	$MT^{-1}Q^{-1}$
Magnetic polarizability	α_m	(meter)^4/henry	$M^{-1}L^2Q^2$
Magnetic scalar potential	V_m	ampere	$T^{-1}Q$
Magnetic susceptibility	χ_m	—	—
Magnetic vector potential	$\mathbf{A}$	weber/meter	$MLT^{-1}Q^{-1}$
Magnetization surface current density	$\mathbf{J}_{ms}$	ampere/meter	$L^{-1}T^{-1}Q$
Magnetization vector	$\mathbf{M}$	ampere/meter	$L^{-1}T^{-1}Q$
Magnetization volume current density	$\mathbf{J}_m$	ampere/square meter	$L^{-2}T^{-1}Q$
Magnetizing field	$\mathbf{B}_m$	weber/square meter (or tesla)	$MT^{-1}Q^{-1}$
Mass	M, m	kilogram	M
Mobility of electron	μ_e	(meter)^2/volt-second	$M^{-1}TQ$
Mobility of hole	μ_h	(meter)^2/volt-second	$M^{-1}TQ$
Molecular polarizability	α	farad-(meter)^2	$M^{-1}T^2Q^2$
Mutual inductance	L_{12}, L_{21}	henry	ML^2Q^{-2}
Mutual inductance per unit length	$\mathcal{L}_{12}, \mathcal{L}_{21}$	henry/meter	MLQ^{-2}
Natural frequency of oscillation	f_n	hertz	T^{-1}
Normalized admittance	$\bar{y}$	—	—
Normalized impedance	$\bar{z}$	—	—
Normalized radiation intensity	U_n	—	—
Normalized reactance	x	—	—

TABLE A.1 Cont'd

Quantity	Symbol	Unit	Dimensions
Normalized resistance	r	—	—
Normalized susceptance	b	—	—
Permeability	μ	henry/meter	MLQ^{-2}
Permeability of free space	μ_0	henry/meter	MLQ^{-2}
Permittivity	ϵ	farad/meter	$M^{-1}L^{-3}T^2Q^2$
Permittivity of free space	ϵ_0	farad/meter	$M^{-1}L^{-3}T^2Q^2$
Phase constant	β	radian/meter	L^{-1}
Phase refractive index	μ	—	—
Phase velocity	v_p	meter/second	LT^{-1}
Plasma frequency	f_N	hertz	T^{-1}
Polarization current density	$\mathbf{J}_p$	ampere/square meter	$L^{-2}T^{-1}Q$
Polarization surface charge density	ρ_{ps}	coulomb/square meter	$L^{-2}Q$
Polarization vector	$\mathbf{P}$	coulomb/square meter	$L^{-2}Q$
Polarization volume charge density	ρ_p	coulomb/cubic meter	$L^{-3}Q$
Polarizing electric field	$\mathbf{E}_p$	volt/meter	$MLT^{-2}Q^{-1}$
Position vector of field point	$\mathbf{r}$	meter	L
Position vector of source point	$\mathbf{r}'$	meter	L
Power	P	watt	ML^2T^{-3}
Power density	p	watt/square meter	MT^{-3}
Power dissipation density	p_d	watt/square meter	MT^{-3}
Poynting vector	$\mathbf{P}$	watt/square meter	MT^{-3}
Propagation constant	$\bar{\gamma}$	complex neper/meter	L^{-1}
Propagation vector	$\boldsymbol{\beta}$	radian/meter	L^{-1}
Quality factor	Q	—	—
Radian frequency	ω	radian/second	T^{-1}
Radiated power	P_{rad}	watt	ML^2T^{-3}
Radiation intensity	U	watt/steradian	ML^2T^{-3}
Radiation resistance	R_{rad}	ohm	$ML^2T^{-1}Q^{-2}$
Reactance	X	ohm	$ML^2T^{-1}Q^{-2}$
Reflection coefficient	Γ	—	—
Relative permeability	μ_r	—	—
Relative permittivity	ϵ_r	—	—
Reluctance	$\mathcal{R}$	ampere-turn/weber	$M^{-1}L^{-2}Q^2$
Resistance	R	ohm	$ML^2T^{-1}Q^{-2}$
Resistance per unit length	$\mathcal{R}$	ohm/meter	$MLT^{-1}Q^{-2}$
Skin depth	δ	meter	L
Skin effect resistance	R_s	ohm	$ML^2T^{-1}Q^{-2}$
Skin effect resistance per unit length	$\mathcal{R}_s$	ohm/meter	$MLT^{-1}Q^{-2}$
Solid angle	Ω	steradian	—
Spherical coordinates	$\begin{cases} r, r_s \\ \theta \\ \phi \end{cases}$	meter radian radian	L — —
Surface	S	square meter	L^2
Surface charge density	ρ_s	coulomb/square meter	$L^{-2}Q$
Surface current density	$\mathbf{J}_s$	ampere/meter	$L^{-1}T^{-1}Q$

TABLE A.1. Cont'd

Quantity	Symbol	Unit	Dimensions
Susceptance	B	mho	$M^{-1}L^{-2}TQ^2$
Time	t	second	T
Unit normal vector	$\mathbf{i}_N, \mathbf{i}_n$	—	—
Velocity of light in free space	c	meter/second	LT^{-1}
Voltage	V	volt	$ML^2T^{-2}Q^{-1}$
Voltage standing wave ratio	VSWR	—	—
Voltage transmission coefficient	τ_v	—	—
Volume	V	cubic meter	L^3
Volume charge density	ρ	coulomb/cubic meter	$L^{-3}Q$
Volume current density	$\mathbf{J}$	ampere/square meter	$L^{-2}T^{-1}Q$
Wavelength	λ	meter	L
Work	W	joule	ML^2T^{-2}

We know that the dimension of v is LT^{-1}. Hence, we have to show that the dimension of $1/\sqrt{\mu_0\epsilon_0}$ is also LT^{-1}. To do this, we note from Coulomb's law that

$$\epsilon_0 = \frac{Q_1Q_2}{4\pi FR^2}$$

Hence, the dimension of ϵ_0 is $Q^2/[(MLT^{-2})(L^2)]$ or $M^{-1}L^{-3}T^2Q^2$. We note from Ampere's force law applied to two infinitesimal current elements parallel to each other and normal to the line joining them that

$$\mu_0 = \frac{4\pi FR^2}{(I_1\,dl_1)(I_2\,dl_2)}$$

Hence, the dimension of μ_0 is $[(MLT^{-2})(L^2)]/(QT^{-1}L)^2$ or MLQ^{-2}. We now obtain the dimension of $1/\sqrt{\mu_0\epsilon_0}$ as $1/\sqrt{(M^{-1}L^{-3}T^2Q^2)(MLQ^{-2})}$ or LT^{-1}, which is the same as the dimension of v. It should however be noted that the test for the equality of the dimensions of the two sides of a derived equation is not a sufficient test to establish the equality of the two sides since any dimensionless constants associated with the equation may be in error.

It is not always necessary to refer to the table of dimensions for checking the possible validity of a derived equation. For example, let us assume that we have derived the expression for the characteristic impedance of a transmission line, i.e., $\sqrt{\mathcal{L}/\mathcal{C}}$ and we wish to verify that $\sqrt{\mathcal{L}/\mathcal{C}}$ does indeed have the dimension of impedance. To do this, we write

$$\sqrt{\frac{\mathcal{L}}{\mathcal{C}}} = \sqrt{\frac{\omega\mathcal{L}1}{\omega\mathcal{C}1}} = \sqrt{\frac{\omega L}{\omega C}} = \sqrt{(\omega L)\left(\frac{1}{\omega C}\right)}$$

We now recognize from our knowledge of circuit theory that both ωL and $1/\omega C$, being the reactances of L and C, respectively, have the dimensions of impedance. Hence, we conclude that $\sqrt{\mathcal{L}/\mathcal{C}}$ has the dimension of $\sqrt{(\text{impedance})^2}$ or impedance.

BIBLIOGRAPHY

Adler, R. B., L. J. Chu, and R. M. Fano, *Electromagnetic Energy Transmission and Radiation*, John Wiley & Sons, Inc., New York, 1960.

Bradshaw, M. D. and W. J. Byatt, *Introductory Engineering Field Theory*, Prentice-Hall, Inc., Englewood Cliffs, N. J., 1967.

Bohn, E. V., *Introduction to Electromagnetic Fields and Waves*, Addison-Wesley Publishing Company, Inc., Reading, Mass., 1968.

Della Torre, E. and C. V. Longo, *The Electromagnetic Field*, Allyn and Bacon, Inc., Boston, 1969.

Elliot, R. S., *Electromagnetics*, McGraw-Hill Book Company, Inc., New York, 1966.

Fano, R. M., L. J. Chu, and R. B. Adler, *Electromagnetic Fields, Energy and Forces*, John Wiley & Sons, Inc., New York, 1960.

Hayt, W. H., Jr., *Engineering Electromagnetics*, (2nd ed.), McGraw-Hill Book Company, Inc., New York, 1967.

Javid, M. and P. M. Brown, *Field Analysis and Electromagnetics*, McGraw-Hill Book Company, Inc., New York, 1963.

Jordan, E. C. and K. G. Balmain, *Electromagnetic Waves and Radiating Systems*, (2nd ed.), Prentice-Hall, Inc., Englewood Cliffs, N. J., 1968.

Kraus, J. D., *Electromagnetics*, McGraw-Hill Book Company, Inc., New York, 1953.

Lorrain, P. and D. Corson, *Electromagnetic Fields and Waves*, (2nd ed.), W. H. Freeman and Company, San Francisco, 1970.

Moore, R. K., *Traveling Wave Engineering*, McGraw-Hill Book Company, Inc., New York, 1960.

527

Nussbaum, A., *Electromagnetic and Quantum Properties of Materials*, Prentice-Hall, Inc., Englewood Cliffs, N. J., 1966.

Owen, G. E., *Electromagnetic Theory*, Allyn and Bacon, Inc., Boston, 1963.

Panofsky, W. K. H. and M. Phillips, *Classical Electricity and Magnetism*, (2nd ed.), Addison-Wesley Publishing Company, Inc., Reading, Mass., 1962.

Plonsey, R. and R. E. Collin, *Principles and Applications of Electromagnetic Fields*, McGraw-Hill Book Company, Inc., New York, N. Y., 1961.

Ramo, S., J. R. Whinnery, and R. Van Duzer, *Fields and Waves in Communication Electronics*, John Wiley & Sons, Inc., New York, 1965.

Reitz, J. R. and F. J. Milford, *Foundations of Electromagnetic Theory*, Addison-Wesley Publishing Company, Inc., Reading, Mass., 1960.

Rogers, W. E., *Introduction to Electric Fields*, McGraw-Hill Book Company, Inc., New York, 1954.

Schelkunoff, S. A., *Electromagnetic Fields*, Blaisdell Publishing Company, New York, 1963.

Scott, W. T., *The Physics of Electricity and Magnetism*, (2nd ed.), John Wiley & Sons, Inc., New York, 1966.

Skilling, H. H., *Fundamentals of Electric Waves*, (2nd ed.), John Wiley & Sons, Inc., New York, 1948.

Stratton, J. A., *Electromagnetic Theory*, McGraw-Hill Book Company, Inc., New York, 1941.

ANSWERS TO
ODD-NUMBERED PROBLEMS

Chapter 1

1.1.   (a) $4\sqrt{3}$ units and directed 30° south of east   (b) 0.51764 units and directed 45° north of east   (c) 9.928 units and directed 30° south of east   (d) $-6\sqrt{3}$ (e) 6 units and directed upwards   (f) 0   (g) 0   (h) 2.784   (i) 1.607 units and directed upwards   (j) 0   (k) 0   (l) 24 units and directed towards the north   (m) 24 units and directed 30° north of east

1.3. $C^2 = A^2 + B^2 - 2AB \cos \theta$ where θ is the angle between **A** and **B**.

1.5.   (c) $dl = \sqrt{u^2 + v^2}\, du\, \mathbf{i}_u + \sqrt{u^2 + v^2}\, dv\, \mathbf{i}_v + dz\, \mathbf{i}_z$ (d) $dv = (u^2 + v^2)\, du\, dv\, dz$

1.7.   (c) $\alpha = 129.25°$, $\Delta = 29.38°$, $S = 38673$ Km   (d) 208.97°, 25.17°

1.15. $(\mathbf{i}_x + 2\mathbf{i}_y + 3\mathbf{i}_z)/\sqrt{14}$

1.17.   (a) Surfaces of constant magnitudes, T_0, are ellipsoids with intercepts on the x, y, and z axes at $\pm\sqrt{T_0}$, $\pm\sqrt{T_0/4}$, and $\pm\sqrt{T_0/9}$, respectively.

(b) Surfaces of constant magnitudes, U_0, are cylinders parallel to the z axis, having radii equal to $1/2U_0$ and with their axes passing through $x = \pm1/2U_0, y = 0$.

(c) Surfaces of constant magnitudes, V_0, are toruses obtained by revolving, about the z axis, circles in the $\phi =$ constant plane with centers at $r_c = 1/2V_0$ and $z = 0$ and with radii equal to $1/2V_0$.

1.19. $\mathbf{F} = -(mMG/r^2)\mathbf{i}_r$ in the spherical coordinate system having its origin at the center of the earth. Constant magnitude surfaces are spheres concentric with the earth. Direction lines are radial lines converging towards the center of the earth.

1.23.   (a) $\mathbf{v} = a\mathbf{i}_r + abt\mathbf{i}_\phi$, $\dfrac{d\mathbf{v}}{dt} = -ab^2t\mathbf{i}_r + 2ab\mathbf{i}_\phi$

(b) $\mathbf{v} = -\omega a \sin \omega t\, \mathbf{i}_x + \omega b \cos \omega t\, \mathbf{i}_y + c\mathbf{i}_z$

$\dfrac{d\mathbf{v}}{dt} = -\omega^2 a \cos \omega t\, \mathbf{i}_x - \omega^2 b \sin \omega t\, \mathbf{i}_y$

529

1.25. $(\mathbf{i}_r - \sqrt{3}\,\mathbf{i}_\phi)/2$

1.27.

Scalar function	x	y	z	r_c	ϕ	r_s	θ
Gradient	$\mathbf{i}_x$	$\mathbf{i}_y$	$\mathbf{i}_z$	$\mathbf{i}_{rc}$	$(1/r_c)\mathbf{i}_\phi$	$\mathbf{i}_{rs}$	$(1/r_s)\mathbf{i}_\theta$

1.29. 6.983

1.31. (a) 1/720 (b) $2\pi a l$ (c) $\pi/16$

1.33. (a) 0 (b) $a^2 l/2$

1.35. (a) $\pi/2$ (b) $\pi/2$ (c) $\pi/2$ (d) $\pi/2$

1.37. $-2/3$

1.39. $(e^{-1} - 1)\pi/2$

1.43.

Unit vector	$\mathbf{i}_x$	$\mathbf{i}_y$	$\mathbf{i}_z$	$\mathbf{i}_{rc}$	$\mathbf{i}_\phi$	$\mathbf{i}_{rs}$	$\mathbf{i}_\theta$
Divergence	0	0	0	$1/r_c$	0	$2/r_s$	$(\cot\theta)/r_s$

1.49. (a) 21/16 (b) 1/2 (c) 0 (d) 0

1.53. (a) $-y\mathbf{i}_x - z\mathbf{i}_y - x\mathbf{i}_z$ (b) $-2\mathbf{i}_z$ (c) $(2 + 2\sin\phi)\mathbf{i}_z$ (d) 0 except for $z = 0$
(e) $-(e^{-r}/r)\mathbf{i}_\phi$

1.63. (a) $6xyz^2 + 2x^3y$ (b) 0 (c) e^{-r}/r (d) $2(yz\mathbf{i}_x + zx\mathbf{i}_y + xy\mathbf{i}_z)$

Chapter 2

2.1. $-(mg/q)\mathbf{i}_r$, 55.7×10^{-12} N/C

2.3. (b) $y_L = qE_0L^2/2mv_0^2$, $\mathbf{v}_L = v_0\mathbf{i}_x + (qE_0L/mv_0)\mathbf{i}_y$
(c) $y_d = (qE_0L/mv_0^2)[(L/2) + d]$

2.5. $Q^2/\epsilon_0 l^2 mg = 4\pi/\sqrt{6}$

2.7. (a) $6Qd^2/4\pi\epsilon_0 z^4$ away from the quadrupole
(b) $3Qd^2/4\pi\epsilon_0 r^4$ towards the quadrupole

2.9. (a) $E_x = 0$, $E_y = 0$, $E_z = \rho_{L0}az/2\epsilon_0(a^2 + z^2)^{3/2}$
(b) $E_x = 0$, $E_y = -\rho_{L0}a^2/\pi\epsilon_0(a^2 + z^2)^{3/2}$, $E_z = 0$
(c) $E_x = -\rho_{L0}a^2/4\epsilon_0(a^2 + z^2)^{3/2}$, $E_y = 0$, $E_z = 0$
(d) $E_x = 0$, $E_y = -\rho_{L0}a^2/4\epsilon_0(a^2 + z^2)^{3/2}$, $E_z = 0$

2.11. (a) 0 for $|z| < a$, $\rho_{s0}a^2|z|/\epsilon_0 z^3$ for $|z| > a$
(b) $-\rho_{s0}/3\epsilon_0$ for $|z| < a$, $2\rho_{s0}a^3/3\epsilon_0|z^3|$ for $|z| > a$

2.13. $(\rho_0 r/2\epsilon_0)\mathbf{i}_r$ for $r < a$, and $(\rho_0 a^2/2\epsilon_0 r)\mathbf{i}_r$ for $r > a$, where the axis of the cylindrical charge is the z axis.

2.15. (a) $\mathbf{E} = (\rho_{L0}d/2\pi\epsilon_0 r^2)(\cos\phi\,\mathbf{i}_r + \sin\phi\,\mathbf{i}_\phi)$
(b) $x^2 + (y - c/2)^2 = (c/2)^2$, $z = $ constant; circles in planes normal to the z axis, with centers at $x = 0$ and $y = \pm c/2$, and having radii $c/2$.

2.17. $\mathbf{E} = \dfrac{\rho_{L0}}{4\pi\epsilon_0 r}\left[\left(\dfrac{z+a}{\sqrt{(z+a)^2 + r^2}} - \dfrac{z-a}{\sqrt{(z-a)^2 + r^2}}\right)\mathbf{i}_r \right.$
$\left. + \left(\dfrac{r}{\sqrt{(z-a)^2 + r^2}} - \dfrac{r}{\sqrt{(z+a)^2 + r^2}}\right)\mathbf{i}_z\right]$
where the line charge is located along the z axis between $z = -a$ and $z = a$.
Direction lines are given by
$\sqrt{(z+a)^2 + r^2} - \sqrt{(z-a)^2 + r^2} = $ constant

2.19. (a) π (b) $2\pi/3$ (c) $\pi/6$ (d) 2π (e) $\pi/2$ (f) $\pi/2$

2.21. $Q/8\epsilon_0$

2.23. (a) $\dfrac{\rho_0 z}{\epsilon_0}\mathbf{i}_z$ for $|z| < a$, $\dfrac{\rho_0 a|z|}{\epsilon_0 z}\mathbf{i}_z$ for $|z| > a$

(b) $\dfrac{\rho_0}{\epsilon_0}(|z| - a)\mathbf{i}_z$ for $|z| < a$, 0 for $|z| > a$

(c) $\dfrac{z^3}{2\epsilon_0|z|}\mathbf{i}_z$ for $|z| < a$, $\dfrac{a^2|z|}{2\epsilon_0 z}\mathbf{i}_z$ for $|z| > a$

(d) $\dfrac{z^2 - a^2}{2\epsilon_0}\mathbf{i}_z$ for $|z| < a$, 0 for $|z| > a$

(e) $\dfrac{1}{\epsilon_0}\left(az - \dfrac{z^3}{2|z|}\right)\mathbf{i}_z$ for $|z| < a$, $\dfrac{a^2 z}{2\epsilon_0|z|}\mathbf{i}_z$ for $|z| > a$

2.25. (a) 0 for $r < a$, $\dfrac{\rho_0}{3\epsilon_0 r^2}(r^3 - a^3)\mathbf{i}_r$ for $a < r < b$, $\dfrac{\rho_0}{3\epsilon_0 r^2}(b^3 - a^3)\mathbf{i}_r$ for $r > b$

(b) $\dfrac{\rho_0 r^2}{4\epsilon_0 a}\mathbf{i}_r$ for $r < a$, $\dfrac{\rho_0 a^3}{4\epsilon_0 r^2}\mathbf{i}_r$ for $r > a$

(c) $\dfrac{\rho_0(5a^2 r^3 - 3r^5)}{15\epsilon_0 a^2 r^2}\mathbf{i}_r$ for $r < a$, $\dfrac{2\rho_0 a^3}{15\epsilon_0 r^2}\mathbf{i}_r$ for $r > a$

2.27. $(\rho_0/2\epsilon_0)\mathbf{c}$ where $\mathbf{c}$ is the vector drawn from the axis of the cylindrical surface of radius a to the axis of the cylindrical surface of radius b.

2.31. (a) $\rho_s = \begin{cases} (2/3)\rho_{s0} & z = 0 \\ (4/3)\rho_{s0} & z = a \end{cases}$

(b) $\rho = e^{-r}/r$

(c) $\rho_s = \begin{cases} Q/4\pi a^2 & r = a \\ -Q/4\pi b^2 & r = b \end{cases}$

2.35. 1 unit of work done by the field.

2.39. $\dfrac{3Q(\Delta x)(\Delta z)}{4\pi\epsilon_0 r^3}\sin\theta\cos\theta\cos\phi$

2.41. (a) $\dfrac{5Q}{32\pi\epsilon_0 r} + \dfrac{Q(9\sin\theta\cos\phi + 3\sin\theta\sin\phi)}{32\pi\epsilon_0 r^2}$

(b) $-\dfrac{3}{4\pi\epsilon_0 r} - \dfrac{\sin\theta\cos\phi + \sin\theta\sin\phi + \cos\theta}{\pi\epsilon_0 r^2}$

(c) $\dfrac{Q(\sin\theta\cos\phi + \cos\theta)}{2\pi\epsilon_0 r^2}$

$\quad + \dfrac{Q(3\sin^2\theta\cos^2\phi + 3\cos^2\theta + 3\sin\theta\cos\theta\cos\phi - 2)}{4\pi\epsilon_0 r^3}$

2.43. (a) $\dfrac{\rho_{L0}}{4\pi\epsilon_0}\ln\dfrac{\sqrt{r^2 + z_0^2} + z_0}{\sqrt{r^2 + z_0^2} - z_0}$

(b) $\dfrac{1}{2\pi\epsilon_0}(\sqrt{r^2 + z^2} - r)$

(c) 0

2.45. (a) $(\rho_{s0}/2\epsilon_0)(|z_0| - |z|)$

(b) $(\rho_{s0}/2\epsilon_0)(|\sqrt{r_0^2 + z^2}| - |\sqrt{r_0^2 + z_0^2}| - |z| + |z_0|)$

(c) $(\rho_{s0}/2\epsilon_0)(|\sqrt{r_0^2 + z_0^2}| - |\sqrt{r_0^2 + z^2}|)$

(d) 0

(e) 0

2.47. For the line charge lying along the z axis between $z = -a$ and $z = a$,

$$V = \dfrac{\rho_{L0}}{4\pi\epsilon_0}\ln\dfrac{\sqrt{r^2 + (z + a)^2} + (z + a)}{\sqrt{r^2 + (z - a)^2} + (z - a)}$$

Equipotential surfaces are given by

$$\frac{(c-1)^2}{4c}\left(\frac{r}{a}\right)^2 + \frac{(c-1)^2}{(c+1)^2}\left(\frac{z}{a}\right)^2 = 1$$

where c is constant.

2.49. $\frac{\rho_0}{2\epsilon_0}\left(a^2 - \frac{r^2}{3}\right)$ for $r < a$, $\frac{\rho_0 a^3}{3\epsilon_0 r}$ for $r > a$

2.51. (a) $\frac{\rho_0}{4\epsilon_0}(a^2 - r^2)$ for $r < a$, $\frac{\rho_0 a^2}{2\epsilon_0}\ln\frac{a}{r}$ for $r > a$

(b) 0 for $r < a$, $\frac{\rho_0}{2\epsilon_0}\left(\frac{a^2 - r^2}{2} - a^2\ln\frac{a}{r}\right)$ for $a < r < b$,

$\frac{\rho_0}{2\epsilon^0}\left(\frac{a^2 - b^2}{2} - a^2\ln\frac{a}{b}\right) + \frac{\rho_0(b^2 - a^2)}{2\epsilon_0}\ln\frac{b}{r}$ for $r > b$

(c) $\frac{\rho_0 a^2}{3\epsilon_0}\ln\frac{a}{r}$ for $r < a$, $\frac{\rho_0}{9\epsilon_0 a}(a^3 - r^3)$ for $r > a$

2.53. (a) $\frac{\rho_{s0}z}{\epsilon_0}$ for $|z| < a$, $\frac{\rho_{s0}a|z|}{\epsilon_0 z}$ for $|z| > a$

(b) $\frac{\rho_{s0}a}{\epsilon_0}\ln\frac{b}{a}$ for $r < a$, $\frac{\rho_{s0}a}{\epsilon_0}\ln\frac{b}{r}$ for $a < r < b$, 0 for $r > b$

(c) $\frac{\rho_{s0}a^2}{\epsilon_0}\left(\frac{1}{a} - \frac{1}{b}\right)$ for $r < a$, $\frac{\rho_{s0}a^2}{\epsilon_0}\left(\frac{1}{r} - \frac{1}{b}\right)$ for $a < r < b$, 0 for $r > b$

2.55. (a) $\pi a^2\rho_{L0}\mathbf{i}_x$ (b) 0 (c) $(\rho_{L0}a^2/2)(-\pi\mathbf{i}_x + 2\pi^2\mathbf{i}_y)$

Dipole moments for cases (a) and (b) about any point other than the origin are the same as the respective dipole moments about the origin.

Chapter 3

3.1. $-e\mathbf{i}_x$

3.3. (b) $x_L = \frac{mv_0}{qB_0}\left[1 - \sqrt{1 - \left(\frac{qB_0 L}{mv_0}\right)^2}\right]$

$\mathbf{v}_L = \frac{qB_0 L}{m}\mathbf{i}_x + v_0\sqrt{1 - \left(\frac{qB_0 L}{mv_0}\right)^2}\,\mathbf{i}_y$

(c) $x_d = x_L + \frac{qB_0 Ld}{mv_0}\left[1 - \left(\frac{qB_0 L}{mv_0}\right)^2\right]^{-1/2}$

3.5. mg/LB_0, 9800 amp from west to east

3.9. $\mathbf{F}_{21} = -(\mu_0/4\pi)I_1 I_2\,dl_1\,dl_2\,\mathbf{i}_z$, $\mathbf{F}_{12} = (\mu_0/4\pi)I_1 I_2\,dl_1\,dl_2\,\mathbf{i}_z$

$\mathbf{F}_{31} = (\mu_0/12\sqrt{3}\,\pi)I_1 I_2\,dl_1\,dl_3\,\mathbf{i}_y$, $\mathbf{F}_{13} = (\mu_0/12\sqrt{3}\,\pi)I_1 I_2\,dl_1\,dl_3\,\mathbf{i}_x$

$\mathbf{F}_{32} = (\mu_0/8\sqrt{2}\,\pi)I_2^2\,dl_2\,dl_3\,\mathbf{i}_y$, $\mathbf{F}_{23} = (\mu_0/8\sqrt{2}\,\pi)I_2^2\,dl_2\,dl_3\,\mathbf{i}_x$

3.11. $\frac{\mu_0 I_1 I_2 a}{2\pi}\left(\frac{1}{d} - \frac{1}{d+b}\right)$ towards the loop

$\frac{\mu_0 I_1 I_2 a}{2\pi}\left(\frac{1}{d} - \frac{1}{d+b}\right)$ towards the infinitely long wire

3.13. $\frac{d}{2}\sqrt{\frac{\pi W}{IL}}$

3.15. $\frac{\mu_0 nIa^2\sin(2\pi/n)}{4\pi[(a\cos\pi/n)^2 + z^2](a^2 + z^2)^{1/2}}\,\mathbf{i}_z$

3.17. (a) $\frac{\mu_0 Ia^2}{2}\left\{\frac{1}{[a^2 + (z-b)^2]^{3/2}} + \frac{1}{[a^2 + (z+b)^2]^{3/2}}\right\}\mathbf{i}_z$

3.19. (a) $\dfrac{\mu_0 n_0 I}{2}\left[\ln\dfrac{a+\sqrt{a^2+z^2}}{|z|}-\dfrac{a}{\sqrt{a^2+z^2}}\right]\mathbf{i}_z$

(b) $\dfrac{\mu_0 n_0 I}{2}\left[\dfrac{1}{|z|}-\dfrac{1}{\sqrt{a^2+z^2}}\right]\mathbf{i}_z$

(c) $\dfrac{\mu_0 n_0 I a}{z^2\sqrt{a^2+z^2}}\mathbf{i}_z$

3.21. (a) $(\mu_0 Id/2\pi r^2)(-\sin\phi\,\mathbf{i}_r+\cos\phi\,\mathbf{i}_\phi)$

(b) $(x-c/2)^2+y^2=(c/2)^2$, $z=$ constant; circles in planes normal to the z axis, with centers at $y=0$ and $x=\pm c/2$, and having radii $c/2$.

3.23. (a) $3\mu_0 Ia^2 d/z^4$ away from the quadrupole

(b) $3\mu_0 Ia^2 d/2r^4$ towards the quadrupole

3.27. (a) $\dfrac{\mu_0 J_0 r}{2}\mathbf{i}_\phi$ for $r<a$, $\dfrac{\mu_0 J_0 a^2}{2r}\mathbf{i}_\phi$ for $r>a$

(b) 0 for $r<a$, $\dfrac{\mu_0 J_0}{2r}(r^2-a^2)\mathbf{i}_\phi$ for $a<r<b$, $\dfrac{\mu_0 J_0}{2r}(b^2-a^2)\mathbf{i}_\phi$ for $r>b$

(c) $\dfrac{\mu_0 J_0 a^2}{(n+2)r}\left(\dfrac{r}{a}\right)^{n+2}$ for $r<a$, $\dfrac{\mu_0 J_0 a^2}{(n+2)r}$ for $r>a$

3.29. (a) $-\mu_0 J_0 y\,\mathbf{i}_x$ for $|y|<a$, $-(\mu_0 J_0 a|y|/y)\mathbf{i}_x$ for $|y|>a$

(b) $\mu_0 J_0(|y|-a)\mathbf{i}_x$ for $|y|<a$, 0 for $|y|>a$

(c) $-(\mu_0 y^3/2|y|)\mathbf{i}_x$ for $|y|<a$, $-(\mu_0 a^2|y|/2y)\mathbf{i}_x$ for $|y|>a$

(d) $[\mu_0(a^2-y^2)/2]\mathbf{i}_x$ for $|y|<a$, 0 for $|y|>a$

(e) $-\mu_0[ay-(y^3/2|y|)]\mathbf{i}_x$ for $|y|<a$, $-(\mu_0 a^2|y|/2y)\mathbf{i}_x$ for $|y|>a$

3.31. (a) $\mu_0 J_{s0}\mathbf{i}_x$ for $|y|<a$, 0 for $|y|>a$

(b) 0 for $r<a$, $(\mu_0 J_{s0}a/r)\mathbf{i}_\phi$ for $r>a$

(c) 0 for $r<a$, $(\mu_0 J_{s0}a/r)\mathbf{i}_\phi$ for $a<r<b$, 0 for $r>b$

3.33. 0 inside the sphere, $-(\mu_0 I/2\pi r_c)\mathbf{i}_\phi$ outside the sphere

3.37. (a) $\dfrac{2}{3}J_{s0}\mathbf{i}_z$ for $y=0$, $\dfrac{4}{3}J_{s0}\mathbf{i}_z$ for $y=a$

(b) $3J_0 r\mathbf{i}_z$ for $r<a$, 0 for $a<r<b$, $-(J_0 a^3/b)\mathbf{i}_z$ for $r=b$, 0 for $r>b$

(c) $\dfrac{3}{2}J_{s0}\sin\theta\,\mathbf{i}_\phi$ for $r=a$

3.41. $\mathbf{A}=\dfrac{\mu_0 I}{4\pi}\ln\left[\dfrac{\sqrt{r^2+(z+a)^2}+(z+a)}{\sqrt{r^2+(z-a)^2}+(z-a)}\right]\mathbf{i}_z$

$\mathbf{B}=\dfrac{\mu_0 I}{4\pi r}\left[\dfrac{z+a}{\sqrt{r^2+(z+a)^2}}-\dfrac{z-a}{\sqrt{r^2+(z-a)^2}}\right]\mathbf{i}_\phi$

3.43. $\mathbf{A}=\dfrac{\mu_0 I\pi a^2\sin\theta}{4\pi r^2}\mathbf{i}_\phi$

$\mathbf{B}=\dfrac{\mu_0 I\pi a^2}{4\pi r^3}(2\cos\theta\,\mathbf{i}_r+\sin\theta\,\mathbf{i}_\theta)$

3.45. (a) $-\dfrac{\mu_0 J_0 y^2}{2}\mathbf{i}_z$ for $|y|<a$, $\left[-\dfrac{\mu_0 J_0 a^2}{2}-\mu_0 J_0 a(|y|-a)\right]\mathbf{i}_z$ for $|y|>a$

(b) $\mu_0 J_0\left(ay-\dfrac{y^3}{2|y|}\right)\mathbf{i}_z$ for $|y|<a$, $\dfrac{\mu_0 J_0 a^2 y}{2|y|}\mathbf{i}_z$ for $|y|>a$

(c) $-\dfrac{\mu_0|y^3|}{6}\mathbf{i}_z$ for $|y|<a$, $\dfrac{\mu_0(2a^3-3a^2|y|)}{6}\mathbf{i}_z$ for $|y|>a$

(d) $\dfrac{\mu_0(3a^2 y-y^3)}{6}\mathbf{i}_z$ for $|y|<a$, $\dfrac{\mu_0 a^3 y}{3|y|}\mathbf{i}_z$ for $|y|>a$

(e) $\dfrac{\mu_0(|y^3| - 3ay^2)}{6} \mathbf{i}_z$ for $|y| < a$, $\dfrac{\mu_0(a^3 - 3a^2|y|)}{6} \mathbf{i}_z$ for $|y| > a$

3.47. (a) $\mu_0 J_{s0} y \mathbf{i}_z$ for $|y| < a$, $(\mu_0 J_{s0} |y|/y) \mathbf{i}_z$ for $|y| > a$

(b) $\mu_0 J_{s0} a \ln \dfrac{b}{a}$ for $r < a$, $\mu_0 J_{s0} a \ln \dfrac{b}{r}$ for $a < r < b$, 0 for $r > b$

3.49. (a) $\mathbf{m} = (\pi n_0 I a^3/3) \mathbf{i}_z$ $\mathbf{A} = (\mu_0 n_0 I a^3/12 r^2) \sin \theta \, \mathbf{i}_\phi$

(b) $\mathbf{m} = (\pi n_0 I a^2/2) \mathbf{i}_z$ $\mathbf{A} = (\mu_0 n_0 I a^2/8 r^2) \sin \theta \, \mathbf{i}_\phi$

(c) $\mathbf{m} = \pi n_0 I a \mathbf{i}_z$ $\mathbf{A} = (\mu_0 n_0 I a/4 r^2) \sin \theta \, \mathbf{i}_\phi$

3.51. $(\mu_0 \rho_0 \omega_0 a^5/15 r^2) \sin \theta \, \mathbf{i}_\phi$

3.53. $\left[x - \dfrac{d}{2}\left(\dfrac{c^2+1}{c^2-1}\right)\right]^2 + y^2 = \left(\dfrac{dc}{c^2-1}\right)^2$ where c is constant, $z = 0$

3.57. (a) Group (a)   (b) Group (d)   (c) Group (c)   (d) Group (b)   (e) Group (c)

Chapter 4

4.1. $\mathbf{E} = \mathbf{i}_x + \mathbf{i}_y, \mathbf{B} = \mathbf{i}_z$

4.3. $x = \dfrac{E_0}{\omega_c B_0}(\omega_c t - \sin \omega_c t) + \dfrac{v_0}{\omega_c}(1 - \cos \omega_c t)$

$y = \dfrac{E_0}{\omega_c B_0}(1 - \cos \omega_c t) + \dfrac{v_0}{\omega_c}\sin \omega_c t$

$z = 0$

4.5. $x = \dfrac{qE_0}{m}\dfrac{\omega_c}{\omega_c^2 - \omega^2}\left(\dfrac{\sin \omega t}{\omega} - \dfrac{\sin \omega_c t}{\omega_c}\right)$

$y = \dfrac{E_0}{B_0}\dfrac{\omega_c}{\omega_c^2 - \omega^2}(\cos \omega t - \cos \omega_c t)$

$z = 0$

4.7. $B_0 v_0 ab/y(y + a)$

4.9. $\left(B_0 b \omega \ln \dfrac{y+a}{y}\right) \sin \omega t + \dfrac{B_0 v_0 ab}{y(y+a)} \cos \omega t$

4.11. $\omega a B_0 \mathbf{i}_r$

4.15. $\dfrac{\mu_0 I}{2}\left(\dfrac{z+d}{\sqrt{a^2 + (z+d)^2}} - \dfrac{z-d}{\sqrt{a^2 + (z-d)^2}}\right)$

4.17. $\dfrac{\mu_0 I}{2}\left(\dfrac{z-a}{\sqrt{r^2 + (z-a)^2}} - \dfrac{z+a}{\sqrt{r^2 + (z+a)^2}}\right)$ for C outside the sphere

$\dfrac{\mu_0 I}{2}\left(2 + \dfrac{z-a}{\sqrt{r^2 + (z-a)^2}} - \dfrac{z+a}{\sqrt{r^2 + (z+a)^2}}\right)$ for C outside the sphere

4.19. $(7/8)\mu_0 I$

4.21. $1.0606\mu_0$

4.23. The magnetic field due to the moving charge is given by

$\mathbf{B} = \dfrac{\mu_0 Q_0 v_0}{4\pi}\dfrac{r}{[r^2 + (v_0 t - z)^2]^{3/2}}\mathbf{i}_\phi$

at an arbitrary point (r, ϕ, z).

4.25. $0.1471/\epsilon_0$ N-m

4.27. (a) $\dfrac{2\pi p_0^2}{15\epsilon_0}(2b^5 + 3a^5 - 5a^3 b^2)$ N-m

(b) $\dfrac{\pi \rho_0^2 a^5}{7\epsilon_0}$ N-m

4.31.   (a) $\dfrac{\mu_0 I_0^2}{4\pi}\left[\dfrac{c^4}{(c^2 - b^2)^2} \ln \dfrac{c}{b} + \ln \dfrac{b}{a} - \dfrac{c^2}{2(c^2 - b^2)}\right]$

(b) $\dfrac{\pi \mu_0 J_0^2}{9}\left(a^4 \ln \dfrac{b}{a} + \dfrac{a^4}{6}\right)$

4.33.   The energies associated with the current distributions of Problem 4.32 are
(a) $\mu_0 J_{s0}^2 a$ (b) $2\mu_0/15$

4.35. $\pi \mu_0 (I_1^2 \ln c/a + 2I_1 I_2 \ln c/b + I_2^2 \ln c/b)$

4.37. $(V_0 I_0/4) \sin 2\beta z$

4.39. $\dfrac{8\pi \beta E_0^2}{15\mu_0 \omega} \cos^2 (\omega t - \beta r)$

4.41. $4 \cos (2t - 96.87°)$

4.43.   (a) $\bar{E}_x = 2\underline{/135°}$, $\bar{E}_y = 2\underline{/225°}$

(b) The magnitude of the field vector is constant and equal to 2 units. The angle which the vector makes with the x axis varies as $(-\omega t + 135°)$ with time. Hence, the field is circularly polarized.

4.45.   (a) $\sqrt{3}\,x - 2y - 3z = $ constant.

(c) The direction of polarization makes an angle of $25.67°$ with its projection, on to the xy plane, which makes an angle of $73.9°$ with the x axis.

4.47.   (a) $\sqrt{3}\,x + 3y + 2z = $ constant.

(c) $\mathbf{B} = \dfrac{0.04\pi}{\omega}[(-1 + j2\sqrt{3})\mathbf{i}_x + (-\sqrt{3} - j2)\mathbf{i}_y + 2\sqrt{3}\,\mathbf{i}_z]e^{-j0.02\pi(\sqrt{3}x + 3y + 2z)}$

The field is left circularly polarized.

4.49. $2\epsilon_0$

Chapter 5

5.3. $-50\epsilon_0 y$ for $x = 0$, $y > 0$; $-50\epsilon_0 x$ for $y = 0$, $x > 0$;
$(50\epsilon_0/x)\sqrt{x^4 + 4}$ for $xy = 2$

5.7. 0 for $r = a$, $\rho_{L1}/2\pi b$ for $r = b$, $-\rho_{L1}/2\pi c$ for $r = c$, and $(\rho_{L1} + \rho_{L2})/2\pi d$ for $r = d$

5.11.   (c) $3\epsilon_0 E_0 \cos \theta$

(d) $\mathbf{E}_a = E_0 \mathbf{i}_z$

$\mathbf{E}_s = \begin{cases} -E_0(\cos \theta\, \mathbf{i}_r - \sin \theta\, \mathbf{i}_\theta) \text{ for } r < a \\ \dfrac{E_0 a^3}{r^3}(2 \cos \theta\, \mathbf{i}_r + \sin \theta\, \mathbf{i}_\theta) \text{ for } r > a \end{cases}$

5.17.   (b) $\dfrac{Q}{4\pi\epsilon_0 r^2} \mathbf{i}_r$ for $r < a$, $\dfrac{Q}{4\pi\epsilon_0 (1 + \chi_{e0})r^2} \mathbf{i}_r$ for $a < r < b$, $\dfrac{Q}{4\pi\epsilon_0 r^2} \mathbf{i}_r$ for $r > b$

5.19.   (b) $\mathbf{E}_a = E_0(\cos \theta\, \mathbf{i}_r - \sin \theta\, \mathbf{i}_\theta)$

$\mathbf{E}_s = \begin{cases} -\dfrac{\chi_{e0}}{3 + \chi_{e0}} E_0(\cos \theta\, \mathbf{i}_r - \sin \theta\, \mathbf{i}_\theta) \text{ for } r < a \\ \dfrac{\chi_{e0}}{3 + \chi_{e0}} \dfrac{E_0 a^3}{r^3}(2 \cos \theta\, \mathbf{i}_r + \sin \theta\, \mathbf{i}_\theta) \text{ for } r > a \end{cases}$

(d) $\dfrac{3\chi_{e0}}{3 + \chi_{e0}} \epsilon_0 E_0 \cos \theta$

5.21.　(a) $\rho_{so}d/\epsilon_0$　(b) $\rho_{so}d/4\epsilon_0$　(c) $\rho_{so}(d + t)/4\epsilon_0$　(d) $\dfrac{\rho_{so}d}{\epsilon_2 - \epsilon_1}\ln\dfrac{\epsilon_2}{\epsilon_1}$

5.23.　(a) $\epsilon_0 E_0 \mathbf{i}_z$　(b) $\epsilon_0 E_0 \mathbf{i}_z$　(c) $\dfrac{E_0}{4}\left(1 + \dfrac{z}{d}\right)^2 \mathbf{i}_z$　(d) $\epsilon_0 E_0\left[1 - \dfrac{1}{4}\left(1 + \dfrac{z}{d}\right)^2\right]\mathbf{i}_z$

　　　(e) $-\dfrac{3}{4}\epsilon_0 E_0$ for $z = 0$, 0 for $z = d$　(f) $\dfrac{\epsilon_0 E_0}{2d}\left(1 + \dfrac{z}{d}\right)$

5.29.　(b) $\dfrac{\mu_0 I}{2\pi r}\mathbf{i}_\phi$ for $r < a$, $\mu_0(1 + \chi_{m0})\dfrac{I}{2\pi r}\mathbf{i}_\phi$ for $a < r < b$, $\dfrac{\mu_0 I}{2\pi r}\mathbf{i}_\phi$ for $r > b$

5.31.　(b) $\mathbf{B}_a = B_0(\cos\theta\,\mathbf{i}_r - \sin\theta\,\mathbf{i}_\theta)$

$$\mathbf{B}_s = \begin{cases} \dfrac{2\chi_{m0}}{3 + \chi_{m0}}\,B_0(\cos\theta\,\mathbf{i}_r - \sin\theta\,\mathbf{i}_\theta) \text{ for } r < a \\[3mm] \dfrac{\chi_{m0}}{3 + \chi_{m0}}\,\dfrac{B_0 a^3}{r^3}(2\cos\theta\,\mathbf{i}_r + \sin\theta\,\mathbf{i}_\theta) \text{ for } r > a \end{cases}$$

　　　(d) $\dfrac{3\chi_{m0}}{3 + \chi_{m0}}\dfrac{B_0}{\mu_0}\sin\theta\,\mathbf{i}_\phi$

5.33.　(a) $\mu_0 J_{so}d$　(b) $4\mu_0 J_{so}d$　(c) $\mu_0 J_{so}(4d - 2t)$　(d) $[(\mu_1 + \mu_2)/2]J_{so}d$

5.35.　(a) $(B_0/\mu_0)\mathbf{i}_y$　(b) $(B_0/\mu_0)\mathbf{i}_y$　(c) $\left(1 + \dfrac{z}{d}\right)^2 B_0 \mathbf{i}_y$　(d) $\left[\left(1 + \dfrac{z}{d}\right)^2 - 1\right]\dfrac{B_0}{\mu_0}\mathbf{i}_y$

　　　(e) 0 for $z = 0$, $(3B_0/\mu_0)\mathbf{i}_x$ for $z = d$　(f) $-2\left(1 + \dfrac{z}{d}\right)\dfrac{B_0}{\mu_0 d}\mathbf{i}_x$

5.37.　$\mu_r = kH$, $\mu_{ir} = 2kH$, $\chi_m = kH - 1$, $\mathbf{M} = (kH - 1)\mathbf{H}$

5.39.　(a) $\rho_{so}^2 d/2\epsilon_0$　(b) $\rho_{so}^2 d/8\epsilon_0$

5.41.　(a) $\mu_0 J_{so}^2 d/2$　(b) $2\mu_0 J_{so}^2 d$

5.43.　$\dfrac{2}{3\sqrt{k\mu_0}}B_0^{3/2}$

5.47.　$B_0(5\mathbf{i}_x + 4\mathbf{i}_y + 5\mathbf{i}_z)$ Wb/m^2

5.49.　(a) $\mathbf{H}_1 = \sqrt{\dfrac{\epsilon_0}{\mu_0}}[E_i\cos\omega(t - \sqrt{\mu_0\epsilon_0}\,z) - E_r\cos\omega(t + \sqrt{\mu_0\epsilon_0}\,z)]\mathbf{i}_y$

　　　$\mathbf{H}_2 = 2\sqrt{\dfrac{\epsilon_0}{\mu_0}}\,E_t\cos\omega(t - 2\sqrt{\mu_0\epsilon_0}\,z)\,\mathbf{i}_y$

　　　(b) $\dfrac{E_r}{E_i} = -\dfrac{1}{3}$　　$\dfrac{E_t}{E_i} = \dfrac{2}{3}$

Chapter 6

6.1.　(a) $\dfrac{3a^2 z - z^3}{6\epsilon}$ for $|z| < a$, $\dfrac{a^3|z|}{3\epsilon z}$ for $|z| > a$

　　　(b) $\dfrac{\rho_0}{4\epsilon}(a^2 - r^2)$ for $r < a$, $-\dfrac{\rho_0 a^2}{2\epsilon}\ln\dfrac{r}{a}$ for $r > a$

　　　(c) $\dfrac{\rho_0}{2\epsilon}(b^2 - a^2)$ for $r < a$, $-\dfrac{\rho_0}{6\epsilon}\left(r^2 + \dfrac{2a^3}{r} - 3b^2\right)$ for $a < r < b$,

　　　　$\dfrac{\rho_0}{3\epsilon r}(b^3 - a^3)$ for $r > b$

　　　(d) $-\dfrac{\rho_0}{\epsilon}\left(\dfrac{r^2}{6} - \dfrac{r^4}{20a^2} - \dfrac{a^2}{4}\right)$ for $r < a$, $\dfrac{2\rho_0 a^3}{15\epsilon r}$ for $r > a$

6.5.　$V = \dfrac{V_0}{\ln\epsilon_2/\epsilon_1}\ln\dfrac{\epsilon_1 d + (\epsilon_2 - \epsilon_1)x}{\epsilon_1 d}$

$$\mathbf{E} = -\frac{V_0}{\ln \epsilon_2/\epsilon_1} \frac{\epsilon_2 - \epsilon_1}{\epsilon_1 d + (\epsilon_2 - \epsilon_1)x} \mathbf{i}_x$$

6.7. (a) $\rho_s = \epsilon M_0 \cos \theta$ for $r = a$

(b) $\mathbf{E} = \begin{cases} -(M_0/3)\mathbf{i}_z & \text{for } r < a \\ (M_0 a^3/3r^3)(2 \cos \theta \, \mathbf{i}_r + \sin \theta \, \mathbf{i}_\theta) & \text{for } r > a \end{cases}$

(c) $\mathbf{H} = \begin{cases} -(M_0/3)\mathbf{i}_z & \text{for } r < a \\ (M_0 a^3/3r^3)(2 \cos \theta \, \mathbf{i}_r + \sin \theta \, \mathbf{i}_\theta) & \text{for } r > a \end{cases}$

$\mathbf{B} = \begin{cases} (2\mu_0 M_0/3)\mathbf{i}_z & \text{for } r < a \\ (\mu_0 M_0 a^3/3r^3)(2 \cos \theta \, \mathbf{i}_r + \sin \theta \, \mathbf{i}_\theta) & \text{for } r > a \end{cases}$

6.9. $V = V_1 \dfrac{\sinh (\pi x/b)}{\sinh (\pi a/b)} \sin (\pi y/b) + V_2 \dfrac{\sinh (3\pi x/b)}{\sinh (3\pi a/b)} \sin (3\pi y/b)$

$V = \dfrac{3V_1}{4} \dfrac{\sinh (\pi x/b)}{\sinh (\pi a/b)} \sin (\pi y/b) - \dfrac{V_1}{4} \dfrac{\sinh (3\pi x/b)}{\sinh (3\pi a/b)} \sin (3\pi y/b)$

6.11. (a) $V = \displaystyle\sum_{n=1,3,5,\ldots}^{\infty} \frac{4V_0}{n\pi} \frac{\sinh [n\pi(x - a/2)/b]}{\sinh (n\pi a/2b)} \sin (n\pi y/b)$

$\mathbf{J}_c = -\dfrac{4V_0\sigma_0}{b} \displaystyle\sum_{n=1,3,5,\ldots}^{\infty} \dfrac{1}{\sinh (n\pi a/2b)}$

$\times \left[\cosh \dfrac{n\pi}{b}\left(x - \dfrac{a}{2}\right) \sin \dfrac{n\pi y}{b} \mathbf{i}_x + \sinh \dfrac{n\pi}{b}\left(x - \dfrac{a}{2}\right) \cos \dfrac{n\pi y}{b} \mathbf{i}_y \right]$

(b) $V = \displaystyle\sum_{n=1,3,5,\ldots}^{\infty} \frac{4V_0}{n\pi} \frac{\cosh [n\pi(x - a/2)/b]}{\cosh (n\pi a/2b)} \sin (n\pi y/b)$

$\mathbf{J}_c = -\dfrac{4V_0\sigma_0}{b} \displaystyle\sum_{n=1,3,5,\ldots}^{\infty} \dfrac{1}{\cosh (n\pi a/2b)}$

$\times \left[\sinh \dfrac{n\pi}{b}\left(x - \dfrac{a}{2}\right) \sin \dfrac{n\pi y}{b} \mathbf{i}_x + \cosh \dfrac{n\pi}{b}\left(x - \dfrac{a}{2}\right) \cos \dfrac{n\pi y}{b} \mathbf{i}_y \right]$

(c) $V = \displaystyle\sum_{n=1,3,5,\ldots}^{\infty} \left[\dfrac{4V_1}{n\pi} \dfrac{\sinh (n\pi x/b)}{\sinh (n\pi a/b)} \sin (n\pi y/b) + \dfrac{4V_2}{n\pi} \dfrac{\sinh (n\pi y/a)}{\sinh (n\pi b/a)} \cos (n\pi x/a) \right]$

$\mathbf{J}_c = -\sigma_0 \displaystyle\sum_{n=1,3,5,\ldots}^{\infty}$

$\left\{ \left[\dfrac{4V_1}{b} \dfrac{\cosh (n\pi x/b)}{\sinh (n\pi a/b)} \sin (n\pi y/b) + \dfrac{4V_2}{a} \dfrac{\sinh (n\pi y/a)}{\sinh (n\pi b/a)} \cos (n\pi x/a) \right] \mathbf{i}_x \right.$

$\left. + \left[\dfrac{4V_1}{b} \dfrac{\sinh (n\pi x/b)}{\sinh (n\pi a/b)} \cos (n\pi y/b) + \dfrac{4V_2}{b} \dfrac{\cosh (n\pi y/a)}{\sinh (n\pi b/a)} \sin (n\pi x/a) \right] \mathbf{i}_y \right\}$

6.13. The image charge is an infinitely long line charge of uniform density $-\rho_{L0}$ C/m situated parallel to the actual line charge and at a distance d from the grounded conductor on the side opposite to that of the actual line charge.

6.17. (a) $-\dfrac{\epsilon_2 V_0}{\epsilon_2 t + \epsilon_1(d - t)} \mathbf{i}_x$ for $0 < x < t$, $-\dfrac{\epsilon_1 V_0}{\epsilon_2 t + \epsilon_1(d - t)} \mathbf{i}_x$ for $t < x < d$

(b) $-\dfrac{\epsilon_1 \epsilon_2 V_0}{\epsilon_2 t + \epsilon_1(d - t)}$ for $x = 0$, $\dfrac{\epsilon_1 \epsilon_2 V_0}{\epsilon_2 t + \epsilon_1(d - t)}$ for $x = d$

6.21. $\mu/12\pi$

6.23. $\pi a^2 \mu N^2$

6.25. $\pi a^2 \mu N_1 N_2$

6.27. 1257

6.31. For f slightly larger than $1/2\pi l\sqrt{\mu\epsilon}$, the input behaviour of the structure is equivalent to a series combination of $C = \epsilon wl/d$ and $\frac{1}{3}L$ where $L = \mu dl/w$. For still higher frequencies, the input behaviour of the structure is equivalent to C in series with the parallel combination of $\frac{1}{3}L$ and $\frac{1}{5}C$.

6.33. (a) $-480\pi^2 \times 10^{-7} \sin 300\pi t$ volts   (b) 0 volts

6.39. (a) For the region $r < (a - b)$, $\mathbf{E} = 0$ for all t.
For the region $(a - b) < r < (a + b)$,
$\mathbf{E}$ is zero for $\cos 2\pi t > (r - a)/b$ and $(Q/4\pi\epsilon_0 r^2)\mathbf{i}_r$ for $\cos 2\pi t < (r - a)/b$.
For the region $r > (a + b)$, $\mathbf{E} = (Q/4\pi\epsilon_0 r^2)\mathbf{i}_r$ for all t.
(b) No wave propagation

6.41. (a) $10 \cos 2\pi \times 10^7 t$, $10 \sin 2\pi \times 10^7 t$, 10^7 Hz
(b) $10 \cos 0.1\pi z$, $10 \sin 0.1\pi z$, 20 m
(c) 2×10^8 m/sec
(d) $\dfrac{1}{8\pi} \cos (2\pi \times 10^7 t - 0.1\pi z)$

6.43. (a) The given $\bar{\mathbf{E}}$ represents the electric field of a uniform plane wave
(b) Direction of propagation is along the unit vector $\frac{1}{4}(\sqrt{3}\,\mathbf{i}_x + 3\mathbf{i}_y + 2\mathbf{i}_z)$.
$\lambda = 25$ m
$f = 12$ MHz
$\lambda_x = 57.7$ m, $\lambda_y = 33.3$ m, $\lambda_z = 50$ m
$v_{px} = 6.928 \times 10^8$ m/sec, $v_{py} = 4 \times 10^8$ m/sec, $v_{pz} = 6 \times 10^8$ m/sec
The polarization is left circular
$$\bar{\mathbf{H}} = \frac{1}{240\pi}[(-1 + j2\sqrt{3})\mathbf{i}_x + (-\sqrt{3} - j2)\mathbf{i}_y + 2\sqrt{3}\,\mathbf{i}_z]e^{-j0.02\pi(\sqrt{3}x + 3y + 2z)}$$

6.47. (a) $[E_x]_{t=0.015} = \begin{cases} 2/3 & -1.5 < z < 0.75 \\ 0 & \text{otherwise} \end{cases}$

$[E_x]_{t=0.035} = \begin{cases} -1/3 & -7.5 < z < 4.5 \\ -8/45 & -1.5 < z < 0.75 \\ 8/15 & 2 < z < 3 \\ 0 & \text{otherwise} \end{cases}$

(b) $\eta_0[H_y]_{z=-3} = \begin{cases} 1 & 0 < t < 0.01 \\ 1/3 & 0.02 < t < 0.03 \\ 8/45 & 0.04 < t < 0.05 \\ -8/675 & 0.06 < t < 0.07 \\ 8/1025 & 0.08 < t < 0.09 \\ \cdots \end{cases}$

$\eta_0[H_y]_{z=2.5} = \begin{cases} 24/15 & 0.03 < t < 0.04 \\ -8/75 & 0.05 < t < 0.06 \\ 8/1125 & 0.07 < t < 0.08 \\ \cdots \end{cases}$

$\eta_0[H_y]_{t=0.035} = \begin{cases} 4/3 & -1.5 < z < 0.75 \\ 0 & \text{otherwise} \end{cases}$

$\eta_0[H_y]_{t=0.035} = \begin{cases} 1/3 & -7.5 < z < 4.5 \\ 8/45 & -1.5 < z < 0.75 \\ 24/15 & 2 < z < 3 \\ 0 & \text{otherwise} \end{cases}$

E_x in volts/m, H_y in amps/m, t in μ sec, and z in m.

6.51. $V(0, t) = 0$ $\qquad\qquad$ $I(0, t) = 2(|\bar{V}^+|/Z_0) \cos \omega t$
$V(\lambda/8, t) = -\sqrt{2}\,|\bar{V}^+| \sin \omega t$ $\qquad$ $I(\lambda/8, t) = \sqrt{2}(|\bar{V}^+|/Z_0) \cos \omega t$
$V(\lambda/4, t) = -2|\bar{V}_+| \sin \omega t$ $\qquad$ $I(\lambda/4, t) = 0$

$$V(3\lambda/8, t) = -\sqrt{2}\,|\,\bar{V}^+\,|\sin \omega t \qquad\qquad I(3\lambda/8, t) = -\sqrt{2}\,(|\,\bar{V}^+\,|/Z_0)\cos \omega t$$
$$V(\lambda/2, t) = 0 \qquad\qquad\qquad\qquad I(\lambda/2, t) = -2(|\,\bar{V}^+\,|/Z_0)\cos \omega t$$

6.53. (a) ∞, $|\,\bar{V}(d)\,| = 10 \sin (\pi d/2l)$, $|\,\bar{I}(d)\,| = 0.2 \cos (\pi d/2l)$
(b) 0, $|\,\bar{V}(d)\,| = 2.5 \sin (\pi d/l)$, $|\,\bar{I}(d)\,| = 0.05\,|\cos (\pi d/l)\,|$

(c)	d	0	$l/2$	l
Voltage, volts		0	5.303	7.07
Current, amps		0.1458	0.1	0.03535

6.57. No standing waves in medium 3.
In medium 2, standing wave ratio is 1.5. Standing wave patterns consist of minima for $|\,\bar{E}_x\,|$ and maxima for $|\,\bar{H}_y\,|$ at either end of medium 2. Both patterns contain three wavelengths.
In medium 1, standing wave ratio is 3. Standing wave patterns consist of minimum for $|\,\bar{E}_x\,|$ and maximum for $|\,\bar{H}_y\,|$ at the right end of medium 1.
Fraction of incident power transmitted into medium 3 = $\frac{3}{4}$.

6.59. 2.5 cm, $4\epsilon_0$

6.61.

Z_0 of stub	stub location	stub length
50 ohms	0.033λ	0.0811λ
50 ohms	0.167λ	0.4189λ
100 ohms	0.033λ	0.0434λ
100 ohms	0.167λ	0.4566λ

6.63. (a) $0.75e^{j0.264\pi}$ (b) 7 (c) 0.316λ (d) $(220 - j310)$ ohms
(e) $(0.0015 + j0.00215)$ mhos (f) 0.258λ

6.67.

Propagating mode	$TE_{1,0}$	$TE_{2,0}$	$TE_{3,0}$
f_c, MHz	1875	3750	5625
θ_i, degrees	71.79	51.32	20.37
λ_g, cm	2.631	3.203	7.184
v_{pz}, m/sec	1.5783×10^8	1.9215×10^8	4.3104×10^8
η_g, ohms	198.35	241.47	541.68

6.69. (a) $\bar{H}_y = 2\bar{H}_0 \cos (\beta x \cos \theta_i)\, e^{-j\beta z \sin \theta_i}$
$\bar{E}_x = 2\sqrt{\mu/\epsilon}\,\bar{H}_0 \sin \theta_i \cos (\beta x \cos \theta_i)\, e^{-j\beta z \sin \theta_i}$
$\bar{E}_z = j2\sqrt{\mu/\epsilon}\,\bar{H}_0 \cos \theta_i \sin (\beta x \cos \theta_i)\, e^{-j\beta z \sin \theta_i}$
(c) $\lambda_c = 2a/m$, $f_c = m/2a\sqrt{\mu\epsilon}$, $\lambda_g = \lambda/\sqrt{1 - (\lambda/\lambda_c)^2}$,
$v_{pz} = v_p/\sqrt{1 - (\lambda/\lambda_c)^2}$, $m = 1, 2, 3, \ldots$
(d) $\bar{H}_y = 2\bar{H}_0 \cos (m\pi x/a)\, e^{-j(2\pi/\lambda_g)z}$
$\bar{E}_x = 2\eta\bar{H}_0(\lambda/\lambda_g) \cos (m\pi x/a)\, e^{-j(2\pi/\lambda_g)z}$
$\bar{E}_z = j2\eta\bar{H}_0(\lambda/\lambda_c) \sin (m\pi x/a)\, e^{-j(2\pi/\lambda_g)z}$
(e) $\eta_g = \bar{E}_x/\bar{H}_y = \eta\sqrt{1 - (\lambda/\lambda_c)^2}$
Transmission-line equivalent consists of $\bar{V} \leftrightarrow \bar{E}_x$, $\bar{I} \leftrightarrow \bar{H}_y$, $Z_0 \leftrightarrow \eta_g$, and $v_p \leftrightarrow v_{pz}$.

6.73. (b) $\bar{V}(z) = e^{-(1/2)az}[\bar{A}e^{(1/2)\sqrt{a^2 - 4\omega^2\mathcal{L}_0\mathcal{C}_0}\,z} + \bar{B}e^{-(1/2)\sqrt{a^2 - 4\omega^2\mathcal{L}_0\mathcal{C}_0}\,z}]$
$$\bar{I}(z) = \frac{e^{(1/2)az}}{j\omega\mathcal{L}_0}\Big[\frac{1}{2}(a - \sqrt{a^2 - 4\omega^2\mathcal{L}_0\mathcal{C}_0})\bar{A}e^{(1/2)\sqrt{a^2 - 4\omega^2\mathcal{L}_0\mathcal{C}_0}\,z}$$
$$+ \frac{1}{2}(a + \sqrt{a^2 - 4\omega^2\mathcal{L}_0\mathcal{C}_0})\bar{B}e^{-(1/2)\sqrt{a^2 - 4\omega^2\mathcal{L}_0\mathcal{C}_0}\,z}]$$
$f_c = a/4\pi\sqrt{\mathcal{L}_0\mathcal{C}_0}$

6.75.

Frequency	α, nepers/m	β, rad/m	$\bar{\eta}$, ohms	λ, m
100 MHz	0.0211	18.73	42.4	0.3354
10 KHz	$2\pi \times 10^{-3}$	$2\pi \times 10^{-3}$	$(1+j)2\pi$	1000

6.77. (a) $\bar{E}_x(d) = \bar{E}_x^+ e^{\bar{\gamma}d} + \bar{E}_x^- e^{-\bar{\gamma}d}$

$$\bar{H}_y(d) = \frac{1}{\bar{\eta}}[\bar{E}_x^+ e^{\bar{\gamma}d} - \bar{E}_x^- e^{-\bar{\gamma}d}]$$

6.79. (a) $\dfrac{\partial \bar{V}}{\partial z} = -[2\mathcal{R}_i + j\omega(2\mathcal{L}_i + \mathcal{L})]\bar{I}(z)$

$$\frac{\partial \bar{I}}{\partial z} = -(\mathcal{G} + j\omega\mathcal{C})\bar{V}(z)$$

(b) $\bar{\gamma} = [2\mathcal{R}_i + j\omega(2\mathcal{L}_i + \mathcal{L})](\mathcal{G} + j\omega\mathcal{C})$

$$Z_0 = \sqrt{\frac{2\mathcal{R}_i + j\omega(2\mathcal{L}_i + \mathcal{L})}{\mathcal{G} + j\omega\mathcal{C}}}$$

(c) $\sqrt{2\mathcal{R}_i\mathcal{G}}$

6.81. (b) 1824.42, 4175.58, 7824.42 MHz

6.83. (a) $f_{1,0,1} = 5303.4$ MHz, $f_{2,0,1} = f_{1,0,2} = 8385.4$ MHz, $f_{2,0,2} = 10606.5$ MHz.

(b) $\bar{E}_y = \bar{E}_0 \sin\dfrac{\pi x}{a} \sin\dfrac{\pi z}{d}$

$$\bar{H}_x = -j\frac{\bar{E}_0}{\eta}\frac{\lambda}{2d}\sin\frac{\pi x}{a}\cos\frac{\pi z}{d}$$

$$\bar{H}_z = j\frac{\bar{E}_0}{\eta}\frac{\lambda}{2a}\cos\frac{\pi x}{a}\sin\frac{\pi z}{d}$$

$$Q = \frac{\pi\sigma\delta\eta}{4}\frac{(a^2 + d^2)^{3/2}}{(a^3 + d^3)}$$

6.87. $56.3°$

6.89. For currents equal in magnitude and phase: (a) linearly polarized in the z direction, (b) linearly polarized in the x direction, (c) linearly polarized parallel to the vector $(\mathbf{i}_x + \mathbf{i}_z)$, and (d) linearly polarized parallel to the vector $(-\mathbf{i}_x + \frac{1}{2}\mathbf{i}_y - \frac{1}{2}\mathbf{i}_z)$.
For currents equal in magnitude but different in phase by $\pi/2$: (a) linearly polarized in the z direction, (b) linearly polarized in the x direction, (c) circularly polarized normal to the y axis, and (d) elliptically polarized with major axis along $\mathbf{i}_x$ and minor axis along $(\mathbf{i}_y - \mathbf{i}_z)$ and with the ratio of the major to the minor axis equal to $\sqrt{2}$.

6.91. (a) $I_1(t) = I_0 \cos \omega t$, $I_2(t) = -I_0 \cos \omega t$, where $I_0 = \omega Q_0$.

(b) $\bar{\mathbf{A}} = \dfrac{\mu\omega Q_0(dl)^2 \cos\theta}{4\pi r}\left(j\dfrac{\omega}{v} + \dfrac{1}{r}\right)e^{-j\omega r/v}\mathbf{i}_z$

(c) $\bar{\mathbf{E}} = -\dfrac{Q_0(dl)^2 \sin\theta \cos\theta}{4\pi\epsilon r}\left[\left(\dfrac{\omega}{v}\right)^3 - j\dfrac{3}{r}\left(\dfrac{\omega}{v}\right)^2 - \dfrac{6\omega}{vr^2} + j\dfrac{6}{r^3}\right]e^{-j\omega r/v}\mathbf{i}_\theta$

$\qquad + \dfrac{Q_0(dl)^2(3\cos^2\theta - 1)}{4\pi\epsilon r^2}\left[j\left(\dfrac{\omega}{v}\right)^2 + \dfrac{3\omega}{vr} - j\dfrac{3}{r^2}\right]e^{-j\omega r/v}\mathbf{i}_r$

$\bar{\mathbf{H}} = \dfrac{\omega Q_0(dl)^2 \sin\theta \cos\theta}{4\pi r}\left[-\left(\dfrac{\omega}{v}\right)^2 + j\dfrac{3\omega}{vr} + \dfrac{3}{r^2}\right]e^{-j\omega r/v}\mathbf{i}_\phi$

(d) $\bar{\mathbf{E}} = -\dfrac{Q_0(dl)^2\omega^3 \sin\theta \cos\theta}{4\pi\epsilon r v^3}e^{-j\omega r/v}\mathbf{i}_\theta$

$\bar{\mathbf{H}} = -\dfrac{Q_0(dl)^2\omega^3 \sin\theta \cos\theta}{4\pi\epsilon\eta r v^3}e^{-j\omega r/v}\mathbf{i}_\phi$

6.93. (c) $U_n = \dfrac{\cos^2[(\pi/2)\cos\theta]}{\sin^2\theta}$

INDEX